THE BIOLOGY AND IDENTIFICATION OF THE COCCIDIA (APICOMPLEXA) OF RABBITS OF THE WORLD

THE BIOLOGY AND IDENTIFICATION OF THE COCCIDIA (APICOMPLEXA) OF RABBITS OF THE WORLD

DONALD W. DUSZYNSKI

University of New Mexico, Albuquerque, NM

LEE COUCH

University of New Mexico, Albuquerque, NM

ELSEVIER

AMSTERDAM • BOSTON • HEIDELBERG • LONDON
NEW YORK • OXFORD • PARIS • SAN DIEGO
SAN FRANCISCO • SINGAPORE • SYDNEY • TOKYO

Academic Press is an Imprint of Elsevier

Academic Press is an imprint of Elsevier
32 Jamestown Road, London NW1 7BY, UK
225 Wyman Street, Waltham, MA 02451, USA
525 B Street, Suite 1800, San Diego, CA 92101-4495, USA

Notice

British Library Cataloguing-in-Publication Data
A catalogue record for this book is available from the British Library.

Library of Congress Cataloging-in-Publication Data
A catalog record for this book is available from the Library of Congress.

ISBN: 978-0-12-397899-8

For information on all Academic Press publications
visit our website at elsevierdirect.com

Typeset by TNQ Books and Journals Pvt Ltd.
www.tnq.co.in

Dedication

This book is dedicated to Dr. Steve J. Upton (June 14, 1953—July 29, 2010), who finished his M.S. degree in our laboratory in 1981. Steve went on to complete his Ph.D. (1983) in Zoology-Entomology, under the direction of Dr. W.L. Current, Auburn University, Auburn, Alabama. From 1983 to 1984, he was a postdoctoral fellow in parasitology in Bill Current's laboratory. He spent 1984–1986, as a Visiting Assistant Professor at the University of Texas-El Paso, Texas, where he taught many classes (General Biology, Medical Parasitology, Human Parasitology, Marine Invertebrate Zoology, Protozoology, Human Anatomy and Physiology) and worked closely with Drs. Jack Bristol and Lillian Mayberry, with whom he developed a close and life-long friendship. In the fall of 1986, he began as an Assistant Professor in Biological Sciences at Kansas State University (KSU), where he progressed rapidly through the professorial ranks and developed a powerful international reputation as the expert on the basic biology, taxonomy, systematics, life history, *in vitro* cultivation, and immunology of numerous coccidia (e.g., *Caryospora*, *Eimeria*, *Isospora*, etc.) and *Cryptosporidium* species. His contributions toward our knowledge and understanding of the parasites he loved were prodigious and included: 225 original research papers in refereed journals; 11 book chapters; three books (including the 2009 seminal work, *The Coccidia of Snakes of the World*, with D.W. Duszynski). Unfortunately, he was taken from us much too soon, after a courageous, decade-long battle against cancer, in Manhattan, Kansas, where he had lived the past 24 years. Steve was an extraordinary collector of hard-copy reprints throughout his life. One of his great friends and colleagues at KSU was Bob La Hew, also in Biological Sciences, who helped Steve in the last decade of his life, and who preserved Steve's reprint collection so that one of us (DWD) could transport it to New Mexico; a collection without which this book could not have been completed in its current detail.

Steve's work ethic and accomplishments were the stuff of which legends are made. Nonetheless, Steve remained low-key, humble, and very human. We never met a person who did not like or love Steve Upton. Steve would have been proud to be a part of this undertaking. His passing has created a noticeable void and a large tear in the fabric of coccidian parasitology. He is, and always will be, deeply missed by both of us and by so many others.

Contents

6. Coccidia (Eimeriidae) of the Family Leporidae: Genus *Oryctolagus*

7. Coccidia (Eimeriidae) of the Family Leporidae: Genus *Sylvilagus*

8. Sarcocystidae Poche, 1913, the Predator-Prey Coccidia in Rabbits: *Besnoitia, Sarcocystis, Toxoplasma*

9. *Cryptosporidium* and Cryptosporidiosis in Rabbits

10. Strategies for Management, Control, and Chemotherapy

11. Summary and Conclusions

Preface and Acknowledgments

Earlier we (Duszynski and Upton, 2009) reported that the Coccidia were among the first one-celled animals visualized by Antonie van Leeuwenhoek in 1674 when he saw what we now know were oocysts of *Eimeria stiedai* in the bile of a rabbit. The literature that summarizes all of our knowledge about coccidia of rabbits since 1674 is dispersed in hundreds of journals and dozens of languages. After his retirement from the University of Illinois, Dr. Norman D. Levine was kind enough to share many of his original reprints with one of us (DWD). Among the documents he sent was a very preliminary manuscript (circa 1990), hand-typed on a typewriter on yellow paper, giving a clear indication that he previously had intended to produce a monograph on rabbit coccidia. The idea for this book came from the inspiration provided by Dr. Levine's very primitive and incomplete manuscript, and because there is no taxonomic summation currently available for those species that infect rabbits. Dr. Levine was the world leader of taxonomic and systematic knowledge for at least the last four decades of his professional life. And he was a quiet, gentle, and unassuming person. I (DWD) am a better person for having known him.

We are grateful to Dr. Lynn Hertel (deceased), formerly of the Department of Biology, The University of New Mexico (UNM), for her creative interpretation and skillful depiction and production of the life-cycle line drawings used on our cover. This book would not be accurate in defining the host rabbits without the seminal book by Wilson and Reeder (2005), which we consider the current bible for the mammal species of the world. We are grateful to many editors and/or publishers for giving us permission to utilize certain line drawings and photomicrographs from the original papers in which they appeared. This book also would not have been possible without the dedicated work done by our numerous colleagues worldwide who study the coccidia of rabbits and publish their findings. Finally, we are sincerely grateful to a subset of these colleagues, who generously donated their time and creative input to help us. Dr. Terry L. Carpenter, Lieutenant Colonel USAF (retired), was invaluable in helping us obtain some of the more obscure Russian journal articles and translations. Lynda Crispino, *That's It! Creative*, Rio Rancho, New Mexico, helped us with artwork on five of our line drawings. Dr. Brent B. Parker, Santa Fe Animal Hospital, helped us retrieve some of the drug therapy literature and regimens currently used by veterinarians to treat rabbit coccidiosis. Jana Kvićerová (DVM, Ph.D.), Institute of Parasitology, Biology Centre of the Academy of Sciences, České Budějovice, South Bohemia, Czech Republic, provided many helpful discussions on the evolutionary significance of morphological features to help understand the phylogenetic relationships among rabbit eimerians.

Donald W. Duszynski
Professor Emeritus of Biology
The University of New Mexico
Albuquerque, NM 87131

Lee Couch, Principal Lecturer
Department of Biology
The University of New Mexico
Albuquerque, NM 87131

Introduction

Like its predecessor, *The Coccidia of Snakes of the World* (Duszynski and Upton, 2009), this book is intended to be the most comprehensive treatise, to date, describing the structural and biological knowledge of all species in the most pervasive group of **protist** (formerly protozoa) parasites that infect rabbits, a group called the Coccidia. These protists (Phylum **Apicomplexa**) are common in rabbits and are represented by about 87 species that fit taxonomically into six genera in three families that include Cryptosporidiidae Léger, 1911 (*Cryptosporidium*), Eimeriidae Minchin, 1903 (*Eimeria, Isospora*), and Sarcocystidae Poche, 1913 (*Besnoitia, Sarcocystis, Toxoplasma*). An overview of the general biology,

taxonomy, life cycles, and numbers of species of eimeriid and cryptosporid coccidia from wild mammals was published a decade ago (Duszynski and Upton, 2001), and monographic works on the coccidia of certain selected vertebrate groups also are available; these include Amphibia (Duszynski et al., 2007); Chiroptera (Duszynski, 2002); Insectivora (Duszynski and Upton, 2000); Marmotine squirrels in the Rodentia (Wilber et al., 1998); Primates and Scandentia (Duszynski et al., 1999); and the Serpentes of the Reptilia (Duszynski and Upton, 2009). No such review exists for the coccidia of rabbits. Here we strive to resolve that void for rabbits (Lagomorpha), because "bunnies" have a long and important history shared with humans, perhaps more so than any other domesticated animals.

RABBITS ARE FOOD, PETS, LAB ANIMALS, AND PESTS

Rabbits are unique animals that are revered as an important source of protein, both commercially and as objects for hunting; treated as trusted and loved pets; docile and easily adapted to laboratory conditions as important experimental animals in biological research; and horrific pests at times, all qualities deeply imbedded within the human culture. The wild, Old World European rabbit (*Oryctolagus cuniculus*) has been transported and introduced globally, often with serious consequences on local biodiversity (vegetation and wildlife), especially when local, natural predators are missing. In the most classic example, 24 *O. cuniculus* were introduced into Victoria, Australia, in 1859. Because of the lack of any natural predators, the presence of a farming habitat that was nearly ideal for their reproducing, and a climate that allowed year-round breeding, their population expanded exponentially. By the turn of the century, the Australian government built an immense "rabbit-proof fence" in a futile attempt to halt the westward expansion of the ever-increasing rabbit population, but because the rabbits could both jump high and burrow, the fence was useless, and while their population continued to expand, they pushed some native herbivores to near-extinction. During the 1950s, introduction of the myxoma virus (a prototype of *Leporipoxvirus*), which causes the disease myxomatosis in rabbits, provided some relief in Australia, but not in New Zealand, where the insect vectors necessary for spread of the disease were not present. Today, wild rabbits in Australia are largely immune to myxomatosis.

Besides their very important and necessary use as laboratory animals, rabbits are raised for a variety of commercial purposes that include wool, meat, and fur (Bhat et al., 1996). They are efficient converters of vegetable protein into high-quality animal protein. In India, rabbit farming for wool and meat has developed into an important industry, and it has brought handsome returns to rabbit breeders, but those returns are affected by outbreaks of coccidiosis and major losses in the rabbit industry caused by multiple *Eimeria* species (Leysen et al., 1989; Bhat et al., 1996). Thus, coccidia present an emerging disease problem of increasing importance in commercial rabbitries (Licois et al., 1990).

RABBITS AND THE HISTORY OF DISCOVERY OF THE COCCIDIA

Norman D. Levine (1973a, 1974) was the first to attempt a scholarly study of the historical aspects of research on the coccidia. At first, he believed this task would be easy because early research on coccidia was done by some of the most prominent biologists of their times, but when he searched the literature, he found that very little had been written about even the most important and well-known scholars. Two examples will serve to illustrate his frustration. First, he (1974) found nothing about the life of Alphonse Labbé, "who did a great deal of taxonomic work on the coccidia and who assembled all the known information about them around the turn of the century." Second, all he (Levine, 1974) found about Aimé Schneider, "who established the genus *Eimeria*, was a death notice in *Nature* in which his last name was misspelled." What he did find is that "the history of the coccidia is badly tangled," but his (always) positive nature let him conclude "that biographical research is a legitimate type of research for scientists—it puts flesh on the bones of knowledge." Here we try to briefly untangle some of the history.

Coccidia were among the first single-celled animals seen by Antonie van Leeuwenhoek in

1674, when he saw "bodies" in the bile of a domestic rabbit (see Wenyon, 1926; Levine, 1973a, 1974), which Dobell (1922) believed were oocysts of *Eimeria stiedai*. Since then, the literature that records and documents the development of our knowledge about coccidian parasites of vertebrates in general, and of rabbits in particular, has become widely dispersed in hundreds of journals and dozens of languages. The "active" study of rabbit coccidia began in earnest after the work of Hake (1839), who first described pathology in the liver and duodenum of the domestic rabbit (Wenyon, 1926), although he believed the oocysts he saw were "pus globules" associated with a carcinoma of the liver.

Following Hake's (1839) work, much of our early knowledge on the biology of the coccidia was done using domestic rabbits. Unfortunately, early investigators never thought to differentiate the species infecting the liver from the form(s) infecting the intestine, assuming they were all the same species, and this led to many misinterpretations. A second reason for confusion was that it took a long time for early workers to understand that the **endogenous** (sexual and asexual) and **exogenous** (oocyst) stages were part of the same life cycle. And a third reason, which entangled the terminology in the early development of our knowledge, is that several different generic names were given to the coccidia when they were confused with **myxosporans** and **gregarines**; these included *Psorospermium, Coccidium,* and *Monocystis* (for an overview, see Levine, 1973a, 1974).

In 1845, Remak was the first to find oocysts in the intestinal mucosa of rabbits, and Kauffmann (1847) observed that oocysts segmented into four separate sporocysts when kept in water. Two decades later, Stieda (1865) verified Kauffmann's observation and also noted that two elongated structures (i.e., sporozoites) developed within each sporocyst; Lindemann (1865) recognized the parasitic nature of the liver stages and named the organism *Monocystis*

stiedae, thinking it was a gregarine. A decade after this, Schneider (1875) named the genus *Eimeria* for a mouse coccidium and a few years later, apparently unaware of Schneider's paper, Rivolta (1878) named a rabbit coccidium *Psorospermium cuniculi* and Leuckart (1879) proposed the name *Coccidium oviforme* for the same form he found in the rabbit intestine; we now know both are junior synonyms of *Eimeria perforans.* Two decades after Stieda's report, Balbiani (1884) confirmed that sporocysts each produced two sporozoites along with a small residual body. According to Wenyon (1926), L. Pfeiffer (1890, 1891) and R. Pfeiffer (1892) were the first to report that an endogenous multiplication occurred in the rabbit's intestine resulting in the development and production of oocysts. In 1897, Simond was the first to illustrate **Koch's postulates** by feeding oocysts to young rabbits reared from birth and to find both the intermediate stages of merogony, and later, newly formed oocysts that resembled the original ones fed to the rabbits. In spite of all this important early work, the coccidia were considered of no practical importance until the beginning of the twentieth century.

The early days of the development of our knowledge about how many eimerians infected rabbits and how to separate them were confusing, at best. Perard (1924a) and Waworuntu (1924) described some developmental stages of *E. perforans* and *E. stiedai,* and Perard (1924a) also reported a new intestinal eimerian that he named *E. magna* (1925b). Prior to their work, all other workers with rabbit coccidia had considered this to be a larger form of *E. perforans* or had confused it with *E. stiedai.* Waworuntu (1924), unfortunately, didn't (or couldn't) distinguish between these species and said that he produced intestinal coccidiosis with what he thought was *E. stiedai,* when he was really using *E. magna* (1924, see his Fig. 9, Plate II). He was not the only person to make this mistake. Both Krijgsman (1926) and Wenyon (1926) also confused oocysts of *E. magna* and

E. stiedai. Finally, Waworuntu (1924) reported a third kind of oocyst that he said was intermediate in size between *E. magna* and *E. stiedai*, which Kessel and Jankiewicz (1931) later named *E. media*. They (Kessel and Jankiewicz, 1931) also named another rabbit intestinal species, *E. irresidua*, the sporulated oocysts of which can easily be distinguished from those of *E. stiedai*. Our book tries to help sort out the details of the discovery and our total current knowledge of each species known to infect all rabbit species, both domesticated and wild.

HIGH PREVALENCE AND MULTIPLE SPECIES

Mention must be made that almost all rabbits sampled, regardless of genus or species, and from both wild populations and domestic rabbitries, seem to be infected with coccidia, and these infections usually are composed of multiple coccidian species. Nieschultz (1923), working on hares in Holland, found that over 90% were infected with coccidia. Cooper (1927), reporting on the incidence of coccidiosis in rabbits at Mukteswar, India (probably hares, *Lepus nigricollis*), found an infection rate > 83%. Boughton (1932), who surveyed the snowshoe rabbit (*Lepus americanus*) in western Canada, found 336 of 420 (80%) infected with eimerians of various species. Gill and Ray (1960) said that, "out of 855 consecutive postmortem examinations performed during a year or so, 99.4%" of the hares they examined in India were infected with coccidia. Duszynski and Marquardt (1969) found all of 100 cottontail rabbits (*Sylvilagus audubonii*) collected near Fort Collins, Colorado, USA, to be infected with from four to six species of *Eimeria*. Catchpole and Norton (1979) examined 596 fecal samples from *Oryctolagus cuniculus* from three commercial rabbitries in southeast England. In those rabbits managed conventionally in wire cages over dropping pits, 96% contained oocysts,

and mixed infections were most common, with 67% of the rabbits carrying two to four different eimerians. Moreno-Montañez et al. (1979) found 27 of 42 (64%) *L. capensis* from different localities in Spain to be multiply infected with three eimerian species. Tasan and Özer (1989), in Turkey, found 32 of 40 (80%) *L. europaeus* hunted between 1985 and 1987 in the rural districts of two Provinces to be infected with from one to four *Eimeria* species. Pakandl (1990) reported a study done from 1983 to 1985, in which he surveyed 33 digestive tracts and 317 fecal samples of the European hare (*L. europaeus*) from various localities in the Czech Republic, and found 337 of 350 (96%) hares surveyed were positive for coccidia infections representing various combinations of nine different species of *Eimeria*. Polozowski (1993) examined feces from 246 rabbits in six farm rabbitries in the Wroclaw District of Poland and found 234 (95%) to be infected with from one to nine species of *Eimeria*; their individual prevalences ranged from 21% to 85% in the infected hosts. Rabbits 1 to 3 months old always had five to nine species of coccidia concurrently, while rabbits > 24 months old were infected by only one to three species concurrently (Polozowski, 1993). Of the age groups examined, 183 of 246 (74%) were < 3 months old and 15 of 246 (6%) were 4 to 12 months old; these two age groups harbored all nine eimerians identified, while 37 of 246 (15%) were 13 to 24 months old and were infected with eight species. Additionally, 11 of 246 (4.5%) in which six eimerians were identified were < 24 months old. Darwish and Golemansky (1991) found 10 *Eimeria* species in 58/75 (73%) domestic rabbits from four localities in Syria. The 58 infected rabbits harbored from one to seven *Eimeria* species each.

As clearly documented above, mixed infections are the rule, with single species infections seen rarely in the wild or usually only under laboratory conditions. In terms of pathology, the most important rabbit coccidium is *E. stiedai*, which occurs in the liver of the Old World wild rabbit (*O. cuniculus*) and its domesticated subspecies, the common (albino) laboratory rabbit; all other *Eimeria* (and two *Isospora*; see Chapter 3) species in rabbits are found in the intestines, as far as we currently know. Interestingly, *E. stiedai* has been transmitted experimentally both to cottontail rabbits (*Sylvilagus* spp.) and to hares (*Lepus* spp.), and it has been reported many times from both in surveys of wild rabbits. It is the only rabbit eimerian that has reliable (experimental) evidence that it can infect multiple rabbit species in three different genera. Including *E. stiedai*, there are about 74 *Eimeria* and two *Isospora* species now known to parasitize five rabbit genera (see Chapters 3–7). Of these intestinal forms from the domestic (lab) and Old World wild rabbits, some are important pathogens (e.g., *E. flavescens*), some seem to be only slightly pathogenic in young rabbits (e.g., *E. irresidua*, *E. matsubayashii*), and for other species (e.g., *E. nagpurensis*, *E. oryctolagi*), we don't know anything about how they affect the health of rabbits infected with them. Some of the intestinal coccidia of the domestic rabbit have been reported in cottontails, and a lesser number have been reported from hares; most of these records should be looked at critically and await confirmation with cross-transmission studies and molecular evidence. Coccidiosis (the manifestation of disease symptoms) is entirely unknown in all other rabbit genera and species (*Ochotona, Lepus, Sylvilagus* spp.), with one exception. We do know that the only eimerian reported to date from endangered pygmy rabbits (*Brachylagus idahoensis*) can be highly pathogenic and may be responsible for recent declines in some of their isolated populations.

AN EMERGING DISEASE PROBLEM

It is difficult to find uninfected rabbits, and probably no rabbit remains uninfected for life. The uninfected condition, which is likely transient, may be more common in wild rabbits in their natural environments (however, see above), but coccidiosis in rabbits is an emerging disease of growing importance in commercial rabbits throughout the world (Licois et al., 1990) because of the increasing commercial production of rabbits as a source of protein, for medical research, and as pets. For example, in India, where the incidence of disease is reported to be 13–64%, it has a major impact on rabbit production and is responsible for a high incidence of morbidity and mortality. Similarly, the incidence of disease in European countries is 21–60% (Bhat et al., 1996). Al-Mathal (2008) found 158/490 (32%) domestic rabbits, 1 to 4 months old, on three farms in Saudi Arabia, to be infected with *E. stiedai* and demonstrating symptoms of infection (and see Toula and Ramadan, 1998). Coccidia also are an important contributing factor to diseased rabbits in Iran (Yakhchali and Tehrani, 2007; Razavi et al., 2010), Nigeria (Musongong and Fakae, 1999), Syria (Darwish and Golemansky, 1991), southwestern Australia (Hobbs and Twigg, 1998), Brazil (Mundin and Barbon, 1990), and elsewhere throughout the world.

COCCIDIOSIS: DISEASE, SYMPTOMS, PERPETRATORS

Coccidiosis in rabbits primarily is a disease of young animals, while adults most often act as carriers by discharging oocysts into the environment. Rabbits become infected by ingesting sporulated oocysts with their food or water. The severity of the disease depends upon the number of infective oocysts ingested, upon the coccidian species involved and its preferred habitat and cell type within the rabbit, and upon each rabbit's own immune and nutritional status. Disease occurs most often in intensively managed animals, but it appears also in well-cared-for rabbits (van Praag, 2011). Feces may contain blood and threads of mucus. Young rabbits present with retarded growth, due mainly to side effects on the kidney and the liver. In general, during both intestinal and hepatic coccidiosis, the normal function of infected cells is inhibited, the cells are hypertrophied, and they eventually die.

Intestinal coccidiosis mainly affects youngsters from age 6 wks to 5 mo. Clinical symptoms may include a rough coat, dullness, depression, diminished appetite and poor feed conversion, abdominal pain, dehydration, loss of weight, pale watery mucous membranes, and (sometimes profuse) diarrhea, 4 to 6 days post-infection (PI). Within intestinal enterocytes, when numerous cells are infected the villi will atrophy, leading to malabsorption of nutrients, electrolyte imbalance, anemia, **hypoproteinemia**, and dehydration due to erosion and ulceration of the epithelium. If weight loss reaches 20%, death may follow within 24 hr and can be preceded by convulsions or paralysis. In intestinal coccidiosis, disturbances in water and electrolyte balance occur in parasitized parts of the intestine before the appearance of the macroscopic lesions and are characterized by a loss of water and sodium (Bhat et al., 1996). The loss of sodium is compensated by the exchange of potassium from the blood, thereby leading to **hypokalemia**, and may cause death (Lebas et al., 1986). The majority of intestinal species develop in the small intestine, while only *E. flavescens* and *E. piriformis* complete their development in the caecum and colon, respectively. Upon necropsy, inflammation and **edema** will be found in the most heavily infected portions of the gastrointestinal tract (dependent on the coccidian species), sometimes accompanied by bleeding and mucosal ulcerations (Oncel et al., 2011; van Praag, 2011). Intestinal coccidia species can be classified into four groups based on clinical parameters (weight loss, diarrhea, mortality): non-pathogenic to slightly pathogenic: *E. exigua*, *E. irresidua*, *E. matsubayashi*, *E. perforans*, *E. piriformis*; moderately pathogenic: *E. coecicola*,

E. media; very pathogenic: *E. flavenscens*, *E. intestinalis*, *E. magna*; and unknown pathology: *E. nagpurensis*, *E. oryctolagi*, *E. roobroucki*, *E. vejdovskyi* (numerous authors; see Chapter 6).

Hepatic coccidiosis, caused only by *E. stiedai*, affects rabbits of all ages when the parasite develops in the bile ducts of the liver, which become enormously enlarged, and thereby interferes with liver function. In hepatic coccidiosis, white nodules or cords develop on the liver, which later tend to coalesce; infected animals may have diarrhea and their mucous membranes may be **icteric**. Other symptoms include listlessness, thirst, wasting of the back and hindquarters, and enlargement of the abdomen caused by an enlarged liver and gall bladder. This infection can take either a chronic course during several weeks, or it may turn acute and end in death within 10 days, preceded by coma and sometimes diarrhea. At necropsy, liver, gall bladder, and the bile duct will be found to be distended and enlarged with white nodules covering the surface. Secondary bacterial infection, in particular with *Escherichia coli*, can lead to bacterial presence in the nervous system (Abdel-Ghaffar et al., 1990; Bhat et al., 1996; Yakchali and Tehrani, 2007; Freitas et al., 2010; van Praag, 2011).

Finally, hematological studies of rabbits infected with either intestinal or hepatic eimerian species (or both) demonstrate numerous changes in blood parameters including, but not limited to: reduced hemoglobin and RBC count, accompanied by a significant increase in packed cell volume (PCV) and total white blood cell (WBC) count; decreased levels of liver lipids and of sodium and chloride, but increased levels of total protein, globulin, cholesterol, low-density lipoproteins (LDL-c), **triacylglycerols**, and potassium. Electrolyte imbalance is likely attributed to diarrhea. Serum calcium, iron, copper, zinc, and glucose are usually slightly lower than in healthy animals and may indicate malnutrition due to intestinal damage or secondary bacterial infection. Liver coccidiosis is accompanied by significant elevation of serum **bilirubin**, **alkaline phosphatase (ALP)**, **alanine aminotransferase (ALT)** and **aspartate aminotransferase (AST)**, and **gamma glutamyl transpeptidase (GGT)** (Licois et al., 1978a, b; Peeters et al., 1984; Abdel-Ghaffar et al., 1990; Bhat and Jithendran, 1995; Jithendran and Bhat, 1996; Kulišić et al., 1998, 2006; Tambur et al., 1998a, b, c, 1999; Freitas et al., 2010). When animals return to normal after appropriate treatment, values return to normal levels (van Praag, 2011).

EPIDEMIOLOGY OF DISEASE IN RABBITS AND ITS LIMITATIONS

All populations of organisms are limited partially or completely by diseases in their ecosystems (Real, 1996; Pimentel et al., 1998), but the complexity of variables that must be considered when studying the epidemiology of any disease condition is daunting. Disease prevalence in populations and ecosystems is influenced by numerous biological and environmental factors, including infectious organisms (viruses, bacteria, fungi, protists, helminths, etc.), presence or absence of intermediate hosts (in **heteroxenous** life cycles) and of mechanical vectors (e.g., filth flies, earthworms, soil nematodes, etc.), air, water and soil pollutants (chemical and biological wastes), and shortages of food and nutrients (Dubos, 1965). In addition, selection that exists in natural populations is inhibited by domestication; this was demonstrated early by Osipovskiy (1955), who showed the importance of selection with regard to resistance to coccidiosis. While working with chinchilla rabbits from 60 to 65 days old, he showed that, through selection, mortality due to artificial infections could be reduced from 45% in the first generation to 18% in the second, and to 9% in the third generation. This complex of factors and their interactions makes tracking and assessing the causes and effects of individual diseases extremely difficult (McMichael, 1993). Disease dynamics are further complicated by the increased density of the animals during crowding (loss of habitat in the wild; domesticated breeding facilities), because high densities facilitate both the increase and spread of infectious organisms. Thus, it would be ideal if we could assess the relationship between high population density and increased environmental degradation and be able to quantify and correlate such variables as: population growth and disease transmission; water pollution and diseases; atmospheric pollution and diseases; pesticide pollution and diseases, especially those chemicals that affect nervous systems by inhibiting **cholinesterase** (e.g., **organophosphate** and **carbamate** classes); the effect of land degradation on disease incidence; food contamination, disease, and malnutrition; drug resistance; and changes in ecosystem biological diversity, evolution of parasites, and invasion by exotic species that all can result in disease outbreaks. Clearly, the puzzle is complex and our limitations are great.

Mykytowycz (1962) noted that coccidia frequently have been held responsible for the death of free-living rabbits and hares. However, very few such studies were carried out systematically and in sufficient detail to warrant those conclusions, including those by Ritchie (1926), who reported coccidiosis among mountain hares in Scotland; Clapham (1954), who pointed out the association of coccidia with mortality of wild rabbits in England; Rieck (1956), who attributed death from coccidiosis among free-living hares and rabbits in Germany; Naumov (1939), who drew the same conclusion of hares in Russia; and Bull (1953, 1960), who, like others before him, came to the conclusion that "*E. stiedae* is well suited to control rabbit numbers in New Zealand."

Mykytowycz's (1962) own study of the epidemiology of coccidiosis involved six *Eimeria* species he studied in a free-living population of wild rabbits (*O. cuniculus*) confined to a 2-acre experimental enclosure near Canberra, A.C.T., Australia. Eight wild rabbits (three does, five bucks) born in captivity, and infected naturally with the six *Eimeria* species, were introduced into the enclosed pasture and the infections were monitored. Over 35 months, 2,326 fecal samples were examined from these rabbits and only 3.6% were found to be coccidia-free. The proportion of infected rabbits increased as the population increased. Mykyto-wycz (1962) felt that the high percentage of infected rabbits was not an artifact from the enclosure conditions, since separate examina-tion of field-caught rabbits showed an 89% infection rate. He also reported that multiple infections were common: only one *Eimeria* species was found in ~7% of the rabbit samples, two species in 29%, three in 38%, four in 17%, five in 4.5% and six in 1%. Other interesting conclusions from this study were: (1) the increased discharge of oocysts by infected rabbits over time was due to the breakdown of resistance rather than to exposure to new infec-tions; (2) once three eimerians (*E. perforans, E. piriformis, E. irresidua*) had established them-selves in the population, they were persistent and tended to maintain high levels of infection; (3) in contrast, infections with three others (*E. media, E. magna, E. stiedai*) were subject to pronounced fluctuations; (4) resistance to coccidia increased with age, as noted by the decrease in the discharge of oocysts by older animals; (5) development of resistance to these eimerian parasites appeared at the age of 3 mo; (6) three eimerians (*E. perforans, E. irresidua, E. media*) showed no tendency to infect rabbits of any particular age group; (7) in contrast, two species (*E. magna, E. stiedai*) were most frequently found in rabbits of the youngest age groups, 1 to 3 mo; (8) one species (*E. piriformis*), on the other hand, was less common in younger age groups and infected a high proportion of older rabbits; (9) two species (*E. piriformis, E. perforans*) were the most common species found, whereas infections with two others

(*E. media, E. magna*) were intermittent; (10) *Eime-ria piriformis* affected all age groups with equal severity, while *E. stiedai* affected the youngest rabbits most severely; (11) large numbers of oocysts in the feces did not indicate the presence of demonstrable disease; and (12) there was little difference in the level of infection between the sexes, even during the breeding season.

THE SPECIES CONCEPT

One of the stated objectives of this book is to document, in one place, the number of nominal species of coccidia that infect rabbits. But what is a species? Darwin said that species are related by the physical connection of genealogical descent, but questioned the reality of species, saying they were artificial combinations made for convenience (Gould, 1992). Gould (1992), however, argued that there was a rationale that could be used to give a historical definition of species as "unique and separate branches on nature's bush." His rationale generally works for most **eukaryotes**, but **prokaryotes** (bacteria, **archaea**), viruses, and **prions** undermine all old definitions by the ways they can share, incorpo-rate, and exchange DNA.

Defining any species is a challenging task, particularly in groups of microorganisms that are not well known, and species concepts, there-fore, are necessarily tailored for specific groups. Schipani (2011) correctly stated that "Imposing separations on the fluid process of evolution inevitably breeds disagreement on what exactly constitutes a species." Delimiting and then defining species has challenged biologists since Carolus Linneaus first began to sort and name organisms 250 years ago using only macro-scopic physical characteristics, and Mayden (1997) identified at least 22 different species concepts. Only since the advent of molecular sequencing techniques have species become less confounding to researchers, because now two or more "cryptic" organisms once thought to be identical can be split into separate species due to differences in their genetic code. Since DNA is the only thing that unites all species, gene sequencing and DNA analysis are helpful in many ways, but they do not replace classical scholarship, and, as we will see with the coccidia in general and rabbit coccidia in particular, gene sequencing data are simply not available for the vast majority of species. Thus, structural (sporulated oocysts) and biological (endogenous developmental stages, host specificity) characters must still play a more prominent role in defining and distinguishing between most species of coccidia, including those from rabbits. And we must always keep in mind that alpha taxonomy (describing and naming new species based on the appropriate and available character states at the time) is always a work in progress. Those people (taxonomists) who delimit a species will always be followed by someone who does a better job as better tools are available.

One of the fundamental questions of biology is, "What is the full nature of biodiversity on Earth?" In the time since Linnaeus introduced the practice of **binomial nomenclature** and articulated the goal of identifying all species of organisms, we still have accounted for only a tiny fraction of the whole (Wilson, 2003, 2004, 2011). The number of living species discovered and named to date is about 2 million, but the true number of all living things (including viruses, archaeans, and bacteria) is minimally 10 times that number (Reaka-Kudla et al., 1997; Stork, 1997) and may be > 100 million species (Wilson, 2004)! For example, the insects, with about 1 million named species, easily could have five times that number. And the nematodes, with about 25,000 named species, might be the most abundant eukaryotes on Earth if we consider that virtually every insect that has been studied in detail may have at least one nematode species that is unique to it. In fact, it is estimated that, "If all the matter in the universe except the nematodes were

swept away … we should find [our world's] mountains, hills, vales, rivers, lakes, and oceans represented by a thin film of nematodes" (Cobb, 1914; and quoted by Roberts and Janovy Jr., 2009, p. 369). Some say that four of every five animals on earth are nematodes. Similar comparisons of magnitude can be made for mites, gregarines, and viruses, and it is estimated that microbes make up 90% of the ocean's biomass, including perhaps 20 million bacterial species (Wilson, 2011). Wilson (2011) further predicts that as we look more carefully at protists, fungi, and small invertebrates, we will discover numbers "even larger than expected, as species are added that are very rare, seasonal, limited in distribution, or specialized for niches seldom examined. Others will be found that are 'cryptic,' forming genetically separate populations, but so similar in anatomical traits conventionally used by taxonomists as to be overlooked until their DNA is sequenced." Clearly, we only know a small fraction of the number of species that share the Earth with us, and defining these species is certainly the most challenging part of assessing our planet's biodiversity. So what about the coccidia?

TAXONOMY AND SYSTEMATICS CRISES AFFECT BIODIVERSITY

Science journalist Carl Zimmer (2012, p. 181) asked outspoken, visionary astrophysicist Dr. Neil deGrasse Tyson, Director of the Hayden Planetarium, about our current knowledge of the universe: he said, "Everything we know and love (about the universe)—electrons, protons, neutrons, light, black holes, planets, stars—everything we know and understand occupies four percent of the universe. So we're just dumb—stupid—about what's driving the cosmos." How does knowing and understanding about 4% of the universe compare to what we know and understand about the coccidian parasites of vertebrates? About the same!

Thorough, revisionary studies of coccidia from vertebrates have been completed for only a few host groups. For example, only 45 of ~6,009 amphibian species (< 1%; Duszynski et al., 2007), 208 of ~3,108 snake species (7%; Duszynski and Upton, 2009), 37 of 428 soricomorph (insectivore) species (9%; Duszynski and Upton, 2000), 18 of 233 primate species (8%; Duszynski et al., 1999), 86 of 925 chiroptera (bat) species (9%; Duszynski, 2002), and 23 of 92 lagomorph species (25%; this work) ever have been examined for coccidia (that is, 417/10,795; **4%**); from these few vertebrate groups, 10 genera and about 620 valid coccidia species have been documented. We can only imagine what remains to be learned when the other 96% of these vertebrate hosts are examined and their coccidia discovered. And these numbers don't include the birds, the fish, the rodents and other mammals, the lizards and other reptiles, etc. We think this is the perfect prelude to our discussion on taxonomy, systematics, and biodiversity, because these are sciences in crises.

Taxonomy

Taxonomy is the science of naming organisms, and it is arguably the oldest of all biological disciplines, going as far back in written form to the time of Aristotle (Luoma, 1991). Carolus Linneaus set out to catalogue all the plants and animals known in the mid-1700's and described about 7,300 plants (Linneaus, 1753) and 4,200 animals (Linneaus, 1758) in his 2 treatises, *Systema Plantarum* and *Systema Naturae*, respectively (Luoma, 1991). But taxonomy is much more than merely a descriptive science that inventories the Earth's biodiversity; given the extreme particularity of all species groups, and how little we know about the species in most of them, taxonomy can justly be called "the pioneering exploration of life on a little known planet" (Wilson, 2004). Taxonomy provides the database for ecology

and conservation science, and it lays the foundation both for the phylogenetic tree of life (Wilson, 2004) and the emerging encyclopedia of life (Wilson, 2003; Blaustein, 2009). Taxonomists interpret and integrate millions of facts about the myriad of species now known; they make the identification of species available to others, they provide the vocabulary to talk about them, they test the evolutionary units of biological diversity, and they make accessible all that we know about life on this planet (Wheeler, 2004). Taxonomy makes testable hypotheses and its results are as rigorously scientific as any; it is *not* just an identification service for other biologists, although the need for *reliable* species identifications is a major justification for all biologists to support it.

As humans, we are innately curious about other species. There will always be both professionals and amateurs who appreciate aesthetic beauty in nature, and who will want to go into new places to identify species based on observable morphology (Wilson, 2011). What critics of taxonomy as a valid discipline fail to acknowledge is that it is in no way different from genomics or proteomics, also descriptive disciplines, which are involved in endeavors such as bacterial artificial chromosome libraries and the Human Genome Project (Wheeler, 2004). But taxonomy is not *just* descriptive, it is hypothesis-driven and tests hypotheses on multiple levels. For example, if a taxonomist claims that one (new) species differs from all (previously) known species, that claim is a hypothesis about the discontinuous distribution of its unique combinations of characters. Every division (clade), at every Linnaean rank, is a hypothesis, and all of these hypotheses represent generalizations from which specific future observations can be predicted (Wheeler, 2004). When those future observations are made, the initial hypothesis is either corroborated or refuted. When one studies the coccidia, and each time a new host specimen is examined—particularly one in a previously unexamined population—new species are described establishing hypotheses that can be tested. When new species and characters are added to a data matrix, higher taxa are opened to testing anew. As additional specimens and populations are studied, these hypothesized boundaries are tested. Given that many species have wide and complex ranges, truly rigorous species testing is neither automatic nor easy (Williamson and Day, 2007). But before we begin species testing, we must know what has been done and what yet needs to be done. Thus, a book such as this one makes an immediate and lasting contribution toward understanding the diversity of less conspicuous species, like the coccidia, which will never be as popular with professionals or amateurs as are mammals or birds. Is it important that the identification of protists, and knowledge of their biology, is unknown to most biologists? Yes! Does it seem to some that process-based ecological research and laboratory-based genomic sequencing take a more modern approach to species identification? Yes! Yet the latter are of little value, and likely unrepeatable, unless they are closely linked to the scientific identities of the organisms concerned. Such identities are needed to connect biodiversity and ecosystem processes, information that must be at hand to effectively diagnose and control diseases caused by parasitic protists (Williamson and Day, 2007). Without taxonomy, phylogeny is impoverished (Korf, 2005), ecology is deprived of one of its fundamental units of currency (Gotelli, 2004), and conservation biology loses focus and aim (Godfray and Knapp, 2004; Mace, 2004).

Systematics

Systematics is the science that looks beyond the naming of species and more deeply into the evolutionary and ecological relationships among organisms (Luoma, 1991). The conundrum of non-comparable species concepts (see Mayden, 1997) stresses the importance of, and

can be resolved by, revisionary systematics. In taxonomic work, it is critical that well-trained systematists revisit the earlier work of taxonomists. That is, the best approach is to stand not on the shoulders of a few giants, but upon a pyramid that continues to grow by ordinary researchers revising each other's work and building upon it. Ecologists who study biodiversity rarely consider revisionary systematics as important to their studies because it's dismissed as merely descriptive. But counts of species very well may be unreliable without frequent reevaluation and expansion of the character and distributional data upon which taxonomic concepts are based. Such revisionary work adjusts earlier concepts, sometimes dramatically. For example, Wilber et al. (1998) revised the taxonomy of *Eimeria* species from rodents in the mammalian Tribe Marmotini (Sciuridae) and reduced the number of named eimerians from 40 to 26 species. With successive efforts, a general community standard will eventually take shape, and the tally of taxa should converge on a reliable estimate. This means that large-scale ecological agendas need to support not only alpha taxonomy, but also revisionary systematics on taxa of special interest to them. There are important practical consequences implicit in the recognition that revisionary taxonomy also is hypothesis-driven. The confidence one has in a species hypothesis is directly proportional to the extent to which that hypothesis has been tested. Unfortunately, hypothesis testing is rarely done in taxonomy, especially in the case of little-studied microorganisms like the coccidia, where new observations of related species have the potential to alter the hypothesis. Without well-trained taxonomists in these disciplines there will be no revisions (e.g., monographs), and without them, most species hypotheses remain untested and no real progress in taxonomy or in phylogenetic biology can occur. Revisions, the summation of scattered reports of species and subordinate taxa, when carefully reconsidered and revisited, are the essential tools for progress in understanding biodiversity.

It is widely known, and unfortunate, that we have a crisis in the training of new taxonomists and systematists (Wheeler and Cracraft, 1997; Brooks and Hoberg, 2000; Wheeler, 2004; Wheeler et al., 2004; Wilson, 2004; Ebach and Holdrege, 2005a, b; Korf, 2005; Agnarsson and Kuntner, 2007; Raczkowski and Wenzel, 2007; Grant, 2009). The worldwide total of specialists with sufficient skills to identify microorganisms like protists is likely less than a few hundred individuals (Williamson and Day, 2007). Worldwide, as few as 6,000 biologists work in all of taxonomy, and this number is dwindling fast as traditional taxonomists retire and universities fail to train new people to replace them (Wilson, 2003). According to Korf (2005), in the last 30 years, the decline in taxonomic work has been "catastrophic, to the point that taxonomy is today potentially bleeding to death," and Ebach and Holdrege (2005a, b) offer a similar cataclysmic view: "Taxonomists soon will become fossils in the strata of scientific evolution themselves." Clearly, not enough people are qualified currently to interpret classical taxonomic scholarship, and that number is shrinking, which means that in the very near future we will not be able to revise our taxonomy, leaving us with whatever baseline the original authors described.

Biodiversity

The guiding agenda of understanding biodiversity is to fully discover and describe the species of an entire planet (Wilson, 2003, 2011; Wheeler, 2004; Blaustein, 2009). This taxonomic imperative has theoretical, practical, and ethical implications. Even with conservative estimates, it is likely that as few as 50% or as many as 90—98% of the Earth's species remain to be discovered and described. The majority of the lay public seem surprised to learn that most of biodiversity is still entirely unknown. Our lack of knowledge compromises any hope of resolving the phylogeny of life on Earth, unless there is a truly ambitious effort to advance descriptive taxonomy. In the face of this biodiversity crisis, the need to rebuild expertise and strengthen the infrastructure for taxonomy is paramount (Wheeler and Cracraft 1997). Although the precise impact of species excluded from any systematic analysis varies from case to case, there is general agreement that multiple missing taxa are a serious impediment to the recovery of true phylogenetic patterns. For all but a few relatively well-known (and small) clades, this ignorance of species diversity poses an impediment to resolving phylogenetic relationships. And what is the result of this ignorance? Biologists in the future will have little faith that proposed phylogenies are even approximately correct and phylogenies will be subject to frequent and major reorganizations. Expertise on many taxa, especially within the Protista, is disappearing entirely, and what passes for morphology in phylogenetic studies is sometimes just a literature review added as a few lines to a predominantly molecular data matrix. Our book builds on reconciling the taxonomic work on rabbit coccidia of all who went before us. The older literature is still pertinent to the biodiversity enterprise, especially for our lesser-studied species. In these groups, our knowledge may exist in the form of one article or one photomicrograph of one life-history stage describing and representing the entire species content. Therefore, current research will necessarily rely on older literature in a way that is not paralleled by other scientific fields (e.g., particle physics) where it is not necessary to consult eighteenth- and nineteenth-century works to do current research.

MODERN DNA TAXONOMY

Wilson (2003, 2004, 2011) emphasized that each species is a universe unto itself: its own genetic code, anatomy, behavior, life cycle, and environmental role, "a self-perpetuating system created during an almost unimaginably complicated evolutionary history." It wasn't until the advent of DNA technology, about 2 decades ago, that taxonomists began to incorporate parts of an organism's genetic code into their taxonomic descriptions. However, we must be cautious; DNA taxonomy is useful for labeling specimens whose names are already known (Raczkowski and Wenzel, 2007), but such data are not likely to be a substitute for formal revisionary systematics for several fundamental reasons. First, molecular studies rely on the standards of species that were described through traditional means, but they do not reexamine the original descriptions, nor do they often include many individuals (Lipscomb et al., 2003; Raczkowski and Wenzel, 2007). Indeed, as we progressively use DNA sequencing to enhance our power to discover small differences among species, it is likely we also will amplify problems associated with "over-splitting" and lose track of the more universal species concepts upon which the rest of biology is founded. Thus, it cannot be just about DNA sequencing; the best future approach lies in advancing classical taxonomy and systematics hand-in-hand with modern DNA methods (Raczkowski and Wenzel, 2007). In the past, molecular biologists have criticized taxonomists for conducting descriptive work without themselves recognizing that much of molecular biology is descriptive and *not* hypothesis-driven. DNA is simply more species data and Lipscomb et al. (2003) felt that "it is clearly impossible to equate DNA sequences with taxonomic insights." We need to remember that good science is, unequivocally, always hypothesis-driven.

TAXONOMY VS. BARCODING

It's been stated both implicitly and explicitly that taxonomy has little intellectual content and is seen as a descriptive science whose primary function is identification. Many molecular biologists have mistakenly expressed the view that such a discipline could, and should, be replaced by mitochondrial DNA (mtDNA) barcoding (Ebach and Holdrege, 2005a, b). Early barcoders (Herbert et al., 2003) proposed a universal animal barcode: a segment of ~650 basepairs of the mitochondrial gene cytochrome *c* oxidase subunit 1 (CO1), which is attractive because primers used to amplify that gene fragment worked across many animal groups, including birds (Herbert et al., 2004), fish (Ward et al., 2005), leeches (Siddall and Budinoff, 2005), and mosquitoes (Grant, 2007; Kumar et al., 2007). These studies took species that were already well delineated through morphology, ecology, and other characters, collected their CO1 sequences and determined how closely traditional classifications matched with those derived from the barcode DNA. Barcoding is an extremely exciting and potent tool for helping make species identifications easier in some protist taxa, like the coccidia, with complex and multiple life-cycle stages. For example, when one stage of the life history is readily identified by morphology (e.g., a sporulated oocyst), DNA barcodes can enable associations with other life-cycle stages (e.g., endogenous developmental stages). In the clinical setting, it also has the potential to rapidly identify biological pathogens in ailing hosts (Grant, 2007; Williamson and Day, 2007).

However, barcoding is not a good approach to species discovery and description. Very often, the CO1 gene sequence is used as the animal barcode, but it does not differentiate accurately between all animal species, and for some taxonomic groups (some Diptera, Scleractinia corals), it presumably doesn't work at all (Grant, 2007). In rabbits, for example, mtDNA sequences gave misleading signals that proved unreliable and impractical in attempted phylogenetic reconstruction of hares (*Lepus* spp.) (Slimen et al., 2008; Kriegs et al., 2010).

This leaves some taxonomists unconvinced of the utility of trying to barcode every species, especially because it is generally accepted that any single-character system is unlikely to be the panacea for anything. Plant barcoders have not yet identified a suitable, standardized alternative to CO1 because it doesn't work in plants; their evolution rates are much more variable than in animal species, meaning that divergence in plant mitochondrial genomes is virtually nil (Grant, 2007).

Barcoding could indeed be valuable in quick species identification for rapid environmental assessment and in allowing a rapid survey of the biodiversity in a polluted lake or in a planned development site near an environmentally sensitive area (Ebach and Holdrege, 2005a, b). But we must remember that DNA barcoding generates information, not knowledge, and that information will prove meaningful only if it can be placed within the context of morphological, physiological, and behavioral knowledge; that is, it needs to be put into context to be useful (Ebach and Holdrege, 2005a, b). On its own, a barcode cannot confer new species status; it is just another technique in the taxonomist's toolbox and should not be taken for more than that. However, when used with an appropriate bioinformatics database, barcodes promise to provide rapid and effective identification with good species separation. Such an evaluation process requires traditional taxonomic standards in order to connect a barcode to an authoritatively named, museum-preserved sample or culture (Ebach and Holdrege, 2005a, b; Williamson and Day, 2007).

PARASITES, SHRINKING ECOSYSTEMS, AND DISEASE

Every living organism is host to many kinds of **commensals** and parasites during part or all of its lifetime, but in most instances and most of the time they do little or no harm. Under our current paradigm of host-parasite co-evolution, parasites are continually evolving more effective methods to exploit their host, and the host is continually evolving more effective methods to control that exploitation (e.g., immunological, behavioral); that is, each partner in this association is exerting a selective pressure on the other one. Under what circumstances, then, do parasites become pathogens? Holmes (1996) suggested the answer involves a triad consisting of the host, its parasites, and their environment. All parasites, and especially obligate intracellular parasites like the coccidia, are dependent upon their hosts for survival, and both host and parasite survival is dependent upon the environment in which they both exist. It is not within the scope of this book to belabor either parasite ecology or how parasites can (and do) affect biodiversity; those topics are best left to the experts (see Combes, 1996; Holmes, 1996; Renaud et al., 1996). We think it is relevant, however, to superficially highlight a few parts of the complexity involved that have influenced the host-parasite patterns we see in the coccidia of some rabbits.

Shrinking ecosystems (habitat destruction), the introduction of exotic animals and their parasites, overkilling of certain host species, environmental toxins (e.g., acid rain, industrial wastes, pesticides, etc.), and climate change have important consequences on the emergence of disease, especially in shrinking ecosystems (Wilson, 1992; Agnarsson and Kuntner, 2007). Indeed, Wilson (1992) identified these components as the "mindless horsemen of the environmental apocalypse." Shrinking ecosystems, for example, will concentrate both species and individuals into restricted areas, and this can only promote both the transmission and exchange of parasites among individuals and host species, especially for parasites with direct life cycles, like most coccidia. As ecosystems shrink, their habitats become fragmented, which increases their edge effect and can bring an influx of new host species into the disturbed habitats between fragments (Holmes, 1996). If new hosts introduce their parasites upon invasion, these may be new species and/or more pathogenic strains than those in the resident host population; thus, as habitats become compressed, we should witness the emergence of new diseases and increasing numbers of epidemics (Scott, 1988; Holmes, 1996). Each of these effects increases the potential for a host's usual parasites to become pathogenic, so the importance of disease is expected to increase in shrinking ecosystems. Some of these changes may contribute to what we are witnessing in pygmy rabbit populations in the Pacific northwestern USA (see Chapter 4).

1988; Holmes, 1996). Each of these events increase the potential for a host's usual parasites to become pathogenic, so the importance of disease is expected to increase in shrinking ecosystems. Some of these changes may contribute to whatever are witnessing in pygmy rabbit populations in the Pacific northwestern USA (see Chapter 4).

new host species into the disturbed habitats between fragments (Holmes, 1996). In new hosts introduce their parasites upon invasion, these may be host species and/or more pathogenic strains than those in the resident host population: thus, as habitats become compressed, we should witness the emergence of new diseases and increasing numbers of epidemics (Scott

2

Lagomorpha Origins and Diversification

There are differing opinions as to the evolutionary relationships of the lagomorphs relative to the rest of the **eutherian** (placental) mammals (see McKenna, 1982). Li et al. (1990), using only the coding regions of up to 14 protein-coding genes, constructed molecular phylogenies to try to determine the branching order of five major mammal orders: Artiodactyla, Carnivora, Lagomorpha, Primates, and Rodents, because the evolutionary position of the Lagomorpha always has been in controversy, with some paleontologists arguing that rodents and lagomorphs are most closely related to each other, while others argue that lagomorphs separated from the rodents at the time of eutherian radiation (Li et al., 1990). They interpreted their data to provide "strong evidence" for their four major conclusions: (1) Rodentia is an outgroup to the other four mammalian orders; (2) there is not a close relationship between the orders Rodentia and Lagomorpha; (3) the Lagomorpha branched

off from the major eutherian lineage after the Rodentia, but before Artiodactyla, Carnivora, and Primates; and (4) the Artiodactyla-Carnivora formed a single, "superordinal" clade. Later work by Rose et al. (2008), based on recent fossil discoveries, returned to the concept of a mammalian superorder, Gilres, which includes both the Rodentia and Lagomorpha as closest relatives (also see Meng and Wyss, 2005; Missiaen et al., 2006).

The order Lagomorpha is generally agreed upon to be a homogeneous grouping that comprises two extant families, Ochotonidae (pikas) and Leporidae (rabbit and hares). Although rabbits and hares are fairly similar to one another in body size, when compared with the diversity of body forms and sizes in the closely related rodents, the phylogenetic relationships among leporids are still controversial and may become more so. Recently, paleontologists unearthed the bones of a giant rabbit, the largest ever found, on the Mediterranean island of Minorca (Millius, 2011). There, about 3 to 5 million years ago (MYA), a rabbit species now called *Nuralagus rex*, grew to about 0.5 m high and had an estimated weight of 12 kg, six times the size of today's European wild rabbit (*Oryctolagus cuniculus*); this shows the kind of "unusual turn" evolution can take on islands. When pioneer animals arrive and begin colonizing an island, rates of evolution can speed up, and initially small creatures can become very large and/or big ones can shrink. To date, no potential carnivores were found among the rabbit fossils from the same time period on Minorca, suggesting the rabbits could have evolved to a larger body size with no evolutionary pressures to escape predators by maintaining agility and speed. In fact, the short spinal column documented from these fossils suggests that *N. rex* probably didn't even hop much (Millius, 2011).

LAGOMORPH BIODIVERSITY AND FAMILIAL RELATIONSHIPS

When compared with other mammalian orders, the Lagomorpha is not diverse and contains just three families, Prolagidae (Sardinian pika, extinct), Ochotonidae (pikas), and Leporidae (rabbits and hares).

The monotypic Prolagidae (type species, *Prolagus sardus* [Wagner, 1832]) was found on a number of Mediterranean islands (e.g., Sardinia, Italy; Corsica, France), but it is now considered extinct by the IUCN. It is described from fossils, but actually may have survived until historic times, perhaps as late as 1774 (Hoffmann and Smith, 2005).

Species in the second family, Ochotonidae, most commonly are called pikas, but individuals also have been called cony, mouse hare, rock rabbit, and whistling hare by various authors (Holtcamp, 2010; Chapter 3). All pika species share a single genus, *Ochotona*, and show little morphological variation or diversity,

which makes them one of the most complex and problematic groups for mammalogists. They are **Holarctic** lagomorphs comprising about 30 species (Hoffmann and Smith, 2005) and numerous sub-species. Most ochotonid species are found in Asia, mainly on the Qinghai-Tibetan Plateau of (West) China, but also in Afghanistan, Burma, India, Iran, Japan, Kazakhstan, Korea, Nepal, Pakistan, and Russia, while only two species are found in North America (Chapman and Flux, 1990; Hoffmann and Smith, 2005). Yu et al. (2000) divided the genus into three groups based on molecular sequencing of mitochondrial cytochrome b and ND4 genes: "shrub-steppe," "mountain," and "northern," but the placement of species within these groups is still a matter of debate among mammalogists.

The Leporidae consists of 11 genera, which share a total of 61 species; seven of these 11 genera are monotypic. For the purpose of this book, the four leporid genera with coccidia described from them bear mentioning. *Lepus*, with 32 species, has nearly worldwide distribution (Chapter 5), but *Sylvilagus*, with 17 species, is limited to the Americas (Chapter 7). The monotypic *Oryctolagus cuniculus* may be considered the most important lagomorph species, with worldwide distribution and multiple human associations as lab animals, pets, pests, and sources of food and fur (see Chapters 1, 6), while the monotypic *Brachylagus idahoensis* (Chapter 4) may have the most restricted distribution (Pacific Northwest, USA) and also may be the most endangered lagomorph species (Hoffmann and Smith, 2005).

Several *Lepus* species have a long historical association with humans, and this has fostered their economic and ecological importance worldwide. Despite this, the evolution of hares and jackrabbits has been poorly studied. They lack a lot of morphological variation between species, and this has resulted in numerous taxonomic revisions throughout the last two centuries (Flux, 1983; Halanych et al., 1999). Prior to a phylogenetic analysis (mitochondrial cytochrome b gene data) of 11 of 32 (34%) *Lepus* species by Halanych et al. (1999), only two prior studies used molecular or biochemical analyses to address the evolutionary history of *Lepus*; Pérez-Suárez et al. (1994) used mitochondrial DNA to look at speciation of four Mediterranean hares, and Robinson and Osterhoff (1983) documented protein variation and its systematic implications in three South African hares. Halanych et al. (1999) provided the first phylogenetic framework for hares, but many of their conclusions seem equivocal based on work by subsequent authors. Their analysis consistently demonstrated two species groups, a western American clade and an arctic clade that allied *L. townsendii* with the arctic species; these results support previous groupings based on morphology. Their data also supported the view that species of *Lepus* within North America (including Canada and Mexico) do not form a monophyletic group, and the status of some of these species needs to be reassessed.

Sylvilagus is the other large genus in the Leporidae; these are called "cottontail" rabbits and include 17 species that are distributed throughout North (including Canada), Central and South America, as far as northern Argentina (Hoffmann and Smith, 2005). Several of these species have become commercially important (e.g., game animals) throughout their range, and some may be considered threatened or endangered (Halanych and Robinson, 1997). In appearance, the majority of cottontails somewhat resemble the wild European rabbit, *O. cuniculus*, the pygmy rabbit, *B. idahoensis*, and several other rabbit genera; at one time, several of these were synonymized by Hall (1981) and treated as subgenera of *Sylvilagus*. However, more rigorous morphological and recent molecular studies have separated these latter monotypic genera from *Sylvilagus*. Most *Sylvilagus* species have a short stub tail, called a **scut**, that is brown above, but with a white underside that shows prominently when they

are fleeing, thus giving them their nickname "cottontails." The diversity cottontails lack in morphology is compensated by great diversity in habitat selection. Some, like *S. floridanus*, are **euryoecious** because they can be found in deserts, farmlands, prairies, swamps, woodlands, and in hardwood, rain and boreal forests, while other species in the genus, for example *S. palustris* and *S. aquaticus*, must be considered **stenoecious**, because they are narrowly restricted to marshes or swamps and are limited by the availability of water. All species are terrestrial, although a few, such as *S. nuttallii*, are sometimes said to be semiarboreal because they climb into juniper trees to feed. At least one species, *S. palustris*, swims extensively, but all species probably can swim (Nowak, 1991).

Of the 12 extant lagomorph genera, only the five identified above have been examined for coccidia (Chapters 3–7).

PHYLOGENETIC RELATIONSHIPS AND HISTORICAL BIOGEOGRAPHY

Biogeography is the study of past and present animal and plant distributions and diversity. It is not our intent to cover the biogeography of lagomorphs worldwide, but there are several studies available on this subject that should set the framework for much of what we know. Also, there are a handful of modern studies on their evolutionary history and/or molecular phylogenetic analyses between and within various genera that seem relevant. These will allow us to provide a brief overview of some of the patterns and processes across time and space at the scales for which they are available, in spite of the fact that the data are contradictory in some instances. One should keep in mind that all organisms may become distributed temporally and geographically by two general mechanisms, dispersal and vicariance. When organisms disperse it means that they have moved across pre-existing barriers to create their present distributions. Vicariant events are those natural historic occurrences in Earth's history (e.g., uplifting of a mountain range) that introduce barriers to an environment that fragment or separate ancestral ranges of once continuously distributed organisms.

The oldest eutherian mammals now are estimated to have originated about 125 MYA, about 40–50 MY older than previously thought (Qiang et al., 2002). From a paleontological perspective, it was not clear whether the Lagomorpha originated in Asia or North America, and prior to the work by Rose et al. (2008), the lagomorph fossil record for the early and middle Eocene (54–45 MYA) was completely lacking. Dawson (1981) was the first to postulate that the expansion of Leporidae occurred in North America during the Miocene, an idea supported by some fossil discoveries also suggesting the family originated here (White, 1991; Voorhies and Timperley, 1997; Matthee et al., 2004). However, Rose et al. (2008) reported what may be the oldest known record of Lagomorpha fossils from early Eocene deposits (about 53 MYA) in west-central India, a timeframe that is in agreement with an earlier molecular study (Springer et al., 2003). Thus, the lagomorphs have a long evolutionary history, with the origin of the Ochotonidae and Leporidae from a common lagomorph ancestor extending back to at least the early Eocene in Asia (Springer et al., 2003; Rose et al., 2008). Phylogenetic analysis of molecular evidence supports the fossil record suggesting such a split (McKenna 1982; Douzery et al. 2003; Asher et al. 2005).

Ochotonidae

Although Ochotonidae are thought to have been more diverse during the Miocene (23–25 MYA), they currently are characterized by a single genus, *Ochotona*, that includes only about 30 extant species (Angermann et al., 1990; Nowak, 1999; Yu et al., 2000; Hoffmann and Smith, 2005); however, some of these species are thought to have one or more subspecies, with *O. princeps*, the American pika, consisting of 36 (!) subspecies (Yu et al., 2000). There is evidence in the fossil record that shows that *Ochotona*-like forms first originated in Asia (Rose et al., 2008), most likely spreading subsequently to Europe and the **Nearctic,** via **Beringia**, and undergoing a complex diversification in the Nearctic, with representatives of some clades transferring back and forth across Beringia during times of glacial maxima (2.5 MYA–11,000 years before present [YBP]). One of these lines, the monotypic *Prolagidae*, persisted until the late Pleistocene or early modern times (~1 MYA) on Corsica and Sardinia. Other ochotonid branches reached Africa and North America. The extant genus *Ochotona* appeared in Asia in the Pliocene (5–1.8 MYA) and spread from there, reaching western Europe and eastern North America in the Pleistocene, about 1.8 MYA to 11,000 YBP (Yu et al., 2000). The current range of the genus represents a considerable reduction from that during the Pleistocene. Approximately 28 *Ochotona* species occur throughout Asia while only two, *O. princeps and O. collaris*, are recognized from the Nearctic.

Leporidae

The hares and rabbits have a nearly worldwide distribution, but more than 70% of the genera have geographically restricted distributions (Matthee et al., 2004). Despite a number of attempts using morphological, cytogenetic, and mitochondrial (mt) DNA evidence, a rigorous phylogeny for the family remains obscure. In an attempt to reach some resolution for familial relationships, Matthee et al. (2004) constructed a "molecular supermatrix" using 27 taxa, including all 11 leporid genera, that analyzed five nuclear and two mitochondrial gene fragments. One of the interesting observations from their study was that their analysis of each gene fragment separately, as well as the combined mtDNA data, failed to provide good statistical support for intergeneric relationships, whereas the combined nuclear DNA topology greatly increased phylogenetic resolution among the leporid genera. Their analysis of dispersal-vicariance events revealed a parsimonious solution suggesting the current geographic distribution of leporids involved an Asian or North American origin (their work was before the discovery by Rose et al., 2008), followed by at least nine dispersals and five vicariant events. Of the dispersal events, at least three were intercontinental exchanges between North America and Asia (or vice versa) via the Bering land bridge, three were independent dispersals into Africa, and at least two dispersals into South America (one by *Lepus*, another by *Sylvilagus*) were identified (Matthee et al., 2004). However, as far as we know, there is no fossil or other evidence that species of *Lepus* made it into South America in the past. Their (2004) molecular clock analysis said these dispersal events occurred ~14 to 8 MYA (Miocene), when all the modern leporid genera are thought to have originated. It is noteworthy that tectonic events during the middle Miocene epoch (~20 MYA) resulted in the formation of land bridge connections between Eurasia and North America (Kummel, 1970). Prior to that time, global ice volume was relatively stable and low from 26 to 15 MYA, thus isolating Asia from North America (Zachos et al., 2001). According to Matthee et al. (2004), the opening up of the Bering land bridge during the middle Miocene, ~15 MYA, allowed a leporid lineage to be established in either North America (descendants: *Romerolagus, Lepus, Sylvilagus, Brachylagus, Pentalagus, Caprolagus, Bunolagus, Oryctolagus*) or in Asia (descendants: *Nesolagus, Pronolagus, Poelagus*). Within North America, the leporid ancestor gave rise to two lineages that diverged ~12.8 MYA: (1) *Romerolagus*; and (2) the lineage comprising *Lepus, Sylvilagus, Brachylagus, Pentalagus, Caprolagus, Bunolagus,*

and *Oryctolagus*. Between 11.8 and 10.3 MYA, a second intercontinental exchange occurred between Asia and North America (or vice versa) and was followed by a vicariant event leaving the ancestor of *Sylvilagus* and *Brachylagus* in North America and causing the establishment of ancestral *Pentalagus, Caprolagus, Bunolagus,* and *Oryctolagus* in Asia (Matthee et al., 2004). The "cottontail lineage" in North America later colonized South America. It is noteworthy that the pygmy rabbit, *B. idahoensis*, at the root of this clade, clearly is distinct from the cottontails (*Sylvilagus*), giving credence to its separate generic status (Halanych and Robinson, 1997; Matthee et al., 2004).

Resolving leporid evolutionary relationships and history among genera is difficult with conventional phylogenetic approaches, because anatomical characters converge and **chromosomal synapomorphies** are absent (Matthee et al., 2004). Early attempts at phylogenetic resolution based on premolar tooth patterns were the most definitive at that time (Dawson, 1958; Hibbard, 1963). Later, Dawson (1981) incorporated nine of the 11 leporid genera (*Poelagus, Bunolagus* not included) and concluded that *Nesolagus* and *Brachylagus* were basal (most primitive) sister taxa and the remaining leporid genera were subdivided into three more recent and contemporaneous evolutionary lineages: (1) *Pronolagus* and *Pentalagus* as sister taxa; (2) *Lepus, Oryctolagus, Caprolagus,* and *Sylvilagus*; and (3) the monotypic *Romerolagus diazi* (Dawson, 1981). However, Kriegs et al. (2010), who did phylogenetic reconstructions with **retroposon insertions**, said that only *Pronolagus* was the sister group to the remaining leporids. Following Dawson (1981), Corbet (1983) examined 21 morphological character states for 22 leporid species (including all genera) and concluded there is considerable **homoplasy** in leporid morphology. Likewise, the analysis of partial sequences of two mitochondrial genes (cytochrome b and 12S rRNA; Halanych and Robinson, 1999) failed to fully resolve the phylogenetic relationships among most genera, leading those authors to conclude that most genera originated contemporaneously, approximately 12 to 16 MYA. About the same time, Su and Nei (1999) suggested that this date was closer to 20 MYA. Cross-species **chromosome painting** by Robinson et al. (2002) showed that the numerous chromosomal rearrangements that characterize the karyotypes of many of the extant leporid genera are **autapomorphic**, and these probably arose as a result of a combination of low population numbers and founder events. The failure of the chromosomal data to provide markers that track descent from common ancestry, particularly at the intergeneric level, was thought to reflect the group's rapid radiation, thus mirroring the pattern suggested by the mtDNA sequence data.

Lepus

Hibbard (1963) proposed that *Lepus* first arose in North America based on the fossil evidence of his time, implying that hares radiated to other continents (Halanych et al., 1999). However, the recent fossil data by Rose et al. (2008) confounds Hibbard's argument, suggesting that the ancestors of modern-day lagomorphs had an Asian origin. Hibbard (1963) also wrote, "the genus *Lepus* has arisen from both pro-*Sylvilagus* and *Oryctolagus* instead of from just one of the stocks," but the molecular phylogeny of Matthee et al. (2004) suggested the divergence of *Lepus* predated the development of the *Sylvilagus/Oryctolagus* clade, ~10.3 MYA, placing the divergence of *Lepus* at ~11.8 MYA. It's been suggested that the *Lepus* radiation coincided with the opening of temperate grasslands, which facilitated the development of **precocial** young and other adaptations suitable for an open habitat (larger body size, strong hind legs; Corbet, 1986; Yamada et al., 2002).

The limited data provided by Halanych et al. (1999) suggest that North American *Lepus* species do not form a monophyletic clade to the exclusion of other hares, whereas Kriegs

et al. (2010), using retroposon insertions, argued that since "one orthologous CSINE3 element and one 17-nt deletion were found in all *Lepus* species, but not in other Leporidae," *Lepus* was a monophyletic genus. Halanych et al.'s (1999) conclusions were based on phylogenies constructed from mitochondrial cytochrome b genes which showed *L. townsendii*, *L. arcticus*, and *L. othus* in a clade separate from the remaining North American taxa. The placement of Old World taxa at the base of and within the arctic clade suggested to them that some hares invaded North America secondarily, perhaps via the Bering land connection. In contrast, *L. americanus*, which ranges across the high latitudes of the continental USA, Canada, and Alaska, is more closely related to taxa from the southwestern USA and Mexico (*L. alleni*, *L. californicus*, *L. callotis*) than to the arctic clade (Matthee et al., 2004). In addition, Halanych et al. (1999) noted deep divergence between *Lepus* species pairs that hybridize in the wild (e.g., *L. timidus*, *L. europaeus*). Hybridization between more distantly related species allowed Halanych et al. (1999) to argue that isolation mechanisms (e.g., geographic, behavioral, or ecological) may be driving speciation within *Lepus* and that the known lack of chromosomal diversity within *Lepus* points to mechanisms of speciation that do not invoke the chromosomal models suggested for other mammals (Halanych et al., 1999; Matthee et al., 2004).

Sylvilagus

Halanych and Robinson (1997) examined the evolutionary history of the genus *Sylvilagus* using sequence data from the mitochondrial 12S rDNA gene, which they felt was particularly reliable and effective for determining interspecific relationships and relatively recent evolutionary events in these mammals. Their 12S data supported earlier arguments by Green and Flanders (1980) and Corbet (1983) for the separate generic status of *Sylvilagus* and also indicated separate generic status (sister taxon) for the pygmy rabbit, *B. idahoensis*, based on sequence divergence values and genetic distance. They also noted that taxa that are geographically adjacent also are phylogenetically closely related (e.g., *S. palustris*, marsh rabbit and *S. aquaticus*, swamp rabbit). Their finding suggests that recent vicariance events might explain the diversification of several *Sylvilagus* lineages.

Oryctolagus

Research based on mitochondrial DNA polymorphism (Monnerot et al., 1994) of *O. cuniculus* has shown that all European rabbits carry common genetic markers and descend from one of two maternal lines. According to the theory proposed by Monnerot et al. (1994), these lines originated between approximately 6.5 MYA and 12,000 YBP when glaciers isolated them into two groups; one on the Iberian Peninsula and the other in Mediterranean France. Paleontologic data support the origin of the genus *Oryctolagus* in southern Spain, 6–6.5 MYA (Lopez-Martinez, 1989). However, it would be useful to have a more exhaustive analysis of genetic diversity using different markers to completely support their proposed hypothesis. Thus, it is believed that *O. cuniculus* is a native of southwestern Europe and northwestern Africa.

Coccidia (Eimeriidae) of the Family Ochotonidae: Genus *Ochotona*

The Biology and Identification of the Coccidia (Apicomplexa) of Rabbits of the World
http://dx.doi.org/10.1016/B978-0-12-397899-8.00003-2

FIGURE 3.1 The American pika, *Ochotona princeps* (Richardson, 1828).

INTRODUCTION

According to Hoffmann and Smith (2005), the Ochotonidae is monogeneric, consisting only of the genus *Ochotona*, and its species have been called a variety of common names including pikas, rock rabbits, and others by many different authors (Yu et al., 2000; Holtcamp, 2010). Pikas are Holarctic lagomorphs consisting of about 30 species, but only two are found in North America, while the remainder occur in Asia (see Chapter 2; Chapman and Flux, 1990; Hoffmann and Smith, 2005). Although they are remarkably similar in size, morphology, and body mass, Yu et al. (2000) divided the genus into three groups based on molecular sequencing, but the placement of species within these groups is still a matter of debate among mammalogists. Even less is known about the parasites of these hosts, especially their coccidian parasites. To wit, only about a dozen papers are published documenting the presence of coccidia in pikas and these described 17 *Eimeria* and two *Isospora* species, some of questionable validity, in seven *Ochotona* species (see Yi-Fan et al., 2009). All the information known about these 19 coccidian "species" is presented, below.

Oocyst and sporocyst structural characters used in this chapter follow the convention established earlier (Wilber et al., 1998; Duszynski and Upton, 2009) and now used by most journals for species descriptions; these characters are abbreviated throughout as follows: oocyst and sporocyst length (L), width (W) and ratio (shape index) (L/W), oocyst (OR) and/or sporocyst (SR) residuum, polar granule (PG), micropyle (M), micropyle cap (MC), Stieda body (SB), substieda body (SSB), parastieda body (PSB), refractile body (RB), nucleus (N), and sporozoites (SZ).

FAMILY OCHOTONIDAE THOMAS, 1897

HOST GENUS OCHOTONA LINK, 1795

EIMERIA BALCHANICA GLEBEZDIN, 1978

Type host: *Ochotona rufescens* (Gray, 1842) (syn. *Lagomys rufescens*), Afghan pika.

Type locality: ASIA: Russia: Turkmen, Kara-Kalinski region, Bolshoi Balkhan.

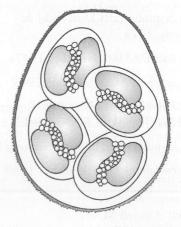

FIGURE 3.2 Line drawing of the sporulated oocyst of *Eimeria balchanica* (original) redrawn from Glebezdin, 1978.

Other hosts: None reported to date.

Geographic distribution: ASIA: Russia: Turkmen, Kara-Kalinski region, Bolshoi Balkhan.

Description of sporulated oocyst: Oocyst shape: ovoidal, flattened on top; number of walls: 1; wall characteristics: rough, yellowish; L × W: 26.5 × 20.9 (25−31 × 20−25); L/W ratio: 1.3 (1.2−1.3); M: present; OR, PG: both absent. Distinctive features of oocyst: shape and lack of both OR and PG.

Description of sporocyst and sporozoites: Sporocyst shape: ovoidal; L × W: 11.5 × 8.7 (11−14 × 8−11); L/W ratio: 1.3; SB: not mentioned, but absent in original drawing; SSB, PSB: both absent; SR: present; SR characteristics: small granules lying between SZ; SZ not described, but in Glebezdin's (1978) line drawing they appear blunt and lie at opposite ends of the sporocyst. Distinctive features of sporocyst: blunt-appearing SZ.

Prevalence: In 3/28 (11%) specimens of the type host.

Sporulation: Sporulation time unknown.

Prepatent and patent periods: Unknown.

Site of infection: Unknown. Oocysts recovered from feces.

Exogenous stages: Unknown.

Cross-transmission: None.

Pathology: Unknown.

Material deposited: None.

Remarks: There are 11 *Eimeria* species described from pikas with oocysts that have a M, in addition to this species: *E. barretti, E. calentinei, E. circumborealis, E. cryptobarretti* (?), *E. erschovi, E. haibeiensis, E. klondikensis, E. metelkini, E. ochotona, E. princepsis* (?), and *E. qinghaiensis*. Of these, sporulated oocysts of *E. balchanica* are closest in size to those of *E. calentinei* and *E. circumborealis*, but differ from the former by having a rough outer wall, a less defined M, lacking an OR, which *E. calentinei* possesses, and by having sporocysts without distinct SBs; they differ from the latter by having a rough outer wall, which *E. circumborealis* lacks, a less well-defined M, and by having sporocysts without distinct SBs. This species has not been seen since its original description.

EIMERIA BANFFENSIS LEPP, TODD & SAMUEL, 1973

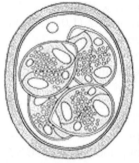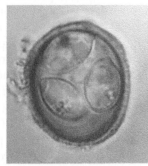

FIGURES 3.3, 3.4 Line drawing of the sporulated oocyst of *Eimeria banffensis* from Lepp et al., 1973, with permission of John Wiley & Sons. Photomicrograph of a sporulated oocyst of *E. banffensis* copied from Duszynski and Brunson, 1973, with permission of the *Journal of Parasitology*.

Type host: *Ochotona princeps* (Richardson, 1828) (syn. *Lepus* [*Lagomys*] *princeps*), American pika.

Type locality: NORTH AMERICA: Canada: Alberta, Banff, Sheep River area, 51°N, 115°W.

Other hosts: *Ochotona collaris* (Nelson, 1893) (syn. *Lagomys collaris*), Collared pika; *Ochotona curzoniae* (Hudgson, 1858), Plateau pika; *Ochotona hyperborea* (Pallas, 1811), Northern pika.

Geographic distribution: NORTH AMERICA: Canada: Alberta, Banff, Sheep River area, near Jumpinground and Sibbald Creeks, 51°N, 115°W (Hobbs and Samuel, 1974); Yukon Territory, Ogilvie Mountains, 64°N, 138°W (Hobbs and Samuel, 1974); USA: Colorado, Clear Creek and Larimer Counties (Duszynski and Brunson, 1972; Duszynski, 1974); ASIA: Japan, Hokkaido, Daisetzusan National Park (Hobbs and Samuel, 1974); Russia: Siberia, Providenya Oblast, Anadyr River (Lynch et al., 2007); People's Republic of China: Qinghai Province, Haibei Alpine Meadow Ecosystem Research Station, Chinese Academy of Sciences, 37°36′N, 101°18′E; altitude 3,205 m (Yi-Fan et al., 2009).

Description of sporulated oocyst: Oocyst shape: spheroidal to subspheroidal; number of walls: 2; wall thickness: 2.0–2.5; wall characteristics: outer is rough, pitted, brown, ~$^3/_4$ total thickness; inner is dark green; L × W: 29.8 × 25.2 (27–32 × 24–28); L/W ratio: 1.2 (1.0–1.4); M, OR: both absent; PG: present, one, ~2.0–2.5 wide. Distinctive features of oocyst: lack of OR, thick rough wall, and relatively large PG.

Description of sporocyst and sporozoites: Sporocyst shape: ovoidal; L × W: 13.0 × 9.6 (10–15 × 8–12); L/W ratio: 1.4 (1.0–1.7); SB, SSB: both present; PSB: absent; SR: present; SR characteristics: granular, spread throughout sporocyst; SZ: arranged lengthwise in sporocyst, with one large posterior RB and a smaller one anterior, and arranged transverse to long axis of SZ. Distinctive features of sporocyst: presence of SB and SSB, and arrangement of RB in SZ.

Prevalence: In 6/34 (17%) *O. princeps* from Alberta (Lepp et al., 1973), and Hobbs and Samuel (1974) found it in 9/111 (8%) *O. princeps* on talus slopes in southwestern Alberta (51°N, 115°W); 5/92 (5%) *O. collaris* in Yukon Territory (Hobbs and Samuel, 1974); 21/52 (40%) *O. curzoniae* in China (Yi-Fan et al., 2009); 3/14 (21%) *O. hyperborea* in Japan (Hobbs and Samuel, 1974); 5/35 (14%) *O. hyperborea* in Siberia (Lynch et al., 2007); and in 40/167 (24%) *O. princeps* in Colorado (Duszynski and Brunson, 1973; Duszynski, 1974).

Sporulation: Exogenous. Oocysts sporulated within 72 hr in 2.5% (w/v) aqueous $K_2Cr_2O_7$ solution at 30°C (Duszynski and Brunson, 1973).

Prepatent and patent periods: Unknown.

Site of infection: Unknown. Oocysts recovered from feces and intestinal contents.

Endogenous stages: Unknown.

Cross-transmission: None.

Pathology: Unknown.

Excystation: SZs of this species did not excyst when excysting fluid (trypsin-sodium taurocholate, pH 7.5) was added to oocyst/sporocyst suspensions. The optical density of the sporocyst wall changed, but then the SB and sporocyst contents became indistinct and apparently dissolved within 60 min after addition of the excysting fluid (Duszynski and Brunson, 1973).

Material deposited: Skull, skeleton and tissues of a symbiotype host of *O. hyperborea* are preserved in the University of Alaska Museum of the North (UAM), No. 84368 (IF 5252), male, 11 August 2002, collected by N.E. Dokuchaev and A.A. Tsvetkova. Photosyntype of sporulated oocysts are in the U.S. National Parasite Collection (USNPC), Beltsville, MD, No. 87390 (Lynch et al., 2007). Skull and skin of the symbiotype host of *O. curzoniae* are preserved in the Qinghai-Tibet Plateau Biological Specimen Museum (QPBSM), No. 0006870, male, 16 September 2008. Photosyntypes of a sporulated oocyst in the Key Laboratory of Adaptation and Evolution of Plateau Biota (KLAEPB), No. 08094 (Yi-Fan et al., 2009).

Remarks: This species was first described from *O. princeps* by Lepp et al. (1973). Duszynski and Brunson (1973) believed that oocysts collected from *O. princeps* in Colorado also were this species, but their oocysts differed slightly by being ~2 μm smaller overall and by lacking a PG; the oocysts they measured were 27.6 × 23.3 (23–34 × 18–25), but the L/W ratio was still 1.2. Their sporocysts also were slightly smaller, 12.5 × 8.5 (10–16 × 7–10), but with a L/W ratio of 1.5 vs. 1.35 in the original description. The size of the sporulated oocysts of *E. banffensis* from *O. curzoniae* in China (Yi-Fan et al., 2009) also were slightly smaller than those in the original description, 25.6 × 21.2 (21–29 × 14.5–28) vs. 29.8 × 25.2 (27–32 × 24–28), respectively; however, the L/W ratios were the same (1.2). Similarly, the sporocysts in the Plateau pika also were slightly smaller, 11.9 × 8.3 (8–15 × 6–10) vs. 13.0 × 9.6 (10–15 × 8–12), respectively, but the L/W was the same (1.4 vs. 1.35, respectively). We attribute these minor size variations to the difference between host species. However, oocysts collected from *O. collaris* in the Yukon Territory of Canada

(Hobbs and Samuel, 1974) and those from *O. hyperborea* in Siberia, Russia (Lynch et al., 2007), did not differ much when compared with the original description of Lepp et al. (1973).

EIMERIA BARRETTI LEPP, TODD & SAMUEL, 1972

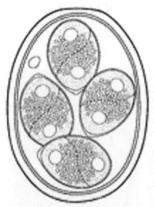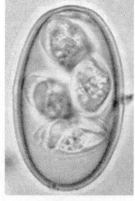

FIGURES 3.5, 3.6 Line drawing of the sporulated oocyst of *Eimeria barretti* from Lepp et al., 1972, with permission of the *Journal of Protozoology.* Photomicrograph of a sporulated oocyst of *E. barretti* from Hobbs and Samuel, 1974, with permission of the *Canadian Journal of Zoology.*

Type host: *Ochotona princeps* (Richardson, 1828) (syn. *Lepus* [*Lagomys*] *princeps*), American pika.

Type locality: NORTH AMERICA: Canada: Alberta, Plateau Mountain in the Crowsnest Forest, 50°12′N, 114°30′W.

Other hosts: *Ochotona collaris* (Nelson, 1893) (syn. *Lagomys collaris*), Collared pika.

Geographic distribution: NORTH AMERICA: Canada: Alberta, Plateau Mountain in the Crowsnest Forest, 50°12′N, 114°30′W; Sheep River area in Bow River Forest, 50°36′N, 114°50′W (Lepp et al., 1972) and talus slopes in SW Alberta, 51°N, 115°W; Yukon Territory, Ogilvie Mountains 64°N, 138°W (Hobbs and Samuel, 1974).

Description of sporulated oocyst: Oocyst shape: ellipsoidal to slightly ovoidal; number of walls: 2; wall thickness: 3; wall characteristics: outer is smooth, light brown, ~$2/3$ of total thickness; inner is dark brown; L × W: 32.9 × 23.8 (27–36 × 21–27); L/W ratio: 1.4 (1.2–1.6); M: present; M characteristics: a thinning of outer oocyst wall, 5–7 wide, slightly flattened; MC: reported in a few oocysts, ~3 × 3, but not shown in original line drawing by Lepp et al. (1972) nor in the line drawing and photomicrograph by Hobbs and Samuel (1974); OR: absent; PG: present, one, ~4 wide. Distinctive features of oocyst: lack of OR and thick smooth wall with M and, presumably, a MC.

Description of sporocyst and sporozoites: Sporocyst shape: ovoidal; L × W: 12.6 × 8.8 (9–15 × 6–12); L/W ratio: 1.5 (1.3–1.7); SB: present as small knob-like structure; SSB, PSB: both absent; SR: present; SR characteristics: diffuse and granular; SZ: arranged diagonally, each with one RB. Distinctive features of sporocyst: diffuse SR and arrangement of SZ in sporocyst.

Prevalence: Not stated in original description, but in 5/92 (5%) *O. collaris* and 8/111 (7%) *O. princeps* (Hobbs and Samuel, 1974).

Sporulation: Oocysts were sporulated when examined, but sporulation time is unknown because of the method used. Feces collected from the rectum or cloaca were kept in 2.5% (w/v) aqueous $K_2Cr_2O_7$ solution at air temperature (1–25°C) for 1–5 hr in the field, then at 4°C in the refrigerator for an unspecified time; finally, they were later placed in Petri dishes at 19°C in the dark for at least 10 days before they were examined.

Prepatent and patent periods: Unknown.

Site of infection: Unknown. Oocysts recovered from feces and colon contents.

Endogenous stages: Unknown.

Cross-transmission: None.

Pathology: Unknown.

Material deposited: None.

Etymology: This species was named for Richard E. Barrett, Alaska State Federal

Laboratory, Animal Health Division, Palmer, Alaska, USA.

Remarks: In their original description of *E. barretti*, Lepp et al. (1972) archived a line drawing of a sporulated oocyst (their Fig. 2 and the one we use here) that is slightly ellipsoidal and slightly flattened on one end, but it did not show either a M or a MC. However, their written description said, "The micropyle was characterized by a thinning of the oocyst wall and was 5–7 μm in diameter. A micropyle cap about 3 by 3 μm was present in a few oocysts" and, "The oocyst wall was smooth, approximately 3 μm thick." They also said that a prominent PG, ~4 μm, was present in sporulated oocysts. Unfortunately, they did not provide a photomicrograph of a sporulated oocyst.

In their redescription of *E. barretti*, however, Hobbs and Samuel (1974) reported the oocysts they measured were 29.0 × 20.1 (23–32 × 17–24), with L/W 1.4 (1.2–1.6) and sporocysts were 14.5 × 7.6 (12–20 × 7–8), with L/W 1.9 (1.6–2.5). They also stated that a PG was absent. Finally, their line drawing (their Fig. 4) showed a *distinct* M, which their description said was "indistinct," they did not mention the presence of a MC, and their photomicrograph (their Fig. 8) was distinctly ellipsoidal vs. the "slightly ovoid" reported by Lepp et al. (1972). Given these discrepancies, it seems to us that Hobbs and Samuel (1974) would have been prudent to give the form they redescribed as *E. barretti* a new species name, rather than comingle the mensural data they reported with the parameters that defined the original *E. barretti* of Lepp et al. (1972). We considered naming their form (Hobbs and Samuel, 1974) as a new species, but there is so much discrepancy between the line drawings, written descriptions, and the one photomicrograph presented in the redescription of *E. barretti* (Lepp et al., 1972; Hobbs and Samuel, 1974), that sorting out what they said and what they meant was impossible for us to determine.

Of the 11 eimerian species described from pikas with oocysts that have a M, in addition to this species (see *Remarks* under *E. balchanica*), the sporulated oocysts of *E. barretti* are closest in size to those of *E. cryptobarretti*, but differ by having a smooth outer wall vs. the rough outer wall of the latter.

EIMERIA CALENTINEI DUSZYNSKI & BRUNSON, 1973

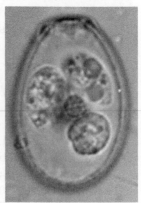

FIGURES 3.7, 3.8 Line drawing of the sporulated oocyst of *Eimeria calentinei*, with permission of the *Journal of Parasitology*. Photomicrograph of a sporulated oocyst of *E. calentinei* (original).

Type host: *Ochotona princeps* (Richardson, 1828) (syn. *Lepus* [*Lagomys*] *princeps*), American pika.

Type locality: NORTH AMERICA: USA: Colorado, Larimer County, Crown Point, alt. 3,666–3,833 m.

Other hosts: *Ochotona collaris* (Nelson, 1893) (syn. *Lagomys collaris*), Collared pika; *Ochotona curzoniae* (Hudgson, 1858), Plateau pika; *Ochotona hyperborea* (Pallas, 1811), Northern pika.

Geographic distribution: NORTH AMERICA: Canada: Yukon Territory, Ogilvie Mountains, 64°N, 138°W (Hobbs and Samuel, 1974); Alberta, 51°N, 115°W (Hobbs and Samuel, 1974); USA:

Alaska, Yukon-Charley Rivers National Preserve (Lynch et al., 2007); Colorado, Clear Creek and Larimer Counties (Duszynski and Brunson, 1973; Duszynski, 1974); ASIA: Japan, Hokkaido, Daisetzusan National Park (Hobbs and Samuel, 1974); Russia: Siberia, Omolon River basin, Providenya Oblast (Lynch et al., 2007); People's Republic of China: Qinghai Province, Haibei Alpine Meadow Ecosystem Research Station, Chinese Academy of Sciences, 37°36′N, 101°18′E; alt. 3,205 m (Yi-Fan et al., 2009).

Description of sporulated oocyst: Oocyst shape: ovoidal, widest in the middle and tapering toward the M; number of walls: 2; wall thickness ~2; wall characteristics: outer is yellow, smooth, of irregular thickness, ~1, thinnest at end opposite M and thickest on sides and around margin of M; inner is orange, of uniform thickness; L × W ($N = 50$): 28.5 × 20.0 (26—32 × 18—23); L/W ratio: 1.4 (1.3—1.7); M: present; M characteristics: distinct, 4.4 (3—7) wide; OR: present; OR characteristics: well-defined, spheroidal, 3—6 wide; PG: absent. Distinctive features of oocyst: distinct M and a well-defined OR.

Description of sporocyst and sporozoites: Sporocyst shape: ovoidal; L × W: 12.0 × 7.4 (10—14 × 6—9); L/W ratio: 1.6 (1.4—2.0); SB: present as distinct knob; SSB, PSB: both absent; SR: present; SR characteristics: granular, diffuse to spheroidal; SZ: banana-shaped ($N = 10$), 18—21 × 2, folded and tightly packed in sporocyst with one large posterior RB and N located anterior to RB in center of SZ. Distinctive features of sporocyst: knob-shaped SB and structure of SZ.

Prevalence: In 5/53 (9%) *O. collaris* in Alaska (Lynch et al., 2007); 8/92 (9%) *O. collaris* in the Yukon Territory (Hobbs and Samuel, 1974); 21/52 (40%) *O. curzoniae* in China (Yi-Fan et al., 2009); 2/35 (6%) *O. hyperborea* in Siberia (Lynch et al., 2007); 1/14 (7%) *O. hyperborea* in Japan (Hobbs and Samuel, 1974); 2/111 (2%) *O. princeps* in Alberta (Hobbs and Samuel, 1974); and in 39/167 (23%) *O. princeps* in Colorado (Duszynski and Brunson, 1973; Duszynski, 1974).

Sporulation: Exogenous. Oocysts sporulated after 72 hr in 2.5% (w/v) aqueous $K_2Cr_2O_7$ solution at 30°C.

Prepatent and patent periods: Unknown.

Site of infection: Unknown. Oocysts recovered from feces.

Endogenous stages: Unknown.

Cross-transmission: None.

Pathology: Unknown.

Excystation: SZ became active from 30 sec to 12 min after excysting solution (trypsin-sodium taurocholate, pH 7.5) reached the sporocyst. After 2 min of exposure, the SB changed in optical density, quickly became indistinct, and disappeared while the sporocyst wall became transparent. About 2 min after activity started, SZs began to excyst. Excystation occurred less rapidly than in *I. marquardti* (Duszynski and Brunson, 1972) and *E. cryptobarretti* (Duszynski and Brunson, 1973). Each SZ would probe the area of the SB, protrude its anterior (pointed) end through the opening and then withdraw into the sporocyst. This process was repeated many times; each time the SZ would progress farther out of its sporocyst. When only the most posterior part of the SZ (containing the large RB) remained in the sporocyst, the SZ would gyrate as if attempting to free itself. In only one instance did Duszynski and Brunson (1973) see a SZ excyst completely.

Material deposited: Skin, skull, skeleton, and tissues of two symbiotype hosts, *O. collaris* from Alaska, and *O. hyperborea* from Russia, are preserved in the University of Alaska Museum of the North (UAM): *O. collaris*, UAM No. 58399 (AF 49333), male, 1 August 2001, collected by H. Henttonen, J. Niemimaa, K. Gamblin, and L.B. Barrelli; and *O. hyperborea*, UAM No. 80824 (AF 38535), 4 September 2000, collected by S.O. MacDonald, N.E. Dokushaev, and K.E. Galbreath. Photosyntype of a sporulated oocyst in the USNPC, Beltsville, MD, No. 87393. Skull and skin of a symbiotype host of *O. curzoniae* in China (Yi-Fan et al., 2009) are preserved in the Qinghai-Tibet Plateau

Biological Specimen Museum (QPBSM), No. 0006871, male, 17 September 2008. A photosyntype of a sporulated oocyst is in the Key Laboratory of Adaptation and Evolution of Plateau Biota (KLAEPB), No. 08096.

Etymology: This species was named for Dr. Robert L. Calentine, Wisconsin State University, River Falls, WI, USA, for his many contributions to parasitology, for mentoring one of us (DWD), and for a life devoted to teaching and research on the cestodes of fish.

Remarks: This species was first described from *O. princeps* by Duszynski and Brunson (1973) in Colorado, USA. Later it was found from other hosts, including *O. collaris* from the Yukon Territory, Canada, *O. hyperborea* from Japan and Russia, and *O. princeps* from Alberta, Canada (see Lynch et al., 2007). The size of the sporulated oocysts of *E. calentinei* from *O. curzoniae* in China (Yi-Fan et al., 2009) was slightly smaller than those in the original description, 25.5 × 17.9 (24–28 × 17–21) vs. 28.5 × 20.0 (26–32 × 18–23), respectively; however, the L/W ratios were the same (1.4). Similarly, the sporocysts in the Plateau pika also were slightly smaller, 10.9 × 6.5 (10–12 × 6–7) vs. 12.0 × 7.4 (10–14 × 6–9), respectively, but their L/Ws were similar (1.6 vs. 1.7, respectively). All other qualitative features of the sporulated oocysts were identical; thus, these minor size variations are attributed to the difference between host species (Yi-Fan et al., 2009). The sporulated oocysts from *O. hyperborea* from Russia, and from *O. collaris* from Alaska, were nearly identical to those described by Duszynski and Brunson (1973) from *O. princeps* in Colorado.

Of the 11 eimerian species described from pikas with oocysts that have a M, in addition to this species (see *Remarks* under *E. balchanica*), the sporulated oocysts of this species differ from those of *E. barretti* in oocyst size and shape, in the structure and thickness of the oocyst wall, in lacking a PG, and in having an OR. They differ from those of *E. erschovi* in oocyst size

and shape, in the structure and thickness of the oocyst wall, and in having larger sporocysts. They differ from those of *E. metelkini* in oocyst size and shape, in having a two-layered wall, and in the size and shape of the sporocysts. They differ from those of *E. ochotona* in oocyst size and shape, in having a two-layered wall, and in having considerably larger sporocysts which have a SR. However, as Hobbs and Samuel (1974) correctly point out, accurate line drawings and photomicrographs of the Asian pikas' coccidia (e.g., *E. erschovi*, *E. daurica*, *E. metelkini*, etc.) are needed before conclusions can be drawn on the validity of these latter species.

EIMERIA CIRCUMBOREALIS HOBBS & SAMUEL, 1974

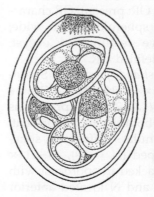 

FIGURES 3.9, 3.10 Line drawing of the sporulated oocyst of *Eimeria circumborealis* from Hobbs and Samuel, 1974, with permission of the *Canadian Journal of Zoology*. Photomicrograph of a sporulated oocyst of *E. circumborealis* from Hobbs and Samuel, 1974, with permission of the *Canadian Journal of Zoology*.

Type host: *Ochotona collaris* (Nelson, 1893) (syn. *Lagomys collaris*), Collared pika.

Type locality: NORTH AMERICA: Canada: Yukon Territory, Ogilvie Mountains, 64°N 138°W.

Other hosts: *Ochotona princeps* (Richardson, 1828) (syn. *Lepus* [*Lagomys*] *princeps*), American pika; *Ochotona hyperborea* (Pallas, 1811), Northern pika.

Geographic distribution: NORTH AMERICA: Canada: Yukon Territory, 64°N 138°W; talus slopes in southwestern Alberta, 15°N 115°W; ASIA: Japan: Hokkaido, Daisetzusan National Park.

Description of sporulated oocyst: Oocyst shape: ellipsoidal to subspheroidal; number of walls: 2; wall thickness: ~2; wall characteristics: smooth; L × W: 26.6 × 20.2 (21−34 × 15−23); L/W ratio: 1.3 (1.1−1.8); M: present; M characteristics: slightly flattened area in outer wall layer with opening in inner wall; OR: absent; PG: present as many, tiny diffuse granules that seem to congregate under M. Distinctive features of oocyst: lack of OR and diffuse, tiny PGs under M.

Description of sporocyst and sporozoites: Sporocyst shape: ovoidal; L × W: 13.4 × 8.2 (10−16 × 6−9); L/W ratio: 1.6 (1.3−1.9); SB: present as nipple-like structure at pointed end of sporocyst; SSB, PSB: both absent; SR: present; SR characteristics: compact mass of granules that appear to be membrane-bound; SZ: with one end narrower than the other and each with two RB, one anterior and the second one posterior to N. Distinctive features of sporocyst: SZ with two RB, one at each end of SZ.

Prevalence: In 7/92 (8%) specimens of the type host; 9/111 (8%) *O. princeps*; and in 10/14 (71%) *O. hyperborea* (Hobbs and Samuel, 1974).

Sporulation: Oocysts were sporulated when examined, but sporulation time is unknown because of the method used. Feces collected from the rectum or colon were kept in 2.5% (w/v) aqueous $K_2Cr_2O_7$ solution at air temperature (1−25°C) for 1−5 hr in the field, then at 4°C in the refrigerator for an unspecified time; finally they were later placed in Petri dishes at 19°C in the dark for at least 10 days before they were examined.

Prepatent and patent periods: Unknown.

Site of infection: Unknown. Oocysts recovered from feces.

Endogenous stages: Unknown.

Cross transmission: None.

Pathology: Unknown.

Material deposited: None.

Remarks: Of the 11 eimerian species described from pikas with oocysts that have a M, in addition to this species (see *Remarks* under *E. balchanica*), the sporulated oocysts of *E. circumborealis* are similar in size only to those of *E. barretti*, *E. calentinei*, *E. erschovi*, and *E. metelkini*, but differ from them by having the characteristic diffuse PGs (diffuse OR granules?) concentrated near the M. It differs from *E. erschovi* only in sporocyst size and the presence of a PG. Since the PG (OR?) granules here is/are diffuse, they may have been overlooked, or regarded as wrinkles of the inner oocyst wall in *E. erschovi* described earlier (Hobbs and Samuel, 1974).

EIMERIA CRYPTOBARRETTI DUSZYNSKI & BRUNSON, 1973

Type host: *Ochotona princeps* (Richardson, 1828) (syn. *Lepus* [*Lagomys*] *princeps*), American pika.

Type locality: NORTH AMERICA: USA: Colorado, Larimer County, Crown Point, altitude 3,666−3,833 m.

Other hosts: *Ochotona collaris* (Nelson, 1893) (syn. *Lagomys collaris*), Collared pika; *Ochotona curzoniae* (Hudgson, 1858), Plateau pika; *Ochotona hyperborea* (Pallas, 1811), Northern pika.

Geographic distribution: NORTH AMERICA: USA: Colorado, Clear Creek and Larimer Counties (Duszynski and Brunson, 1973; Duszynski, 1974); Alaska, Wrangell-St. Elias National Park (Lynch et al., 2007), Yukon-Charley Rivers National Preserve, mountainside NW of Headwater Lake of Crescent Creek, 64°82'N, 143°75'W (Lynch et al., 2007); ASIA: Russia: Siberia, Magadanskaya Oblast, mouth of Kegali River, 64°26'N, 161°47'E (Lynch et al., 2007); People's Republic of China: Qinghai Province, Haibei Alpine Meadow Ecosystem Research Station, Chinese Academy of Sciences,

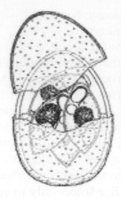
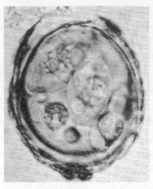
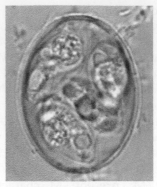

FIGURES 3.11, 3.12, 3.13 Line drawing of the sporulated oocyst of *Eimeria cryptobarretti* from Duszynski and Brunson, 1973, with permission of the *Journal of Parasitology*. Photomicrograph of a sporulated oocyst of *E. cryptobarretti* with the rough outer wall beginning to separate from the smooth inner wall (original). Photomicrograph of a sporulated oocyst of *E. cryptobarretti* showing smooth wall after rough outer wall has been removed (original).

37°36′N, 101°18′E; alt. 3,205 m (Yi-Fan et al., 2009).

Description of sporulated oocyst: Oocyst shape: ellipsoidal; number of walls: 2; wall thickness: ~2.1; wall characteristics: outer is brown, rough, ~1 thick, and usually has debris clinging to it; inner is smooth; L × W (N = 200, 21—27 from each of eight hosts): 29.8 × 20.6 (24—35 × 18—22); L/W ratio: 1.5 (1.3—1.8); M: may be present; M characteristics: when outer layer of wall is broken, the smooth inner layer is slightly flattened and thinner at one end, giving the appearance of a M; OR, PG: both absent. Distinctive features of oocyst: lack of OR and PG and possibility of an indistinct M.

Description of sporocyst and sporozoites: Sporocyst shape: ellipsoidal, but pointed at end opposite SB; L × W: 15.2 × 8.4 (13—18 × 7—11); L/W ratio: 1.8 (1.5—2.1); SB: present; SSB, PSB: both absent; SR: present; SR characteristics: a compact, spheroidal mass, but sometimes as diffuse granules; SZ (N = 10): banana-shaped, 15.2 × 2.5 (13.5—17 × 2—3.5), with one large posterior RB and a smaller one just anterior to it. Distinctive features of sporocyst: shape of sporocyst with pointed end opposite the SB.

Prevalence: In 17/30 (57%) and 90/137 (67%) *O. princeps* in Colorado (Duszynski and Brunson, 1973; Duszynski, 1974, respectively); 6/53 (11%) *O. collaris* in Alaska (Lynch et al., 2007); 14/52 (27%) *O. curzoniae* in China (Yi-Fan et al., 2009); and in 5/35 (14%) *O. hyperborea* in Siberia (Lynch et al., 2007).

Sporulation: Exogenous. Oocysts sporulated within 72 hr in 2.5% aqueous (w/v) $K_2Cr_2O_7$ solution at 30°C.

Prepatent and patent periods: Unknown.

Site of infection: Unknown. Oocysts recovered from feces.

Endogenous stages: Unknown.

Cross-transmission: None.

Pathology: Unknown.

Excystation: The SZ began active movement 1—10 min after the excysting fluid (trypsin-sodium taurocholate, pH 7.5) reached the sporocysts. This activity varied in intensity from rapid to slow and was interrupted by moments of complete inactivity. Before SZ activity began, the SB disappeared. As the SZs began moving, the sporocyst wall became increasingly transparent. Just before SZs excysted, the SR "exploded" and, immediately thereafter, each SZ rapidly left the sporocyst. It is uncertain whether excystation was triggered by the fragmenting SR or whether this fragmentation resulted from a rapid change in pressure within the sporocyst forcing the SZs out. After excystation, only the transparent wall of the sporocyst remained. Total time to complete excystation was 15—60 min after addition of the excysting fluid.

Material deposited: Skull, skin, skeleton, and tissues of two symbiotype hosts, *O. collaris* and *O. hyperborea*, are preserved in the University of Alaska Museum of the North (UAM): *O. collaris*, UAM No. 58213 (AF 49535), 18 July 2001, collected by H. Henttonen, J. Niemimaa, K. Gamblin, and L.B. Barrelli; and *O. hyperborea*, UAM No. 80603 (AF 38233), male, 19 August 2000, collected by S.O. MacDonald, N.E. Dokuchaev, and K.E. Galbreath. Photosyntype and photoparatype of sporulated oocysts are in the USNPC, Nos. 87480 and 88170, respectively (Lynch et al., 2007). Skull and skin of one *O. curzoniae*, another symbiotype host, are preserved in the Qinghai-Tibet Plateau Biological Specimen Museum (QPBSM), No. 0006873, female, 16 September 2008. Photosyntype of a sporulated oocyst is in the Key Laboratory of Adaptation and Evolution of Plateau Biota (KLAEP), No.08098 (Yi-Fan et al., 2009).

Etymology: When the rough, outer oocyst wall was removed, Duszynski and Brunson (1973) noticed that the oocysts of this species closely resemble those of *E. barretti*, as drawn by Lepp et al. (1972). Thus, the prefix *crypto-* (L., hidden) was added to *barretti* when they named their new species.

Remarks: This species was first described from *O. princeps* in Colorado, USA, by Duszynski and Brunson (1973) and Duszynski (1974). Later it was found in *O. collaris* from Alaska, USA, and in *O. hyperborea* from Siberia, Russia, and Lynch et al. (2007) said that oocysts from both of these hosts were similar to those in the original description (Duszynski and Brunson, 1973). The sporulated oocysts of *E. cryptobarretti* from *O. curzoniae* (Yi-Fan et al., 2009) were smaller than those in the original description from *O. princeps*, 26.0 × 18.7 (25—31 × 16—26) vs. 29.8 × 20.6 (24—35 × 18—22), respectively, but the L/W ratios were the same (1.4). Likewise, the

sporocysts from oocysts in *O. curzoniae* also were smaller, 12.3 × 7.2 (11—14 × 6—8) vs. 15.2 × 8.4 (13—18 × 7—11), respectively, as were their L/W indices, 1.7 vs. 1.8, respectively. We attribute these minor size variations to the difference between host species. Sporulated oocysts, with their rough outer wall removed, were 28.9 × 20.7 (27—33 × 19—22) with a L/W ratio 1.4 (1.3—1.6), figures similar to measurements for *E. barretti* (Lepp et al., 1972).

EIMERIA DAURICA MATSCHOULSKY, 1947a

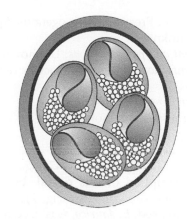

FIGURE 3.14 Line drawing of the sporulated oocyst of *Eimeria daurica* (original) redrawn from Matschoulsky, 1947a.

Synonym: *Eimeria matschoulskyi* Pellérdy, 1974, in *Cricetulus barabensis* (striped hamster), pro parte.

Type host: *Ochotona dauurica* (Pallas, 1776) (syn. *Lepus dauuricus*), Daurian pika.

Type locality: ASIA: Asiatic Russia (former Soviet Union): Mongolia, southern regions of Buryat-Mongol.

Other hosts: None reported to date.

Geographic distribution: ASIA: Asiatic Russia (former Soviet Union): Mongolia, southern regions of Buryat-Mongol.

Description of sporulated oocyst: Oocyst shape: cylindroidal or ellipsoidal; number of walls: two (?), described as double-contoured; wall characteristics: membranous to translucent, ~1 thick; L × W: 23.8 × 16.8 (20—30 × 15—18); L/W ratio: 1.4 (1.3—1.6); M, OR, PG: all absent. Distinctive features of oocyst: none.

Description of sporocyst and sporozoites: Sporocyst shape: subspheroidal to slightly ovoidal (line drawing); L × W: 7—10 × 6.6; SB, SSB, PSB: not given, but presumably absent; SR: present; SZ: not described. Distinctive features of sporocyst: insufficient information provided.

Prevalence: In 1/3 (33%) specimens of the type host.

Sporulation: Presumably exogenous. Sporulation time unknown.

Prepatent and patent periods: Unknown.

Site of infection: Unknown. Oocysts recovered from feces.

Endogenous stages: Unknown.

Cross-transmission: None.

Pathology: Unknown.

Material deposited: None.

Remarks: Matschoulsky (1947a) gave a very limited description of this oocyst, but he did provide a line drawing (his Fig. 7). However, the name he used, *E. daurica*, was used a second time for oocysts he described from the striped hamster, *Cricetulus barabensis*, in the same paper. This prompted Pellérdy (1974) to rename the oocyst from the hamster and retain this name for the *Ochotona* species. However, in his 1963 alphabetical check list of species of Eimeriidae (published 2 years prior to his first edition of *Coccidia and Coccidiosis*) Pellérdy listed oocyst measurements from Matschoulsky's (1947a) description of *E. daurica* as 20.6 × 14.1 (17—23 × 13—15), an error repeated in both editions (1965, 1974) of his classic tome, and in his second edition (1974) he added that the sporocysts were 6.7—8.1 × 4.7, also an error. In fact, almost everyone since then has used Pellérdy's measurements instead of those from the original

description by Matschoulsky (1947a), which are given above. Musaev and Veisov (1965), citing Matschoulsky (1947a), used the same oocyst and sporocyst measurements repeated in Pellérdy (1974), but not actually those of Matschoulsky's (1947a) original description.

EIMERIA ERSCHOVI MATSCHOULSKY, 1949

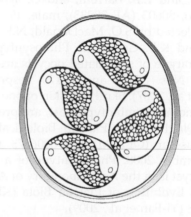

FIGURE 3.15 Line drawing of the sporulated oocyst of *Eimeria erschovi* (original) redrawn from Matschoulsky, 1949.

Type host: *Ochotona dauurica* (Pallas, 1776) (syn. *Lepus dauuricus*), Daurian pika.

Type locality: ASIA: Asiatic Russia (former Soviet Union): Mongolia, southern regions of Buryat-Mongol; Central Kazakhstan.

Other hosts: *Ochotona pallasi* Gray, 1867 (syn. *Ochotona pricei* Thomas, 1911), Mongolian pika.

Geographic distribution: ASIA: Asiatic Russia (former Soviet Union): Mongolia, southern regions of Buryat-Mongol (Matschoulsky, 1949); Kazakhstan, Karaganda Region, Chetak District (Svanbaev, 1958).

Description of sporulated oocyst: Oocyst shape: ovoidal to subspheroidal; number of walls: 2; wall thickness: 1.2—1.8; wall characteristics: smooth, dark yellow; L × W: 21.1 × 18.5 (21—23 × 17—19); L/W ratio: 1.15 (1.1—1.2); M: present; M characteristics: a thinning of the oocyst wall at one end, with only a small, raised portion; OR, PG: both absent. Distinctive features of oocyst: mostly ovoidal shape, only slightly raised near M, and lack of both OR and PG.

Description of sporocyst and sporozoites: Sporocyst shape: ellipsoidal (in original line drawing); L × W: 7.6–9.5 × 5.7; SB, SSB, PSB: all absent; SR: present as scattered delicate granules; SZ: comma-shaped, each with one small RB at rounded end. Distinctive features of sporocyst: small size and tiny SB.

Prevalence: In 3/14 (21%) of the type host; and in 4/66 (6%) from *O. pallasi* (syn. *O. pricei*) (Svanbaev, 1958).

Sporulation: Presumably exogenous. Sporulation time unknown.

Prepatent and patent periods: Unknown.

Site of infection: Unknown. Oocysts recovered from feces.

Endogenous stages: Unknown.

Cross-transmission: None.

Pathology: Unknown.

Material deposited: None.

Etymology: This species was named after Professor Erschov.

Remarks: Svanbaev (1958) reported the dimensions of sporulated oocysts from *O. pallasi* as 21.1 × 18.5 (22–31 × 16–23) with sporocysts 9.7 × 5.3 (8–11 × 4–7), but he did not see a M, while Musaev and Veisov (1965), who identified the host as *O. pricei*, said the oocysts were ovoidal, ellipsoidal or spheroidal, 25.6 × 20.2 (22–31 × 16–23), with a smooth, yellow-brown, double-contoured wall, 1–1.8 thick, and thinner at end with M; presumably, they did not see an OR or PG; sporocysts were reported to be ellipsoidal or broadly ellipsoidal, 10 × 5 (3–11 × 4–7), without SB, SSB, and PSB, but with a SR; SZ were 6.4 × 3.3 (5–7.5 × 2–4).

Of the 11 *Eimeria* species described from pikas with oocysts that have a M, in addition to this species (see *Remarks* under *E. balchanica*), the sporulated oocysts of four of them have oocysts with a distinct to modest collar-like feature around the M, and they are easily distinguished from each other by oocyst size, shape of the collar around the M, and by other quantitative and qualitative features (see *Remarks* section under *E. ochotona*).

EIMERIA HAIBEIENSIS YI-FAN, RUN-ROUNG, JIAN-HUA, JIANG-HUI & DUSZYNSKI, 2009

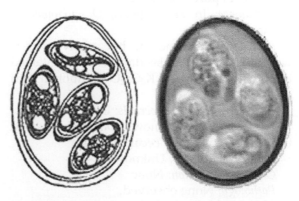

FIGURES 3.16, 3.17 Line drawing of the sporulated oocyst of *Eimeria haibeiensis* from Yi-Fan et al., 2009, with permission of the authors and the *Journal of Parasitology*. Photomicrograph of a sporulated oocyst of *E. haibeiensis* from Yi-Fan et al., 2009, with permission of the authors and the *Journal of Parasitology*.

Type host: *Ochotona curzoniae* (Hudgson, 1858), Plateau pika.

Type locality: ASIA: People's Republic of China (PRC): Qinghai Province, Haibei Alpine Meadow Ecosystem Research Station, Chinese Academy of Sciences, 37°36′N, 101°18′E; alt. 3,205 m.

Other hosts: None to date.

Geographic distribution: ASIA: PRC, Qinghai Province, 37°36′N, 101°18′E; alt. 3,205 m.

Description of sporulated oocyst: Oocyst shape: ellipsoidal to ovoidal; number of walls: 2; wall thickness: ~2; wall characteristics: outer is smooth; L × W ($N = 38$): 22.2 × 16.2 (20–24 × 15–18); L/W ratio: 1.4 (1.3–1.6); M: present; M characteristics: ~3.9 (3.5–4.5) wide; PG, OR: both absent.

Distinctive features of oocyst: M, 2-layered smooth outer wall and lack of PG and OR.

Description of sporocyst and sporozoites: Sporocyst shape: ovoidal; L × W ($N = 38$): 11.6 × 6.6 (10–13 × 5–7); L/W ratio: 1.8 (1.4–2.4); SB: present at slightly pointed end; SSB, PSB: both absent; SR: present; SR characteristics: compact mass of granules; SZ with two RBs, one anterior and the second one posterior to N. Distinctive features of sporocyst: none.

Prevalence: In 21/52 (40%) *O. curzoniae* in China.

Sporulation: Unknown, but many oocysts were sporulated after six days when maintained in 2.5% (w/v) aqueous $K_2Cr_2O_7$ solution in Petri dishes at 25°C.

Prepatent and patent periods: Unknown.

Site of infection: Unknown. Oocysts recovered from feces, colon and cecal contents.

Endogenous stages: Unknown.

Cross-transmission: None.

Pathology: None observed.

Material deposited: Skull and skin of the symbiotype host are preserved in the Qinghai-Tibet Plateau Biological Specimen Museum (QPBSM), No. 0006872 male, 16 September 2008. Photosyntypes of sporulated oocysts in the Key Laboratory of Adaptation and Evolution of Plateau Biota (KLAEPB), No. 08097.

Etymology: The *nomen trivale* is derived from the name of the Research Station in the PRC in which the host animal was collected and *−ensis* (L., belonging to).

Remarks: This species somewhat resembles *E. barretti*, first described from *O. princeps* in Alberta, Canada (Lepp et al., 1972), and later reported from *O. collaris* from the Yukon Territory, Canada (Hobbs and Samuel, 1974). However, even though W.M. Samuel was an author on both papers, there are several discrepancies between the description and line drawing in the original description (Fig. 2 of Lepp et al., 1972) and the description, line drawing, and photomicrograph in their second report of this eimerian (Figs. 4, 8, of Hobbs and Samuel, 1974).

As we noted in the *Remarks* section for *E. barretti*, Lepp et al.'s (1972) original line drawing of the sporulated oocyst of *E. barretti* showed a slightly ellipsoidal oocyst flattened on one end, but without M and MC, even though their written description said the M was 5–7 μm wide, with a 3 × 3 μm MC seen in a few oocysts. Mensural data for their (1972) oocysts and sporocysts were 32.9 × 23.8 (27–36 × 21–27), with L/W 1.4 (1.2–1.6) and 12.6 × 8.8 (9–15 × 6–12) with L/W 1.5 (1.3–1.7), respectively. When Hobbs and Samuel (1974) redescribed *E. barretti*, however, they said those oocysts were 29.0 × 20.1 (23–32 × 17–24), with sporocysts 14.5 × 7.6 (12–20 × 7–8), while their new line drawing (1974, Fig. 4) showed a *distinct* M, which their description said was "indistinct"; they also did not mention the presence (or absence) of a MC, and their photomicrograph (their Fig. 8) was distinctly ellipsoidal vs. the "slightly ovoid" reported by Lepp et al. (1972). As we noted above (*Remarks* under *E. barretti*), Hobbs and Samuel (1974) might have been wiser to give the oocysts they redescribed as *E. barretti* a new species name, rather than confuse the mensural parameters that defined the original *E. barretti* of Lepp et al. (1972). We reiterate these details here because the morphology of *E. haibeiensis* from *O. curzoniae* is similar to the redescription of *E. barretti* provided by Hobbs and Samuel (1974) from *O. collaris*. However, the oocyst and sporocyst dimensions, the absence of both OR and PG, and the L/W ratios of both oocysts and sporocysts give the form described by Yi-Fan et al. (2009) a sufficiently unique suite of characters that justifies their naming it as distinct from *E. barretti*.

EIMERIA KLONDIKENSIS HOBBS & SAMUEL, 1974

Type host: *Ochotona collaris* (Nelson, 1893) (syn. *Lagomys collaris*), Collared pika.

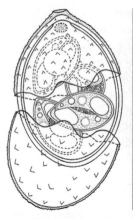
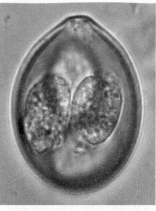

FIGURES 3.18, 3.19 Line drawing of the sporulated oocyst of *Eimeria klondikensis* from Hobbs and Samuel, 1974, with permission of the *Canadian Journal of Zoology*. Photomicrograph of a sporulated oocyst of *E. klondikensis* from Hobbs and Samuel, 1974, with permission of the *Canadian Journal of Zoology*.

Type locality: NORTH AMERICA: Canada: Yukon Territory, Ogilvie Mountains, 64°N, 138°W.

Other hosts: *Ochotona princeps* (Richardson, 1828) (syn. *Lepus* [*Lagomys*] *princeps*), American pika; *Ochotona hyperborea* (Pallas, 1811), Northern pika.

Geographic distribution: NORTH AMERICA: Canada: Yukon Territory, Ogilvie Mountains, 64°N 138°W (Hobbs and Samuel, 1974); Alberta, 51°N, 115°W (Hobbs and Samuel, 1974); USA: Colorado, Clear Creek County (Duszynski, 1974); Alaska: Wrangell-St. Elias National Park and Preserve, SE of Rock Lake, 21 July, 2001, 61°47′N, 141°12′W (Lynch et al., 2007); Yukon-Charley Rivers National Preserve (Lynch et al., 2007); ASIA: Japan: Hokkaido, Daisetzusan National Park (Hobbs and Samuel, 1974); Russia: Siberia, Chukotka, 3 km SSE of confluence of Volchya River and Liman Sea, 64°48′N, 177°33′E (Lynch et al., 2007).

Description of sporulated oocyst: Oocyst shape: ovoidal; number of walls: 3; wall thickness: 2; wall characteristics: outer is rough and of variable thickness, inner is smooth; L ×

W: 35.2 × 24.6 (32–38 × 23–36); L/W ratio: 1.4 (1.3–1.5); M: present; M characteristics: raised, neck-like structure that makes it very prominent; OR: absent; PG: present; number of PGs: one, about 2 wide. Distinctive features of oocyst: prominent M and three walls.

Description of sporocyst and sporozoites: Sporocyst shape: ovoidal; L × W: 15.5 × 9.2 (14–16 × 8–10); L/W ratio: 1.7 (1.6–1.8); SB: present at narrow end; SSB, PSB: both absent; SR: present; SR characteristics: compact mass of granules; SZ: with 2–3 RBs, one anterior and the second one posterior to the N, occasionally with a third present, also anterior to N. Distinctive features of sporocyst: 2–3 RB in SZ.

Prevalence: In 3/92 (3%) *O. collaris* (type host) in the Yukon Territory (Hobbs and Samuel, 1974); 2/53 (4%) *O. collaris* in Alaska (Lynch et al., 2007); 1/35 (3%) *O. hyperborea* in Siberia, Russia (Lynch et al., 2007); 2/14 (14%) *O. hyperborea* in Japan (Hobbs and Samuel, 1974); 7/111 (6%) *O. princeps* in Alberta (Hobbs and Samuel, 1974); and in 31/137 (23%) *O. princeps* in Colorado (Duszynski, 1974).

Sporulation: Presumably exogenous. Oocysts were sporulated when examined, but sporulation time is unknown because of the method used. Feces collected from the rectum or colon were kept in 2.5% (w/v) aqueous $K_2Cr_2O_7$ solution at air temperature (1–25°C) for 1–5 hr in the field, then at 4°C in the refrigerator for an unspecified time; finally, they were later placed in Petri dishes at 19°C in the dark for at least 10 days before they were examined.

Prepatent and patent periods: Unknown.

Site of infection: Unknown. Oocysts recovered from feces.

Endogenous stages: Unknown.

Cross-transmission: None.

Pathology: Unknown.

Material deposited: None from the type host, but preservation of host and parasite materials was done in later studies. Skull, skin, skeleton, and tissues of two symbiotype hosts, *O. collaris* and *O. hyperborea*, are preserved in the

University of Alaska Museum of the North (UAM): *O. collaris*, UAM No. 56067 (AF 54551), female, 21 July 2001, collected by S. Kutz, A. Tsvetkova, A.A. Eddingaas, and M. McCain; and *O. hyperborea*, UAM No. 84369 (IF 5352), male, 11 August 2002, collected by N.E. Dokuchaev and A.A. Tsvetkova. (Lynch et al., 2007). Lynch et al. (2007) also deposited a photoneotype of a sporulated oocyst in the USNPC, No. 99671, because no previous authors had archived a type specimen of this parasite.

Remarks: There are 11 eimerian species described from pikas with oocysts that have a M, in addition to this species: *E. balchanica, E. barretti, E. calentinei, E. circumborealis, E. cryptobarretti* (?), *E. erschovi, E. haibeiensis, E. metelkini, E. ochotona, E. princepsis* (?), and *E. qinghaiensis*. Of these, only one the oocysts of *E. qinghaiensis* are as large, but the presence of the raised, neck-like structure that supports the M distinguishes the oocysts of *E. klondikensis* from all *Eimeria* species from pikas.

The morphology and mensural features of sporulated oocysts of *E. klondikensis* from *O. collaris* in Alaska and *O. hyperborea* in Russia (Lynch et al., 2007) were similar to those in the original description provided by Hobbs and Samuel (1974) for oocysts collected from the same host species in Canada and Japan, respectively.

EIMERIA METELKINI MATSCHOULSKY, 1949

Type host: *Ochotona dauurica* (Pallas, 1776) (syn. *Lepus dauuricus*), Daurian pika.

Type locality: ASIA: Asiatic Russia (former Soviet Union): Mongolia, southern regions of Buryat-Mongol.

Other hosts: None reported to date.

Geographic distribution: ASIA: Asiatic Russia (former Soviet Union): Mongolia, southern regions of Buryat-Mongol.

Description of sporulated oocyst: Oocyst shape: ovoidal with a small neck supporting the M and MC; number of walls: 2, but appears as

FIGURE 3.20 Line drawing of the sporulated oocyst of *Eimeria metelkini* (original) redrawn from Matschoulsky, 1949.

only one in original line drawing; wall thickness: 0.9–1.2; wall characteristics: dark pink; L × W: 24.3 × 19.0 (23–27 × 19); L/W: 1.3 (1.2–1.4); M: present; M characteristics: thinning of the wall at narrow end of oocyst, 3.8–5.7 wide; MC: present, 1.9–2.7 high; OR: present; OR characteristics: separate granules; PG: absent. Distinctive features of oocyst: distinct M with MC and presence of OR.

Description of sporocyst and sporozoites: Sporocyst shape: ellipsoidal (line drawing); L × W: 9–11 × 6–7; SB, SSB, PSB: all absent; SR: present; SR characteristics: large granules between SZ; SZ: elongate with one end broader than the other containing a clear RB and lie end-to-end with SR in between. Distinctive features of sporocyst: ellipsoidal shape and lack of SB.

Prevalence: In 2/14 (14%) specimens of the type host.

Sporulation: Presumably exogenous. Sporulation time is unknown.

Prepatent and patent periods: Unknown.

Site of infection: Unknown. Oocysts recovered from feces.

Endogenous stages: Unknown.

Cross-transmission: None.

Pathology: Unknown.

Material deposited: None.

Etymology: This species was named after Professor Metelkin.

Remarks: Only four of the 11 eimerian species described from pikas that have oocysts reported with a M have a distinct to modest collar-like feature around the M and they are easily distinguished from each other by oocyst size, shape of the collar around the M, and by other quantitative and qualitative features (see *Remarks* under *E. ochotona*).

EIMERIA OCHOTONA MATSCHOULSKY, 1949

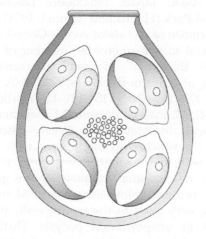

FIGURE 3.21 Line drawing of the sporulated oocyst of *Eimeria ochotona* (original) redrawn from Matschoulsky, 1949.

Type host: *Ochotona dauurica* (Pallas, 1776) (syn. *Lepus dauuricus*), Daurian pika.

Type locality: ASIA: Asiatic Russia (former Soviet Union): Mongolia, southern regions of Buryat-Mongol.

Other hosts: None reported to date.

Geographic distribution: ASIA: Asiatic Russia (former Soviet Union): Mongolia, southern regions of Buryat-Mongol.

Description of sporulated oocyst: Oocyst shape: ovoidal to vase-shaped, but flattened at one end; number of walls: 2; wall thickness: 1; wall

characteristics: light pink; L × W: 19.9 × 14.2 (19—21 × 13—15); L/W ratio: 1.4 (1.4—1.5); M: present; M characteristics: a thinning of the wall at one end of oocyst, 4.6—5.1 wide (line drawing shows M to be wide, flattened, and pronounced, much like a vase would be flat on top and surrounded by a slight neck); OR: present as a few small granules in center of oocyst; PG: absent. Distinctive features of oocyst: vase-like shape and distinct appearance of M neck in original drawing.

Description of sporocyst and sporozoites: Sporocyst shape: ovoidal to lemon-shaped; L × W: 5.7 × 3.8; L/W ratio: 1.5; SB: present (in original line drawing) as a small nipple-like structure; SSB, PSB: both absent; SR: absent; SZ: lie end-to-end in original line drawing, each with a small posterior RB at rounded end. Distinctive features of sporocyst: small size, lemon-shape and lack of SR.

Prevalence: In 2/14 (14%) specimens of the type host.

Sporulation: Unknown, but presumably exogenous.

Prepatent and patent periods: Unknown.

Site of infection: Unknown. Oocysts recovered from feces.

Endogenous stages: Unknown.

Cross-transmission: None.

Pathology: Unknown.

Material deposited: None.

Remarks: This species has only been seen once since its original description. Dr. Norman Levine (deceased), formerly of the University of Illinois, sent one of us (DWD) a portion of his reprint library prior to his death. Within this treasure trove of taxonomic reprints was a preliminary, hand-typed and hand-edited partial manuscript (on yellow paper) on the coccidia of rabbits. Levine noted that he did not have access to Matschoulsky's (1949) original description, so he used the descriptive data for *E. ochotona* from Musaev and Veisov (1965), who redescribed *E. ochotona*. The data given by Musaev and Veisov (1965), according to Levine's notes, were identical to information (above) in Matschoulsky's (1949) original paper. We have compared the papers and we agree the oocysts seen by Musaev and Veisov (1965) were identical in size (19.9 × 14.2 [19−21 × 13−15]) to those in the original description, as were the sporocysts (6 × 4).

Of the 11 eimerians known from pikas to have oocysts with a M, in addition to *E. ochotona*, only four species have a distinct collar-like feature around the M: *E. ochotona*, *E. metelkini*, *E. erschovi* (all described by Matschoulsky, 1949), and *E. klondikensis*. Oocysts of *E. ochotona* clearly differ in size from those of *E. klondikensis* by being considerably smaller (19.9 × 14.2 [19−21 × 13−15] vs. 35 × 24 [32−38 × 23−36]). Its oocysts are only slightly smaller than those of *E. metelkini* (24 × 19 [23−27 × 19]), but the latter has a MC that is ~2 × 3, which *E. ochotona* lacks. Finally, its oocysts also are only slightly smaller than those of *E. erschovi* (21.1 × 18.5 [20−23 × 17−19]), but the latter lacks an OR and its oocysts are more rounded with a less-pronounced collar around its M.

EIMERIA PRINCEPSIS DUSZYNSKI & BRUNSON, 1973

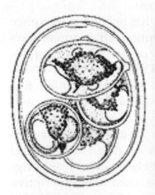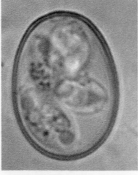

FIGURES 3.22, 3.23 Line drawing of the sporulated oocyst of *Eimeria princepsis* from Duszynski and Brunsen, 1973, with permission of the *Journal of Parasitology*. Photomicrograph of a sporulated oocyst of *E. princepsis* (original).

Type host: *Ochotona princeps* (Richardson, 1828) (syn. *Lepus* [*Lagomys*] *princeps*), American pika.

Type locality: NORTH AMERICA: USA: Colorado, Larimer and Clear Creek Counties.

Other hosts: *Ochotona collaris* (Nelson, 1893) (syn. *Lagomys collaris*), Collared pika; *Ochotona hyperborea* (Pallas, 1811), Northern pika.

Geographic distribution: NORTH AMERICA: USA: Colorado, Larimer and Clear Creek Counties (Duszynski and Brunson, 1973; Duszynski, 1974); Canada: Yukon Territory, Ogilvie Mountains, 69°N 138°W and southwestern Alberta, 51°N, 115°W (Hobbs and Samuel, 1974); ASIA: Japan: Hokkaido, Daisetzusan National Park (Hobbs and Samuel, 1974).

Description of sporulated oocyst: Oocyst shape: ellipsoidal to subspheroidal; number of walls: 1; wall thickness: ~1; wall characteristics: smooth, of uniform thickness; L × W (*N* = 33): 21.5 × 17.3 (19−24 × 15−19); L/W ratio: 1.25 (1.1−1.4); M: may be present; M characteristics: if present, indistinct, but closely associated with a PG attached to the inner surface of the oocyst wall; OR: absent; PG: present; number of PGs: one, a small, almost square granule usually present below area where M may be located, very close to the oocyst wall, and not present in unsporulated oocysts. Distinctive features of oocyst: unique, square/rectangular PG attached to the inner layer of oocyst wall just below the indistinct M.

Description of sporocyst and sporozoites: Sporocyst shape: ovoidal; L × W: 12.3 × 7.3 (10−14 × 5−8); L/W ratio: 1.6 (1.5−2.0); SB: present; SB characteristics: distinct knob at apex of sporocyst; SSB, PSB: both absent; SR: present; SR characteristics: granular, usually scattered throughout sporocyst, but sometimes present as a compact spherical mass; SZ: arranged side-by-side, tightly within sporocyst, each with one posterior RB and probably a second one just anterior to it. Distinctive features of sporocyst: distinct, knoblike SB.

Prevalence: In 11/30 (37%) *O. princeps* (type host; Duszynski and Brunson, 1973); 82/137 (60%) *O. princeps* (Duszynski, 1974); 24/92 (26%) *O. collaris*; 8/111 (7%) *O. princeps*; and in 2/14 (14%) *O. hyperborea* (Hobbs and Samuel, 1974).

Sporulation: Exogenous. Oocysts sporulated within 72 hr in 2.5% (w/v) aqueous $K_2Cr_2O_7$ solution at 30°C.

Prepatent and patent periods: Unknown.

Site of infection: Unknown. Oocysts recovered from feces and intestinal contents.

Endogenous stages: Unknown.

Cross-transmission: None.

Pathology: Unknown.

Excystation: SZs of this species did not excyst when sporocysts were exposed to a trypsin-sodium taurocholate (pH 7.5) excysting fluid after 1 hr.

Material deposited: None.

Remarks: Only sporulated oocysts of *E. daurica* and *E. erschovi* somewhat resemble oocysts of this species because all three have similar oocyst dimensions, but those of *E. princepsis* differ from the other two in oocyst shape, being narrower than *E. erschovi* and broader than *E. daurica*. *Eimeria erschovi* has a distinct M, not seen in *E. princepsis*, and neither *E. erschovi* nor *E. daurica* were reported to have a PG. Also, the sporocysts of the latter two apparently do not possess a knob-like SB and are 20–40% smaller than the sporocysts of *E. princepsis*. These structural differences, plus geographic isolation and host differences, led Duszynski and Brunson (1973) to call this a new species at the time. According to Hobbs and Samuel (1974), this species differs from *E. daurica* only slightly in shape, with other characters being similar.

EIMERIA QINGHAIENSIS YI-FAN, RUN-ROUNG, JIAN-HUA, JIANG-HUI & DUSZYNSKI, 2009

Type host: *Ochotona curzoniae* (Hudgson, 1858), Plateau pika.

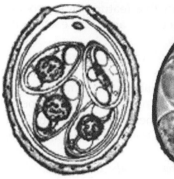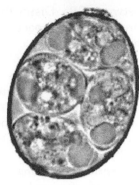

FIGURES 3.24, 3.25 Line drawing of the sporulated oocyst of *Eimeria qinghaiensis* from Yi-Fan et al., 2009, with permission of the authors and the *Journal of Parasitology*. Photomicrograph of a sporulated oocyst of *E. qinghaiensis* from Yi-Fan et al., 2009, with permission of the authors and the *Journal of Parasitology*.

Type locality: ASIA: People's Republic of China (PRC): Qinghai Province, Haibei Alpine Meadow Ecosystem Research Station, Chinese Academy of Sciences, 37°36′N, 101°18′E; alt. 3205 m.

Other hosts: None to date.

Geographic distribution: ASIA: PRC: Qinghai Province, 37°36′N, 101°18′E; alt. 3,205 m.

Description of sporulated oocyst: Oocyst shape: ovoidal; number of walls: 3; wall thickness: ~2; wall characteristics: outer is rough, with prominent, sunken M; inner is smooth with M barely visible; L × W ($N = 70$): 37.2 × 27.2 (34–41 × 24–32); L/W ratio: 1.4 (1.1–1.6); M: present; M characteristics: 9 (7.5–12.1) wide, most prominent in outermost layer; PG: present, 2.6 (2.4–3.2) wide; OR: absent. Distinctive features of oocyst: prominent M and 3-layered wall.

Description of sporocyst and sporozoites: Sporocyst shape: ovoidal; L × W ($N = 50$): 16.6 × 9.8 (14–19 × 9–11); L/W ratio: 1.7 (1.4–1.9); SB: present at pointed end; SSB, PSB: both absent; SR: present; SR characteristics: compact spheroidal mass of tiny granules, 6.8 × 5.7 (5–9 × 5–7); SZ with anterior RB, 2 × 3 (2–4 × 2–3) and posterior RB, 6 × 4 (4–6 × 3–4); N visible between RBs. Distinctive features of sporocyst: prominent compact SR and SZ each with two RBs.

Prevalence: In 18/52 (35%) *O. curzoniae* from China.

Sporulation: Presumably exogenous, but the time is unknown; many oocysts were sporulated after 6 days when maintained in 2.5% (w/v) aqueous $K_2Cr_2O_7$ solution in Petri dishes at 25°C.

Prepatent and patent periods: Unknown.

Site of infection: Unknown. Oocysts recovered from feces, colon, and cecal contents.

Endogenous stages: Unknown.

Cross-transmission: None.

Pathology: None observed.

Material deposited: Skull and skin of the symbiotype host are preserved in the Qinghai-Tibet Plateau Biological Specimen Museum (QPBSM), No. 0006869, female, 15 September 2008. Photosyntypes of sporulated oocysts in the Key Laboratory of Adaptation and Evolution of Plateau Biota (KLAEPB), No. 08091.

Etymology: The *nomen trivale* was derived from the name of the Province in the PRC in which the host animal was collected and *−ensis* (L., belonging to).

Remarks: Mensural features of the sporulated oocysts of this species are similar in size to, but slightly larger than, those of *E. klondikensis* described from *O. collaris* in the Yukon Territory, Canada (Hobbs and Samuel, 1974); their oocysts are 37.2 × 27.2 (34−41 × 24−32) vs. 35.2 × 24.6 (32−38 × 23−36), respectively, and their sporocysts are 16.6 × 9.8 (14−19 × 9−11) vs. 15.5 × 9.2 (14−16 × 8−10), respectively. Qualitative features also are similar, with OR, PG, and SB present in both descriptions. However, there is one distinct structural difference between the oocysts found by Yi-Fan et al. (2009) and those of *E. klondikensis* (Duszynski, 1974; Hobbs and Samuel, 1974; Lynch et al., 2007), and that is the part of the oocyst wall with the M. In previous descriptions, line drawings, and

photomicrographs, the M on the oocysts of *E. klondikensis* is on a distinct raised structure, like the raised neck of a turtleneck sweater, that surrounds the M separating it from the oocyst wall per se. The M on the oocysts of *E. qinghaiensis* is a distinct part of the outermost oocyst wall and slightly sunken into it, which the authors who described it felt was a unique enough difference to separate it from *E. klondikensis*.

EIMERIA WORLEYI LEPP, TODD & SAMUEL, 1972

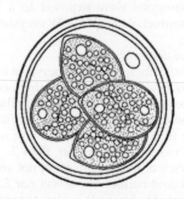

FIGURE 3.26 Line drawing of the sporulated oocyst of *Eimeria worleyi* from Lepp et al., 1972, with permission of the *Journal of Protozoology*.

Type host: *Ochotona princeps* (Richardson, 1828) (syn. *Lepus* [*Lagomys*] *princeps*), American pika.

Type locality: NORTH AMERICA: Canada: Alberta, Blairmore, Frank Slide 49°37′N, 114°25′W.

Other hosts: *Ochotona hyperborea* (Pallas, 1811), Northern pika.

Geographic distribution: NORTH AMERICA: Canada: southwestern Alberta, the Blairmore, Frank Slide area (Lepp et al., 1972); USA: Colorado, Larimer and Clear Creek Counties (Duszynski, 1974); ASIA: Japan: Hokkaido,

Daisetzusan National Park (Hobbs and Samuel, 1974).

Description of sporulated oocyst: Oocyst shape: spheroidal to subspheroidal; number of walls: 2; wall thickness: ~1.5; wall characteristics: outer is smooth, light brown, ~$^2/_3$ total thickness; inner is light green; L × W (N = 50): 13.5 × 12.5 (12−16 × 10−15); L/W ratio: 1.1 (1.0−1.3); M, OR: both absent; PG: present; number of PGs: 1, ~1−2 wide. Distinctive features of oocyst: small size, lack of OR, and presence of a PG.

Description of sporocyst and sporozoites: Sporocyst shape: ovoidal; L × W: 5.6 × 3.7 (4−6 × 3−5); L/W ratio: 1.5 (1.3−1.7); SB: present, small, at pointed end of sporocyst; SSB, PSB: both absent; SR: present; SR characteristics: granular, diffuse, and spread throughout sporocyst; SZ: lie lengthwise in sporocyst, each with a small, posterior RB. Distinctive features of sporocyst: diffuse, granular SR.

Prevalence: Not given for the type host in the original description (Lepp et al., 1972), but in 1/14 (7%) *O. hyperborea* (Hobbs and Samuel, 1974).

Sporulation: Presumably exogenous. Oocysts were sporulated when examined, but sporulation time is unknown because of the method used. Feces collected from the rectum or colon were kept in 2.5% (w/v) aqueous $K_2Cr_2O_7$ solution at air temperature (1−25°C) for 1−5 hr in the field, then at 4°C in the refrigerator for an unspecified time; finally, they were later placed in Petri dishes at 19°C in the dark for at least 10 days before they were examined.

Prepatent and patent periods: Unknown.

Site of infection: Unknown. Oocysts recovered from feces.

Endogenous stages: Unknown.

Cross-transmission: None.

Pathology: Unknown.

Material deposited: None.

Etymology: The species was named for Dr. David E. Worley, Montana State University, Bozeman, MT, USA.

Remarks: Comparing the oocyst species that lack a M, this species differs from all others from pikas in having the smallest sporulated oocyst and sporocysts.

ISOSPORA MARQUARDTI DUSZYNSKI & BRUNSON, 1972

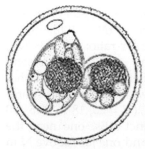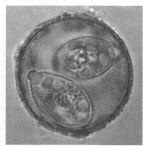

FIGURES 3.27, 3.28 Line drawing of the sporulated oocyst of *Isospora marquardti* from Duszynski and Brunson, 1972, with permission of the *Journal of Protozoology*. Photomicrograph of a sporulated oocyst of *I. marquardti* from Duszynski and Brunson 1972, with permission of the *Journal of Protozoology*.

Type host: *Ochotona princeps* (Richardson, 1828) (syn. *Lepus* [*Lagomys*] *princeps*), American pika.

Type locality: NORTH AMERICA: USA: Colorado, Larimer County, Ft. Collins.

Other hosts: *Ochotona collaris* (Nelson, 1893) (syn. *Lagomys collaris*), Collared pika; *Ochotona hyperborea* (Pallas, 1811), Northern pika.

Geographic distribution: NORTH AMERICA: USA: Colorado, Larimer and Clear Creek Counties (Duszynski and Brunson, 1972; Duszynski, 1974); Canada: Yukon Territory, Ogilvie Mountains, 64°N, 138°W (Lynch et al., 2007); Alberta, 51°N, 115°W (Hobbs and Samuel, 1974); ASIA: Russia: Siberia, Chukotka, Ulhum River, 15 km W of Chaplino Village, 64°25′N, 172°32′E (Lynch et al., 2007).

Description of sporulated oocyst: Oocyst shape: spheroidal to subspheroidal; number of walls:

questionable, but there appear to be two in optical cross section—these could not be demonstrated by crushing oocysts by using friction of the coverslip; wall thickness: ~1.5; wall characteristics: lightly pitted and pale yellow; L × W (*N* = 100): 30.8 × 30.1 (23−36 × 23−36); L/W ratio: 1.0 (1.0−1.2); M, OR: both absent; PG: present; number of PGs: 1, 2 × 4. Distinctive features of oocyst: relatively large size, lightly pitted outer wall, and lack of OR.

Description of sporocyst and sporozoites: Sporocyst shape: ovoidal; L × W (*N* = 100): 19.3 × 12.0 (17−22 × 10−14); L/W ratio: 1.6 (1.3−1.8); SB: present; SSB: present, ~3 × 3; PSB: absent; SR: present; SR characteristics: distinct spheroidal mass, 8.3 (5−10) wide, in center of sporocyst; SZ: banana-shaped, (*N* = 20) 18 × 3 (15−20 × 2−4) with two distinct RBs, one near the posterior rounded end and one just above N in center of SZ; N with prominent nucleolus. Distinctive features of sporocyst: presence of SSB, SR, two distinct RBs in SZ, and N with distinct nucleolus.

Prevalence: In 9/30 (30%) *O. princeps* (type host) including 4/14 (29%) collected on Mt. Evans (alt. 4,267−4,400 m) and 1/4 (25%) collected on Goliath Peak (alt. 3,867−4,000 m), both in Clear Creek County, and 4/12 (33%) collected from Crown Point (elev. 3,667−3,833 m) in Larimer County (Duszynski and Brunson, 1972); 1/92 (1%) *O. collaris* from the Yukon Territory, Alaska (Lynch et al., 2007); 1/35 (3%) *O. hyperborea* from Siberia, Russia (Lynch et al., 2007); 1/111 (<1%) *O. princeps* from Alberta, Canada (Hobbs and Samuel, 1974); and 16/137 (12%) *O. princeps* from Colorado, (Duszynski, 1974).

Sporulation: Exogenous. Oocysts sporulated after 72 hr in 2.5% (w/v) aqueous $K_2Cr_2O_7$ solution at 30°C.

Prepatent and patent periods: Unknown.

Site of infection: Unknown. Oocysts recovered from feces.

Endogenous stages: Unknown.

Cross-transmission: None.

Pathology: Unknown.

Excystation: Excystation occurred 45 sec to 4 min after the excysting fluid (trypsin-sodium taurocholate, pH 7.5) reached the sporocysts. Rapid SZ activity began the excystation process and was accompanied by a change in optical density of the SB. The rapidity with which the first SZ followed the SSB exiting the sporocyst suggested that either the exit of the SSB was aided by the SZs or a rapid change in pressure within the sporocyst forced contents out of it once the SB had disappeared. Both factors may be operating in the excystation of *I. marquardti* SZs. Once the first SZ excysted, the remaining three took an additional 2−8 min to leave and sometimes two excysted simultaneously. Total excystation was completed 7−14 min after exposure to the excysting fluid, leaving the SR intact in the empty sporocyst.

Material deposited: Skull, skin, skeleton, and tissues of a symbiotype host, *O. hyperborea*, are preserved in the University of Alaska Museum of the North (UAM), UAM No. 83846 (IF 7569), female, 28 July 2002, collected by V.F. Fedorov and K.E. Galbreath (Lynch et al., 2007). A photosyntype of a sporulated oocyst is in the USNPC, No. 87408.

Etymology: This species was named in honor of Dr. William C. Marquardt, Colorado State University, Fort Collins, CO, USA, for his contributions mentoring one of us (DWD) and for his lifetime of teaching and research excellence.

Remarks: Citing a previous observation by Duszynski (1971) that the time during patency when oocysts are measured may be important in accurately determining their size, Duszynski and Brunson (1972) measured oocysts from five host specimens to obtain more representative structural parameters of *I. marquardti*, on the assumption that oocysts were collected on different days of the patent period.

The morphology of sporulated oocysts from *O. hyperborea* in Russia differs slightly from those in the original description from *O. princeps* in Colorado; the latter had oocysts and sporocysts that are larger in both length and width (31 × 30 and

19 × 12 vs. 28 × 27 and 17 × 11, respectively) than those from Russia (Lynch et al., 2007). Nonetheless, both oocysts and sporocysts from Russian *O. hyperborea* were larger than those measured by Hobbs and Samuel (1974) from *O. collaris* from the Yukon Territory, Canada (23 × 22 and 15 × 9). Oocysts of some species are known to exhibit phenotypic plasticity (see Duszynski et al., 1992) and, given the similarity of qualitative data, we assume all these oocysts represent *I. marquardti*.

ISOSPORA YUKONENSIS HOBBS & SAMUEL, 1974

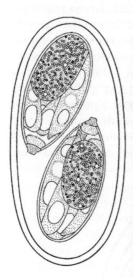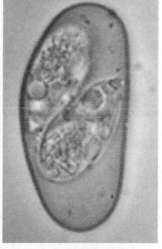

FIGURES 3.29, 3.30 Line drawing of the sporulated oocyst of *Isospora yukonensis* from Hobbs and Samuel, 1974, with permission of the *Canadian Journal of Zoology*. Photomicrograph of a sporulated oocyst of *I. yukonensis* from Hobbs and Samuel, 1974, with permission of the *Canadian Journal of Zoology*.

Type host: *Ochotona collaris* (Nelson, 1893) (syn. *Lagomys collaris*), Collared pika.

Type locality: NORTH AMERICA: Canada: Yukon Territory, Ogilvie Mountains, 64°N, 138°W.

Other hosts: None reported to date.

Geographic distribution: NORTH AMERICA: Canada: Yukon Territory, Ogilvie Mountains, 64°N, 138°W.

Description of sporulated oocyst: Oocyst shape: elongate-ellipsoidal, often asymmetrical; number of walls: 1; wall thickness: < 2; wall characteristics: smooth; L × W: 52.4 × 23.3 (48−57 × 21−27); L/W ratio: 2.25 (1.9−2.6); M, OR, PG: all absent. Distinctive features of oocyst: large size, asymmetrical shape, lack of M, OR, and PG.

Description of sporocyst and sporozoites: Sporocyst shape: elongate-ovoidal; L × W: 27.6 × 12.3 (26−29 × 12−14); L/W ratio: 2.2 (1.9−2.4); SB: present; SB characteristics: flat-ended; SSB: present, slightly wider than SB; PSB: absent; SR: present; SR characteristics: large, compact mass of granules filling lower half of sporocyst; SZ: with two RBs, one anterior and one posterior to N. Distinctive features of sporocyst: large elongate-ovoidal size and presence of SB and SSB.

Prevalence: In 1/92 (1%) specimens of the type host.

Sporulation: Presumably exogenous. Oocysts were sporulated when examined, but sporulation time is unknown because of the method used. Feces collected from the rectum or colon were kept in 2.5% (w/v) aqueous $K_2Cr_2O_7$ solution at air temperature (1−25°C) for 1−5 hr in the field, then at 4°C in the refrigerator for an unspecified time; finally, they were later placed in Petri dishes at 19°C in the dark for at least 10 days before they were examined.

Prepatent and patent periods: Unknown.

Site of infection: Unknown. Oocysts recovered from feces and intestinal contents.

Endogenous stages: Unknown.

Cross-transmission: None.

Pathology: Unknown.

Material deposited: None.

Remarks: Only one other species of *Isospora* has been described from lagomorphs, *I. marquardti* Duszynski and Brunson, 1972 from *O. princeps*. This species differs considerably from

I. marquardti in shape (elongate ellipsoidal vs. spheroidal) and size (52.4 × 23.3 vs. 30.8 × 30.1).

Collared pikas share talus habitats with hoary marmots (*Marmota caligaata*) and arctic ground squirrels, *Urocitellus* (syn. *Spermophilus*) *undulates* (Pallas, 1778) Helgen et al., 2009 (Helgen et al., 2009). Hobbs and Samuel (1974) speculated it is possible that the true host of *I. yukonensis* is the marmot or ground squirrel, since pikas are coprophagous and only one infection was recorded from *O. collaris*. However, to our knowledge, oocysts resembling *I. yukonensis* have not yet been reported from these rodents.

SPECIES INQUIRENDAE (5+)

Eimeria pallasi (Svanbaev, 1958) Lepp, Todd & Samuel, 1972

Synonyms: *E. kriygsmanni* Yakimoff and Gousseff, 1938 of Svanbaev (1958) in *Ochotona pallasi*; non *E. krijgsmanni* Yakimoff and Gousseff, 1938 in *Mus musculus*.

Original host: *Ochotona pallasi* (Gray, 1867), Mongolian pika.

Remarks: Yakimoff and Gousseff (1938) first described *E. krijgsmanni* in *Mus musculus* and included a line drawing (their Fig. 3). Svanbaev (1958) later described what he thought were oocysts of this same species from 6/66 (9%) specimens of *O. pallasi*, but misspelled the name as *E. kriygsmanni*. Svanbaev's (1958) oocysts were described as ellipsoidal to ovoidal with a two-layered wall, ~1.4–2 thick; the oocysts measured 26.3 × 21.3 (19–34 × 17–26) with a L/W ratio of 1.2 (1.1–1.3) and lacked both M and OR, but had one PG. Sporocysts were subspheroidal to ellipsoidal, 11.0 × 7.8 (9–13 × 7–9) with a L/W ratio of 1.4; they lacked SB, SSB, PSB and SR and the SZ were 7.9 × 3.8 (7–10 × 3.5 × 4). Lepp et al. (1972) felt that because rodent coccidia have a relatively high degree of host specificity, the oocysts described by Svanbaev (1958) most likely represented a new species and named it *E. pallasi*. However, neither provided a line drawing or a photomicrograph. Thus, using the criterion established by Duszynski and Upton (2009) for making decisions on the validity of species of snake coccidia, that a species described without existence of a type specimen of any kind (e.g.,

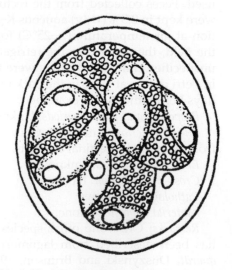 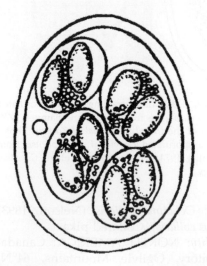

FIGURES 3.31, 3.32 Line drawings of sporulated oocysts of *Eimeria pallasi*, both from Musaev and Veisov, 1965.

line drawing, photomicrograph, stages in tissue sections, etc.) must be relegated to a status of *species inquirenda*, we feel it prudent to do that here for the name *E. pallasi*.

Musaev and Veisov (1965) provided two line drawings (their Fig. 41), which we have included here (Figures 3.31 and 3.32), to approximate what sporulated oocysts of *E. pallasi* may look like, if it actually exists. The reader must understand that Musaev and Veisov (1965) said their line drawings represented oocysts of *E. kriygsmanni* from *Mus*, citing Yakimoff and Gousseff (1938). However, since Svanbaev (1958) considered the oocysts he saw in *O. pallasi* to be the same as those of *E. kriygsmanni* from *M. musculus*, their sporulated oocysts may be quite similar. This form has not been found since it was originally described by Svanbaev (1958).

Eimeria shubini (Svanbaev, 1958) Lepp, Todd & Samuel 1972

FIGURE 3.33 Line drawing of the sporulated oocyst of *Eimeria shubini* from Yakimoff and Gousseff 1938, with permission of *Parasitology*.

Synonyms: *Eimeria musculi* Yakimoff and Gousseff, 1938 of Svanbaev (1958) in *Ochotona pallasi*, non *Eimeria musculi* Yakimoff and Gousseff, 1938 in *Mus musculus*.

Original host: *Ochotona pallasi* (Gray, 1867), Mongolian pika.

Remarks: Yakimoff and Gousseff (1938) first described *E. musculi* in *Mus musculus* and included a line drawing (their Fig. 1). Svanbaev (1958) later described what he thought were oocysts of this same species in 1/66 (1.5%) *O. pallasi*. He described oocysts to be spheroidal with two walls, ~1.3—1.6 thick; the oocysts were 22 × 22 and lacked M, OR and PG. Sporocysts were ellipsoidal, 8.6 × 7.7 (8—9 × 7—8) with a L/W ratio of 1.1 and all lacked SB, SSB, PSB and SR, while their SZ were 6.2 × 3.8 (6—7 × 3—5). Lepp et al. (1972) felt that because rodent coccidia have a relatively high degree of host specificity, the oocysts described by Svanbaev (1958) most likely represented a new species from pikas and named it *E. shubini*. However, neither provided a line drawing or a photomicrograph. Thus, using the criterion established by Duszynski and Upton (2009), as noted above, we feel it prudent to relegate the name *E. pallasi* to a *species inquirenda*.

Since Yakimoff and Gousseff (1938) gave a line drawing for the form they found in the mouse, we have included it here (left) to approximate what sporulated oocysts of *E. shubini* may look like, if it actually exists. The reader must understand that Yakimoff and Gousseff (1938) used this line drawing to represent *E. musculi* from *M. musculus*. However, since Svanbaev (1958) considered the oocysts he saw in *O. pallasi* to be the same as those of *E. musculi*, their sporulated oocysts may be quite similar. This form has not been found since it was originally described by Svanbaev (1958).

Eimeria sp. of Svanbaev, 1958

Original host: *Ochotona pallasi* (Gray, 1867), Mongolian pika.

Remarks: This host was later called *O. pricei* Thomas, 1911, as was common in the (former) Soviet Union literature by Musaev and Veisov (1965) and Svanbaev (1979), but Wilson and Reeder (2005) list *O. pricei* as a synonym of *O. pallasi*. Svanbaev (1958) found this organism in the feces of 16/66 (24%) pikas in the Chetsk District of Karaganda Region, central Kazakhstan (former Soviet Union). Oocysts were ovoidal to elongate-ellipsoidal and measured 103.2 × 46.1 (94—111 × 36—52), L/W ratio, 2.2 and had a smooth, yellow-green or yellow-brown, two-layered wall with the outer (?) layer radially striated, ~3—6 thick, without a M or OR, but sometimes with a PG. The oocysts he saw and described never sporulated completely, so he did not give it a name, an action with which we agree. Svanbaev (1958), however, did see four sporoblasts in each oocyst that were 33 × 20 (31—35 × 18—21), which allowed him to call it an *Eimeria* sp.

Eimeria spp. of Barrett & Worley, 1970

Original host: *Ochotona princeps* (Richardson, 1828) (syn. *Lepus* [*Lagomys*] *princeps*), American pika.

Remarks: Barrett and Worley (1970) examined 54 pikas for helminth and protist parasites from two counties in Montana, USA. They said their fecal exams revealed six *Eimeria* species in four pikas, all from Gallatin County, but they made no attempt to name the species they identified.

Isospora sp. of Barrett & Worley, 1970

Original host: *Ochotona princeps* (Richardson, 1828) (syn. *Lepus* [*Lagomys*] *princeps*), American pika.

Remarks: Barrett and Worley (1970) examined 54 pikas for helminth and protist parasites from two counties in Montana, USA. They said their fecal exams revealed one *Isospora* species in one pika, from Gallatin County, but they did not name or identify the species they saw.

DISCUSSION AND SUMMARY

Living pikas are distributed only in the Northern Hemisphere, and all species are so remarkably homogeneous in morphology and size that they are placed in a single genus, *Ochotona*. Most *Ochotona* species occupy one of two different habitat types, although there are a few intermediate species.

Burrowing forms, with straighter, more powerful claws, are found in steppe, shrub, and forest environments. They tend to be highly social, rather short-lived, and may have high, but fluctuating, population densities and high fecundity rates. Their non-burrowing relatives, with longer vibrissae, occupy (alpine) talus rock fields that lie adjacent to meadows or other vegetation so that they have access to plants for food and hay farming. The talus-dwelling pikas tend toward being asocial, are longer-lived, and have relatively stable, low population densities and low fecundity rates (Yu et al., 2000). They are unique among alpine mammals because they gather vegetation (flowers, grasses, leaves, evergreen needles, pine cones) throughout the summer months and then live off their "hay pile" throughout the winter, instead of moving down-slope and/or hibernating. They are highly territorial throughout the winter and guard their hay piles with their lives. As snow-pack accumulates at high altitudes, it insulates their talus environment by maintaining temperatures just below freezing, wherein pikas can survive.

Because of this unique habitat and behavior, talus-dwelling pikas may be threatened by global warming. The interstices between their talus rocks (the rocks must be of a certain size) provide both a cool, moist micro-climate where they cool down during hot summer days and a stable, near-freezing temperature where they stay during long winter nights. This stable winter temperature is critical to them, because they do not huddle together like many other mammals, as far as anyone knows. Thus, with warming temperatures, the snowpack and their stable underground temperature is reduced in many mountainous regions of the world and "in a strange twist of fate, global warming can cause pikas to freeze to death" (Holtcamp, 2010). Likewise, when the ambient temperature begins to exceed 25°C, these high-energy animals will die if they cannot regulate their body temperature by moving into a cooler micro-climate under the talus rocks. But since they already live near mountain tops, when their particular talus field's micro-climate becomes too hot (summer) or too cold (winter), they simply have nowhere to go and will die (Holtcamp, 2010).

Matschoulsky (1947a) published the first paper on a coccidium in pikas. Since then only 10 additional papers have described and named a total of 19 coccidians (17 *Eimeria*, two *Isospora*) in seven pika species: two from North America and five from Asia (Table 3.1). The coccidia reported from three of those seven hosts, *O. collaris* and *O. princeps* (from North America) and *O. hyperborea* (China, Russia, Japan), which are the best-studied hosts, are remarkably similar (Lynch et al., 2007). These hosts have all been studied on multiple occasions and 10 coccidia species are known from them; 6/10 (60%) of these coccidia species are reported from all three host species, while three others are reported from at least two host species. One coccidium, *I. yukonensis*, has been reported only from a single individual of *O. collaris* (Hobbs and Samuel, 1974). The overlap of coccidia species among *O. collaris*, *O. hyperborea*, and *O. princeps* suggests the possibility that these coccidia may have evolved from a common ancestor; that is, shared coccidia faunas in three closely related *Ochotona* species may reflect a single origin for the parasites in their common ancestor. On the other hand, this pattern may indicate that each coccidium had a common ancestor in the ancestor of the pika species. Thus, the parasite community may have a recent origin, but this doesn't say anything about relationships among these coccidia (Lynch et al., 2007).

With the exception of *E. erschovi*, seven coccidia from three Asian pikas, *O. dauurica* (*E. daurica*, *E. metelkini*, *E. ochotona* in 1949),

O. pallasi (*E. pallasi*, *E. shubini*, *E.* sp., in 1958), and *O. rufescens* (*E. balchanica*, in 1978), have been identified only once from their single host species. Initially, the lack of overlap indicates that these coccidia (if they actually are valid species from pikas) may be more host-species specific, but nothing is known about host specificity of pika coccidia. However, given the known distributions of these three host species, it is not likely their coccidia would ever come into contact with an *Ochotona* species different from the one in which it was first described: *Ochotona rufescens*, the Afghan pika, is geographically separated from the other two species, and although the ranges of *O. dauurica* and *O. pallasi* overlap (e.g., Russia, Mongolia), these species are separated by both altitude and biome (high mountain vs. desert, respectively). Unfortunately, the sampling bias doesn't allow meaningful comparisons, and the question can be asked whether some of these seven coccidia even still exist, since at least some populations of *O. pallasi* are either threatened or critically endangered (IUCN).

Focusing on domestic, cottontail, and jack rabbits, Samoil and Samuel (1981) were among the first to try to use coccidia as indicators of phylogenetic relationships of members of the Lagomorpha. Their summary of the morphological traits of the eimerians from these three rabbit groups, along with results of cross-transmission work among rabbits to that date, led them to suggest that (1) cottontails (*Sylvilagus* spp.) and domestic rabbits (*Oryctolagus* spp.) were more closely related to each other than they are to jackrabbits (*Lepus* spp.), and (2) the 13 *Eimeria* spp. known in 1981 from ochotonids "appear to be more similar to those of rodents and artiodactyls than to those of the Leporidae." More recently, Lynch et al. (2007) approached the same idea, that the similarity, or disparity, of coccidia infecting pika species might reflect the systematics and phylogenetics of their hosts. Work was done by Yu et al. (2000) on the phylogeny of 19 pika species, including five of the seven from which *Eimeria* and *Isospora* species have been described: *O. curzoniae*, *O. dauurica*, *O. hyperborea*, *O. princeps*, and *O. pallasi*, but sequences from *O. collaris* and *O. rufescens* were not incorporated. Their data identified three pika clades: a shrub-steppe group of seven species (including *O. curzoniae* and *O. dauurica*), a northern group of five species (including *O. hyperborea*, *O. pallasi*, and *O. princeps*), and a mountain group of seven species (none of which have yet been examined for coccidia). Based on their host-parasite data, Lynch et al. (2007) supported the notion that *O. hyperborea* and *O. princeps* may be infected by the same coccidia, because they have descended from a recent common ancestor. If true, this would predict that the same or similar coccidian species will be found in other species from the northern group of Yu et al. (2000). In other words, the morphological similarity of the coccidia in the study by Lynch et al. (2007) may reflect close phylogenetic relationships that are a consequence of the close relationship between their hosts.

The data of Yu et al. (2000) suggested that *O. princeps* is the most basal member of the northern group. Interestingly, *O. pallasi*, also a member of the northern clade, is infected by an entirely different set of coccidia from other hosts in that clade (although most of these species are currently suspect). In fact, *O. pallasi* is infected by *E. erschovi*, a coccidium first identified from *O. dauurica*, a member of the shrub-steppe group of pikas. Despite an older association between *O. dauurica* and *O. pallasi*, it is possible that *E. erschovi* is a generalist parasite capable of a broad co-accommodation of hosts (Brooks, 1979); that is, the association between *E. erschovi* and two deeply divergent pika lineages may suggest the generalist nature of this coccidium. These hosts are in relatively close contact, as the range of *O. pallasi* overlaps that of *O. dauurica*, although they occupy different habitats (Chapman and Flux, 1990); unfortunately, it is not known if any burrowing

or talus-dwelling pikas live in enough proximity to connect these lineages. Perhaps the other *Eimeria* spp. identified from *O. pallasi* (*E. pallasi*, *E. shubini*, and *E.* sp.) are more derived than those infecting *O. princeps* and are results of recent speciation. This could mean that co-speciation is occurring, but it is also possible that a host switch could have led to this association. Only phylogenetic sequence data for these coccidia will resolve the relationships among them.

The dichotomy seen by Lynch et al. (2007) where three hosts overlap in eimeriid fauna and the other three hosts have divergent fauna, could be the result of poor species descriptions as suggested by earlier authors (Lepp et al., 1972; Hobbs and Samuel, 1974), who noted the strong similarity between many of the "continental Asian" and the "North American" coccidia. In all cases, however, there were enough differences among those coccidia to prevent the synonymy of species. Before conclusions can be made regarding the validity of species descriptions, more Asian pikas must be surveyed. Finding evidence for cryptic species of coccidia in pikas also could be an important issue for sorting out the origins of their host-parasite associations. Finally, when interpreting their work in light of all previous studies, Lynch et al. (2007) made three points: (1) the similarity in coccidian fauna among *O. princeps*, *O. collaris*, and *O. hyperborea*; (2) the different and more diverse coccidian parasites in Asian hosts; and (3) the apparent widespread species of coccidia found in pikas representing two different host clades.

To date, only 7/30 (23%) pika species worldwide have been examined for coccidia and only the 19 species or forms, all described above, are known. For several of these host species, only small numbers of individuals were sampled, so it is likely that other new *Eimeria* and *Isospora* species exist in them, still to be discovered, especially cryptic species (see Lynch et al., 2007). Based on the information given here, we believe it safe to assume that each pika species is capable of harboring at least four coccidia species that are unique to it. This means that there should be, minimally, 100 new coccidia yet to be discovered from the 23 pika species that have not yet been examined for these parasites. Put another way, to date, only about 17% of the total coccidia species from pikas have been discovered and described!

Finally, of the 30 extant pika species on Earth, 23 (77%) are found in China, several with overlapping host ranges, but only four (17%) of these hosts (*O. curzoniae, O. dauurica, O. hyperborea, O. pallasi*) have been examined for coccidia. From these four pika species, 17 coccidial forms have been described: *O. curzoniae* (*E. banffensis,*

E. calentinei, E. cryptobarretti, E. qinghaiensis, E. haibeiensis), *O. dauurica* (*E. daurica, E. erschovi, E. metelkini, E. ochotona*), *O. hyperborea* (*E. banffensis, E. calentinei, E. circumborealis, E. cryptobarretti, E. klondikensis, E. princepsis, E. worleyi, I. marquardti*), and *O. pallasi* (*E. erschovi, E. pallasi, E. shubini, E.* sp.). Not until the remaining Asian pika species are examined and their coccidia discovered and described will we be able to construct rigorous and testable hypotheses about whether their coccidia would reflect the systematics and phylogenetics of their hosts. Of course, we must rely on our colleagues who study mammals to clarify and elaborate on the systematics of the host genus in the future (Yi-Fan et al., 2009).

TABLE 3.1 All Reported Coccidia Species (Apicomplexa: Eimeriidae) from Members of the Genus *Ochotona* Link, 1795 (Family Ochotonidae)

Ochotona spp.	Coccidia (Eimeriidae)[1]	References
collaris[2]	*Eimeria banffensis*	Hobbs & Samuel, 1974
	E. barretti	Hobbs & Samuel, 1974
	E. calentinei	Hobbs & Samuel, 1974; Lynch et al., 2007
	E. circumborealis	Hobbs & Samuel, 1974
	E. cryptobarretti	Lynch et al., 2007
	E. klondikensis	Hobbs & Samuel, 1974; Lynch et al., 2007
	E. princepsis	Hobbs & Samuel, 1974
	Isospora marquardti	Lynch et al., 2007
	I. yukonensis	Hobbs & Samuel, 1974
curzoniae	*E. banffensis*	Yi-Fan et al., 2009
	E. calentinei	Yi-Fan et al., 2009
	E. cryptobarretti	Yi-Fan et al., 2009
	E. haibeiensis	Yi-Fan et al., 2009
	E. qinghaiensis	Yi-Fan et al., 2009
dauurica	*E. daurica*	Matschoulsky, 1947a
	E. erschovi	Matschoulsky, 1949
	E. metelkini	Matschoulsky, 1949
	E. ochotona	Matschoulsky, 1949

(Continued)

TABLE 3.1   All Reported Coccidia Species (Apicomplexa: Eimeriidae) from Members of the Genus *Ochotona* Link, 1795 (Family Ochotonidae) (*cont'd*)

Ochotona spp.	Coccidia (Eimeriidae)[1]	References
hyperborea	*E. banffensis*	Hobbs & Samuel, 1974; Lynch et al., 2007
	E. calentinei	Hobbs & Samuel, 1974; Lynch et al., 2007
	E. circumborealis	Hobbs & Samuel, 1974
	E. cryptobarretti	Lynch et al., 2007
	E. klondikensis	Hobbs & Samuel, 1974; Lynch et al., 2007
	E. princepsis	Hobbs & Samuel, 1974
	E. worleyi	Hobbs & Samuel, 1974
	I. marquardti	Lynch et al., 2007
princeps[2]	*E. banffensis*	Lepp & Todd, 1973; Hobbs & Samuel, 1974; Duszynski & Brunson, 1973; Duszynski, 1974
	E. barretti	Lepp et al., 1972
	E. calentinei	Duszynski & Brunson, 1973; Hobbs & Samuel, 1974; Duszynski, 1974
	E. circumborealis	Hobbs & Samuel, 1974
	E. cryptobarretti	Duszynski & Brunson, 1973; Duszynski, 1974
	E. klondikensis	Hobbs & Samuel, 1974; Duszynski, 1974
	E. princepsis	Duszynski & Brunson, 1973; Duszynski, 1974; Hobbs & Samuel, 1974
	E. worleyi	Lepp et al., 1972
	I. marquardti	Duszynski & Brunson, 1972; Duszynski, 1974; Hobbs & Samuel, 1974
	Eimeria spp.	Barrett & Worley, 1970
	Isospora sp.	Barrett & Worley, 1970
pallasi	*E. erschovi*	Svanbaev, 1958
	E. pallasi (?)	Svanbaev, 1956; Lepp et al., 1972
	E. shubini (?)	Svanbaev, 1958; Lepp et al., 1972
	Eimeria sp. (?)	Svanbaev, 1958
rufescens	*E. balchanica*	Glebezdin, 1978
Seven host species	17 *Eimeria* (15 valid), two *Isospora*	

[1] *All reported as natural infections. No experimental transmissions have been performed, to date, in* Ochotona *spp.*
[2] *North American species; all other pika species are from Asia.*
(?) Not considered a valid species of *Ochotona*.

CHAPTER

4

Coccidia (Eimeriidae) from the Family Leporidae: Genus *Brachylagus*

The mammalian order Lagomorpha contains two extant families, Ochotonidae (Chapter 3) and Leporidae. Within the Leporidae there are 11 genera with about 60 distinct species (Hoffmann and Smith, 2005) including *Brachylagus* (one species), *Bunolagus* (one species), *Caprolagus* (one species), *Lepus* (32 species), *Nesolagus* (one species), *Oryctolagus* (one species), *Pentalagus* (one species), *Poelagus* (one species), *Pronolagus* (three species), *Romerolagus* (one species), and *Sylvilagus* (17 species). To date, only four of the 11 (36%) leporid genera (*Brachylagus*, *Lepus*, *Oryctolagus*, *Sylvilagus*) and 17 of the 60 (28%) leporid species have been examined for and have had coccidia (all *Eimeria* species) described from them (Duszynski and Upton, 2001).

Relatively little is known about the biology of the pygmy rabbit, *Brachylagus idahoensis* (Merriam, 1891). Weighing ~400 g as adults, they are the only species in the genus *Brachylagus*, they are restricted to the Great Basin of the western United States, and they are found only in isolated populations in northeastern California, southern Idaho, southwestern Montana, northern Nevada, eastern Oregon, western Utah, western Wyoming, and southeastern Washington (Green and Flinders, 1980; Campbell et al., 1982; Nowak, 1991). The population in Washington State is confined to the Columbia Basin, and is thought

FIGURE 4.1 The pygmy rabbit, *Brachylagus idahoensis* (Merriam, 1891).

to have been geographically isolated from other populations of the species for thousands of years (Lyman, 1991, 2004; K. Warheit, pers. comm.).

Pygmy rabbits display several traits that set them apart from species of cottontails (*Sylvilagus*), jackrabbits (*Lepus*), and Old World rabbits (*Oryctolagus*): (1) they are the only leporids in the continental United States that dig their own extensive and intercommunicating burrows; (2) they give alarm calls and other vocalizations; (3) they move by a low scampering gait (not leaping); and (4) they are almost totally dependent upon sagebrush (*Artemisia tridentata*) for their diet, especially during winter months, when it may constitute 98–99% of their food intake (Nowak, 1991).

In the last decade, populations have been declining in Washington, Oregon, and California, where sagebrush habitat has been burned, converted to agriculture, or cleared from large areas and replaced with bunch grasses to improve livestock forage. Studies by the Washington Department of Fish and Wildlife (WDFW) in 2001 determined that the Columbia Basin population of pygmy rabbits (Douglas County, Washington) is genetically distinct and has been isolated from Idaho and Oregon populations for at least 7,000 years. The population has declined precipitously from about 150 pygmy rabbits in 1995 (WDFW, 1995) to fewer than 30 in 2001 (Hays, 2001). Since no one had successfully bred this species in captivity, the WDFW initiated a captive breeding program in cooperation with Washington State University and the Oregon Zoo. It started in December 2000, when four pygmy rabbits were brought from the Lemhi Valley, Idaho, to the Oregon Zoo to help develop pygmy rabbit husbandry protocols. Later, 16 Columbia Basin pygmy rabbits were captured as an initial source for captive breeding efforts, and in 2002, both facilities began breeding the endangered Columbia Basin rabbits for eventual reintroduction to a protected habitat in central Washington. The U.S. Fish and Wildlife Service listed the Columbia Basin pygmy rabbit population under emergency provisions of the Endangered Species Act on November 30, 2001 (USFWS, 2001) and provided the population full listing status on March 5, 2003 (USFWS, 2003).

There are now about two dozen Columbia Basin pygmy rabbits in captive breeding programs in Washington and Oregon, but the program did not produce many offspring in the first 2 years. In addition, in May 2002, four captive-bred young rabbits died in 5 days, apparently due to gastrointestinal coccidiosis. The coccidian species responsible for these deaths was reported as new by Duszynski et al. (2005). This, so far, is the only documentation of a potentially pathogenic eimerian in pygmy rabbits, and the structure of the sporulated oocyst and some of the endogenous stages and histopathology are all the information available on this parasite (Duszynski et al., 2005).

FAMILY LEPORIDAE FISCHER, 1817

HOST GENUS *BRACHYLAGUS* MILLER, 1900

EIMERIA BRACHYLAGIA DUSZYNSKI, HARRENSTIEN, COUCH & GARNER, 2005

Type host: *Brachylagus idahoensis* (Merriam, 1891), Pygmy rabbit.

Type locality: NORTH AMERICA: USA, Washington State, Douglas County (47°67′N, 119°89′W).

Other hosts: None to date.

Geographic distribution: NORTH AMERICA: USA, Washington, Oregon, Idaho.

Description of sporulated oocyst: Oocyst shape: subspheroidal; number of walls: 2, outer ~$^3/_4$ of total thickness; wall thickness: 0.8–1.1 (1.0); wall characteristics: smooth; L × W ($N = 50$): 25.6 × 23.8 (22–28.5 × 21–27); L/W: 1.1

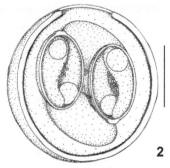
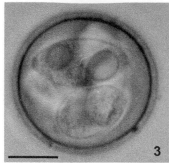
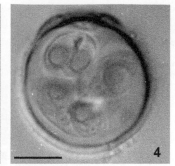

FIGURES 4.2, 4.3, 4.4 **2.** Line drawing of a sporulated oocyst of *E. brachylagia*. **3.** Photomicrograph of sporulated oocysts of *E. brachylagia*. Note the large refractile bodies in each sporozoite and the Stieda body on the pointed end of the sporocyst at the top of the oocyst. **4.** Photomicrograph of sporulated oocysts of *E. brachylagia*. Notice the micropyle at the top of the oocyst wall. All figures from Duszynski et al. (2005) with permission of the *Journal of Parasitology*. Scale bar = 10 μm in all figures.

(1.0—1.2); M: present as a thin membranous area at one end of oocyst with wall ending bluntly on each side of M; OR: absent; PG: 0—1. Distinctive features of oocyst: presence of a M and lack of an OR.

Description of sporocyst and sporozoites: Sporocyst shape: ellipsoidal; L × W: 13.4 × 8.1 (11—16.5 × 7.5—9); L/W: 1.7 (1.3—2.2); SB: present as a slightly rounded extension of the sporocyst wall; SSB and PSB: absent; SR: present; SR characteristics: many fine granules evenly dispersed between SZ, usually in middle of sporocyst; SZ: lie head-to-tail and have one large, posterior RB. Distinctive features of sporocysts: large posterior RB in each SZ.

Prevalence: 11/20 (55%); 8/17 in feces, 3/3 in tissue sections of intestinal epithelium.

Sporulation: Exogenous.

Prepatent and patent periods: Unknown.

Site of infection: Epithelial cells of the proximal- and mid-small intestine. No coccidial endogenous stages were found scanning sections of liver parenchyma for 20+ min.

Endogenous stages: Several different merogonous stages were seen. These may represent a progression in development of one stage or they may represent two or more different merogonous generations (Figures 4.11—4.15 and see Duszynski et al., 2005). If the former, early merozoites seem to develop and bud from a central globule in the meront (Figure 4.11) and then enlarge until they become what appear to be fully developed meronts (Figure 4.6); these (N = 11) were subspheroidal, 14.8 × 13.9 (13—18 × 10.5—16.5), with ~46 (26—70) merozoites. Early developing macro- and microgamonts were indistinguishable from each other (Figure 4.7). They (N = 20) were spheroidal to subspheroidal, 10.4 × 9.5 (9—11 × 7.5—10.5), and stained light purple with the N staining somewhat darker. The cytoplasm varied from homogeneous to having a slight frothy appearance. The N (~2.0) was usually centric and only rarely was a small, non-centric nucleolus visible. Mature macrogamonts, with clearly defined wall-forming bodies that had not yet coalesced at the margins (Figure 4.8), were spheroidal to subspheroidal, 14.2 × 13.7 (12—17 × 11—16). Young and mature microgamonts (Figures 4.9, 4.10), with a central N surrounded by homogeneous cytoplasm and many scattered, small N that eventually end up at the periphery, were subspheroidal, 11.9 × 10.8 (10.5—13 × 9—12). During the course of an infection, the majority of epithelial cells on the villi are infected with one or more endogenous developmental stages of this parasite (Figure 4.5).

Cross-transmission: None to date.

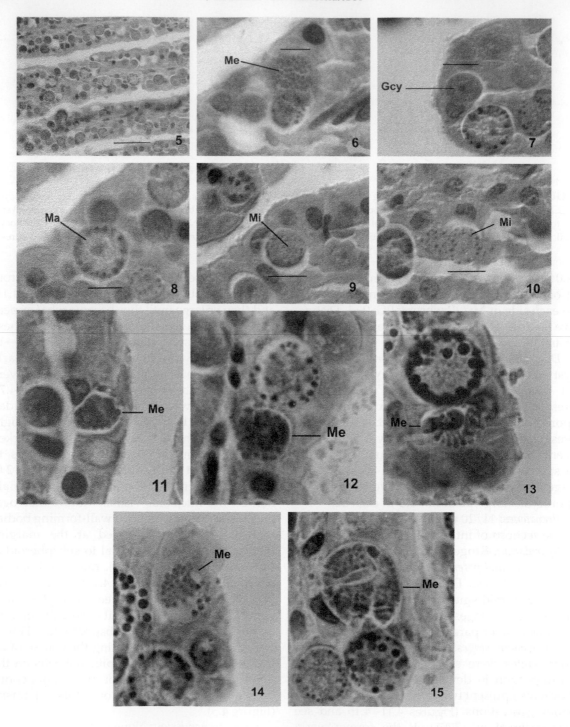

Pathology: Case History No. 1: A 3-wk-old male, captive-bred rabbit was ill for less than one day and was found dead. No gross abnormalities were noted other than absence of pelleted stool in the distal colon. Massive numbers of developing gamonts and a lesser number of meronts were seen in the villus enterocytes of the mid-small intestine associated with apical necrosis and mild chronic inflammation in the lamina propria. Bacterial overgrowth was noted in the central lumen of this section of the intestine. The animal was emaciated with marked **atrophy** of adipose stores, and the severe intestinal coccidiosis likely contributed to this condition. The endogenous stages were succinctly segmental in one convoluted section of small intestine, while the cecum, colon, and distal part of the small intestine near the ileum were unaffected.

Case History No. 2: A 26-day-old male, captive-bred pygmy rabbit was found in **moribund** condition with moderate seizure-like muscle movements. The rabbit died within 25 min despite treatment with oxygen, warmth, fluid, dextrose, **doxapram hydrochloride**, and epinephrine. Necropsy revealed reddening of the **calvarium** and ingesta throughout the gut, and formed fecal pellets in the distal colon also were reddened, but it is not known whether this was fresh blood or post-mortem artifact. The rabbit had pneumonia, and its lungs were consolidated with **clostridial overgrowth**. The proximal- and mid-small intestine had large numbers of endogenous developmental stages (mostly gamonts) in villus enterocytes that seemed to be associated with mild elongation of villi (Figure 4.5). The presence of lymphocytes and plasma cells in the **lamina propria** is normal, but in infected villi there was a mild to moderate increase in the number of these infiltrates. Large numbers of unsporulated oocysts were in the central lumen. Adipose stores were markedly atrophic.

Case History No. 3: A wild-caught, adult female presented with a mass in the neck region that was diagnosed as a **hemangiosarcoma**. The rabbit was humanely euthanized and microscopic examination of the small intestine noted many intraepithelial coccidial meronts and gamonts. There were increased numbers of lymphocytes, plasma cells, macrophages, and hemosiderin-laden macrophages in the lamina propria, beyond background infiltrates.

Material deposited: Photosyntypes of sporulated oocysts, No. 095050.00, and tissue sections of endogenous stages, No. 095051.00, are deposited in the U.S. National Parasite Collection (USNPC), Beltsville, MD, USA.

Remarks: Historically, the monotypic genus *Brachylagus* was included as a subgenus of *Sylvilagus*, but as more data accumulated from paleontology, morphology, serology, ecology, and behavior, it is now accepted as a distinct genus by most mammalogists (Green and Flinders, 1980; Nowak, 1991). Thus, it is most logical to compare sporulated oocysts of *E. brachylagia* with those found in *Sylvilagus* spp., still the closest phylogenetic relative, which may be the

FIGURES 4.5—4.15 Photomicrographs of endogenous stages of *E. brachylagia* in villus epithelial cells of the small intestine from a pygmy rabbit that died of coccidiosis. **5.** Section through three villi; note the large number of cells with developing meronts and/or gamonts in them. **6.** Presumably mature meront (Me) with ~46 merozoites in cross-section. **7.** Young developing gametocyte (Gcy); note that it is not possible at this stage to distinguish between macro- and microgametocytes. **8.** Macrogametocyte (Ma) with unfused wall forming bodies near the periphery of the cell. **9.** Young microgametocyte (Mi) with a few nuclei beginning to accumulate at the periphery of the cell. **10.** Two microgametocytes (Mi) in the same cell. **11, 12.** Young meronts with ~10—12 merozoites (cross-section) budding from a central mass. **13.** Young meront with ~12 merozoites (long-section) budding from a central mass. **14, 15.** Nearly mature **(14)** and presumably mature **(15)** meronts with ~35 and 50+ merozoites, respectively. Scale bar = 50 μm in Figure 5 and = 10 μm in Figs. **6—1 5**. All figures from Duszynski et al. (2005) with permission of the *Journal of Parasitology*.

most similar. Of the dozen *Eimeria* species known to infect only *Sylvilagus* species, sporulated oocysts of *E. brachylagia* are similar in size only to those of *E. environ* described by Honess (1939), from the mountain cottontail *Sylvilagus nuttallii* (Bachman) (26 × 24 [22–28.5 × 21–27] vs. 27 × 21 [22–30 × 16–23]), but their L/W ratios are different (1.1 vs. 1.3), making *E. environ* more elongate and not subspheroidal. Oocysts of the two species also are similar by having a M and by lacking an OR, but the M of *E. environ* often has "a cap which slightly protrudes beyond the curvature of the oocyst wall" (Honess, 1939), which *E. brachylagia* lacks. The size of the sporocysts of *E. brachylagia* also are similar (13 × 8 [11–16.5 × 7.5 × 9]) to those of *E. environ* (16 × 8 [13–18 × 7–10]), and both species possess a SB. However, the sporocyst L/W ratios are, again, different (1.7 vs. 2.0) with those from *E. brachylagia* being more subspheroidal, while those from *E. environ* are more ellipsoidal. Finally, sporocysts of *E. brachylagia* possess a diffuse SR, which those of *E. environ* lack.

The pathogenic nature of this coccidium seems surprising. In natural populations of other leporids with which we are familiar, especially *Lepus* and *Sylvilagus* species, it is not uncommon for an individual to be infected with up to six *Eimeria* species at a single time and not exhibit demonstrable pathology (DWD, pers. obs.). Yet, when young pygmy rabbits get infected with sporulated oocysts of *E. brachylagia*, it appears that virtually every epithelial cell on an infected villus can harbor one or more developmental stages of the parasite. This suggests to us that there may be several successive generations of meronts, each producing many merozoites, that eventually lead to the massive concentration of the gamogonous stages reported infecting epithelial cells (Duszynski et al., 2005). Whether the merogony reported represents one or possibly two different generations can only be a matter of speculation due to the temporal nature of endogenous development and the time at which the animals were examined. One question that arises is whether it is the nature of *E. brachylagia* to be pathogenic (as is *E. bovis* in cattle, *E. tenella* in chickens, *E. stiedai* in other leporids) or whether the pathology reported was exacerbated by the stress of captivity. Other questions on the biology of *E. brachylagia* need to be answered soon if, indeed, this coccidium is a contributing factor to the decline of pygmy rabbits in the wild. Does the burrowing nature of these rabbits contribute to the concentration of oocysts in their natural habitat such that animals are more likely to encounter massive infections early in life? Are there other *Eimeria* species that infect *Brachylagus* and may contribute to the pathology reported? To our knowledge, none of the pygmy rabbits of the northwestern USA have been surveyed for coccidia or any other parasites. What are the mechanisms of pathogenicity? Is it simple destruction of epithelium, in the mid-small intestine, due to large numbers of merogonous stages, or is some yet unknown mechanism in operation?

Given the severe lesions associated with the presence of *E. brachylagia* and the precarious nature of this unique genetic population of rabbits, it appears that this is an emerging pathogen that certainly deserves immediate further study.

Coccidia (Eimeriidae) from the Family Leporidae: Genus *Lepus*

The Biology and Identification of the Coccidia (Apicomplexa) of Rabbits of the World
http://dx.doi.org/10.1016/B978-0-12-397899-8.00005-6

INTRODUCTION

When compared with other mammalian orders, the Lagomorpha is not diverse and contains just three families, Ochotonidae, Prolagidae, and Leporidae; the first two are monogeneric, while the Leporidae is divided into 11 genera with 61 species. Interestingly, of the 13 genera in the Lagomorpha, eight are monotypic, while *Lepus*, with 32 species, is the most speciose

and has worldwide distribution (Wilson and Reeder, 2005). The long historical association of several species of *Lepus* with humans has fostered their economic and ecological importance worldwide. Despite this, the evolution of hares and jackrabbits has been poorly studied. A lack of morphological variation between species has resulted in numerous taxonomic revisions throughout the last two centuries (Flux, 1983; Halanych et al., 1999). Prior to the

FIGURE 5.1 The blacktail jackrabbit, *Lepus californicus* Gray, 1837.

recent molecular analysis (mitochondrial cytochrome b) of 11 North American *Lepus* species by Halanych et al. (1999), only two previous studies have used molecular or biochemical analyses to address the evolution and historical biogeography of *Lepus*; Pérez-Suárez et al. (1994) used mitochondrial DNA to look at speciation in four species of Mediterranean hares, and Robinson and Osterhoff (1983) documented protein variation and its systematic implications in three South African hares.

Based on fossil evidence, Hibbard (1963) proposed that *Lepus* first arose in North America, implying that hares radiated to other continents. Given the limited scope of molecular work to look at evolution and historical biogeography within the Lagomorpha (Robinson and Osterhoff, 1983; Pérez-Suárez et al., 1994; Halanych et al., 1999), it is not possible to

confirm or refute this hypothesis. However, the limited data provided by Halanych et al. (1999) suggest that North American *Lepus* are not a monophyletic clade to the exclusion of other hares, because *L. townsendii*, *L. arcticus*, and *L. othus* are in a clade separate from the remaining North American taxa. The placement of Old World taxa at the base of, and within the arctic clade, suggested to them that some hares invaded North America secondarily, perhaps via the Beringian Asian-American land connection. In contrast, *L. americanus*, which ranges across the high latitudes of the continental USA, Canada, and Alaska, is more closely related to taxa from the southwestern USA and Mexico (*L. alleni*, *L. californicus*, *L. callotis*) than to the arctic clade. In addition, Halanych et al. (1999) noted deep divergence between *Lepus* species pairs that hybridize in the wild (e.g., *L. timidus*, *L. europaeus*). Hybridization between more distantly related species allowed Halanych et al. (1999) to argue that isolation mechanisms (e.g., geographic, behavioral, or ecological) may be driving speciation within *Lepus* and that the known lack of chromosomal diversity within *Lepus* points to mechanisms of speciation that do not invoke the chromosomal models suggested for other mammals (Halanych et al., 1999).

Only 10 of 32 (31%) species in this genus have been examined for coccidia, and combined, they have been reported to harbor 43 *Eimeria* species (Table 5.1); however, nine of 43 (21%) were described originally from the domestic rabbit, *Orcytolagus cuniculus* (*E. exigua*, *E. intestinalis*, *E. irresidua*, *E. magna*, *E. matsubayashii*, *E. media*, *E. perforans*, *E. piriformis*, *E. stiedai*) and four of 43 (9%) were originally described from cottontail rabbits, *Sylvilagus* spp. (*E. audubonii*, *E. minima*, *E. neoleporis*, *E. sylvilagi*). The validity of 12 of these eimerians as bona fide parasites of *Lepus* species needs to be confirmed by molecular and/or cross-transmission work. Only *E. stiedai* is known to parasitize *Lepus* species for certain. It is also of interest to note that species from other, related coccidian genera (e.g., *Isospora, Caryospora, Cyclospora*) have never been observed in *Lepus* species despite the many surveys that have been done on the members of this host genus.

HOST GENUS *LEPUS* LINNAEUS, 1758

EIMERIA AMERICANA CARVALHO, 1943

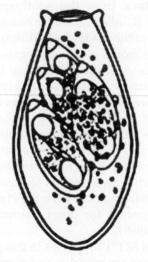

FIGURE 5.2 Line drawing of the sporulated oocyst of *Eimeria americana* from Carvalho, 1943, with permission of John Wiley & Sons.

Type host: *Lepus townsendii* Bachman, 1839, White-tailed jackrabbit.

Type locality: NORTH AMERICA: USA: Iowa, near Ames.

Other hosts: None reported to date.

Geographic distribution: NORTH AMERICA: USA: Iowa.

Description of sporulated oocyst: Oocyst shape: ovoidal to elongate-ellipsoidal, tapering toward the ends; number of walls: 1; wall

characteristics: delicate, of uniform thickness, pinkish, with a distinct dark-brownish marginal ridge around the M; L × W: 38 × 21 (34—43 × 18—25); L/W ratio: 1.8; M: present; M characteristics: distinct, ~ 6 wide, with a collar-like thickening around it; OR: present; OR characteristics: distinct, granular, and spread over and among the sporocysts; PG: absent. Distinctive features of oocyst: marginal lip or collar around M.

Description of sporocyst and sporozoites: Sporocyst shape: ovoidal; L × W: 17 × 7—8; SB: present; SSB, PSB, SR: all absent; SZ: comma-shaped, with a central N and large RB, ~5.5 wide. Distinctive features of sporocyst: absence of SR and presence of large RB.

Prevalence: 5/12 (42%) in the type host (Carvalho, 1943, p. 122).

Sporulation: Exogenous. Oocysts sporulated within 55—65 (mean, 60) hr in 3% aqueous (w/v) potassium dichromate solution ($K_2Cr_2O_7$) at 30° C (Carvalho, 1943).

Prepatent and patent periods: Unknown.

Site of infection: Unknown, but probably the intestine. Oocysts recovered from feces and intestinal contents.

Endogenous stages: Unknown.

Cross-transmission: Carvalho (1943) was not able to transmit this species either to the domestic rabbit, *O. cuniculus*, or to the Iowa cottontail, *S. floridanus*.

Pathology: Unknown.

Material deposited: None.

Remarks: This species has not been reported since it was first described (Carvalho, 1943), which raises some skepticism about its validity.

EIMERIA ATHABASCENSIS SAMOIL & SAMUEL, 1977a

Type host: *Lepus americanus* Erxleben, 1777, Snowshoe hare.

Type locality: NORTH AMERICA: Canada: north-central Alberta, near Rochester.

Other hosts: None reported to date.

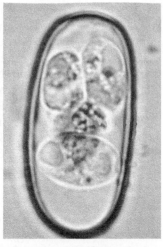

FIGURES 5.3, 5.4 Line drawing of the sporulated oocyst of *Eimeria athabascensis* from Samoil and Samuel, 1977a, with permission of the *Canadian Journal of Zoology*. Photomicrograph of a sporulated oocyst of *E. athabascensis* from Samoil and Samuel, 1977a, with permission of the *Canadian Journal of Zoology*.

Geographic distribution: NORTH AMERICA: Canada: Alberta.

Description of sporulated oocyst: Oocyst shape: cylindroidal; number of walls: 2; wall thickness: ~1; wall characteristics: outer is smooth and uniformly thick, several times thicker than inner; L × W: 33.8 × 15.6 (24–38 × 13–17); L/W ratio: 2.2; M: absent; OR: present: OR characteristics: compact, granular, 3.5 (2–5) wide; PG: absent. Distinctive features of oocyst: cylindroidal shape and presence of compact OR.

Description of sporocyst and sporozoites: Sporocyst shape: elongated spindle-shape, tapering toward one end; L × W: 15.1 × 6.2 (12–18 × 6–8); L/W ratio: 2.5; SB: present; SB characteristics: indistinct, light-refracting area at narrow end; SSB, PSB, SR: all absent; SZ: with a single, clear RB posterior to the N. Distinctive feature of sporocyst: shape of sporocyst and its light-refracting SB.

Prevalence: 156/629 (25%) in the type host in Alberta, Canada (Samoil and Samuel, 1977a).

Sporulation: Exogenous. Oocysts sporulated within 6 days in 2.5% aqueous (w/v) $K_2Cr_2O_7$ solution at 20°C.

Prepatent and patent periods: Unknown.

Site of infection: Unknown. Oocysts recovered from feces.

Endogenous stages: Unknown.

Cross-transmission: None to date.

Pathology: Unknown.

Material deposited: None.

Remarks: Samoil and Samuel (1977a) reported that sporulated oocysts of *E. athabascensis* resemble those of *E. leporis*, *E. ruficaudati*, and *E. neoleporis*. Oocysts of the first two species have a M and/or SR, which those of *E. athabascensis* lack, while oocysts of *E. neoleporis* lack an OR and the oocysts are larger than those of *E. athabascensis*.

EIMERIA AUDUBONII DUSZYNSKI & MARQUARDT, 1969 (FIGURES 7.2, 7.3)

Type host: *Sylvilagus audubonii* (Baird, 1858), Desert cottontail.

Remarks: Aoutil et al. (2005) studied the *Eimeria* species in *L. granatensis* and *L. europaeus* in France, and recorded what they provisionally called *E. audubonii* in 6/46 (13%) *L. granatensis*, but apparently did not find it in the 18 *L. europaeus* they examined. The short ellipsoidal oocysts they measured were 21 × 17 (19–24 × 14–18), without either an OR or M, and the sporocysts were 12 × 7, without a SR, but with SZs that each had two RB, one small and one larger. They noted that all qualitative and quantitative features of the sporulated oocysts they found were identical to those detailed by Duszynski and Marquardt (1969) in their original description of *E. audubonii* while they wrote, "It is doubtful that the species we have observed is indeed identical to that found in an American *Sylvilagus*, but morphological features that would allow to distinguish them are lacking.

Moreover, we are aware that some *Sylvilagus* have been introduced in France by hunters on a number of occasions." We believe that this may be a new species from *Lepus*, but it is not *E. audubonii* from *Sylvilagus*. We await molecular sequence evidence to confirm that the form from *Lepus* is indeed distinct from that found in *Sylvilagus* species. See Chapter 7 (Figures 7.2, 7.3) for the complete description of *E. audubonii*.

EIMERIA BAINAE AOUTIL, BERTANI, BORDES, SNOUNOU, CHABAUD & LANDAU, 2005

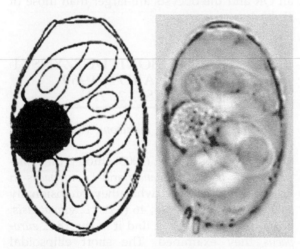

FIGURES 5.5, 5.6 Line drawing of the sporulated oocyst of *Eimeria bainae* from Aoutil et al., 2005, with permission of the authors and from (the journal) *Parasite*. Photomicrograph of a sporulated oocyst of *E. bainae* from Aoutil et al., 2005, with permission of the authors and from (the journal) *Parasite*.

Type host: *Lepus granatensis* Rosenhauer, 1956, Granada hare.

Type locality: EUROPE: France: Pyrénées-Orientales.

Other hosts: None reported to date.

Geographic distribution: EUROPE: France.

Description of sporulated oocyst: Oocyst shape: ellipsoidal, slightly flattened at end with M; number of walls: 1 (line drawing), of uneven thickness; wall characteristics: yellowish, thickens slightly around M; L × W: 29 × 18 (26–33 × 15–21); L/W ratio: 1.6; M: present; M characteristics: ~6 wide, with oocyst wall thicker near margin; OR: present; OR characteristics: a large spheroid, ~8 wide, composed of tightly-packed granules; PG: absent. Distinctive features of oocyst: large, round OR of tightly-packed granules.

Description of sporocyst and sporozoites: Sporocyst shape: spindle-shaped, slightly flattened at one end (line drawing); L × W: 15 × 8 (ranges not given); L/W ratio: 1.9; SB: small, at flattened end; SSB, PSB: both absent; SR: absent; SZ: with one large RB in middle of their body (line drawing). Distinctive features of sporocyst: small, flat SB and a large RB mid-body in SZ.

Prevalence: Apparently found in 4/46 (9%) *L. granatensis*.

Sporulation: Presumably exogenous, but the time and temperature are not known.

Prepatent and patent periods: Unknown.

Site of infection: Unknown. Oocysts recovered from feces.

Endogenous stages: Unknown.

Cross-transmission: None to date.

Pathology: Unknown.

Material deposited: None, although the authors designated their photomicrograph (their Fig. 1.10) as the holotype and said "the photos of the oocysts … have been deposited at the laboratory."

Remarks: Aoutil et al. (2005) stated that this species was closest morphologically to *E. honessi* from *S. floridanus*, the Eastern cottontail of North America, and said that because of host specificity and geographic separation they are different species, an opinion with which we concur. Aoutil et al. (2005) provided a line drawing and a good photomicrograph in their original description, but they did not deposit either of these in an <u>accredited</u> museum. They also did not provide all of the quantitative and qualitative information needed/expected in a modern, standardized species description (sporocyst L, W ranges; sizes of SZ, RB, etc.). In presenting their description of this new species, Aoutil et al. (2005) listed the name of this species as "*Eimeria bainae* n. sp. Aoutil & Landau"; this is not, and cannot be, the correct authority, since there is no separate published paper by Aoutil and Landau describing and naming it. Thus, the full scientific name (genus, species epithet, authority) must be as we listed it above. Finally, Aoutil et al. (2005) raise some doubt regarding the naming and true identity of this species. When we compared their photomicrograph representing *E. bainae* (their Fig. 1.10) to the photomicrograph they presented for another "species" named as new, *E. reniai* (their Fig. 3.14, below), the oocysts look nearly identical. The authors also found these "species" in at least one common animal (*L. granatensis* #20), and we wonder how they were able to distinguish between them. When we measured the images in their Fig. 1.10 vs. their Fig. 3.14, the oocyst L and W were the same for both.

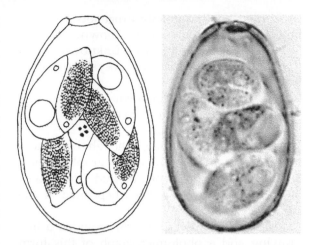

FIGURES 5.7, 5.8 Line drawing of the sporulated oocyst of *Eimeria cabareti* from Aoutil et al., 2005, with permission of the authors and from (the journal) *Parasite*. Photomicrograph of a sporulated oocyst of *E. cabareti* from Aoutil et al., 2005, with permission of the authors and from (the journal) *Parasite*.

EIMERIA CABARETI AOUTIL, BERTANI, CORDES, SNOUNOU, CHABAUD & LANDAU, 2005

Type host: *Lepus granatensis* Rosenhauer, 1956, Granada hare.

Type locality: EUROPE: France: Pyrénées-Orientales.

Other hosts: *Lepus europaeus* Pallas, 1778, European hare.

Geographic distribution: EUROPE: France.

Description of sporulated oocyst: Oocyst shape: elongate-ovoidal, slightly flattened at end with M; number of walls: 1 (line drawing), but probably 2 as the authors said, "a space is frequently observed between the inner and outer walls"; wall characteristics: thin, smooth, light brown, and it thickens around M; L × W: 35 × 22 (32–38 × 20–24); L/W ratio: 1.6; M: present; M characteristics: 5 wide, "barely convex and slightly concave at the base," surrounded by oocyst wall that is thicker near the margin; OR: present; OR characteristics: formed by 3–5 grains; PG: absent. Distinctive features of oocyst: elongate shape that narrows slightly at end with M and an OR consisting only of 3–5 small grains.

Description of sporocyst and sporozoites: Sporocyst shape: ovoidal, slightly pointed at one end (line drawing); L × W: 17 × 9 (ranges not given); L/W ratio: 1.9; SB: small, at pointed end; SSB, PSB: both absent; SR: present; SR characteristics: well-defined, ovoidal, compact mass, composed of small granules; SZ: with one large, elongate RB near center of body (line drawing), but the authors said, "two round RB, one large and one small." Distinctive features of sporocyst: small SB and SZ with 1–2 RBs of different sizes.

Prevalence: Found in 4/46 (4%) *L. granatensis* and in 1/9 (11%) *L. europaeus*.

Sporulation: Presumably exogenous, but the time and temperature are not known.

Prepatent and patent periods: Unknown.

Site of infection: Unknown. Oocysts recovered from feces.

Endogenous stages: Unknown.

Cross-transmission: None to date.

Pathology: Unknown.

Material deposited: None, although the authors designated their photomicrograph (their Fig. 3.16) as the holotype and said "the photos of the oocysts ... have been deposited at the laboratory."

Remarks: Aoutil et al. (2005) provided a line drawing and a photomicrograph of this form in their original description, but they did not deposit either of these in an <u>accredited</u> museum. Nor did they provide all of the quantitative and qualitative information needed and expected in a modern, standardized species description (e.g., age of the oocysts when studied, sporulation time and temperature, sporocyst L, W ranges; measurements of SZ, RB, etc.). The authors say that the sporulated oocysts of this species can be compared to those of *E. coecicola* from *O. cuniculus*, but differ by having an OR of only 2—3 grains vs. one that is well-defined in *E. coecicola* and that there are slight differences in the shape of the SR. They noted that their oocysts also resemble those of *E. stiedai*, but differ by "its ovoid rather than ellipsoid shape, by the presence of a more marked M and a jutting micropyle ring." They also remarked that oocysts of this species may be the same as those first seen in the Czech Republic by Pakandl (1990), who identified it as *E. townsendi*. In presenting their description of this species, Aoutil et al. (2005) listed the name as "*Eimeria cabareti* n. sp. Aoutil & Landau"; this is not, and cannot be, the correct authority for this species, because there is no separate published paper by Aoutil and Landau describing and naming it. Thus, the full scientific name (genus, specific epithet, authority) must be as we listed it above.

EIMERIA CAMPANIA (CARVALHO, 1943) LEVINE & IVENS, 1972

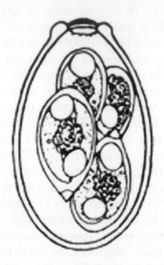

FIGURE 5.9 Line drawing of the sporulated oocyst of *Eimeria campania* from Levine and Ivens, 1972, with permission of the *Journal of Protozoology*.

Synonyms: *Eimeria irresidua* forma *campanius* Carvalho, 1943; *Eimeria irresidua* Kessel and Jankiewicz, 1931, of Gill and Ray (1960); *Eimeria semisculpta* Madsen, 1938, of Pellérdy, 1956, Pastuszko (1961a, b), and Gräfner et al. (1967), *pro parte*; *non Eimeria semisculpta* Madsen, 1938.

Type host: *Lepus townsendii* Bachman, 1839, White-tailed jackrabbit.

Type locality: NORTH AMERICA: USA: Iowa, Ames.

Other hosts: *Lepus europaeus*, Pallas, 1778, European hare; *Lepus nigricollis* F. Cuvier, 1823, Indian hare (syn. *Lepus ruficaudatus*).

Geographic distribution: NORTH AMERICA: USA: Iowa; EUROPE: (former East) Germany, Hungary, Italy, Poland; ASIA: India, Punjab, District Ludhiana, Uttar Pradesh, Kashipur.

Description of sporulated oocyst: Oocyst shape: cylindroidal to ellipsoidal with an elongated neck; number of walls: 2; wall characteristics: outer is thin, brownish, and markedly granulated on one end after being in $K_2Cr_2O_7$ solution for several days; inner is thicker and colorless ; L × W: 38 × 25 (35–45 × 22–27); L/W ratio: 1.5; M: present; M characteristics: 6–8 wide, atop an elongated neck, but without a ridge around the outside; OR, PG: both absent. Distinctive features of oocyst: its unique shape with an elongated neck and the lack of an OR.

Description of sporocyst and sporozoites: Sporocyst shape: oblong-fusiform; L × W: 18 × 9 (ranges not given); L/W ratio: 2.0; SB: present; SSB, PSB: both absent; SR: present; SR characteristics: consists of large granules; SZ: elongate with one end wider than the other, lying lengthwise head-to-tail and having a clear RB at the wide end. Distinctive features of sporocyst: fusiform shape and a SR of large granules.

Prevalence: 2/12 (17%) in the type host (Carvalho, 1943); Pastuszko (1961a, b) found it in 46/462 (10%) *L. europaeus* in Poland; Gräfner et al. (1967) found it in 3/176 (1.6%) *L. europaeus* in (former East) Germany.

Sporulation: Exogenous. Oocysts sporulated within 3–4 days at room temperature (Pellérdy, 1956).

Prepatent and patent periods: Unknown.

Site of infection: Unknown, presumably in the intestine. Oocysts recovered from feces.

Endogenous stages: Unknown.

Cross-transmission: Carvalho (1943) was not able to transmit this species either to the domestic rabbit, *O. cuniculus*, or to the Iowa cottontail, *S. floridanus*. Pellérdy (1956) could not infect the domestic rabbit with it.

Pathology: Unknown.

Material deposited: None.

Remarks: Carvalho (1943) did not describe this form, saying that its oocysts had the same structure and dimensions as those of *E. irresidua* from the domestic rabbit, and he called it *E. irresidua* forma *campanius*. He illustrated the oocyst, however, with a ridge around the M, a structure which *E. irresidua* does not have, so Levine and Ivens (1972) emended the name. The oocyst and sporocyst descriptions (above) are based on the observations of Carvalho (1943) and

Pellérdy (1956, 1965). Pellérdy (1956) reported it from Hungary, Pastuszko (1961a, b) from Poland, and Gräfner et al. (1967) from (former East) Germany, but they called it *E. semisculpta* Madsen, 1938. However, it differs from *E. semisculpta* in lacking an OR. Further, Pellérdy (1956) commented that Carvalho (1943) had not seen the anterior "granulations" that he saw; nevertheless, Pellérdy used the same name for the two forms. The form reported as *E. irresidua* from *L. nigricollis* (syn. *L. ruficaudatus*) by Gill and Ray (1960) was more likely *E. campania*. Tacconi et al. (1995) found this species (which they called *E. semisculpta*) in some breeding farms raising *L. europaeus* in Italy; their ellipsoidal oocysts were 35–45 × 22–27 with a M, and without an OR; this species was always found concurrently with *E. leporis*. Terracciano et al. (1988) identified oocysts of this species (which they also called *E. semisculpta*) in *L. europaeus* from three protected areas in the Province of Pisa, Italy. The ovoidal to cylindroidal oocysts they studied were 35–45 × 22–27 with a M 6–8 wide, and they lacked an OR, while their sporocysts had a distinct SR.

EIMERIA COQUELINAE AOUTIL, BERTANI, BORDES, SNOUNOU, CHABAUD & LANDAU, 2005

Synonym: *Eimeria gresae* of Aoutil et al., 2005 (see *Remarks* for *E. gresae* under *Species Inquirendae*, below).

Type host: *Lepus europaeus* Pallas, 1778, European hare.

Type locality: EUROPE: France: specific locality not stated.

Other hosts: *Lepus granatensis* Rosenhauer, 1956, Granada hare.

Geographic distribution: EUROPE: France.

Description of sporulated oocyst: Oocyst shape: short ellipsoidal, slightly flattened at end with M; number of walls: 1 (line drawing), but probably 2, of uneven thickness; wall characteristics:

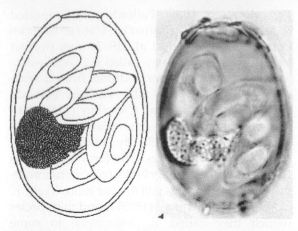

FIGURES 5.10, 5.11 Line drawing of the sporulated oocyst of *Eimeria coquelinae* from Aoutil et al., 2005, with permission of the authors and from (the journal) *Parasite*. Photomicrograph of a sporulated oocyst of *E. coquelinae* from Aoutil et al., 2005, with permission of the authors and from (the journal) *Parasite*.

outer is brownish, thickens slightly around M, and is finely granular, but this is best seen only after the oocyst wall(s) has broken; L × W: 35 × 23 (32–39 × 20–26); L/W ratio: 1.5; M: present; M characteristics: ~7–8 wide, with oocyst wall thicker near margin; OR: present; OR characteristics: a large mass of irregular contour, ~11 wide, composed of medium, tightly packed, granules; PG: absent. Distinctive features of oocyst: large M, ~7–8 wide, and large, OR, ~11 wide, and of irregular shape.

Description of sporocyst and sporozoites: Sporocyst shape: spindle-shaped, slightly pointed at one end (line drawing); L × W: 19 × 9 (ranges not given); L/W ratio: 2.1; SB: small, at pointed end; SSB, PSB: both absent; SR: absent; SZ: with one large RB in middle of their body (line drawing). Distinctive features of sporocyst: small, pointed SB and a large, round to ellipsoid RB mid-body in SZ.

Prevalence: Apparently found in 2/9 (22%) *L. europaeus* (which must be the type host since their figure 1.11, from *L. europaeus*, was designated the holotype, Aoutil et al., 2005) and in 2/46 (4%) *L. granatensis*.

Sporulation: Presumably exogenous, but the time and temperature are not known.

Prepatent and patent periods: Unknown.

Site of infection: Unknown. Oocysts recovered from feces.

Endogenous stages: Unknown.

Cross-transmission: None to date.

Pathology: Unknown.

Material deposited: None, although the authors designated their photomicrograph (their Fig. 1.11) as the holotype and said "the photos of the oocysts … have been deposited at the laboratory."

Remarks: Aoutil et al. (2005) stated that oocysts of this species were closest morphologically to those "wrongly referred to by most authors as *E. robertsoni* Madsen, 1938." Their argument centered on the differences between the forms described by Madsen (1938) as *E. magna* var. *robertsoni* from *L. arcticus* in Greenland, the oocysts Carvalho (1943) called *E. robertsoni* from *L. townsendii* collected in Iowa, USA, and the oocysts Pellérdy (1956) also called *E. robertsoni* from *L. europaeus* in Hungary. They remarked that Bouvier (1967) found "the species of Pellérdy in Switzerland," and said it was a parasite of the cecum, while the *E. robertsoni* from North America was located in the duodenum. Thus, they argued, "we think that the name '*robertsoni*' which appears to refer to three different species, should be abandoned." Unfortunately, they proposed no solution as to how the forms now relegated to *E. robertsoni* by these and other authors should be named (also see *Remarks* under *E. robertsoni* below). They only offered that this form could be distinguished from the Greenland species by being significantly more rounded, and from the Hungarian and North American species by being smaller and broader. Aoutil et al. (2005) provided a line drawing and a good photomicrograph of this form in their original description, but they did not deposit either of these in an <u>accredited</u> museum. Nor did they provide all of the quantitative and qualitative information needed and expected in a modern, standardized species description (e.g., age of the oocysts when studied, sporulation time and temperature, sporocyst L, W ranges; measurements of SZ, RB, etc.). This was disappointing to us as it seems below acceptable standards. In presenting their description of this species as new, Aoutil et al. (2005) listed the name as "*Eimeria coquelinae* n. sp. Aoutil & Landau"; this is not, and cannot be, the correct authority for this species, since there is no separate published paper by Aoutil and Landau describing and naming it. Thus, the full scientific name must be as we have listed it above. Finally, Aoutil et al. (2005) described the oocysts of what they said was another form, which they named *E. gresae* (which they spelled '*grease*' earlier in their paper, thus creating a *nomen nudum*), that they recorded in 6/46 (13%) *L. granatensis* and in 3/9 (33%) *L. europaeus*. Given their measurements, (incomplete) descriptions, photomicrographs (their Figs. 1.11 vs. 1.12) and line drawings (their Figs. 2.11 vs. 2.12), there is no question in our minds that these are variations on a theme and should be considered synonyms. Once again, the authors listed the name of this other species as "*Eimeria gresae* n. sp. Aoutil & Landau," which is not, and cannot be, the correct authority for this species, since there is no separate published paper by Aoutil and Landau describing and naming it.

EIMERIA EUROPAEA PELLÉRDY, 1956

Synonyms: *Eimeria perforans* of Bouvier (1967); *Eimeria belorussica* Litvenkova, 1969.

Type host: *Lepus europaeus* Pallas, 1778, European hare.

Type locality: EUROPE: Hungary.

Other hosts: *Lepus capensis* L., 1758, Cape hare; *Lepus timidus* L., 1758, Mountain hare.

Geographic distribution: ASIA: Byelorussia (of former USSR); Lithuania (of former USSR); EUROPE: Austria, Bulgaria, Czech Republic,

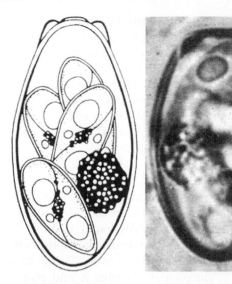

FIGURES 5.12, 5.13 Line drawing of the sporulated oocyst of *Eimeria europaea* from Levine and Ivens, 1972, with permission of the *Journal of Protozoology*. Photomicrograph of a sporulated oocyst of *E. europaea* from Pakandl (1990) with permission of *Folia Parasitologica*.

France, (former East) Germany, Hungary, Italy, Poland, Spain, Switzerland; MIDDLE/NEAR EAST: Turkey.

Description of sporulated oocyst: Oocyst shape: oblong-ellipsoidal or fusiform, tapering at both ends (although illustrated as ovoidal with a flat top); number of walls: 2; wall characteristics: thin, of equal thickness throughout except at the end with a M, inner is dark, outer is pale; L × W: 32 × 18 (26–34 × 15–20); L/W ratio: 1.8; M: present; M characteristics: darker in color than the rest of the wall, 6–9 wide, with a slightly protrusive marginal ring ("lappet") that forms a distinct collar or auricle-like projection around edges; OR: present: OR characteristics: coarse granules in a spheroidal or angular mass; PG: absent. Distinctive feature of oocyst: its distinct shape with auricle-like edges around the M and a large, granular OR.

Description of sporocyst and sporozoites: Sporocyst shape: spindle-shaped or elongate-ovoidal, slightly pointed at one end (line drawing); L ×

W: 9 × 6 (ranges not given); L/W ratio: 1.5; SB: reported as absent in original description, but it is likely present, but very small at the slightly pointed end of sporocyst; SSB, PSB: both absent; SR: occasionally seen; SR characteristics: small, centrally placed granules between SZs; SZ: comma-shaped with RB at thicker end. Distinctive features of sporocyst: tiny SB and a SR consisting only of a few granules.

Prevalence: Pastuszko (1961a) found this species in 61/462 (13%) *L. europaeus* in Poland; Gräfner et al. (1967) found it in 9/176 (5%) *L. europaeus* in (former East) Germany; Arnastauskene and Kazlauskas (1970) found it in 4/41 (10%) *L. europaeus* in Lithuania; Gottschalk (1973) reported a prevalence of 46.7% for this species in *L. europaeus* in Germany; and Sugár et al. (1978) found it in 52/374 (14%) in Hungary. Moreno Montañez et al. (1979) found this species in 27/42 (64%) of *L. capensis* in Spain. In the Czech Republic, Chroust (1984) reported a prevalence of 9.4%, while Pakandl (1990), also in the Czech Republic, surveyed *L. europaeus* from 1983–1985 and reported finding 85/350 (24%) infected with this species. Taşan and Özer (1989), in Turkey, said it occurred in 69% of the 40 hares they surveyed between 1985 and 1987.

Sporulation: Exogenous. Oocysts sporulated within 3–4 days (Pellérdy, 1956); 48–56 hr at 23°C in 3% K$_2$Cr$_2$O$_7$ solution (Golemanski, 1975), or 4 days at 20°C (Romero-Rodriguez, 1976).

Prepatent and patent periods: Unknown.

Site of infection: Unknown, presumably the intestine. Oocysts recovered from feces.

Endogenous stages: Unknown.

Cross-transmission: Pellérdy (1956) was not able to infect the domestic rabbit, *O. cuniculus*, with this species isolated from *L. europaeus*.

Pathology: Tacconi et al. (1995) found this species in the intestinal contents of four dead juvenile *L. europaeus* in Italy. The oocysts were always found with *E. hungarica*, *E. robertsoni*, and *E. townsendi* in these dead hares, which all showed severe enteritis as the cause of death.

Material deposited: None.

Remarks: Mensural data on the description of sporulated oocysts (above) is from Pellérdy (1956), who thought these oocysts resembled those of *E. americana* in shape and structure of the M, but they differed by being smaller in length and by having a compact OR rather than one that is dispersed. Bouvier (1967) called the form that he found in *L. europaeus* in Switzerland *E. perforans*. He said that it differed from *E. robertsoni* essentially only in the smaller size of its oocysts. However, genuine oocysts of *E. perforans* have never been found in *Lepus* and it does not have a conspicuous M. Bouvier's description (1967) fits that of *E. europaea* more closely than that of any other species of *Eimeria* from *Lepus*, and therefore, we assign it to this species. Litvenkova (1969) examined both *L. europaeus* and *L. timidus* in Byelorussia, but it is not certain from her description whether she found this species in both or only one host species. Pastuszko (1961a) said the ellipsoidal oocysts were 30 × 17 (25–30.5 × 15–20) with a M and OR. Golemanski (1975) said that the oocysts were 33 × 18 (26–36 × 14–20) with sporocysts 8–11 × 5–8. Romero-Rodriguez (1976) found this species in the Cape hare in Granada, Spain, and said the ellipsoidal oocysts were 32.8 × 18.2 (26–35 × 15–21) with the outer wall thickened around the M (9.5 wide), and sporocysts 8.7 × 5.9 (7.5–10 × 4–8), with a SB. Oocysts studied by Moreno Montañez et al. (1979) in *L. capensis* from Spain were 28–35 × 17–22 with sporocysts about 15 × 8. Terracciano et al. (1988) identified this species (which they called *E. belorussica*) in *L. europaeus* from three protected areas in the Province of Pisa, Italy; the ellipsoidal oocysts they studied were 26–28 × 14–16 with a M and a SR, but without an OR. The oocysts measured by Pakandl (1990) were 29.3 × 18.1 (26–33 × 14.5–20) with a M 4.0 (3–4.5) wide, an OR 7 (3.5–9) wide, and sporocysts that were 15.7 × 6.6 (13–17 × 5.5–7). Tacconi et al. (1995) found this species in the intestinal contents of four dead juvenile

L. europaeus from a breeding farm in Italy. Their ellipsoidal oocysts were 26—34 × 15—20, with both a M and an OR and this species was always found with *E. hungarica*, *E. robertsoni*, and *E. townsendi* in these dead hares. Aoutil et al. (2005) studied the *Eimeria* species in *L. granatensis* and *L. europaeus* in France, and recorded this species in 1/18 (5.5%) *L. europaeus*, but apparently did not find it in 46 *L. granatensis* they examined. The elongate oocysts they measured were 31 × 19 (30—35 × 17—21), with both an OR ~9 wide and a dome-like M (~3—5 wide); the sporocysts were 19 × 9, with a SR formed by a poorly defined aggregation of small granules. The SZ each had one elongated RB.

EIMERIA EXIGUA YAKIMOFF, 1934 (FIGURES 6.4, 6.5)

Type host: *Oryctolagus cuniculus* (Linnaeus, 1758) (syn. *Lepus cuniculus*), European (domestic) rabbit.

Remarks: This species was originally described from the domestic rabbit in Russia. It may be a synonym of *E. perforans* (Cheissin, 1947a, 1967). Matschoulsky (1941) said he found oocysts that resembled this species in *L. timidus* from Buryat-Mongol (former USSR) and that the spheroidal oocysts were 15.7 × 13.6 (15—18 × 13—15) with sporocysts 7—8 × 4—5. Ryšavý (1954) said he found this species in *L. europaeus* in the Czech Republic; its oocysts were 18.1 × 12.9 (16.5—21 × 10—17.5), with sporocysts 6.5 × 3.0, and these oocysts sporulated in 36—48 hr. The parasite was found in the duodenum and small intestine of hares and rabbits, but only in a very small percentage (Ryšavý, 1954). In an unpublished Master's thesis, Ogedengbe (1991) said he found this species in 19% of the rabbits surveyed in Kaduna State, Nigeria. However, he (1991) neglected to state how many rabbits were surveyed, nor did he mention the host species; Anonymous (2012) lists the Cape hare, *L. capensis*, as the only

lagomorph species found in Nigeria. See Chapter 6 (Figures 6.4, 6.5) for the complete description of this species.

EIMERIA GANTIERI AOUTIL, BERTANI, BORDES, SNOUNOU, CHABAUD & LANDAU, 2005

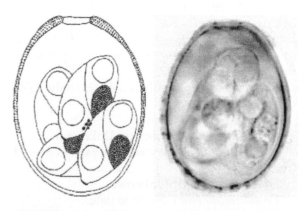

FIGURES 5.14, 5.15 Line drawing of the sporulated oocyst of *Eimeria gantieri* from Aoutil et al., 2005, with permission of the authors and from (the journal) *Parasite*. Photomicrograph of a sporulated oocyst of *E. gantieri* from Aoutil et al., 2005, with permission of the authors and from (the journal) *Parasite*.

Type host: *Lepus granatensis* Rosenhauer, 1956, Granada hare.

Type locality: EUROPE: France: Pyrénées-Orientales.

Other hosts: None reported to date.

Geographic distribution: EUROPE: France.

Description of sporulated oocyst: Oocyst shape: ovoidal, slightly flattened at end with M; number of walls: 1 (line drawing), of uneven thickness; wall characteristics: light brown, described as sculptured, with the top $^2/_3$ appearing finely granular (Aoutil et al., 2005), while their photomicrograph shows that the entire oocyst appears to be rough and sculptured (their Fig. 1.7); L × W: 30 × 22 (29—33 × 20—24); L/W ratio: 1.4; M: present; M characteristics: ~4 wide and wall thickens near margin surrounding it; OR: present; OR characteristics: a few small granules often hidden by sporocysts (line drawing); PG: absent. Distinctive features of oocyst: sculptured wall and presence of a M and a tiny OR.

Description of sporocyst and sporozoites: Sporocyst shape: ellipsoidal or elongate-ovoidal, slightly tapered at one end; L × W: 16 × 9 (ranges not given); SB: tiny, present at tapered end; SSB, PSB: both absent; SR: a compact ovoidal mass of small granules; SZ: comma-shaped, with one large, RB at widest end (line drawing). Distinctive features of sporocyst: tiny SB, large RB in SZ.

Prevalence: Apparently found in only 1/46 (2%) *L. granatensis*.

Sporulation: Unknown.

Prepatent and patent periods: Unknown.

Site of infection: Unknown. Oocysts recovered from feces.

Endogenous stages: Unknown.

Cross-transmission: None to date.

Pathology: Unknown.

Material deposited: None, although the authors designated their photomicrograph (their Fig. 1.7) as the holotype and said "the photos of the oocysts … have been deposited at the laboratory."

Remarks: Aoutil et al. (2005) stated that the ornate anterior half of the oocyst wall, the small OR of only a few granules, the "more roundish shape," and its relatively small size distinguish this form as a new species. They (2005) provided a line drawing and a good photomicrograph in their original description, but they did not deposit either of these in an <u>accredited</u> museum, nor did they provide all of the quantitative and qualitative information needed and expected in a modern, standardized species description (sporocyst L, W ranges; measurements of OR, SR, SZ, RB, etc.). This is disappointing and below normal standards. Finally, in presenting their description of this new species Aoutil et al. (2005) listed the name as "*Eimeria gantieri* n. sp. Aoutil & Landau"; this is not, and cannot

be, the correct authority for this species, since there is no separate published paper by Aoutil and Landau describing and naming it. Thus, the full scientific name must be as we have listed it above.

EIMERIA GOBIENSIS GARDNER, SAGGERMAN, BATSAIKAN, GANZORIG, TINNIN & DUSZYNSKI, 2009

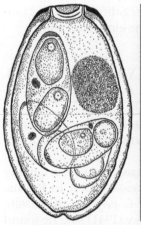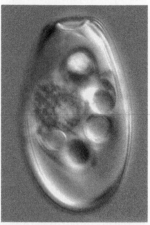

FIGURES 5.16, 5.17 Line drawing of the sporulated oocyst of *Eimeria gobiensis* from Gardner et al., 2009, with permission of the authors and the *Journal of Parasitology*. Photomicrograph of a sporulated oocyst of *E. gobiensis* from Gardner et al., 2009, with permission of the authors and the *Journal of Parasitology*.

Type host: *Lepus tolai* Pallas, 1778, Tolai hare.

Type locality: ASIA: Mongolia: Ulziyt Uul, 1,640 m alt., 44°41′09″N, 102°00′57″E.

Other hosts: None reported to date.

Geographic distribution: ASIA: Mongolia.

Description of sporulated oocyst: Oocyst shape: elongate-ellipsoidal; number of walls: 2; wall thickness: ~2.1 of uneven thickness; wall characteristics: outer is pale blue to transparent, smooth, ~$^3/_4$ of total thickness, inner is yellow; L × W: 38.6 × 24.2 (27−49 × 19−32.5); L/W ratio: 1.6 (1—3); M: present; M characteristics: wall thickens near margin making it appear triple-layered, ~3 wide; OR: present; OR characteristics: large compact mass of tiny granules, 12.1 × 10.8 (10—18 × 9—17); PG: absent. Distinctive features of oocyst: distinct M at tapered end and large, compact OR.

Description of sporocyst and sporozoites: Sporocyst shape: ovoidal, slightly tapered at one end; L × W: 15.0 × 7.7 (9—21 × 5—12); SB: present at tapered end; SSB, PSB, SR: all absent; SZ: with one large, posterior RB, 6.1 × 4.8 (4—11 × 3—8). Distinctive features of sporocyst: no SR and large RB in SZ.

Prevalence: 1/1 (100%) in type host.

Sporulation: Unknown. Oocysts were 3,343 days old when measured and photographed.

Prepatent and patent periods: Unknown.

Site of infection: Unknown. Oocysts recovered from intestinal contents.

Endogenous stages: Unknown.

Cross-transmission: None to date.

Pathology: Unknown.

Material deposited: Photosyntypes of sporulated oocysts in the Harold W. Manter Laboratory of Parasitology, phototype collection, Lincoln, Nebraska, USA, No. 49155. Skin, skull, skeleton, and tissues of the symbiotype host are in the Mammal Division, Museum of Southwestern Biology, University of New Mexico, Albuquerque, New Mexico, USA (accession number not given).

Remarks: Sporulated oocysts of this species do not closely resemble those from other leporids of Eurasia. They can be easily distinguished from *E. hungarica* Pellérdy, 1956, *E. leporis* Nieschulz, 1923, *E. punjabensis* Gill and Ray, 1960, *E. robertsoni* Madsen, 1938, *E. ruficaudatus* Gill and Ray, 1960, *E. septentrionalis* Yakimoff et al. 1936, and *E. stefanskii* Pastuszko, 1961, by having two distinct oocyst wall layers that the others lack. They differ from *E. campania* Carvalho, 1943 and *E. townsendi* Pellérdy, 1956, by having a massive, well-developed OR, which these others lack. And they differ from *E. europaea*

Pellérdy, 1956, by having much larger oocysts, 38.6 × 24.2 vs. 26—34, respectively.

EIMERIA GROENLANDICA MADSEN, 1938 EMEND. LEVINE & IVENS, 1972

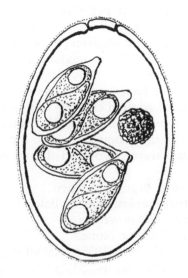

FIGURE 5.18 Line drawing of the sporulated oocyst of *Eimeria groenlandica* from Levine and Ivens, 1972, with permission of the *Journal of Protozoology*.

Synonym: *Eimeria perforans* var. *groenlandica* Madsen, 1938, *pro parte*.

Type host: *Lepus arcticus* Ross, 1819, Arctic hare.

Type locality: GREENLAND: East side, near Eskimonaes, 74°05'N lat.

Other hosts: None reported to date.

Geographic distribution: GREENLAND.

Description of sporulated oocyst: Oocyst shape: broadly ellipsoidal; number of walls: 1; wall thickness: ~1; wall characteristics: colorless and smooth; L × W: ~28 × 21 (23—33 × 17—26); L/W ratio: 1.3; M: narrow, plug-shaped, if present; OR: present; OR characteristics: small, 4—7 wide; PG: absent. Distinctive features of oocyst: small M (if present).

Description of sporocyst and sporozoites: Sporocyst shape: spindle-shaped; L × W: 13—9 × 7—8; SB: present; SB characteristics: peg-shaped; SSB, PSB: both absent; SR: present in some sporocysts; SR characteristics: when visible, present only as a few granules; SZ: elongate with one end broader than the other, lying lengthwise head-to-tail with a clear RB at the large end. Distinctive features of sporocyst: peg-shape of SB.

Prevalence: Reported in 11/22 (50%) of the hares examined by Madsen (1938), but he may have been dealing with a mixture of this species and *E. robertsoni*.

Sporulation: Exogenous. Oocysts sporulated in a saturated sodium chloride solution.

Prepatent and patent periods: Unknown.

Site of infection: Thought to be in the intestine as oocysts were recovered from intestinal contents.

Endogenous stages: Unknown.

Cross-transmission: None to date.

Pathology: Unknown.

Material deposited: None.

Remarks: Madsen (1938) originally gave the oocyst dimensions as 33 × 22 (23—40 × 17—27), but Carvalho (1948) and Pellérdy (1956) said that the larger oocysts were actually those of *E. robertsoni*. To our knowledge, this species has not been seen or reported since its original description, which raises some skepticism about its validity.

EIMERIA HOLMESI SAMOIL & SAMUEL, 1977a

Type host: *Lepus americanus* Erxleben, 1777, Snowshoe hare.

Type locality: NORTH AMERICA: Canada: Alberta, Rochester.

Other hosts: None reported to date.

Geographic distribution: NORTH AMERICA: Canada: Alberta.

Description of sporulated oocyst: Oocyst shape: spheroidal to subspheroidal; number of walls:

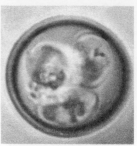

FIGURES 5.19, 5.20 Line drawing of the sporulated oocyst of *Eimeria holmesi* from Samoil and Samuel, 1977a, with permission of the *Canadian Journal of Zoology*. Photomicrograph of a sporulated oocyst of *E. holmesi* from Samoil and Samuel, 1977a, with permission of the *Canadian Journal of Zoology*.

2; wall thickness: ~1; wall characteristics: outer is smooth and uniformly thick; L × W: 15.1 × 14.8 (13—19 × 12—19); L/W ratio: 1.0; M: absent; OR: present; OR characteristics: small, compact, granular; PG: absent. Distinctive features of oocyst: spheroidal shape, small size, and lack of a M.

Description of sporocyst and sporozoites: Sporocyst shape: ellipsoidal, tapering toward one end; L × W: 10.4 × 5.2 (10—11 × 5—6); L/W ratio: 2.0; SB: present; SB characteristics: an indistinct light refracting area at narrow end; SSB, PSB: both absent; SR: present; SR characteristics: small, compact and granular; SZ: with one RB that is lateral or posterior to N. Distinctive features of sporocyst: unusual SB on sporocyst and RB in SZ.

Prevalence: 30/629 (5%) in the type host.

Sporulation: Exogenous. Oocysts sporulated within 6 days in 2.5% aqueous (w/v) $K_2Cr_2O_7$ solution at 20°C.

Prepatent and patent periods: Unknown.

Site of infection: Unknown. Oocysts recovered from feces.

Endogenous stages: Unknown.

Cross-transmission: None to date.

Pathology: Unknown.

Material deposited: None.

Remarks: This species most closely resembles *E. hungarica* and *E. punjabensis*, but it is smaller than *E. punjabensis* and possesses a SR, which is lacking in both.

EIMERIA HUNGARICA PELLÉRDY, 1956

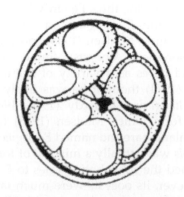

FIGURE 5.21 Line drawing of the sporulated oocyst of *Eimeria hungarica* from Pellérdy, 1956, in *Acta Veterinaria Academiae Scientiarum Hungaricae* with permission of the Editor, *Acta Veterinaria Hungarica* (legal successor to the former).

Synonyms: *Eimeria exigua* Yakimoff, 1934 of Pellérdy, 1954, *pro parte*; *Eimeria minima* Carvalho, 1943, *pro parte*; *Eimeria minima* Yakimoff, 1943 of Gill and Ray, 1960; *Eimeria orbiculata* Lucas, Laroche and Durand, 1959; *Eimeria orbiculata* of Aoutil et al., 2005.

Type host: *Lepus europaeus* Pallas, 1778, European hare.

Type locality: EUROPE: Hungary.

Other hosts: *Lepus nigricollis* F. Cuvier, 1823 (syn. *Lepus ruficaudatus*), Indian hare; *Lepus capensis* L., 1758, Cape hare; *Lepus granatensis* Rosenhauer, 1956, Granada hare; *Lepus timidus* L., 1758, Mountain hare (?).

Geographic distribution: ASIA: India, Punjab, District Ludhiana; Uttar Pradesh, Kashipur; Republic of Belarus (Byelorussia of the former USSR) (?) Lithuania (of former USSR); EUROPE: Austria, Bulgaria, Czech Republic, France, (former East) Germany, Hungary, Italy, Norway, Poland, Portugal, Spain, Sweeden, Switzerland.

Description of sporulated oocyst: Oocyst shape: spheroidal to subspheroidal; number of walls: 1; wall characteristics: thin, smooth, colorless; L × W: 14 × 13 (12−15 × 11−14); L/W ratio: 1.1; M: absent; OR: 1−3 granules ("grains") between sporocysts, which according to Pellérdy (1974) comprise an OR, but these are sometimes called a PG by others; PG: may or may not be present − the distinction between the OR and the PG in this species is not clear-cut. Distinctive features of oocyst: small size, absence of M, and the small grains/granules that compose the OR.

Description of sporocyst and sporozoites: Sporocyst shape: short and "stocky," but no mensural data given by Pellérdy (1956); Pakandl (1990) said they were 8.5 × 5.0 (7.5−10 × 4.5−6); SB: apparently absent, but sporocysts are stout and tightly confined within the walls of the sporocyst, so a small SB may have been missed; SSB, PSB, SR: all absent; SZ: contain a single large, clear RB. Distinctive features of sporocyst: stout shape, packed within the confines of the oocyst wall, and lack of SR.

Prevalence: Pastuszko (1961a) found this species in 22/462 (5%) *L. europaeus* in Poland; Gräfner et al. (1967) found it in 58/176 (33%) *L. europaeus* in (former East) Germany; and Bouvier (1967) found this species rarely in Switzerland and thought that it had been imported quite recently at the time. Litvenkova (1969) said, "Of 52 hares (*Lepus europaeus* L. and *L. timidus* L.) coccidian oocysts were found in 22 (42.3%)," but she did not say whether *E. hungarica* was found in one or both of the *Lepus* species. Arnastauskene and Kazlauskas (1970) found it in 10/41 (24%) *L. europaeus* in Lithuania; and Sugár et al. (1978) found it in 10/374 (<3%) *L. europaeus* in Hungary. Moreno Montañez et al. (1979) found this species in 27/42 (64%) *L. capensis* in Spain. In the Czech Republic, Pakandl (1990) surveyed *L. europaeus* from 1983−1985 and found only 24/350 (7%) infected with this species.

Sporulation: Exogenous. Oocysts sporulated within 2—3 days at room temperature (Pellérdy, 1956; Pastuszko, 1961a).

Prepatent and patent periods: Unknown.

Site of infection: Pellérdy (1974) speculated this parasite localized in the small intestine, and Pastuszko (1961a) and Pakandl (1990) confirmed this observation; however, Golemanski (1975) said he found oocysts in the cecum.

Endogenous stages: Unknown.

Cross-transmission: Pellérdy (1954a, b) failed to infect 11 4-wk-old domestic rabbits, *O. cuniculus*, with this species isolated from *L. europaeus*. He (1956) also could not infect five young domestic rabbits, nor was he able to infect 15 young rabbits with a mixture of "all *Eimeria* species known to occur in the hare in Hungary"; this mixture may have included *E. hungarica*. Burgaz (1973) was unable to transmit *E. hungarica* oocysts from *L. timidus* to *O. cuniculus*, but her paper must be viewed with caution (see Chapter 6).

Pathology: Bouvier (1967) recovered this species in the European hare in Switzerland and said that the hare had severe enteritis. Post-mortem examination of the **cachectic** carcass revealed extensive epithelial desquamation and hemorrhagic inflammation in the intestine. Masses of endogenous stages were seen microscopically in the affected areas. The cause of the enteritis was purported to be a mixed infection with *E. leporis*, *E. robertsoni*, and, chiefly, *E. hungarica*, but since the endogenous stages of this species are unknown, we don't know how Bouvier (1967) made this decision. Tacconi et al. (1995) found this species in the intestinal contents of four dead, juvenile *L. europaeus* in Italy. The oocysts were always found with *E. europaea*, *E. robertsoni*, and *E. townsendi* in these dead hares, which all showed severe enteritis as the cause of death.

Material deposited: None.

Remarks: The mensural data provided for the sporulated oocyst (above) are based on Pellérdy

(1956) and Pastuszko (1961a). This eimerian resembles *E. exigua* from *Oryctolagus* and *E. minima* from *Sylvilagus*, but because his previous cross-infection studies failed Pellérdy (1956) determined that this species is host specific and named it for the first time. Lucas et al. (1959) found what they thought was a new species, *E. orbiculata*, from *L. europaeus*, but the description was essentially identical to that of *E. hungarica*; thus, Pellérdy (1974) synonymized it with *E. hungarica*, a decision with which we agree. Madsen (1938) considered the small, spherical oocysts found in East Greenland hares as a variety of *E. exigua*, but Pellérdy (1974) thought he was likely dealing with *E. hungarica*. Pellérdy (1956, 1965) said that the form which Madsen (1938) found in the Greenland hare and named *E. exigua* var. *septentrionalis* was actually a mixture of forms and he assigned the subspherical ones to *E. hungarica*. However, its oocysts were much larger (27 × 33) and they had sporocysts with SB and SR, so it cannot be this species. Gill and Ray (1960) described a form from *L. nigricollis* (syn. *L. ruficaudatus*) under the name *E. minima*, but it was probably *E. hungarica*, and the form that Lucas et al. (1959) saw in France and named *E. orbiculata* is a junior synonym of *E. hungarica* (Bouvier, 1967). Pastuszko (1961a) said the small, spheroidal oocysts were 12—15 × 11—14, with a thin wall and no M. Golemanski (1975) said the oocysts were spheroidal, 15 (13—16) wide. The oocysts measured by Pakandl (1990) were 14.3 × 13.7 (12.5—17 × 12.5—15) without a M, but with an OR, and sporocysts were 8.5 × 5.0 (7.5—10—17 × 4.5—6). Berg (1981) reported this species in hares from Norway, but gave no other information. Tacconi et al. (1995) found this species in the intestinal contents of four dead juvenile *L. europaeus* from a breeding farm in Italy. The round oocysts were 13—15 × 12—14, without a M, but with an OR; this species was always found with *E. europaea*, *E. robertsoni*, and *E. townsendi* in these dead hares. Oocysts studied by Moreno Montañez et al. (1979) in

L. capensis from Spain were 13—15 × 13—14 with sporocysts about 8 × 5. Terracciano et al. (1988) identified this species in *L. europaeus* from one of three protected areas in the Province of Pisa, Italy. The spheroidal oocysts they studied were 13—15 × 12—14 without a M, OR, or SR. Vila-Viçosa and Caeiro (1997) reported this species in *L. capensis* in Portugal, with the spheroidal oocysts being 13.2 × 12.3. Aoutil et al. (2005) studied the *Eimeria* species in *L. granatensis* and *L. europaeus* in France, and recorded this species in 2/46 (4%) *L. granatensis*, but apparently did not find it in the 18 *L. europaeus* they examined. The subspheroidal oocysts they measured were 13 × 13 (12—16 × 12—15), with an OR, but without a M, and the sporocysts were 9 × 5, without a SR, and had SZ which each had two RB. They described these oocysts under the name *E. orbiculata*, which we and others (Pellérdy, 1956; Bouvier, 1967) consider to be a junior synonym of *E. hungarica*.

EIMERIA INTESTINALIS CHEISSIN, 1948 (FIGURES 6.8, 6.9)

Type host: *Oryctolagus cuniculus* (L., 1758), European (domestic, tame) rabbit.

Remarks: Mandal (1976) said he found oocysts of this species in the intestine of a *Lepus* sp. from Kashipur, India. The oocysts were piriform, 25—30 × 17.5—20, with an OR; he reported the sporocysts to be ovoidal with an SR and sporulation time was 2—3 days. See Chapter 6 (Figures 6.8, 6.9) for a complete species description.

EIMERIA IRRESIDUA KESSEL & JANKIEWICZ, 1931 (FIGURES 6.10, 6.11)

Type host: *Oryctolagus cuniculus* (L., 1758), European (domestic, tame) rabbit.

Remarks: This species was originally described from the domestic rabbit, *O. cuniculus*,

by Kessel and Jankiewicz (1931) from southern California, USA, and it since has been found in rabbits worldwide. It has been reported from *L. nigricollis* (syn. *L. ruficaudatus*), the Indian hare, but its identity in *Lepus* species needs to be verified. Ryšavý (1954) said he found this species in *L. europaeus* in the Czech Republic with oocysts that were 35.8 × 21.4 (29—43 × 14—27), with sporocysts 11 × 7, and that these oocysts sporulated in 50—60 hr. The parasite was found in the small intestine and was abundant in the hares and rabbits they sampled (Ryšavý, 1954). Gill and Ray (1960) reported finding this species both in the domestic rabbit and in *L. nigricollis* (syn. *L. ruficaudatus*) collected in India. The measurements they gave for sporulated oocysts were only for oocysts collected from domestic rabbits, but they included a line drawing of an oocyst they said was from *L. nigricollis* and four line drawings of oocysts they said were from *O. cuniculus* and the five drawings show definite similarities. Finally, Mandal (1976) said he found this species in the intestine of a *Lepus* sp. from Punjab, India. The ovoidal oocysts were 25—37.5 × 17.5—24.8 and lacked an OR; ovoidal sporocysts were 18 × 9 with a SR. See Chapter 6 (Figures 6.10, 6.11) for the complete description of this species.

EIMERIA KEITHI SAMOIL & SAMUEL, 1977a

Type host: *Lepus americanus* Erxleben, 1777, Snowshoe hare.

Type locality: NORTH AMERICA: Canada: Alberta, Rochester.

Other hosts: None reported to date.

Geographic distribution: NORTH AMERICA: Canada, Alberta.

Description of sporulated oocyst: Oocyst shape: cylindroidal; number of walls: 2; wall thickness: ~1.5; wall characteristics: outer is smooth and uniformly thick, several times thicker than inner; L × W: 32.2 × 16.9 (24—38 × 15—19);

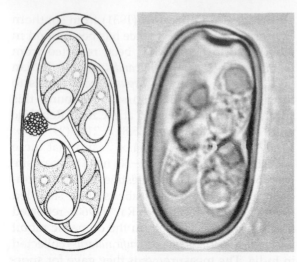

FIGURES 5.22, 5.23 Line drawing of the sporulated oocyst of *Eimeria keithi* from Samoil and Samuel, 1977a, with permission of the *Canadian Journal of Zoology*. Photomicrograph of a sporulated oocyst of *E. keithi* from Samoil and Samuel, 1977a, with permission of the *Canadian Journal of Zoology*.

L/W ratio: 1.9; M: present; M characteristics: distinct, sunken slightly into oocyst wall, 4.9 (2—6); OR: present; OR characteristics: compact, granular, 4.3 (2—7) wide; PG: absent. Distinctive features of oocyst: thick wall and compact OR.

Description of sporocyst and sporozoites: Sporocyst shape: elongated spindle-shaped bodies tapering toward one end; L × W: 14.9 × 6.7 (13—18 × 6—7); L/W ratio: 2.2; SB: present; SB characteristics: an indistinct light-refracting area at narrow end; SSB, PSB, SR: all absent; SZ: with one large RB posterior to N. Distinctive features of sporocyst: spindle-shape and lack of SR.

Prevalence: 54/629 (9%) in the type host.

Sporulation: Exogenous. Oocysts sporulated within 6 days in 2.5% aqueous (w/v) K₂Cr₂O₇ solution at 20° C.

Prepatent and patent periods: Unknown.

Site of infection: Unknown. Oocysts recovered from feces.

Endogenous stages: Unknown.

Cross-transmission: None to date.

Pathology: Unknown.

Material deposited: None.

Remarks: Sporulated oocysts of this species most closely resemble those of *E. ruficaudati*, *E. leporis*, and *E. athabascensis* of *Lepus* spp. and *E. honessi* of *Sylvilagus* spp.; however, they lack the SR of *E. ruficaudati* and *E. leporis* and have a M, which those of *E. leporis* and *E. athabascensis* lack. They differ from those of *E. honessi* in shape and size of the oocyst.

EIMERIA LAPIERREI AOUTIL, BERTANI, BORDES, SNOUNOU, CHABAUD & LANDAU, 2005

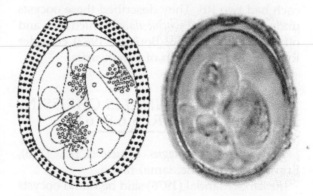

FIGURES 5.24, 5.25 Line drawing of the sporulated oocyst of *Eimeria lapierrei* from Aoutil et al., 2005, with permission of the authors and from (the journal) *Parasite*. Photomicrograph of a sporulated oocyst of *E. lapierrei* from Aoutil et al., 2005, with permission of the authors and from (the journal) *Parasite*.

Type host: *Lepus granatensis* Rosenhauer, 1956, Granada hare.

Type locality: EUROPE: France: Pyrénées-Orientales.

Other hosts: *Lepus europaeus* Pallas, 1778, European hare.

Geographic distribution: EUROPE: France.

Description of sporulated oocyst: Oocyst shape: ovoidal; number of walls: one (line drawing), but actually two (photomicrograph, their Fig. 3.18); wall characteristics: thick, sculptured, dark brown; L × W: 33 × 26 (31–35 × 24–28); L/W ratio: 1.3; M: present; M characteristics: 6 wide; OR, PG: both absent. Distinctive features of oocyst: sculptured outer wall and lacking an OR.

Description of sporocyst and sporozoites: Sporocyst shape: ovoidal, slightly pointed at one end (line drawing); L × W: 17 × 9 (ranges not given); L/W ratio: 1.9; SB: tiny, at pointed end; SSB, PSB: both absent; SR: present; SR characteristics: composed of bunched small granules (line drawing), ~7 × 4; SZ: with one large and one small RBs. Distinctive features of sporocyst: tiny SB, dispersed SR of small granules, and SZ with two RBs of different sizes.

Prevalence: Found in 1/46 (2%) *L. granatensis* and in 1/9 (11%) *L. europaeus*.

Sporulation: Presumably exogenous, but the time and temperature are not known.

Prepatent and patent periods: Unknown.

Site of infection: Unknown. Oocysts recovered from feces.

Endogenous stages: Unknown.

Cross-transmission: None to date.

Pathology: Unknown.

Material deposited: None, although the authors designated their photomicrograph (their Fig. 3.18) as the holotype and said "the photos of the oocysts … have been deposited at the laboratory."

Remarks: Aoutil et al. (2005) provided a line drawing and a photomicrograph of this form in their original description, but they did not deposit either of these in an accredited museum, nor did they provide all of the quantitative and qualitative information needed and expected in a modern, standardized species description (e.g., age of the oocysts when studied, sporulation time and temperature, sporocyst L, W ranges; measurements of SZ, RB, etc.). The authors say that this species can be compared to *E. sculpta* first described from *L. arcticulus groenlandicus* in Greenland (Madsen, 1938), which was described

as 37 × 29, with a very small M (~2 wide), and with a well-defined ellipsoidal SR, all characters which differ from those of *E. lapierrei*. Carvalho (1943) also described *E. sculpta* from *L. townsendii campanius* in Iowa, USA, with piriform oocysts that measured 36 × 29.7, characterized by a large M (7.5 wide), and a SR with an ill-defined shape, characteristics to which this species does not conform (Aoutil et al. 2005). In presenting their description of this species Aoutil et al. (2005) listed the name as "*Eimeria lapierrei* n. sp. Aoutil & Landau"; this is not, and cannot be, the correct authority for this species, since there is no separate published paper by Aoutil and Landau describing and naming it. Thus, the full scientific name (genus, specific epithet, authority) must be as we list it, above.

EIMERIA LEPORIS NIESCHULZ, 1923

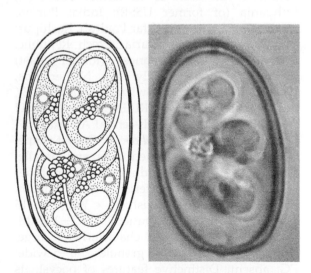

FIGURES 5.26, 5.27 Line drawing of the sporulated oocyst of *Eimeria leporis* from Samoil and Samuel, 1977a, with permission from the *Canadian Journal of Zoology*. Photomicrograph of a sporulated oocyst of *E. leporis* from Samoil and Samuel, 1977a, with permission from the *Canadian Journal of Zoology*.

Synonyms: *non Eimeria leporis* Nieschulz of Morgan and Waller, 1940 and of Waller and Morgan, 1941; *Eimeria leporine* Nieschulz, 1923 of Sugár et al. (1978); *Eimeria leporis leporis* Nieschulz, 1923 of Aoutil et al., 2005; *Eimeria leporis brevis* of Aoutil et al., 2005.

Type host: *Lepus europaeus* Pallas, 1778, European hare.

Type locality: EUROPE: The Netherlands: area near Utrecht.

Other hosts: *Lepus americanus* Erxleben, 1777, Snowshoe hare; *Lepus arcticus* Ross, 1819, Arctic hare; *Lepus capensis* L., 1758, Cape hare; *Lepus granatensis* Rosenhauer, 1956, Granada hare; *Lepus nigricollis* F. Cuvier, 1823 (syn. *Lepus ruficaudatus*), Indian hare; *Lepus timidus* L., 1758, Mountain hare (?); *Lepus tolai* Pallas, 1778, Tolai hare; *Oryctolagus cuniculus* (L., 1758) (syn. *Lepus cuniculus*), European (domestic) rabbit (?), but this is an unlikely host and cross-transmission does not support it as a host for *E. leporis*.

Geographic distribution: ASIA: Republic of Belarus (Byelorussia of the former USSR) (?), Lithuania (of former USSR); India: Punjab, District Ludhiana and Uttar Pradesh, Kashipur; EUROPE: Austria, Bulgaria, Czech Republic, England, Finland, France, (former East) Germany, Hungary, Italy, The Netherlands, Norway, Poland, Spain, Switzerland; MIDDLE/ NEAR EAST: Turkey; GREENLAND; NORTH AMERICA: Canada: north-central Alberta.

Description of sporulated oocyst: Oocyst shape: cylindroidal with rounded ends and parallel sides, often bean- or kidney-shaped; number of walls: 1; wall characteristics: thin, smooth, colorless or pale yellowish; L × W: 32 × 16 (26−38 × 13−20); L/W ratio: 2.0 (1.7−2.6); M: absent: OR: present; OR characteristics: large, compact mass of coarse granules, ~4−5 wide; PG: absent. Distinctive features of oocyst: its shape and size and the presence of compact OR and lack of a M.

Description of sporocyst and sporozoites: Sporocyst shape: elongate-ovoidal, slightly pointed at one end; L × W: 9−10 × 6−7; SB: present, button-like, at slightly pointed end; SSB, PSB: both absent; SR: present; SR characteristics: a compact ellipsoidal mass of granules (original line drawing) or a few scattered granules between SZ; SZ: elongate with one end wider than the other, lying lengthwise head-to-tail and with a clear RB at the large end. Distinctive features of sporocyst: elongate egg-shape and button-like SB.

Prevalence: In > 90% of the type host (Nieschulz, 1923). Pastuszko (1961a) found it in 139/462 (30%) *L. europaeus* in Poland. Gräfner et al. (1967) found it in 99/176 (56%) *L. europaeus* in (former East) Germany, while Bouvier (1967) found it "occasionally" in young hares in Switzerland. Litvenkova (1969) said, "Of 52 hares (*Lepus europaeus* L. and *L. timidus* L.) coccidian oocysts were found in 22 (42.3%)," but she did not say whether *E. leporis* was found in one or both of the *Lepus* species collected in Belarus. Arnastauskene and Kazlauskas (1970) found it in 22/41 (54%) hares in Lithuania. Pellérdy et al. (1974) said that it was 300−500 times as common in Hungary as *E. robertsoni*, *E. semisculpta*, *E. europaea*, and *E. hungarica* put together, apparently in hares which died spontaneously of coccidiosis. Samoil and Samuel (1977a) reported it in 223/629 (35.5%) *L. americanus* in Alberta, Canada, and Sugár et al. (1978) found it in 102/374 (27%) *L. europaeus* in Hungary. Yakimoff et al. (1931) found it in *L. tolai*, Lucas et al. (1959) reported it in *Lepus* sp. in France, and Gill and Ray (1960) found it in *L. nigricollis* (syn. *L. ruficaudatus*) in India. In the Czech Republic, Pakandl (1990) surveyed *L. europaeus* from 1983−1985 and found 110/ 350 (31%) infected with this species, and Ryšavý (1954) also reported it in this host from the Czech Republic. Nickel and Gottwald (1979) reported a prevalence of 97.9% with this species, whereas Gottschalk (1973) found only 18.9% of *L. europaeus* to be infected with it. Moreno Montañez et al. (1979) found this species in 27/42 (64%) *L. capensis* in Spain. Tasan and Özer (1989), in Turkey, said it occurred in 31% of the

40 hares they surveyed between 1985 and 1987. Berg (1981) reported this species in European hares from Norway, and Soveri and Valtonen (1983) reported it in mountain hares in Finland, but neither gave any other information.

Sporulation: Exogenous. Oocysts sporulated within 2–3 days (Pellérdy, 1956); in ~2 days (Lucas et al., 1959); in 48 hr at 23°C when in 3% $K_2Cr_2O_7$ solution (Golemanski, 1975); 4 days at 20°C (Romero-Rodriguez, 1976); in 3 days (Mandal, 1976); in 6 days at 20°C in 2.5% $K_2Cr_2O_7$ solution (Samoil and Samuel, 1977a).

Prepatent and patent periods: Prepatency is 7 days (Pellérdy et al., 1974).

Site of infection: According to Pellérdy et al. (1974), endogenous stages occur throughout the small intestine, usually in epithelial cells of the mucosa and less often subepithelially down to the muscularis mucosae. Sometimes there are three to five organisms in a single intestinal cell.

Endogenous stages: There are three merogonous stages (Pellérdy et al., 1974). The Type A meront was found less frequently than Type B and is probably the first-generation meront. Type A meronts were 6–14 wide and contained 2–8 or more, stout, short merozoites, 5 long × 1.5–3 wide, arranged like the segments of an orange, and they are contained in a distinct **parasitophorous vacuole** (PV). There is no residuum in these meronts, and the merozoite N is centrally located. Type B meronts were 10–20 or more wide and were found in groups or nests. They were arranged in larger groups in the lacteals of the villi, and such groups might contain several hundred merozoites. Many were present along with gamonts. Smaller meronts had 16 and the larger meronts had 30–50 slim merozoites, 8–10 long × 1–1.5 wide, tapering at the ends with their N near one end. Type B meronts also lay in a PV and also had no residuum. Type C meronts were seen only sporadically. They contained two or four stout merozoites, 12–16 long and containing 8–10 dark spots, presumably multiple N. Perhaps

they are predecessors of a gametogonic generation (?). Macrogamonts were found in cells of the subepithelium contained within a PV. A single cell often contained 4–5 macrogamonts. Mature macrogamonts were 20–25 wide and changed gradually into oocysts 22–32 × 12–18 in stained sections. Microgamonts were 15–20 wide, lobular with many N around the periphery of the lobules, and had a central residuum. Microgametes were 2.5–4 × 0.3–0.5. Pellérdy et al. (1974) often saw macrogametes with freshly penetrated microgametes inside them.

Cross-transmission: Nieschulz (1923), Pellérdy (1956), and Lucas et al. (1959) were unable to infect domestic rabbits, *O. cuniculus*, with this species isolated from *L. europaeus*. Burgaz (1973) was unable to transmit *E. leporis* oocysts from *L. timidus* to *O. cuniculus*, but her paper must be viewed with caution (see Chapter 6).

Pathology: Bouvier (1967) said that this species may cause severe enteritis, sometimes hemorrhagic, especially in young hares. Pellérdy et al. (1974) thought this species was responsible for the death of 14/227 (6%) hares in Hungary that died spontaneously of coccidiosis. The main gross lesions consisted of grayish-white foci 1–2 mm wide along the entire length of the small intestine, especially the ileum. The villi were enlarged and the focal lesions were often confluent. The intestinal capillaries also were noted to be dilated and congested (Pellérdy et al., 1974). The same authors killed two young hares, which already had what they said was a "slight" infection with *E. leporis*, by administering 2 million sporulated oocysts to each of them. They concluded that the hare's immune response to this eimerian is very weak, and that this low immunogenicity may account for the high prevalence of this species in hares.

Material deposited: None.

Remarks: This was the first coccidium described from wild hares when Nieschulz (1923) found it in association with *E. stiedai* from *L. europaeus* and the description of sporulated oocysts (above) is based mostly on Nieschulz (1923) and Pellérdy (1956). Morgan and Waller (1940) and Waller and Morgan (1941) reported oocysts from the cottontail, *Sylvilagus floridanus*, in Iowa and Wisconsin, under the name *E. leporis*, but it was actually *E. neoleporis* (see Carvalho, 1943). Ryšavý (1954) said he found this species in both *O. cuniculus* and in *L. europaeus* in the Czech Republic, with oocysts that were 35.8 × 21.4 (29—43 × 14—27) and sporocysts 11 × 7, and that these oocysts sporulated in 50—60 hr; Ryšavý (1954) said the parasite was found in the small intestine and was abundant in the hares and rabbits he sampled. Gill and Ray (1960) gave the dimensions of oocysts they recovered from *L. nigricollis* (syn. *L. ruficaudatus*) as 28 × 18 (22—28 × 15—26) and said that both the OR and SR disappear during preservation. Lucas et al. (1959) said the sporulated oocysts were 32 × 16 (27—37 × 14—18). Pellérdy et al. (1974) described the sporulated oocysts as ellipsoidal, often bean-shaped, 23—38 × 12—20, with a smooth, pale, two-layered wall ~1.5 thick, with a M that was 2—4 wide, and with ovoidal sporocysts that tapered bluntly at one end, were 9—10 × 6—7 when measured inside the oocyst, but were 10—14 × 6—8 when released from the oocyst, and they had a button-like SB and a SR composed of a few scattered granules. These contained sausage-shaped SZs, narrower at one end than the other, that measured 8 × 2 when released. Golemanski (1975) said that the sporulated oocysts are 34 × 16 (28—41 × 15—20) with sporocysts 12—15 × 8—9. Samoil and Samuel (1977a) found oocysts in *L. americanus* that contained sporocysts that were larger (11—18 vs. 9—10) than those described by Pellérdy et al. (1974). The oocysts measured by Pakandl (1990) were 33.3 × 16.6 (27.5—37 × 15—19) with a M 3.5 (2.5—4) and an OR 5.7 (4—7) and sporocysts that were 13.5 × 7.1 (12—15 × 6.5—8). Ryšavý (1954) said he found this species in *L. europaeus* in the Czech Republic with oocysts that were 31.4 × 15.0 (25.5—36 × 14—17.5), with sporocysts

12 × 7, and that these oocysts sporulated in 48—60 hr. The parasite was found in the small intestine and was abundant in the hares and rabbits they sampled (Ryšavý, 1954). Romero-Rodriguez (1976) found this species in the Cape hare in Granada, Spain, and said the oocysts were elongated with round ends and measured 30.4 × 17.1 (27—38 × 14—19) without a M, but with a granular OR 5.6 wide, and sporocysts that were 10 × 5.8; he did not see a SB and said that a SR does not exist, but that there were scanty, fine granules scattered in the sporocyst. Mandal (1976) found it in a *Lepus* sp. in Kashipur, India, and said the cylindrical to slightly ovoidal oocysts were 22—28 × 15—26, with an OR, and the ovoidal sporocysts were 12—13 × 9 and had a SR. Tacconi et al. (1995) reported this species in some breeding farms raising *L. europaeus* in Italy and said elongated oocysts were 28—38 × 16—20 without a M, but with an OR; this species was always found with *E. semisculpta*. Robertson (1933) found *E. leporis* in one *L. europaeus* from England that had oocysts 21—40 × 11—20, without a M and with an OR. Oocysts studied by Moreno Montañez et al. (1979) in *L. capensis* from Spain were 18—36 × 14—19 with sporocysts about 12 × 6. Terracciano et al. (1988) identified this species in *L. europaeus* from three protected areas in the Province of Pisa, Italy. The cylindroidal oocysts they studied were 26—38 × 13—20 without a M, but with both an OR and SR.

We cannot attempt to resolve the differences between these descriptions. If we take the extreme endpoints from all these reports (above) the range of mean oocyst L is 28—34, mean oocyst W is 15—18, and the extreme oocyst L and W ranges are 18—28 to 28—41 × 11—16 to 16—26; thus, there seems to be an enormous plasticity of size of *E. leporis* oocysts discharged during patency. Such variation may be due to host and parasite genetic factors, season, immunological or nutritional differences/deficiencies, the influence of multiple congeneric eimerians in the gut, or the complex of these and multiple unknown factors we are yet to understand. Aoutil et al. (2005), who studied the *Eimeria* species in *L. granatensis* and *L. europaeus* in France, also noticed distinct size differences of oocysts from their rabbits and in the work of others, especially Robertson (1933) in England, whose histogram of oocyst measurements showed small, medium, and large oocysts. The approach they chose was to create two new subspecies. Their larger forms, which conformed closely to those in the original description by Nieschulz (1923), were called *Eimeria leporis leporis* for oocysts 33 × 15 (28–41 × 15–19), with sporocysts 15 × 8, and their smaller forms were called *Eimeria leporis brevis* for oocysts 27 × 14 (23–31 × 13–15). They said that both subspecies were found in the same four *L. granatensis* they examined, but that *E. l. brevis* was found only in two *L. europaeus*. Subspecies designations have not been well accepted by the protist community of scholars, and we believe the creation of numerous subspecies either within the same host species or in congenerics is a slippery slope that only adds further confusion to a literature that is already voluminous, complex, and internally confusing. Subtle differences in oocyst/sporocyst sizes, even gradations from one size to the next, may be used to separate species, but only if some other biological differences can be documented (e.g., location in the host, number of endogenous stages, etc.). There are sufficient examples (e.g., Duszynski, 1971, and many others) of oocysts changing dramatically in size during the patent period. Most field collections, as the one in France, take only small amounts of fecal material (e.g., 2 g feces from the caecum; Aoutil et al., 2005) to examine, so there is no way to know what day of the patent period the oocysts in that sample represent. Until the validity of different eimerians, both from rabbits and from all other hosts, can be confirmed by molecular studies, we prefer not to clutter the literature with subspecies names for similar oocysts produced by the same host animals.

EIMERIA MACROSCULPTA SUGÁR, 1979

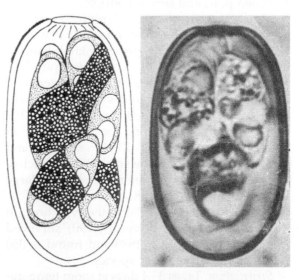

FIGURES 5.28, 5.29 Line drawing of the sporulated oocyst of *Eimeria macrosculpta* from Sugár, 1979. Photomicrograph of a sporulated oocyst of *E. macrosculpta* from Pakandl, 1990, with permission from *Folia Parasitologica*.

Type host: *Lepus europaeus* Pallas, 1778, European hare.

Type locality: EUROPE: Hungary: Gödöllő, Babat Puszta (Com. Pest), Besenyőtelek (Com. Heves), Gyarmatpuszta (Com. Komárom), Komádi (Com. Békés), Biharkeresztes (Com. Hajdú-Bihar).

Other hosts: *Lepus granatensis* Rosenhauer, 1956, Granada hare.

Geographic distribution: EUROPE: France, Hungary.

Description of sporulated oocyst: Oocyst shape: cylindroidal or elongate-ellipsoidal, rarely asymmetrical; number of walls: 2 (although line drawing shows only 1); wall thickness: 1.2–1.4; wall characteristics: outer surface with fine granulation, smooth, uniformly thick; L × W: 46 × 26 (40–50 × 25–33); L/W ratio: 1.8; M: present; M characteristics: distinct, giving one end of oocyst a truncated appearance, 7.5–9 wide; OR, PG: both absent. Distinctive features of oocyst: elongate-ellipsoidal shape, large size with one flattened pole, and lack of both OR and PG.

Description of sporocyst and sporozoites: Sporocyst shape: elongate-ovoidal; L × W: 19.6 × 9.4 (19–20 × 9–10); SB: present, knob-like, at slightly pointed end of each sporocyst; SSB, PSB: both absent; SR: present; SR characteristics: numerous refractile granules that sometimes form a coherent, elongated mass up to 7–12 × 6–8 between and covering the SZ; SZ: said to be "brush-stroke like," 13–15 × 5–6, and enclose one large and one small RB. Distinctive features of sporocyst: elongate shape, large knob-like SB and large SR.

Prevalence: 18/108 (17%) in the type host. In the Czech Republic, Pakandl (1990) surveyed *L. europaeus* from 1983–1985 and found 6/350 (< 2%) infected with this species.

Sporulation: Takes 3–4 days at room temperature in 1% potassium "bichromate" solution.

Prepatent and patent periods: Unknown.

Site of infection: Unknown, but Sugár (1979) said it localizes in the large intestine.

Endogenous stages: Unknown.

Cross-transmission: None to date.

Pathology: Unknown.

Material deposited: None.

Remarks: Sporulated oocysts of this species are most similar to those of *E. sculpta*, *E. semisculpta*, and *E. townsendi*. According to Sugár (1979) this species has larger oocysts (46 × 26 [40–50 × 25–33] vs. 36.8 × 28.7 [32–42 × 23–32]) and sporocysts (19.6 × 9.6 [19–20 × 9–10] vs. 17.1 × 9.5 [15–19 × 9–10]) than those of *E. sculpta*; has longer oocysts (46 × 26 vs. 43 × 26) than those of *E. semisculpta*, and lacks an OR, which *E. semisculpta* possesses; it has longer oocysts (46 [40–50] vs. 40 [37–44]), longer sporocysts (19.6 vs. 17) that possess a prominent knob-like SB, which sporocysts of *E. townsendi* lack. The oocysts measured by Pakandl (1990) were 41.7 × 23.6 (37–44.5 × 22–25) with a M 5.2 (4.5–6.5), but no OR, and its sporocysts were 23.0 × 8.4

(21–25 × 8–10). Aoutil et al. (2005) found this species in 5/46 (11%) *L. granatensis*, but in 0/9 *L. europaeus* in France; the oocysts they measured were 42 × 21 (40–45 × 18–22), L/W ratio 2.0, M 6.5, and sporocysts 22 × 8.5.

EIMERIA MAGNA PÉRARD, 1925b
(FIGURES 6.12, 6.13)

Type host: *Oryctolagus cuniculus* (L., 1758), European (domestic) rabbit.

Remarks: This species was originally described from the domestic rabbit, *O. cuniculus*, by Pérard (1925b) in France, but it has been reported in wild rabbits worldwide, including from three species of *Lepus*: *L. californicus*, the Black-tailed jackrabbit, *L. europaeus*, the European hare, and *L. nigricollis* (syn. *L. ruficaudatus*), the Indian hare. Unfortunately, its validity in *Lepus* species has not yet been verified. Henry (1932) said that she saw this eimerian and three others reported from the domestic rabbit, in one *L. californicus* from San Andreas, California. Ryšavý (1954) said he found this species in both *O. cuniculus* and in *L. europaeus* in the Czech Republic with oocysts that were 36.5 × 23.1 (31–46 × 20–27) and with sporocysts 12 × 7.5, and that these oocysts sporulated in 48–65 hr. The parasite was found in the "blind gut" (cecum?), but was found only rarely in the hares and rabbits they sampled (Ryšavý, 1954). Gill and Ray (1960) reported finding this species both in the domestic rabbit and in *L. nigricollis* collected in Punjab, India. The measurements they gave for sporulated oocysts were only for oocysts collected from domestic rabbits, but they did include a line drawing of an oocyst from *L. nigricollis* along with three line drawings of oocysts from *O. cuniculus*, and they are remarkably similar. Mandal (1976) said he recovered oocysts of this species in *L. nigricollis* from Punjab, India, with ellipsoidal to ovoidal oocysts that were 20.3–31.5 × 18–25.5 with an OR, and had ovoidal sporocysts 15.3–18 ×

7—8.8 with a SR. He (1976) also said that sporulation took 4 days. In an unpublished Master's thesis, Ogedengbe (1991) said he found this species in 19.5% of the rabbits surveyed in Kaduna State, Nigeria. However, he (1991) neglected to state how many rabbits were surveyed, nor did he mention the host species; Anonymous (2012) lists the Cape hare, *L. capensis*, as the only lagomorph species found in Nigeria. Vila-Viçosa and Caeiro (1997) reported this species in both *O. cuniculus* and *L. capensis* in Portugal, with the ovoidal oocysts from *L. capensis* being 35.6 × 22.5—24.3. Its validity as a parasite of *Lepus* species needs to be verified by cross-transmission and/or molecular studies. See Chapter 6 (Figures 6.12, 6.13) for the complete description of this species.

EIMERIA MATSUBAYASHII TSUNODA, 1952 (FIGURE 6.14)

Type host: *Oryctolagus cuniculus* (L., 1758) (syn. *Lepus cuniculus*), European (domestic) rabbit.

Remarks: Mandal (1976) said he recovered oocysts of this species in *Lepus* sp. from Kashipur, India, with ellipsoidal to ovoidal oocysts that were 23.5—29.5 × 14.5—19.3, with an OR, and ovoidal sporocysts 7 × 6 with a SR. Its validity as a parasite of *Lepus* species needs to be verified by cross-transmission and/or molecular studies. See Chapter 6 (Figure 6.14) for the complete description of *E. matsubayashii*.

EIMERIA MEDIA KESSEL, 1929 (FIGURES 6.15, 6.16)

Type host: *Oryctolagus cuniculus* (L., 1758) European (domestic) rabbit.

Remarks: This species was originally described from the domestic rabbit, *O. cuniculus*, by Kessel (1929). A few years later, Henry (1932) said she saw this eimerian and three others reported from the domestic rabbit, in one

Black-tailed jackrabbit from San Andreas, California. Matschoulsky (1941) said he found ovoidal to ellipsoidal oocysts that resembled this species in *L. timidus* from Buryat-Mongol (former USSR) with a M (~5—6 wide), and that were 28.5 × 17.5 (28—29 × 16.5—18.5), with sporocysts 10 × 7. Carvalho (1943) reported it in *Sylvilagus floridanus* from Iowa, and said he transmitted it from *O. cuniculus* to another *S. floridanus*. Gill and Ray (1960) reported finding oocysts of this species both in the domestic rabbit and in *L. nigricollis* (syn. *L. ruficaudatus*) collected in India. They gave measurements for sporulated oocysts collected from wild hares of 32 × 19 (24.5—39.5 × 16—25), with a convex M and both an OR and SR present. They (1960) said that these oocysts collected from the rectal contents of wild hares were "morphologically indistinguishable" from oocysts taken from *O. cuniculus*, and they included two line drawings of oocysts from *L. nigricollis* along with three line drawings of oocysts from *O. cuniculus* that all showed striking similarities to each other. Burgaz (1973) said she was able to transmit *E. media* from *L. timidus* to *O. cuniculus*, but her paper must be viewed with caution because she did not give specific information regarding how she identified the coccidian species, nor did she detail the procedures regarding whether or not the hosts were infected prior to inoculation and we think it likely that she was dealing with another eimerian from *Lepus* (e.g., *E. robertsoni*). In an unpublished Master's thesis, Ogedengbe (1991) said he found this species in 59% of the rabbits surveyed in Kaduna State, Nigeria. However, he (1991) neglected to state how many rabbits were surveyed, nor did he mention the host species; Anonymous (2012) lists the Cape hare, *L. capensis*, as the only lagomorph species found in Nigeria. Vila-Viçosa and Caeiro (1997) reported this species in both *O. cuniculus* and *L. capensis* in Portugal, with the ovoidal oocysts from *L. capensis* being 30—33.7 × 15.7—16.8. It has been confirmed as a valid parasite of *Sylvilagus*, but its true identity and validity in *Lepus* species needs to be verified either by further cross-transmission studies or by gene sequencing or both. See Chapter 6 (Figures 6.15, 6.16) for the complete species description.

EIMERIA MINIMA CARVALHO, 1943 (FIGURE 7.11)

Synonym: *Eimeria exigua* Yakimoff, 1934 of Morgan and Waller, 1940.

Type host: *Sylvilagus floridanus* (J.A. Allen, 1890), Eastern cottontail.

Remarks: Gill and Ray (1960) reported this species in the Indian field hare, *L. nigricollis* (syn. *L. ruficaudatus*), but according to Levine and Ivens (1972) they likely were dealing with *E. hungarica*, an evaluation with which we agree. Later, Mandal (1976) said he recovered oocysts of this species in *Lepus* sp. from Punjab, India, with subspheroidal oocysts that were 10–15.5 × 9–15, without an OR, and with ovoidal sporocysts that were 5 × 2.8 with a SR. There is no way to know what species of *Eimeria* Mandal (1976) actually had. Its true identity and validity in *Lepus* species needs to be verified either by well-controlled cross-transmission studies or by gene sequencing or both. See Chapter 7 (Figure 7. 11) for the complete description of *E. minima*.

EIMERIA NEOLEPORIS CARVALHO, 1942 (FIGURE 7.14)

Synonyms: *E. leporis* Nieschulz of Morgan and Waller, 1940, and Waller and Morgan, 1941; *E. neoleporis* Carvalho, 1943 of Gill and Ray, 1960 (*lapsus calami*); non *E. neoleporis* Nieschulz, 1923; non *E. neoleporis* Carvalho of Pellérdy, 1954, 1965.

Type host: *Sylvilagus floridanus* (J.A. Allen, 1890), Eastern cottontail.

Remarks: This species was originally described from the Eastern cottontail rabbit in Iowa, USA;

it has been transmitted experimentally many times to *O. cuniculus* (see Carvalho, 1943), but not to any *Lepus* species. Mandal (1976) said he recovered oocysts of this species in *Lepus* sp. from Punjab, India, with ellipsoidal oocysts that were 30–54 × 16–22 without an OR, and had ovoidal sporocysts, 16.5 × 8 with a SR. Its validity as a true parasite of any *Lepus* species needs to be verified either by well-controlled cross-transmission studies or by gene sequencing or both. See Chapter 7 (Figure 7.14) for the complete description of *E. neoleporis*.

EIMERIA NICOLEGERAE AOUTIL, BERTANI, BORDES, SNOUNOU, CHABAUD & LANDAU, 2005

Type host: *Lepus granatensis* Rosenhauer, 1956, Granada hare.

Type locality: EUROPE: France: Pyrénées-Orientales.

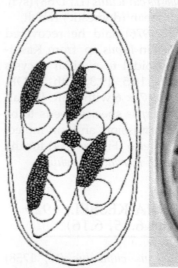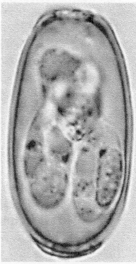

FIGURES 5.30, 5.31 Line drawing of the sporulated oocyst of *Eimeria nicolegerae* from Aoutil et al., 2005, with permission of the authors and from (the journal) *Parasite*. Photomicrograph of a sporulated oocyst of *E. nicolegerae* from Aoutil et al., 2005, with permission of the authors and from (the journal) *Parasite*.

Other hosts: None reported to date.

Geographic distribution: EUROPE: France.

Description of sporulated oocyst: Oocyst shape: elongate-ellipsoidal with parallel edges; number of walls: 1; wall characteristics: smooth, thick, brownish, thickens slightly around M; L × W: 34 × 18 (28–40 × 15–20); L/W ratio: 1.9; M: present; M characteristics: flat, thick, 5–6 wide, surrounded by slightly thickened edge of oocyst wall; OR: present; OR characteristics: small spheroidal mass of large granules, ~4 wide; PG: absent. Distinctive features of oocyst: long-ellipsoidal shape and a small OR with large granules.

Description of sporocyst and sporozoites: Sporocyst shape: ellipsoidal, slightly pointed at one end; L × W: 15 × 8 (ranges not given); L/W ratio: 1.9; SB: apparently present (line drawing) at pointed end, tiny; SSB, PSB: both absent; SR: present as an elongate mass of large granules lying along one side of sporocyst wall, parallel to SZ; SZ: comma-shaped (line drawing), each with one large, round RB. Distinctive features of sporocyst: elongate-ellipsoidal shape with SR as a compact, elongate mass of granules lying along sporocyst wall.

Prevalence: Apparently found in only 1/46 (2%) *L. granatensis*.

Sporulation: Unknown.

Prepatent and patent periods: Unknown.

Site of infection: Unknown. Oocysts recovered from feces.

Endogenous stages: Unknown.

Cross-transmission: None to date.

Pathology: Unknown.

Material deposited: None, although the authors designated their photomicrograph (their Fig. 1.6) as the holotype and said "the photos of the oocysts … have been deposited at the laboratory."

Remarks: Aoutil et al. (2005) noted that the description they provided closely resembles the parameters of the sporulated oocysts of *E. coecicola*, a parasite of *O. cuniculus*, with the oocysts of both being long ellipsoids and possessing a M, OR, and SR. Their first argument for separating the two species is that the OR and SR in the two are different sizes and shapes. It should be noted, however, that there are several reports in the literature documenting that both OR and SR in sporulated oocysts of rabbit eimerians can change size, get smaller, or disappear during storage (e.g., Robertson, 1933; Cheissin, 1959; Kvičerová et al., 2007). Their second argument, which seems more tenable, is that hare coccidia documented to date seem to be reasonably genus specific and are not known to infect *O. cuniculus*. Aoutil et al. (2005) provided a line drawing and a good photomicrograph in their original description, but they did not deposit either of these in an <u>accredited</u> museum, nor did they provide all of the quantitative and qualitative information needed and expected in a modern, standardized species description (range of oocyst L, W measurements; sporocyst mean L, W measurements and their ranges; measurements of SR, SZ, RB, etc.). This is disappointing and below usual standards. Finally, in presenting their description of this species, Aoutil et al. (2005) listed the name of this species as "*Eimeria nicolegerae* n. sp. Aoutil & Landau"; this is not, and cannot be, the correct authority for this species, since there is no separate published paper Aoutil and Landau describing and naming it. Thus, the full scientific name must be as we have listed it above.

EIMERIA PERFORANS (LEUCKART, 1879) SLUITER & SCHWELLENGREBED, 1912 (FIGURES 6.19, 6.20)

Synonyms: *Coccidium perforans* Leuckart, 1879; *Eimeria nana* Marotel and Guilhon, 1941; *Eimeria lugdunumensis* Marotel and Guilhon, 1942.

Type host: *Oryctolagus cuniculus* (L., 1758) (syn. *Lepus cuniculus*), European (domestic) rabbit.

Remarks: This species was originally described from the domestic rabbit, *O. cuniculus*, by Leuckart (1879). Several *Lepus* species have been reported to be infected with this eimerian naturally, but most, if not all, of these reports are probably erroneous. In 1932, Boughton (p. 535) said he found *E. perforans* in the American hare, *L. americanus*, in Western Canada, but gave no facts or evidence in support of his identification, and Henry (1932, p. 282) said that she saw this eimerian and three others reported from the domestic rabbit, in one Black-tailed jackrabbit from California, USA. Robertson (1933) said he found *E. perforans* in one *L. europaeus* from England that had oocysts 18−31 × 13−20, with a M and an OR, and sporocysts 12.5 (10−14) long, with a SB and a SR, which he said, "disappeared fairly rapidly in the majority." Matschoulsky (1941) said he found oocysts that resembled this species in *L. timidus* from Buryat-Mongol (former USSR) and that the ovoidal oocysts were 18.0 × 15.1 (16.5−20 × 15−16.5), with ovoidal sporocysts 7−9 × 4−7. Ryšavý (1954) said he found this species in both *O. cuniculus* and in *L. europaeus* in the Czech Republic with oocysts that were 20.2 × 14.5 (10−23 × 14−29), with sporocysts 8.5 × 4.5, and that these oocysts sporulated in 30−60 hr. The parasite was reported to be in the small intestine and was abundant in the hares and rabbits they sampled (Ryšavý, 1954). Gill and Ray (1960) reported finding this species both in the domestic rabbit and in *L. nigricollis* (syn. *L. ruficaudatus*) collected at Kashipur, India. The measurements they gave for sporulated oocysts were only for oocysts collected from domestic rabbits, but they included four line drawings of oocysts they said were from *L. nigricollis* and two line drawings of oocysts they said were from *O. cuniculus*, and the six drawings were remarkably similar. Mandal (1976) said he recovered oocysts of this species in *L. nigricollis* from Kashipur, India, with ellipsoidal oocysts that were 17−32 × 12.5−19.5, with an OR, and with ovoidal sporocysts 7.8 × 3.8 with a SR. In an unpublished Master's thesis, Ogedengbe (1991) said he found this species in 30% of the rabbits surveyed in Kaduna State,

Nigeria. However, he (1991) neglected to state how many rabbits were surveyed, nor did he mention the host species; Anonymous (2012) lists the Cape hare, *L. capensis*, as the only lagomorph species found in Nigeria. Vila-Viçosa and Caeiro (1997) reported this species in both *O. cuniculus* and *L. capensis* in Portugal, with the ellipsoid oocysts from *L. capensis* being 20.6−22.5 × 13.1−15. While cross-transmission studies verify that this species can infect *Sylvilagus* species, its true identity and validity in *Lepus* species are questionable and need to be verified either by well-controlled cross-transmission studies or by gene sequencing or both. See Chapter 6 (Figures 6.19, 6.20) for the complete description of *E. perforans*.

EIMERIA PIERRECOUDERTI AOUTIL, BERTANI, BORDES, SNOUNOU, CHABAUD & LANDAU, 2005

Type host: *Lepus granatensis* Rosenhauer, 1956, Granada hare.

Type locality: EUROPE: France: Pyrénées-Orientales.

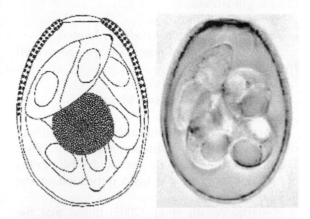

FIGURES 5.32, 5.33 Line drawing of the sporulated oocyst of *Eimeria pierrecouderti* from Aoutil et al., 2005, with permission of the authors and from (the journal) *Parasite*. Photomicrograph of a sporulated oocyst of *E. pierrecouderti* from Aoutil et al., 2005, with permission of the authors and from (the journal) *Parasite*.

Other hosts: None to date.

Geographic distribution: EUROPE: France.

Description of sporulated oocyst: Oocyst shape: ovoidal, slightly flattened at end with M; number of walls: 1 (line drawing), but probably 2, of uneven thickness; wall characteristics: thin, brown, thickens slightly around M, and anterior half of wall is thicker than posterior half and covered by "protruding granules" (i.e., sculptured [?], although this is not visible in the photomicrograph presented); L × W ($N = 12$): 35 × 24 (33–37 × 23–26); L/W ratio: 1.5; M: present; M characteristics: ~8 wide, with a "slight dome"; OR: present; OR characteristics: a large mass of granules, ~10 wide, with some of the edges extending between sporocysts; PG: absent. Distinctive features of oocyst: sculptured anterior half (?) of outer wall and large OR with undefined shape that protrudes between the sporocysts.

Description of sporocyst and sporozoites: Sporocyst shape: elongate-ellipsoidal, slightly pointed at one end (line drawing); L × W: 21 × 9.5 (ranges not given); L/W ratio: 2.2; SB: small, at pointed end; SSB, PSB, SR: all absent; SZ: with one large, elongate RB near rounded, wider end (line drawing). Distinctive features of sporocyst: small SB and lack of a SR.

Prevalence: Found in only 1/46 (2%) *L. granatensis*.

Sporulation: Presumably exogenous, but the time and temperature are not known.

Prepatent and patent periods: Unknown.

Site of infection: Unknown. Oocysts recovered from feces.

Endogenous stages: Unknown.

Cross-transmission: None to date.

Pathology: Unknown.

Material deposited: None, although the authors designated their photomicrograph (their Fig. 3.15) as the holotype and said "the photos of the oocysts … have been deposited at the laboratory."

Remarks: Aoutil et al. (2005) provided a line drawing and a photomicrograph of this form in their original description, but they did not deposit either of these in an <u>accredited</u> museum, nor did they provide all of the quantitative and qualitative information needed and expected in a modern, standardized species description (e.g., age of the oocysts when studied, sporulation time and temperature, sporocyst L, W ranges; measurements of SZ, RB, etc.). The authors argued about the validity of the epithet *"semisculpta"* used by Madsen (1938) and by Pellérdy (1956), stating that, in their opinion, the type from the Greenland hare differs from that of the Hungarian hare by its size and the presence of an OR in the former, but not in the latter (see *Remarks* under *E. semisculpta* to help resolve this dilemma). The oocysts of this species are very similar to those of *E. semisculpta*, but differ by having shorter and stockier sporocysts (L/W ratio 2.2 vs. 3.3 for *E. semisculpta*). Unfortunately, Aoutil et al. (2005) describe their OR with protrusions extending between the sporocysts (thus, hardly compact), but this giant OR is not visible in their photomicrograph (their Fig. 3.15), and they gave no ranges for their sporocysts, which their line drawing shows to be elongate-ellipsoids (their Fig. 4.15). In their description of this species, Aoutil et al. (2005) listed the name as *"Eimeria pierrecouderti* n. sp. Aoutil & Landau"; this is not, and cannot be, the correct authority for this species, since there is no separate published paper by Aoutil and Landau describing and naming it. Thus, the full scientific name (genus, specific epithet, authority) must be as we listed it above. Finally, the photomicrograph of this form (their Fig. 3.15) clearly seems to lack an OR, although the line drawing presented (Fig. 4.15, Aoutil et al., 2005), and their written description, report a massive OR.

EIMERIA PIRIFORMIS KOTLÁN & POSPESCH, 1934 (FIGURES 6.21, 6.22)

Synonyms: *Eimeria piriformis* Marotel and Guilhon, 1941; *Eimeria pyriformis* Kotlán &

Pospesch, 1934 *lapsus calami; non Eimeria piriformis* Lubimov, 1934.

Type host: Oryctolagus cuniculus (L., 1758) (syn. *Lepus cuniculus*), European (domestic) rabbit.

Remarks: Ryšavý (1954) said he found this species in both *O. cuniculus* and in *L. europaeus* in the Czech Republic with oocysts that were 30.2 × 20.6 (29—33 × 18—25), but they did not measure sporocysts nor did they determine sporulation time. The parasite was found in the small intestine of the hares and rabbits they sampled near Prague (Ryšavý, 1954). Vila-Viçosa and Caeiro (1997) reported this species in both *O. cuniculus* and *L. capensis* in Portugal, with the piriform oocysts from *L. capensis* being 37.5 × 18—20.6. Its true identity and validity in *Lepus* species need to be verified either by well-controlled cross-transmission studies or by gene sequencing or both. See Chapter 6 (Figures 6.21, 6.22) for the complete description of *E. piriformis.*

EIMERIA PUNJABENSIS GILL & RAY, 1960

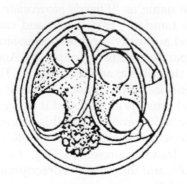

FIGURE 5.34 Line drawing of the sporulated oocyst of *Eimeria punjabensis* from Levine and Ivens, 1972, with permission of the *Journal of Protozoology.*

Type host: Lepus nigricollis F. Cuvier, 1823, (syn. *Lepus ruficaudatus*), Indian hare.

Type locality: ASIA: India: Punjab, District Ludhiana and Uttar Pradesh, Kashipur.

Other hosts: None reported to date.

Geographic distribution: ASIA: India: Punjab.

Description of sporulated oocyst: Oocyst shape: spheroidal to subspheroidal; number of walls: 1; wall characteristics: of even thickness throughout, light yellow or with a salmon tint; L × W: 22.5 × 22 (20—24 × 20—22); L/W ratio: 1.0; M: absent; OR: present; OR characteristics: distinct, ~5 wide; PG: absent. Distinctive features of oocyst: spheroidal shape with an OR, but without a M.

Description of sporocyst and sporozoites: Sporocyst shape: ovoidal; L × W: 12.5 × 8.5 (ranges not given); L/W ratio: 1.5; SB: present (line drawing); SSB, PSB, SR: all absent; SZ: with one large RB, 5 wide, at blunt end and a centrally located N. Distinctive features of sporocyst: lack of a SR and a large RB in SZ.

Prevalence: Unknown.

Sporulation: Exogenous. Oocysts sporulated in 2.5% (w/v) $K_2Cr_2O_7$ solution at 28°C in about 3—4 days (Gill and Ray, 1960; Mandal, 1976).

Prepatent and patent periods: Unknown.

Site of infection: Unknown. Oocysts recovered from feces.

Endogenous stages: Unknown.

Cross-transmission: None to date.

Pathology: Unknown.

Material deposited: None.

Remarks: This species has only been reported once since it was first described in 1960. Mandal (1976) is the only other person to report *E. punjabensis* when he said he recovered oocysts in *L. ruficaudatus*, also from Punjab, India; his spheroidal oocysts were 20—24 × 19.5—22.5, with an OR, and had ovoidal sporocysts that were 12.5 × 3.5 without a SR.

EIMERIA RENIAI AOUTIL, BERTANI, BORDES, SNOUNOU, CHABAUD & LANDAU, 2005

Type host: Lepus granatensis Rosenhauer, 1956, Granada hare.

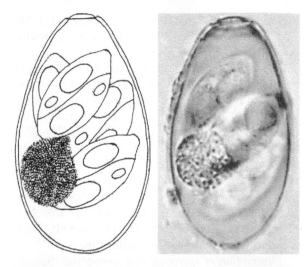

FIGURES 5.35, 5.36 Line drawing of the sporulated oocyst of *Eimeria reniai* from Aoutil et al., 2005, with permission of the authors and from (the journal) *Parasite*. Photomicrograph of a sporulated oocyst of *E. reniai* from Aoutil et al., 2005, with permission of the authors and from (the journal) *Parasite*.

Type locality: EUROPE: France: Pyrénées-Orientales.

Other hosts: *Lepus europaeus* Pallas, 1778, European hare.

Geographic distribution: EUROPE: France.

Description of sporulated oocyst: Oocyst shape: elongate-ellipsoidal, slightly flattened at end with M; number of walls: 1 (line drawing), but probably 2, of uneven thickness; wall characteristics: thin, smooth, brown, thickens slightly around M, and one side sometimes seems flatter than the other; L × W: 37 × 22 (35—40 × 19—24); L/W ratio: 1.7; M: present; M characteristics: ~4—5 wide, with a "jutting dome," surrounded by oocyst wall thicker near margin; OR: present; OR characteristics: a large spheroidal mass of granules, ~10 wide; PG: absent. Distinctive features of oocyst: elongate shape that narrows slightly at end with M and large OR, ~10 wide.

Description of sporocyst and sporozoites: Sporocyst shape: spindle-shaped, slightly pointed at both ends (line drawing); L × W: 19 × 9 (ranges not given); L/W ratio: 2.1; SB: small, at one pointed end; SSB, PSB: both absent; SR: absent; SZ: with one larger, elongate RB near rounded, wider end and a second, smaller spheroidal RB above it (line drawing). Distinctive features of sporocyst: spindle-shape, with a small SB, and SZ with two RBs of different sizes.

Prevalence: Found in 4/46 (4%) *L. granatensis* and in 1/9 (11%) *L. europaeus*.

Sporulation: Presumably exogenous, but the time and temperature are not known.

Prepatent and patent periods: Unknown.

Site of infection: Unknown. Oocysts recovered from feces.

Endogenous stages: Unknown.

Cross-transmission: None to date.

Pathology: Unknown.

Material deposited: None, although the authors designated their photomicrograph (their Fig. 3.14) as the holotype and said "the photos of the oocysts … have been deposited at the laboratory."

Remarks: Aoutil et al. (2005) provided a line drawing and a good photomicrograph of this form in their original description, but they did not deposit either of these in an <u>accredited</u> museum, nor did they provide all of the quantitative and qualitative information needed and expected in a modern, standardized species description (e.g., age of the oocysts when studied, sporulation time and temperature, sporocyst L, W ranges; measurements of SZ, RB, etc.). In presenting their description of this species Aoutil et al. (2005) listed the name as "*Eimeria reniai* n. sp. Aoutil & Landau"; this is not, and cannot be, the correct authority for this species, since there is no separate published paper by Aoutil and Landau describing and naming it. Thus, the full scientific name (genus, specific epithet, authority) must be as we listed it above. Finally, although the oocyst L and W ranges and ratios are slightly different, this form bears a striking resemblance to the morphology of sporulated oocysts from *E. bainae*.

EIMERIA ROBERTSONI (MADSEN, 1938) CARVALHO, 1943

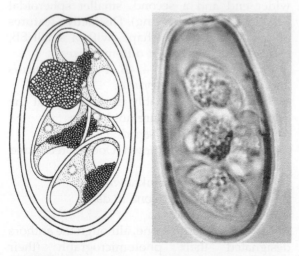

FIGURES 5.37, 5.38 Line drawing of the sporulated oocyst of *Eimeria robertsoni* from Samoil and Samuel, 1977a, with permission of the *Canadian Journal of Zoology*. Photomicrograph of a sporulated oocyst of *E. robertsoni* from Samoil and Samuel, 1977a, with permission of the *Canadian Journal of Zoology*.

Synonyms: *Eimeria magna* var. *robertsoni* Madsen, 1938; *Eimeria robertsoni* Madsen, 1938 of Carvalho, 1943; *Eimeria magna* forma *townsendii* Carvalho, 1943 *pro parte*; *Eimeria perforans* var. *groenlandica* Madsen, 1938, *pro parte*.

Type host: *Lepus arcticus* Ross, 1819, Arctic hare.

Type locality: GREENLAND: Eastern end.

Other hosts: *Lepus americanus* Erxleben, 1777, Snowshoe hare; *Lepus capensis* L., 1758, Cape hare; *Lepus europaeus* Pallas, 1778, European hare; *Lepus nigricollis* (syn. *Lepus ruficaudatus*) F. Cuvier, 1823, Indian hare; *Lepus timidus* L., 1758, Mountain hare; *Lepus townsendii* Bachman, 1839, White-tailed jackrabbit.

Geographic distribution: ASIA: India; Lithuania (of former USSR); EUROPE: Austria, Czech Republic, England, (former East) Germany, Hungry, Italy, Norway, Poland, Portugal,

Switzerland; GREENLAND; MIDDLE/NEAR EAST: Turkey; NORTH AMERICA: Canada: Alberta; USA: Iowa.

Description of sporulated oocyst: Oocyst shape: broadly ellipsoidal to ovoidal; number of walls: 2; wall thickness: 1.9; wall characteristics: outer yellowish brown, finely granular, easily detached, ~1.4; inner is 0.5; L × W: 42.7 × 25.8 (34—52 × 23—32); L/W ratio: 1.7; M: present; M characteristics: broad, 6—8.5 wide, surrounded by a collar-like thickening of the wall; OR: present; OR characteristics: compact, granular, large, 8.1 (6—11, or up to 12—15); PG: absent. Distinctive features of oocyst: with a broad M and a large, granular OR.

Description of sporocyst and sporozoites: Sporocyst shape: elongate spindle-shaped to ovoidal, tapering toward one end; L × W: 20.9 × 8.0 (18—25 × 6—10, Samoil and Samuel, 1977a), 17—22 × 7—8 (Madsen, 1938), 19.5 × 7 (Carvalho, 1943), 18.5 × 7 (Pellérdy, 1956) or 18.5 × 6 (Gill and Ray, 1960); L/W ratio: 2.6; SB: present as an indistinct light-refracting area at narrow end of sporocyst; SSB, PSB: both absent; SR: present; SR characteristics: diffuse, granular; SZ: have a single RB posterior to the N. Distinctive features of sporocyst: long and narrow spindle-shape.

Prevalence: Madsen (1938) found what has now become this species in ~11/22 (50%) *L. arcticus* that he examined from Greenland; he described it as two different types and then variations of those types (his Type B, Figs. 5a, b, c and his Type C, Figs 6a, b) saying that these Types were found in 11/22 arctic hares (but there is no way to tell if both Types were found in the same 11 hares). Carvalho (1943) found it in 9/12 (75%) *L. townsendii* he examined in Iowa. Gräfner et al. (1967) found it in 7/176 (4%) *L. europaeus* from (former East) Germany; Arnastauskene and Kazlauskas (1970) found it in 13/41 (31%) *L. europaeus* from Lithuania; Pastuszko (1961a) reported it in 30/462 (6%) *L. europaeus* in Poland; and Sugár et al. (1978) said it was in 29/374 (8%) in Hungary. Samoil and

Samuel (1977a) found it in 208/629 (33%) *L. americanus* they examined in Canada. In the Czech Republic, Pakandl (1990) surveyed *L. europaeus* from 1983–1985 and found 120/350 (34%) infected with this species. Jirouš (1979), also in the Czech Republic, identified *E. robertsoni* in 67% of fecal samples collected in the field and in 90% of the juveniles and 74% of the adults he shot. Kutzer and Frey (1976) found it in 54% of the hares they collected in Austria. Tasan and Özer (1989) said it occurred in 53% of the 40 hares they surveyed in Turkey between 1985–1987.

Sporulation: Exogenous. Oocysts sporulated after 6 days in 2.5% aqueous (w/v) $K_2Cr_2O_7$ solution at 20°C (Samoil and Samuel, 1977a; Bouvier, 1967), 3–4 days (Pellérdy, 1956), 56 (40–65) hr (Carvalho, 1943), 2–4 days (Levine and Ivens, 1972), 38 hr (Gill and Ray, 1960), or 4–5 days (Mandal, 1976).

Prepatent and patent periods: Prepatency is 6.5 days and the patent period lasts at least 22 days following infection with one sporulated oocyst (Samoil and Samuel, 1977b).

Site of infection: Duodenum (Carvalho, 1943; Samoil and Samuel, 1977a), but Bouvier (1967) said it occurred in the cecum.

Endogenous stages: Unknown.

Cross-transmission: Carvalho (1943) was not able to transmit this species either to the domestic rabbit, *O. cuniculus*, or to the Iowa cottontail, *S. floridanus*. Pellérdy (1956) was not able to infect the domestic rabbit with this species isolated from *L. europaeus*. Burgaz (1973) was unable to transmit *E. robertsoni* from *L. timidus* to the domestic rabbit. Samoil and Samuel (1977b) were not able to infect seven *O. cuniculus* with various numbers of oocysts (100, 30,600, or 50,000) of this species isolated originally from a naturally-infected *L. americanus*.

Pathology: Madsen (1938) said that one *L. arcticus* fed an unspecified number of oocysts died 1 wk later with emaciation, suggestions of diarrhea (whatever that means), a mucosanguine exudate in the posterior small intestine, and an enormously swollen urinary bladder. Bouvier (1967) said that this species may cause serious lesions and hemorrhage in the cecum of young hares, but more discrete lesions in adults. Levine and Ivens (1972) said that this species is "possibly" pathogenic. Tacconi et al. (1995) found this species in the intestinal contents of four dead juvenile *L. europaeus* in Italy; the oocysts they found were always found with those of *E. europaea*, *E. hungarica*, and *E. townsendi* in these dead hares, which all showed severe enteritis as the cause of death.

Material deposited: None.

Remarks: The history that involves the discovery and description(s) of this species is confusing and tortuous, at best. In 1933, Robertson looked at eimerian oocysts from two *L. europaeus* from an estate near Dorchester, England, and mentioned five different types of oocysts. His Type I, which he called *E. stiedai*, resembled oocysts of *E. magna* from *O. cuniculus*, even though the average oocyst size, 42.1 × 25.1 (38–45 × 21–30), was larger than both *E. stiedai* and *E. magna*, and both have sporocysts with a distinct SR, which Robertson's Type I lacked. Madsen (1938) collected feces from *L. arcticus groenlandicus* Rhodes, 1896, in Eastern Greenland (Eskimonæs, 74°05′N) and measured 760 oocysts, which he grouped into four types (A, B, C, D), with the first three each having subtypes (a–c), each represented by a different line drawing. His Type C (Fig. 6a) was nearly identical to Robertson's (1933) Type I, so Madsen named it *E. magna* var. *robertsoni*. Carvalho (1943), working with *L. townsendii campanius* Hollister, 1915, in Iowa, USA, found eimerian oocysts similar to those reported by Madsen (1938) as *E. m.* var. *robertsoni* and elevated the name to *E. robertsoni*. Carvalho (1943) recognized that Madsen's description was probably a complex of two species so he equated and combined the larger of Madsen's (1938) forms, Type B (Figs. 6a–c), which Madsen had named *E. perforans* var. *groenlandica*, into *E. robertsoni*, although doing this aggravated Aoutil et al. (2005), who said, "we think that the name '*robertsoni*' which appears to refer to three different species, should be abandoned." Looking carefully at Madsen's (1938) measurements and figures (his Fig. 5a—c) of *E. p.* var. *groenlandica*, he likely was dealing with a mixture of two, if not three, different eimerians, one of which (his Fig. 5c) could have been *E. robertsoni*. Pellérdy (1956) called this species *E. robertsoni* Carvalho, 1943 and included the synonyms, *E. p.* var. *groenlandica* Madsen, 1938, pro parte and *E. m. forma townsendi* Carvalho, 1943, *pro parte*. Pellérdy (1956) also said that a SR "is always present in the sporocysts, but consists of a few granules only, frequently invisible ...." Finally, Pellérdy (1956) took exception to Carvalho's adding another variety/form to the literature, *E. magna* var. *townsendi*. In Pellérdy's (1956) view, the majority of oocysts identified as *E. m.* var. *townsendi* by Carvalho (1943) were identical with those of *E. robertsoni*, while the others were those of *E. townsendi*, which he described later in that (1956) publication. His later treatises (Pellérdy, 1965, 1974) maintained that view, as did a later redescription of *E. robertsoni* by Samoil and Samuel (1977a); the latter, however, solidified the presence of a distinct, granular SR that was sometimes diffuse and sometimes a compact mass of granules (their Figs. 1, 11). Carvalho (1943) said the oocysts of *E. robertsoni* were 40.5 × 25 (36—47 × 23—33), Pellérdy (1956) said they were 40 × 25 (38—48 × 22—32), Pastuszko (1961a) said they were 40 × 25 (34—52 × 22—35), and Bouvier (1967) said the oocysts were 40 × 25 (34—48 × 22—32). Samoil and Samuel's (1977a) redescription was of oocysts from *L. americanus* in Alberta, Canada, and provided a concise and coherent rendition that was essentially the same as the original of Madsen (1938), except for the number of oocyst walls reported. Madsen (1938) described only one wall, although his drawing showed two. In subsequent reports (Carvalho, 1943; Pellérdy, 1956; Levine and

Ivens, 1972), the wall was described as one-layered, although all but Carvalho (1943) provided drawings showing two walls. Samoil and Samuel (1977a) were able to show that two walls existed by easily removing the outer wall with a 5% sodium hypochlorite solution and said the oocysts were 41.3 × 22.5 (34—52 × 20—25) and an OR 8.1 (6—11), and sporocysts that were 20.9 × 8.0 (18—25 × 6—10). Also, although the size and shape of the oocysts and sporocysts match those of others describing this species, Samoil and Samuel (1977a) noted a SB on the sporocysts which the other authors (apparently) had missed. Finally, there is a question about whether Bouvier (1967) was actually dealing with this species. His description was so incomplete that it is impossible to be sure, and he reported this form from the cecum rather than the duodenum. Samoil and Samuel (1977b) inoculated a juvenile *L. americanus* with one sporulated oocyst of *E. robertsoni* that resulted in a patent infection; oocyst output rose from ~50 oocysts on day 1 PI to ~7 × 10⁶ on day 3 PI, and then declined steadily thereafter. They (1977b) also found that oocyst size increased in both length (13%) and width (17%) during patency. Berg (1981) reported this species in the European hare from Norway, but gave no other information. Tacconi et al. (1995) found this species in the intestinal contents of four dead juvenile *L. europaeus* from a breeding farm in Italy. The ovoidal oocysts were 34—52 × 23—32, with a M and an OR; this species was always found with *E. europaea*, *E. hungarica*, and *E. townsendi* in these dead hares. Terracciano et al. (1988) identified this species in *L. europaeus* from three protected areas in the Province of Pisa, Italy. The ovoidal oocysts they studied were 34—52 × 23—32 with a M 8—9 wide, OR and SR. Gill and Ray (1960) found oocysts of this species in *L. nigricollis* (syn. *L. ruficaudatus*) in India, and said they were 39 × 25.5 (33—45 × 21—33). Mandal (1976) reported this species when he said he recovered oocysts in *L. nigricollis* (syn. *L. ruficaudatus*) from

Kashipur, India, with ovoidal to ellipsoidal oocysts that were 33–44.5 × 21–32.5, with an OR, and had ovoidal sporocysts, 18.5 × 6.2, with a SR. Vila-Viçosa and Caeiro (1997) reported this species in *L. capensis* in Portugal, with ovoidal oocysts 41.8 × 23.4.

EIMERIA ROCHESTERENSIS SAMOIL & SAMUEL, 1977a

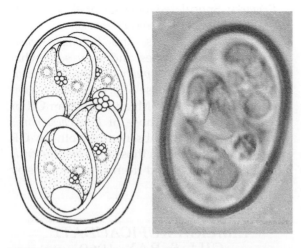

FIGURES 5.39, 5.40 Line drawing of the sporulated oocyst of *Eimeria rochesterensis* from Samoil and Samuel, 1977a, with permission of the *Canadian Journal of Zoology*. Photomicrograph of a sporulated oocyst of *E. rochesterensis* from Samoil and Samuel, 1977a, with permission of the *Canadian Journal of Zoology*.

Type host: *Lepus americanus* Erxleben, 1777, Snowshoe hare.

Type locality: NORTH AMERICA: Canada: Alberta, Rochester.

Other hosts: None reported to date.

Geographic distribution: NORTH AMERICA: Canada: Alberta.

Description of sporulated oocyst: Oocyst shape: cylindroidal; number of walls: 2; wall thickness: ~1; wall characteristics: outer is smooth, uniformly thick; L × W: 22.8 × 14.4 (20–26 × 14–16); L/W ratio: 1.6; M: absent; OR: present; OR characteristics: small, compact mass of a few large granules; PG: absent. Distinctive features of oocyst: cylindroidal shape with an OR of a few large granules.

Description of sporocyst and sporozoites: Sporocyst shape: elongate spindle-shaped, tapering toward one end; L × W: 12.2 × 6.1 (10–13 × 6–7); L/W ratio: 2.0; SB: present; SB characteristics: an indistinct light-refracting area at narrow end of sporocyst; SSB, PSB: both absent; SR: present; SR characteristics: compact, irregular mass of a few large granular; SZ: have one RB posterior to N. Distinctive features of sporocyst: spindle-shape with a large L/W ratio (2.0) and an indistinct light-refracting SB.

Prevalence: 141/629 (22%) in the type host.

Sporulation: Exogenous. Most oocysts were sporulated after 6 days in 2.5% aqueous (w/v) $K_2Cr_2O_7$ solution at 20°C.

Prepatent and patent periods: Unknown.

Site of infection: Unknown. Oocysts recovered from feces.

Endogenous stages: Unknown.

Cross-transmission: None to date.

Pathology: Unknown.

Material deposited: None.

Remarks: The sporulated oocysts of this species resemble those of *E. athabascensis*, *E. keithi*, *E. leporis*, *E. rowani*, and *E. ruficaudati* from other *Lepus* species and those of *E. nagpurensis* described from *O. cuniculus*. It is smaller in size than those of *E. leporis* (22.8 × 14.4 [20–26 × 14–16] vs. 32 × 16 [26–38 × 13–20]) and has an SR which is smaller and compact vs. the large diffuse SR of *E. leporis*. These oocysts also are smaller than those of *E. athabascensis* (33.8 × 15.6 [24–38 × 13–17]) and have a SR. The oocysts of this species lack a M, which is characteristic of *E. ruficaudati* and *E. keithi*. The presence of an OR differentiates these oocysts from those of *E. nagpurensis*, and they differ from those of *E. rowani* by having both an OR and a SR, which *E. rowani* lacks.

EIMERIA ROWANI SAMOIL & SAMUEL, 1977a

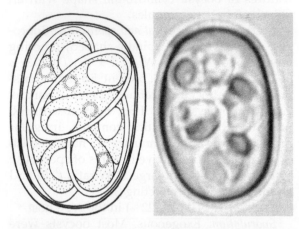

FIGURES 5.41, 5.42 Line drawing of the sporulated oocyst of *Eimeria rowani* from Samoil and Samuel, 1977a, with permission of the *Canadian Journal of Zoology*. Photomicrograph of a sporulated oocyst of *E. rowani* from Samoil and Samuel, 1977a, with permission of the *Canadian Journal of Zoology*.

Type host: *Lepus americanus* Erxleben, 1777, Snowshoe hare.

Type locality: NORTH AMERICA: Canada: Alberta, Rochester.

Other hosts: None reported to date.

Geographic distribution: NORTH AMERICA: Canada: Alberta.

Description of sporulated oocyst: Oocyst shape: cylindroidal; number of walls: 2; wall thickness: ~1; wall characteristics: outer is smooth, uniformly thick; L × W: 22.1 × 15.8 (18−31 × 12−18); L/W ratio: 1.4; M, OR, PG: all absent. Distinctive features of oocyst: lacking a M, OR and PG.

Description of sporocyst and sporozoites: Sporocyst shape: elongate spindle-shaped; L × W: 12.1 × 6.0 (11−17 × 5−7); L/W ratio: 2.0; SB: present; SSB, PSB, SR: all absent; SZ: have one RB posterior to N. Distinctive features of sporocyst: SB an indistinct light-refracting area at narrow end of sporocyst.

Prevalence: 68/629 (11%) in the type host.

Sporulation: Exogenous. Oocysts sporulated after 6 days in 2.5% aqueous (w/v) $K_2Cr_2O_7$ solution at 20°C.

Prepatent and patent periods: Unknown.

Site of infection: Unknown. Oocysts recovered from feces.

Endogenous stages: Unknown.

Cross-transmission: None to date.

Pathology: Unknown.

Material deposited: None.

Remarks: Sporulated oocysts of this species resemble those of *E. rochesterensis*, *E. keithi*, *E. ruficaudati*, *E. leporis*, and *E. athabascensis* from other *Lepus* species and they somewhat resemble those of *E. nagpurensis* from *O. cuniculus*. They differ from those of *E. keithi* and *E. ruficaudati* by not having a M, and either an OR or SR. They differ from those of *E. rochesterensis* and *E. leporis* by the absence of both an OR and SR. They differ from those of *E. athabascensis* by the lack of an OR and from *E. nagpurensis* by the lack of a SR.

EIMERIA RUFICAUDATI GILL & RAY, 1960

Synonyms: *Eimeria deharoi deharoi* Aoutil et al., 2005; *Eimeria deharoi rotunda* Aoutil et al., 2005.

Type host: *Lepus nigricollis* F. Cuvier, 1823, (syn. *Lepus ruficaudatus*), Indian hare.

Type locality: ASIA: India: Punjab, District Ludhiana and Uttar Pradesh, Kashipur.

Other hosts: *Lepus americanus* Erxleben, 1777, Snowshoe hare; *Lepus granatensis* Rosenhauer, 1956, Granada hare.

Geographic distribution: ASIA: India; EUROPE: France; NORTH AMERICA: Canada: Alberta.

Description of sporulated oocyst: Oocyst shape: cylindroidal or ellipsoidal with flat sides; number of walls: 1; wall characteristics: light yellowish-pink, of even thickness throughout, except at M where it becomes thinner; L × W: 31.3 × 17.7

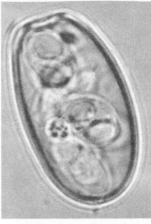

FIGURES 5.43, 5.44 Line drawing of the sporulated oocyst of *Eimeria ruficaudati* from Samoil and Samuel, 1977a, with permission of the *Canadian Journal of Zoology*. Photomicrograph of a sporulated oocyst of *E. ruficaudati* from Samoil and Samuel, 1977a, with permission of the *Canadian Journal of Zoology*.

(28—35 × 12—15 [sic]; Gill and Ray, 1960), 29.9 × 17.5 (23—38 × 14—19, Samoil and Samuel; 1977a); L/W ratio: 1.7; M: present; M characteristics: distinct, ~4.5 (4—6), bounded by a bent line, with a fine, pointed tip in center, which may form a triangular opening; OR: present; OR characteristics: round, composed of compact granules, ~4 (2—6) wide, with a tendency to get smaller upon storage (Gill and Ray, 1960); PG: absent. Distinctive features of oocyst: cylindroidal shape with parallel sides, thin wall especially around the M, and presence of OR.

Description of sporocyst and sporozoites: Sporocyst shape: ovoidal to spindle-shaped, anterior end bluntly pointed; L × W: not given by Gill and Ray (1960), 15.2 × 6.6 (12—18 × 6—8 (Samoil and Samuel, 1977a); SB: present, as an indistinct, light-refracting area at narrow end; SSB, PSB: both absent; SR: present; SR characteristics: round, 2—4 wide or as scattered granules; SZ: medium-sized with one RB at rounded end and N in central position. Distinctive features of sporocyst: the SR is reported to disappear on storage (Gill and Ray, 1960).

Prevalence: Not given in the original description (Gill and Ray, 1960) for the type host; in 101/629 (16%) *L. arcticus* (Samoil and Samuel, 1977a); in 1/46 (2%) *L. granatensis* (Aoutil et al., 2005).

Sporulation: Exogenous. Oocysts sporulated in 66 hr at 28°C (Gill and Ray, 1960), by 6 days at 20°C when maintained in 2.5% aqueous (w/v) $K_2Cr_2O_7$ solution (Samoil and Samuel, 1977a), or in 66 hr (Mandal, 1976).

Prepatent and patent periods: Unknown.

Site of infection: Unknown. Oocysts recovered from feces.

Endogenous stages: Unknown.

Cross-transmission: None to date.

Pathology: Unknown.

Material deposited: None.

Remarks: Gill and Ray (1960) said that the OR has "a tendency to decrease in size" with time and that the SR, usually ~2 wide or as scattered granules, "disappears" with preservation. Mandal (1976) also reported this species when he said he recovered oocysts in *L. nigricollis* (syn. *L. ruficaudatus*) from Kashipur, India, with cylindroidal oocysts that were 28—38 × 12—15, with an OR, and ovoidal sporocysts 13.5—15.5 × 8.5—10.5 with a SR. Aoutil et al. (2005) described two new *Eimeria* subspecies, from *L. granatensis* in France, which they named *E. deharoi deharoi* and *E. d. rotunda*, and compared their form(s) only superficially to *E. media* from *O. cuniculus*. Their larger form had oocysts that were 28 × 16 (26—33 × 14—18), L/W ratio 1.75, with both a M, 4.5 wide, and an OR, 7 wide, and sporocysts 13 × 7, L/W ratio 1.9, with a small SR of "large dispersed grains." However, their larger oocyst forms are similar to those of both *E. europaea* and *E. ruficaudati*, to which they did not compare them. *Eimeria europaea* has been found in multiple *Lepus* species throughout Europe, including France. Its oocysts are 32 × 18 (26—34 × 15—20), L/W ratio 1.8, with a M 6—9 wide, a spheroidal OR, and sporocysts 9 × 6 with a L/W ratio 1.5, and are somewhat smaller than those of *E. deharoi* subspecies, which are 13

× 7, with L/W ratio 1.9. If one compares the line drawings of *E. europaea* given by Pellérdy (both 1956, 1965) with the one given by Aoutil et al. (2005, their Fig. 2.8), they are nearly identical. Likewise, so are the photomicrographs, one of *E. europaea* given by Pakandl (1990) and the one presented by Aoutil et al. (2005, their Fig. 1.8) representing their *E. d. debaroi*. However, given the differences in size and L/W ratio between the sporocysts of *E. europaea* and *E. d. debaroi*, we believe we must defer to *E. ruficaudati* as the appropriate species. Oocysts of *E. ruficaudati* are 30 × 17.5 (23−38 × 14−19), L/W ratio 1.7, with a M 4.5 wide, OR 2−6 wide, and sporocysts 15 × 7 with a L/W ratio 2.1, are quite similar to those of *E. d. debaroi*. The sporulated oocysts of *E. d. rotunda* conform to those of *E. ruficaudati* in every respect except that they are slightly smaller, which we consider to be part of normal size variation characteristic of many eimerians.

EIMERIA SCULPTA MADSEN, 1938

Type host: *Lepus arcticus* Ross, 1819, Arctic hare.

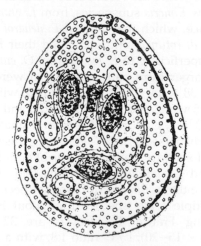

FIGURE 5.45 Line drawing of the sporulated oocyst of *Eimeria sculpta* from Levine and Ivens, 1972, with permission of the *Journal of Protozoology*.

Type locality: GREENLAND: Eskimonaes, 74°05′ N lat.

Other hosts: *Lepus europaeus* Pallas, 1778, European hare; *Lepus townsendii* Bachman, 1839, White-tailed jackrabbit.

Geographic distribution: EUROPE: Czech Republic, Hungary; GREENLAND: Eskimonaes; NORTH AMERICA: USA: Iowa.

Description of sporulated oocyst: Oocyst shape: broadly pear-shaped (piriform) to ovoidal; number of walls: 2; wall thickness: ~2; wall characteristics: outer is rough, sculptured or pitted, dark-brown, uniformly thick except slightly thicker near M; L × W: 36.8 × 28.7 (32−42 × 23−32, Madsen, 1938), 36 × 30 (32−38 × 29−31, Carvalho, 1943); L/W ratio: 1.3−1.4; M: present; M characteristics: plug-shaped, narrow (Madsen, 1938) or a distinct surface structure, 7.5 wide (Carvalho, 1943); OR, PG: both absent. Distinctive features of oocyst: pear-shape, sculptured outer wall, and a distinct M.

Description of sporocyst and sporozoites: Sporocyst shape: ovoidal; L × W: 17.1 × 9.5 (15−19 × 9−10); L/W ratio: 1.8; SB: present; SB characteristics: peg-shaped; SSB, PSB: both absent; SR: present; SR characteristics: sharply defined ellipsoidal body or an irregular shape with sharp contours; SZ: comma-shaped, lie head-to-tail, with one large RB at thicker end. Distinctive features of sporocyst: prominent SB and SR.

Prevalence: 14/22 (64%) from the type host in Greenland; in 9/12 (75%) *L. townsendii* from Iowa (Carvalho, 1943); in 161/374 (43%) *L. europaeus* from Hungary (Sugár et al., 1978). In the Czech Republic, Pakandl (1990) surveyed *L. europaeus* from 1983−1985 and found 32/350 (9%) infected with this species.

Sporulation: Exogenous. Sporulation occurs in 55 (50−60) hr (Carvalho, 1943).

Prepatent and patent periods: Unknown.

Site of infection: Unknown. Oocysts recovered from intestinal canal.

Endogenous stages: Unknown.

Cross-transmission: Carvalho (1943) was unable to transmit this species either to the domestic rabbit, *O. cuniculus*, or to the Iowa cottontail, *S. floridanus*.

Pathology: Unknown.

Material deposited: None.

Remarks: Madsen (1938) distinguished this species from all others reported from hares by two features: the lack of an OR combined "with the sculpture distributed all over the oocyst." Carvalho (1943) said that the sporulated oocysts he observed "possessed a much wider M than Madsen's specimens. The oocyst wall was thicker near the M, and several specimens showed the marginal lappet around it. No other differences were noted." The oocysts measured by Pakandl (1990) were 30.2 × 25.2 (28–35 × 22–27) with a M 4.6 (3.5–6), but no OR, and sporocysts were 15.6 × 8.1 (14–18 × 7.5–9.5).

EIMERIA SEMISCULPTA (MADSEN, 1938) PELLÉRDY, 1956

Synonyms: *Eimeria magna* var. *robertsoni* forma *semisculpta* Madsen, 1938 *pro parte*; *Eimeria*

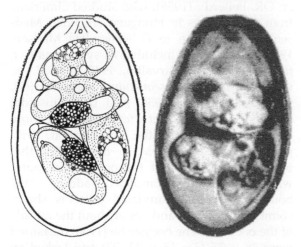

FIGURES 5.46, 5.47 Line drawing of the sporulated oocyst of *Eimeria semisculpta* from Sugár et al., 1978, in *Parasitologica Hungarica* (journal no longer exists). Photomicrograph of a sporulated oocyst of *E. semisculpta* from Pakandl, 1990, with permission of *Folia Parasitologica*.

irresidua forma *campanius* Carvalho, 1943, *pro parte*; *non Eimeria semisculpta* (Madsen, 1938) Pellérdy, 1956, and of Pellérdy, 1965, 1974.

Type host: *Lepus arcticus* Ross, 1819, Arctic hare.

Type locality: GREENLAND: Eastern, near Eskimonaes, 74°05′N.

Other hosts: *Lepus europaeus* Pallas, 1778, European hare; *Lepus timidus* L., 1758, Mountain hare (?); *Lepus townsendii* Bachman, 1839, White-tailed jackrabbit.

Geographic distribution: ASIA: Republic of Belarus (Byelorussia of the former USSR) (?), Lithuania (of former USSR); EUROPE: Czech Republic, Finland, (former East) Germany, Hungary, Norway, Poland; GREENLAND.

Description of sporulated oocyst: Oocyst shape: elongate-fusiform (Madsen, 1938), to broadly ellipsoidal, slightly truncated at one pole (Pellérdy, 1956); number of walls: 2; wall characteristics: ~1 thick, brownish-yellow, uniformly thick, smooth, except over the M hemisphere where it is punctate; the external layer can be easily detached by coverslip compression after the oocyst has been in potassium dichromate for some time; L × W: 42.7 × 25.8 (34–52 × 23–32); L/W ratio: 1.6; M: present; M characteristics: 6–8 wide, plug-shaped; OR: present; OR characteristics: a compact, spheroidal mass; PG: absent. Distinctive features of oocyst: slightly truncated at end with M and compact, spheroidal OR.

Description of sporocyst and sporozoites: Sporocyst shape: long, spindle-shaped or elongate-fusiform; L × W: 22–28 × 7–8; L/W ratio: ~3.3; SB: present; SSB, PSB: both absent; SR: present; SR characteristics: some diffuse granules located centrally between SZ; SZ: comma-shaped and lie head-to-tail with one clear RB at thicker end. Distinctive features of sporocyst: long, spindle-shape with only diffuse SR granules.

Prevalence: 6/22 (27%) in the type host from Greenland; Pastuszko (1961a) reported it in 47/462 (10%) *L. europaeus* from Poland and

Gräfner et al. (1967) in 3/176 (2%) *L. europaeus* from (former East) Germany. Litvenkova (1969) said, "Of 52 hares (*Lepus europaeus* L. and *L. timidus* L.) coccidian oocysts were found in 22 (42.3%)," but she did not say whether *E. semisculpta* was found in one or both of the *Lepus* species. Arnastauskene and Kazlauskas (1970) in 28/41 (68%) *L. europaeus* from Lithuania, and Sugár et al. (1978) in 56/374 (15%) from Hungary. In the Czech Republic, Pakandl (1990) surveyed *L. europaeus* from 1983–1985 and found 48/350 (14%) infected with this species. Also in the Czech Republic, Jirouš (1979) identified *E. semisculpta* in 100% of the hares he examined, while Chroust (1984) found it in only 8.3% and 3.6% of dead European hares.

Sporulation: Exogenous. Oocysts sporulate in 72–80 hr at 23°C in 3% $K_2Cr_2O_7$ solution (Golemanski, 1975).

Prepatent and patent periods: Unknown.

Site of infection: Duodenum.

Endogenous stages: Unknown.

Cross-transmission: Both Carvalho (1943) and Pellérdy (1956) were unable to infect the domestic rabbit with this species isolated from L. europaeus. Burgaz (1973) was unable to transmit *E. semisculpta* from *L. timidus* to the domestic rabbit.

Pathology: Unknown.

Material deposited: None.

Remarks: As with other rabbit eimerian species that began with the publication by Madsen (1938), the history involved in the discovery and description(s) of this species is confusing and tortuous at best. Madsen (1938) had collected feces from *L. arcticus groenlandicus* Rhodes, 1896, in Eastern Greenland and measured 760 oocysts, which he grouped into four types (A, B, C, D), with the first three each having subtypes (a–c) each represented by a different line drawing. He named his Type C (his Fig. 6b) *E. magna* var. *robertsoni* forma *semisculpta*. He cited the features that the oocysts of his form shared with those of

E. magna, but distinguished his by the thickened edge he saw around the M and, "only by the front part of the cyst being finely sculptured." That is, the top half of the oocyst in which the M is located showed a finely sculptured or stippled effect in optical cross section, but was not seen in the other half. In his brief description, he did not mention (p. 14), and then did mention (p. 33), the presence of an OR. His (1938) line drawing (Fig. 6b), showed a large, prominent OR as a ball of granules. On the one hand, the OR in his line drawing may have been a *lapus calami* (slip of the pen), if we are to believe later, credible authors, or these latter authors (Carvalho, 1943; Pellérdy, 1956, 1965, 1974) were remiss in including an OR as one of the distinguishing structural features of this species. To wit, Carvalho (1943), working with *L. townsendii campanius* Hollister, 1915, in Iowa, USA, found eimerian oocysts similar to those reported by Madsen (1938, as *E. m.* var. *robertsoni* forma *semisculpta*), but named them as a new form (forma *campanius*) of *E. irresidua* and gave a line drawing of a sporulated oocyst (his Fig. 15) that lacked an OR. Pellérdy (1956), who studied eimerians from *L. europaeus* in Hungary, emended Madsen's (1938) name to *E. semisculpta* Madsen, 1938 (emend.) and included as synonyms *E. i.* forma *campanius* Carvalho, 1943, and *E. m.* var. *robertsoni* forma *seimsculpta* Madsen, 1938, *pro parte*. Pellérdy (1956) said the ellipsoidal oocysts he studied were reminiscent of those of *E. irresidua* from the tame rabbit, with parallel walls, a broad M ~6–8 wide, and that in the anterior half of the oocysts the outer wall is markedly granulated, adding, "the boundary of the granulated layer is sharp, commonly oblique, and lies at about the middle of the oocysts." The oocysts he (1956) measured were 38×25 ($35–45 \times 22–27$) and lacked an OR, with sporocysts 18×9, each with a large SR. His later treatises (Pellérdy, 1965, 1974) retained that view, as did another redescription of *E. semisculpta* by Pakandl (1990), who said oocysts were 40.7 × 25.3 (37—46 × 23—29) with a M 4.3 (3.5—5.5), but no OR, and sporocysts were 19.8 × 9.4 (18—21 × 9—10). These latter reports may well have solidified the presence of a distinct, granular SR and the absence of a distinct OR in *E. semisculpta*. However, Aoutil et al. (2005) took an exception to Pellérdy (1956, 1965, 1974) and the existence of *E. semisculpta* as a name; they stated, "The taxon *semisculpta* is in our opinion invalid, since the type from Greenland differs from that of Hungary by its size and the presence of OR in the former but not the later. Thus validity of elevating the form "*semisculpta*" to species rank cannot be maintained." Obviously, one cannot just delete a species name that has existed in the literature for eight decades; however, the presence or absence of a large, distinct OR as a structural character has been used hundreds of times in the coccidian literature to help distinguish between species, and this difference, it seems to us, must redefine the distinguishing structures of *E. semisculpta* to be in agreement with Levine and Ivens (1972). They emended the name of the form from Carvalho (1943), "*E. irresidua* form *campanius*," to be "*E. campania* Carvalho, 1943" (see *Remarks* under *E. campania*, above) to include the *E. semisculpta* of Pellérdy, 1956 (a species without an OR), as a synonym, but not the *E. m.* var. *r.* forma *semisculpta* Madson, 1938, which has an OR. Golemanski (1975) said that the oocysts are 38.0 × 24.5 (35—43 × 23—27) with sporocysts 19—21 × 8—10. Unfortunately, some authors added little or no new or useful information. For example, Pastuzko (1961a) used Pellérdy's description (which is now *E. campania*) and Gräfner et al. (1967), Litvenkova (1969), and Arnastauskene and Kazlauskas (1970) all gave no description. The sporulated oocysts of this species also resemble, to a certain extent, those of *E. irresidua*. Berg (1981) reported this species in Norway, and Soveri and Valtonen (1983) reported it in Finland, both in the European hare, but neither gave any other information.

EIMERIA SEPTENTRIONALIS YAKIMOFF, MATSCHOULSKY & SPARTANSKY, 1936

FIGURE 5.48 Line drawing of the sporulated oocyst of *Eimeria septentrionalis* redrawn from Yakimoff et al., 1936 (original).

Synonyms: *Eimeria exigua* var. *septentrionalis* Yakimoff et al. (1936) of Madsen, 1938; *Eimeria babatica* Sugár (1978).

Type host: *Lepus timidus* L., 1758, Mountain hare.

Type locality: ASIA: Russia: Murmansk (Kola Peninsula).

Other hosts: *Lepus arcticus* Ross, 1819, Arctic hare; *Lepus europaeus* Pallas, 1778, European hare; *Lepus townsendii* Bachman, 1839, White-tailed jackrabbit.

Geographic distribution: ASIA: Russia; EUROPE: Czech Republic, Portugal; GREENLAND: near Eskimonaes; MIDDLE/NEAR EAST: Turkey; NORTH AMERICA: USA: Iowa.

Description of sporulated oocyst: Oocyst shape: ovoidal to subspheroidal; number of walls: 2; wall characteristics: ~1, smooth, with horn-like extensions on each side of M; L × W: 26.7 × 21.6 (24—32 × 20—22; Yakimoff et al., 1936) or 23.8 × 20.6 (23—32 × 20—23; Carvalho, 1943);

L/W ratio: 1.2 (1.1–1.5); M: present; M characteristics: 10–12 wide, surrounded by a prominent, thickened margin (lappets); OR, PG: both absent. Distinctive features of oocyst: wide M with horn-like extensions of wall (marginal lappets) on each side of it and lack of both OR and PG.

Description of sporocyst and sporozoites: Sporocyst shape: ovoidal, ellipsoidal to fusiform; L × W: 12–16 × 6–8; SB: likely present (Madsen, 1938; Carvalho, 1943), but not seen by Yakimoff et al. (1936); SSB, PSB, SR: all absent; SZ: sausage-shaped to comma-shaped and lie head-to-tail with one clear RB at thicker end. Distinctive features of sporocyst: absence of a SR.

Prevalence: Not given for the type host in Russia. Carvalho (1943) found it in 4/12 (33%) *L. townsendii* in Iowa. The numbers given by Madsen (1938) in Greenland cannot be used because he was dealing with more than one species. In the Czech Republic, Pakandl (1990) surveyed *L. europaeus* from 1983–1985 and found 98/350 (28%) infected with this species. Tasan and Özer (1989) said it occurred in 2/40 (6%) hares they surveyed in Turkey between 1985–1987.

Sporulation: Exogenous. Oocysts sporulated in 60 (55–65) hr (Carvalho, 1943).

Prepatent and patent periods: Unknown.

Site of infection: Unknown.

Endogenous stages: Unknown.

Cross-transmission: Carvalho (1943) was not able to transmit this species either to the domestic rabbit, *O. cuniculus*, or to the Iowa cottontail, *S. floridanus*.

Pathology: Unknown.

Material deposited: None.

Remarks: The description, above, is based on those of Yakimoff et al. (1936) and Carvalho (1943). Madsen (1938) reported this species under the name *E. exigua* var. *septentrionalis*, but *E. exigua* is an entirely distinct species that is smaller and does not possess a M. Madsen's statement, "Micropyle absent or, if present, broad, more or less indistinct, sometimes surrounded by a thickened margin," led Carvalho (1943) to believe that Madsen (1938) was dealing with two distinct species, an evaluation with which we concur. Matschoulsky (1941) said he found oocysts that resembled this species in *L. timidus* from Buryat-Mongol (former USSR) and that the ovoidal oocysts had a M ~8 wide, and were 27.2 × 21.9 (25–30 × 16.5–26) with sporocysts 10–13 × 5–7. Pakandl (1990) identified the oocysts he saw from hares in the Czech Republic (1983–1985) as *E. babatica*, but, as we point out, this species is actually *E. septentrionalis* (see *Remarks* under *E. babatica*, below). The oocysts measured by Pakandl (1990) were 25.7 × 18.9 (22–30 × 17–20.5) with a M, and an OR was either absent or present (?); sporocysts were 14.4 × 7.0 (12–16 × 5.5–8). Vila-Viçosa and Caeiro (1997) reported this species in *L. capensis* in Portugal, with ovoidal oocysts 22.5–24.3 × 20.6–22.5.

EIMERIA STEFANSKII PASTUSZKO, 1961a

Type host: *Lepus europaeus* Pallas, 178, European hare.

Type locality: EUROPE: Poland.

Other hosts: None reported to date.

Geographic distribution: ASIA: Lithuania (former USSR); EUROPE: Austria, (former) Czechoslovakia, Poland.

Description of sporulated oocyst: Oocyst shape: ellipsoidal to ovoidal, with narrow, tapering ends; number of walls: 1 (line drawing); wall characteristics: light yellow to dark brown with a distinct thickening around M; L × W: 59–68 × 32–37 (means not given); L/W ratio: not given; M: present; M characteristics: 7–9 wide, giving oocyst a slightly flattened appearance; OR: present; OR characteristics: a compact mass of large granules that markedly refracts light, ~16 wide; PG: absent. Distinctive features of oocyst: largest of the rabbit eimerians, elongated shape, and its large OR.

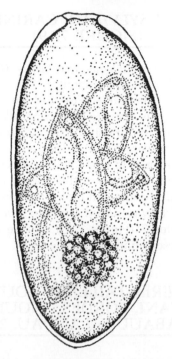

FIGURE 5.49 Line drawing of the sporulated oocyst of *Eimeria stefanskii* from Levine and Ivens, 1972, with permission of the *Journal of Protozoology.*

Description of sporocyst and sporozoites: Sporocyst shape: spindle-shaped to elongate ovoidal; L × W: 15 long (neither W, nor ranges given); SB: reported as absent, but it likely is present at pointed end of sporocyst (line drawing); SSB, PSB, SR: all absent; SZ: comma-shaped and lie head-to-tail in sporocyst with one clear RB at wide end. Distinctive features of sporocyst: spindle-shape and absence of SR.

Prevalence: 3/462 (0.6%) in the type host. Arnastauskene and Kazlauskas (1970) reported it in 1/41 (2%) *L. europaeus* in Lithuania.

Sporulation: Exogenous. Oocysts sporulate in 4–6 wk at 20–25°C.

Prepatent and patent periods: Unknown.

Site of infection: Small intestine.

Endogenous stages: Unknown.

Cross-transmission: None to date.

Pathology: Unknown.

Material deposited: None.

Remarks: This species has the largest sporulated oocysts known from any hare, to date, but the morphology of its sporulated oocyst still remains poorly defined even though it has been reported several times since Pastuszko first described and named it (1961a). It is very rare and, when found, it is always with negligible prevalence (e.g., in Austria by Kutzer and Frey, 1976; in Czechoslovakia by Chroust, 1979; in Lithuania by Arnastauskene, 1982).

EIMERIA STIEDAI (LINDEMANN, 1865) KISSKALT AND HARTMANN, 1907 (FIGURES 6.25, 6.26)

Synonyms: *Psorospermium oviforme* Remack, 1854; *Monocystis stiedae* Lindemann, 1865; *Psorospermium cuniculi* Rivolta, 1878; *Coccidium oviforme* Leuckart, 1879; *Eimeria oviformis* (Leuckart, 1879) Fantham, 1911.

Type host: *Oryctolagus cuniculus* (L., 1758), European (domestic) rabbit.

Remarks: This species was originally described from the domestic rabbit, *O. cuniculus*, by Lindemann (1865). Species of both *Lepus* (hares) and *Sylvilagus* (cottontails) have been reported to be infected with *E. stiedai* either naturally or experimentally or both. Thus, it seems that this is the only eimerian that can infect all three leporid genera, although in *Lepus* and *Sylvilagus* species it seems to be rare in nature. It should also be noted that at least some of these early reports are probably erroneous. For example, in 1932 Boughton (p. 535) said he found *E. stiedai* in the American hare, *L. americanus*, in Western Canada, but gave no facts or evidence to support his identification, and Henry (p. 282) said that she saw this eimerian (and three others reported from the domestic rabbit) in one Black-tailed jackrabbit from California, USA, again with no evidence. Ryšavý (1954) said he found this species in both *O. cuniculus* and in *L. europaeus* in the Czech Republic with oocysts that were 34.4 × 19.3 (30–40 × 15–21) and sporocysts 13–18 × 7.7, and that these oocysts sporulated in 60–75 hr. The parasite was found in the liver and was abundant in the domestic rabbits they sampled (Ryšavý, 1954). Varga (1976) infected nine hares (*L. europaeus*) with *E. stiedai* oocysts from *O. cuniculus* and found that each of them had macroscopic liver lesions after infection. Scholtyseck et al. (1979) transferred sporulated oocysts of *E. stiedai* from *O. cuniculus* to six young, coccidia-free *L. europaeus* and detected all stages of the endogenous cycle from sporozoites to oocysts by both light and electron microscopy. Mandal (1976) said he recovered oocysts of *E. stiedai* in a *Lepus* sp. from Kashipur, India, with ovoidal to ellipsoidal oocysts that were 26–40 × 16–24, without an OR, and that it had ovoidal sporocysts, 17 × 9, with a SR; these oocysts sporulated in 72 hr. Litvenkova (1969) said, "Of 52 hares (*Lepus europaeus* L. and *L. timidus* L.) coccidian oocysts were found in 22 (42.3%)," but she did not say whether *E. stiedai* was found in one or both of the *Lepus* species. There seems to be a sufficient number of observations of *E. stiedai* endogenous stages in the liver of *Lepus* species, but it would be better to have molecular confirmation of its species status in *Lepus* species. Finally, in an unpublished Master's thesis, Ogedengbe (1991) said he isolated this species in the gall bladders of three of the rabbits surveyed in Kaduna State, Nigeria. However, he (1991) neglected to state how many rabbits were surveyed, nor did he mention the host species; Anonymous (2012) lists the Cape hare, *L. capensis*, as the only lagomorph species found in Nigeria. Vila-Viçosa and Caeiro (1997) reported this species in both *O. cuniculus* and *L. capensis* in Portugal, with the ellipsoid oocysts from *O. cuniculus* being 33–37.5 × 16.8–19.5. Aoutil et al. (2005) reported *E. stiedai* in *L. europaeus* in France; their oocysts were 37 × 22 (33–40 × 19–24) with sporocysts, 18 × 10. See Chapter 6 (Figures 6.25, 6.26) for the complete description of *E. stiedai*.

EIMERIA SYLVILAGI CARINI, 1940 (FIGURE 7.19)

Type host: *Sylvilagus brasiliensis* (L., 1758), Tapeti or Brazilian cottontail.

Remarks. Mandal (1976) said he recovered oocysts of this species in *L. nigricollis* (syn. *L. ruficaudatus*) from Kashipur, India, with ovoidal oocysts that were 21–39 × 16.2, without an OR, and had ovoidal to ellipsoidal sporocysts, 16 × 7, with a SR. We suspect this was an erroneous identification and await molecular confirmation that *E. sylvilagi* is indeed found in *Lepus* species. See Chapter 7 (Figure 7.19) for the complete species description of this species.

EIMERIA TAILLIEZI AOUTIL, BERTANI, BORDES, SNOUNOU, CHABAUD & LANDAU, 2005

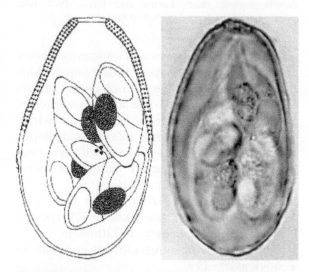

FIGURES 5.50, 5.51 Line drawing of the sporulated oocyst of *Eimeria tailliezi* from Aoutil et al., 2005, with permission of the authors and from (the journal) *Parasite*. Photomicrograph of a sporulated oocyst of *E. tailliezi* from Aoutil et al., 2005, with permission of the authors and from (the journal) *Parasite*.

Type host: *Lepus europaeus* Pallas, 1778, European hare.

Type locality: EUROPE: France: Pyrénées-Orientales.

Other hosts: None to date.

Geographic distribution: EUROPE: France.

Description of sporulated oocyst: Oocyst shape: elongate ellipsoidal; number of walls: 1 (line drawing), but probably 2; wall characteristics: thick, sculptured, dark brown anterior half, often asymmetrical, with "one edge slightly concave and the other more convex;" L × W: 42 × 23 (40–43.5 × 22–24.3); L/W ratio: 1.8; M: present, flat; M characteristics: 6 wide; OR: absent or may be present as only 2–3 small grains; PG: absent. Distinctive features of oocyst: elongate ellipsoidal shape, dark, sculptured outer wall only on anterior half of oocyst, an OR of only 2–3 grains.

Description of sporocyst and sporozoites: Sporocyst shape: elongate-ovoidal, pointed at one end (line drawing); L × W: 19 × 9 (ranges not given); L/W ratio: 2.1; SB: small, at pointed end; SSB, PSB: both absent; SR: present; SR characteristics: ellipsoidal compact mass of small grains (line drawing), ~6 long; SZ: with one round RB at rounded end. Distinctive features of sporocyst: small SB, SR an ellipsoid compact mass of small granule, and SZ with one spheroidal RB.

Prevalence: Found in 1/9 (11%) *L. europaeus*.

Sporulation: Presumably exogenous, but the time and temperature are not known.

Prepatent and patent periods: Unknown.

Site of infection: Unknown. Oocysts recovered from feces.

Endogenous stages: Unknown.

Cross-transmission: None to date.

Pathology: Unknown.

Material deposited: None, although the authors designated their photomicrograph (their Fig. 3.20) as the holotype and said "the photos of the oocysts … have been deposited at the laboratory."

Remarks: Aoutil et al. (2005) found and measured only eight oocysts from one rabbit to describe this species; they provided a line drawing and a photomicrograph in their original description, but they did not deposit either of these in an <u>accredited</u> museum, nor did they provide all of the quantitative and qualitative information needed and expected in a modern, standardized species description (e.g., age of the oocysts when studied, sporulation time and temperature, sporocyst L, W ranges; measurements of SZ, RB, etc.). The authors argue that this species can be compared in size and shape only to *E. neoleporis* from the cottontail rabbit, *S. floridanus*, in Iowa, USA; oocysts of *E. neoleporis* are elongate-ellipsoidal, 38 × 19.8, L/W ratio 2.0, while those of *E. tailliezi* are ovoidal and larger, 42 × 23, L/W ratio 1.8. Aoutil et al. (2005) listed the name as "*Eimeria tailliezi* n. sp. Aoutil & Landau"; this is not, and cannot be, the correct authority for this species, since there is no separate published paper by Aoutil and Landau describing and naming it. Thus, the full scientific name (genus, specific epithet, authority) must be as we listed it above. Finally, based on only eight sporulated oocysts from only one rabbit, and on the shortcomings of the species description noted here, this form might best be relegated to a *species inquirenda* until more information is known about its morphology and biology.

EIMERIA TOWNSENDI (CARVALHO, 1943) PELLÉRDY, 1956

Synonyms: *Eimeria magna* forma *townsendii* Carvalho, 1943 *pro parte*; *Eimeria magna* Pérard, 1925 of Henry, 1932, *pro parte*; *Eimeria townsendi* Pellérdy, 1956; *Eimeria townsendi* (Carvalho, 1943) Pellérdy, 1956 of Pellérdy, 1965, 1974.

Type host: *Lepus townsendii* Bachman 1839, White-tailed jackrabbit.

Type locality: NORTH AMERICA: USA: Iowa.

Other hosts: *Lepus americanus* Erxleben, 1777, Snowshoe hare; *Lepus europaeus* Pallas, 1778,

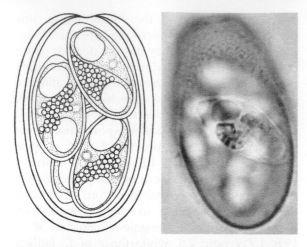

FIGURES 5.52, 5.53 Line drawing of the sporulated oocyst of *Eimeria townsendi* from Samoil and Samuel, 1977a, with permission from the *Canadian Journal of Zoology*. Photomicrograph of a sporulated oocyst of *E. townsendi* from Samoil and Samuel, 1977a, with permission from the *Canadian Journal of Zoology*.

European hare; *Lepus timidus* L., 1758, Mountain hare.

Geographic distribution: EUROPE: Austria, Czech Republic, Finland, (former East) Germany, Hungary, Italy, Norway, Poland, Switzerland; NORTH AMERICA: Canada: Alberta; USA: California, Iowa.

Description of sporulated oocyst: Oocyst shape: broadly ellipsoidal or ovoidal; number of walls: 2; wall characteristics: outer layer is ~1.4 thick, finally granular, yellow-brown, darker and thicker than inner and peels off after a few days in $K_2Cr_2O_7$ solution; inner is ~0.5 thick; L × W: 40 × 23 (36–49 × 19–24) (Samoil and Samuel, 1977a); L/W ratio: 1.7; M: present, 5.6 (5–7) wide; OR: either present (Carvalho, 1943) or "invariably absent" (Pellérdy, 1956; Samoil and Samuel, 1977a); PG: absent. Distinctive features of oocyst: thicker outer wall layer that peels off after storage.

Description of sporocyst and sporozoites: Sporocyst shape: elongate spindle-shaped, but line-drawing shows them to be ellipsoidal, pointed at one end; L × W: 18.4 × 9.1 (17–20 × 8–10); L/W ratio: 2.0; SB: present at slightly pointed end of sporocyst; SSB, PSB: both absent; SR: present; SR characteristics: abundant disbursed granules; SZ: elongate-ovoidal and lie head-to-tail with one central, large RB at wider end. Distinctive features of sporocyst: spindle-shape with SR of abundant disbursed granules.

Prevalence: 3/12 (25%) in the type host; Pastuszko (1961a) found this species in 150/462 (32%) *L. europaeus* in Poland, Gräfner et al. (1967) in 1/176 (0.5%) *L. europaeus* in (former East) Germany, Sugár et al. (1978) in 6/374 (<2%) *L. europaeus* in Hungary, and Bouvier (1967) found it rarely in *L. europaeus* in Switzerland. Kutzer and Frey (1976) found this species in 37% of European hares in Austria, while Jirouš (1979) found it in 33% of rabbit fecal samples collected in the Czech Republic. Samoil and Samuel (1977a) reported it in 314/620 (50%) *L. americanus* in Alberta, Canada. Also in the Czech Republic, Pakandl (1990) surveyed *L. europaeus* from 1983–1985 and found 244/350 (70%) infected with this species.

Sporulation: Exogenous. Sporulation is completed in most oocysts after 6 days at 20°C in 2.5% (w/v) aqueous $K_2Cr_2O_7$ solution (Samoil and Samuel, 1977a).

Prepatent and patent periods: Unknown.

Site of infection: Uncertain, since oocysts were recovered from feces in most studies. Pastuszko (1961a) thought this species was localized in the small intestine and Pellérdy (1974) said "this coccidium probably localizes in the small intestine," but Bouvier (1967) said it occurs in the cecum. Pakandl (1990) looked at three digestive tracts in which he found large numbers of oocysts and gametocytes in the cecum and anterior colon. When he compared these to mucosal scrapings from the same site he concluded these were developmental stages of *E. townsendi*, but it is not clear how he made that decision since the endogenous stages of *E. townsendi* are not known.

Endogenous stages: Unknown.

Cross-transmission: Carvalho (1943) was not able to transmit this species either to the domestic rabbit, *O. cuniculus*, or to the Iowa cottontail, *S. floridanus*. Pellérdy (1954) was not able to infect the domestic rabbit with this species isolated from *L. europaeus*.

Pathology: Unknown, but Bouvier (1967) said it was "null." Tacconi et al. (1995) found oocysts of *E. townsendi* in the intestinal contents of four dead juvenile *L. europaeus* in Italy. The oocysts were always found with those of *E. europaea, E. hungarica*, and *E. robertsoni* in these dead hares, which all showed severe enteritis as the cause of death but, obviously, it was impossible to determine which of these species was the most pathogenic or if the pathology was caused by the synergistic effect(s) of infection with all four species.

Material deposited: None.

Remarks: Carvalho (1943), working with *L. townsendii campanius* Hollister, 1915, in Iowa, USA, found eimerian oocysts in three jack rabbits that were similar to those of *E. magna* known to be a parasite of *O. cuniculus*. He tried to infect six young, tame rabbits and two cottontails (*Sylvilagus*), but could not get patent infections. For this reason, he chose to name the oocysts he found as *E. magna* form *townsendii*, rather than to name a new species. Unfortunately, when he created this new "form" he made two mistakes that confused some later workers. First, he created the "form" based strictly on the negative results of his cross-transmission experiments and failed to mention any morphological differences between his new "form" and the parent species to which it was attached; and second, he referred the reader to a line drawing of *E. magna* (his Plate I, Fig. 1) that he had used earlier to define it when discussing the differences between "The Coccidia of the Tame Rabbit, *Oryctolagus cuniculus* (Linnaeus)." Fortunately, Pellérdy (1956), who studied eimerians from *L. europaeus* in Hungary, emended Carvalho's (1943) name to *E. townsendi* (not *townsendii*), stated that the "extra-residual body is invariably absent" from the ovoidal oocysts, and provided the first line drawing (his Fig. 17) to demonstrate its absence. Later workers (Bouvier, 1967; Levine and Ivens, 1972; Samoil and Samuel, 1977a; Tacconi et al., 1995) have supported Pellérdy's (1956) correction of Carvalho's (1943) omission. However, some authors (e.g., Aoutil et al., 2005) still express confusion, thinking that Carvalho's (1943) reference to a line drawing of *E. magna*, to which he compared his new "form," meant that sporulated oocysts of his "form *townsendi*" literally had an OR, which was never stated in his 1943 paper. Thus, Aoutil et al. (2005) wrote, "The taxon *E. townsendi* must, therefore, be abandoned." Obviously, one cannot just delete a species name that has existed in the literature for more than 50 years, especially given the verification of Pellérdy's (1956) observation by many later authors (Bouvier, 1967; Levine and Ivens, 1972; Pakandl, 1990; Samoil and Samuel, 1977a; Tacconi et al., 1995). In addition to providing both a line drawing and a photomicrograph of this species, Samoil and Samuel (1977a) documented the presence of a SB as "an indistinct light-refracting area at the narrow end" of the sporocyst, a structure that Pellérdy (1956) did not see or describe. Pellérdy (1956) said the oocysts were ovoidal and measured 40×28 ($37-44 \times 28-31$). Pastuszko (1961a) said the ovoidal oocysts were 43×31 ($37-54.5 \times 25-39$) with no OR, but with a M and a SR. The oocysts measured by Pakandl (1990) were 36.0×22.7 ($32.5-39 \times 21-25$) with a M 4.9 (4–6), but no OR, and its sporocysts were 17.0×8.9 ($16-19 \times 8-11$). These oocysts are smaller than those given by other authors (Pellérdy, 1956, 1974; Pastuszko, 1961a; Sugár et al., 1978). Berg (1981) reported this species in Norway, and Soveri and Valtonen (1983) reported it in Finland, both in the European hare, but neither gave any other information. Tacconi et al. (1995) found this species in the intestinal contents of four dead juvenile *L. europaeus* from a breeding farm in Italy. Their ovoidal oocysts were 36–44 × 25–31 and both M and OR were absent; this species was always found with *E. europaea*, *E. hungarica*, and *E. robertsoni* in these dead hares. Terracciano et al. (1988) identified this species in *L. europaeus* from two protected areas in the Province of Pisa, Italy. The ovoidal oocysts they studied were 36–44 × 25–31 with a M and SR, but lacking an OR.

SPECIES INQUIRENDAE (7)

Eimeria babatica Sugár, 1978

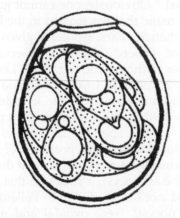

FIGURE 5.54 Line drawing of the sporulated oocyst of *Eimeria babatica* from Sugár, 1978 from *Parasitologida Hungarica* (journal no longer exists); this species is a junior synonym of *E. septentrionalis*.

Synonym: *Eimeria babatica* Aoutil, Bertani, Bordes, Snounou, Chabaud and Landau, 2005

Original host: *Lepus europaeus* Pallas, 1778, European hare.

Remarks: Sugár (1978) described this as a new species from the European hare in Hungary. Sporulated oocysts were broadly- or elongate-ovoidal with a two-layered wall 0.8–0.9, and measured 25 × 19 (20–30 × 16–21), with L/W ratio: 1.3. A distinct M was present ~6–8 wide, with a conspicuous, collar-like projection. Both

an OR and PG were absent. Sporocysts were elongate-ovoidal, 14 × 7 (13–15 × 6–7.5), L/W ratio: 2.0. Both a SB and SR were present, but a SSB and PSB were both absent. SZ were 11–11.5 × 4.2–4.6 with one large and one smaller RB. He compared *E. babatica* with other species of *Eimeria* from *Lepus* species. Initially, in his 1978 publication, he found *E. babatica* to differ from *E. septentrionalis* by being smaller in width, having a smaller M and having a SR. However, in a personal letter he wrote to Dr. Steve J. Upton (formerly of Kansas State University) that was received sometime after 1978 (Sugár's letter was not dated) he wrote, "Now I think the *Eimeria babatica* isn't a new species, as it is identical with *E. septentrionalis*. Sorry, but in Pellérdy's monography the description isn't complete enough and later—after my article (was) published—I read a more complete description by Kutzer." Aoutil et al. (2005) said they found this (non) species in *L. granatensis* from France, but noted that their form differed from the original description by having a smaller M and not having a SR. They, like Sugár (1978), likely were observing oocysts of *E. septentrionalis* and, of course, they had no knowledge of Sugár's personal letter to Dr. Upton that discounted the existence of *E. babatica*.

Eimeria belorussica Litvenkova, 1969

Original host: *Lepus europaeus* Pallas, 1778, European hare.

Remarks: This species was described in an abstract presented in the Third International Congress on Protozoology, Leningrad, by Litvenkova (1969), who said she found it in rabbits in Byelorussia, one of 15 constituent in republics of the former Soviet Union (USSR), now the Republic of Belarus since 1991. Unfortunately, Litvenkova (1969) said, "Of 52 hares (*Lepus europaeus* L. and *L. timidus* L.) coccidian oocysts were found in 22 (42.3%)," but she did not say whether this form was found in one or both of the *Lepus* species. Pellérdy (1974) hints that he also found this species in *L. timidus* in Hungary. The sporulated oocyst was described as ovoidal to ellipsoidal with a two-layered wall that is pale brown; L × W: 26–28 × 14–16; L/W ratio: 1.8; M: present on flattened end; sporocysts were described only as ovoidal and reported to sporulate within 60 hr. To our knowledge, however, neither a line drawing nor a photomicrograph exists for this form, so it must be relegated as a *species inquirenda*.

Eimeria gresae Aoutil, Bertani, Bordes, Snounou, Chabaud & Landau, 2005

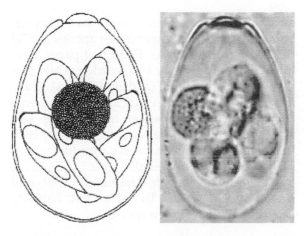

FIGURES 5.55, 5.56 Line drawing of the sporulated oocyst of *Eimeria gresae* from Aoutil et al., 2005, with permission of the authors and from (the journal) *Parasite*. Photomicrograph of a sporulated oocyst of *E. gresae* from Aoutil et al., 2005, with permission of the authors and from (the journal) *Parasite*.

Original host: *Lepus granatensis* Rosenhauer, 1956, Granada hare.

Other hosts: *Lepus europaeus* Pallas, 1778, European hare.

Remarks: This species was described from rabbits collected in France. The oocysts were slightly ovoidal, slightly flattened at the end with the M, and had a thin, smooth, light yellow wall that thickens slightly around the M. They measured 33 × 22 (28–39 × 20–24), L/W ratio: 1.5, had a M ~8 wide, with a slightly bulging dome, and an OR that was a spheroidal mass of coarse granules ~9 wide; there was no PG. Sporocysts were ellipsoidal, 19 × 8.5 (ranges not given), with a small SB at the pointed end, but SSB, PSB, and SR were absent; SZ each have two RBs, one larger than the other. The photomicrographs presented by the authors for this form (their Fig. 1.12) and for *E. coquelinae* (their Fig. 1.11) look somewhat different because of the adherent intestinal material around the M in their Fig. 1.11. However, the oocyst and sporocyst mensural features and L/W ratios, along with all other qualitative features (e.g., large OR present, SR absent, etc.), are identical. Given the natural variability of oocysts of most known eimerians from mammals, especially rabbits, we think these may be the same species and prefer to wait until definitive molecular data can distinguish between them, or not.

Eimeria mazierae Aoutil, Bertani, Bordes, Snounou, Chabaud & Landau, 2005

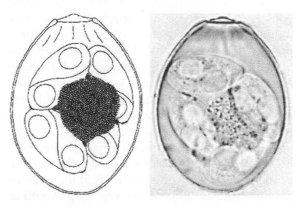

FIGURES 5.57, 5.58 Line drawing of the sporulated oocyst of *Eimeria mazierae* from Aoutil et al., 2005, with permission of the authors and from (the journal) *Parasite*. Photomicrograph of a sporulated oocyst of *E. mazierae* from Aoutil et al., 2005, with permission of the authors and from (the journal) *Parasite*.

Original host: *Lepus granatensis* Rosenhauer, 1956, Granada hare.

Remarks: This species was described from rabbits collected in France. The oocysts were broadly ovoidal, slightly flattened at end with the M, and had a thin, smooth, light yellow wall that thickens slightly around the M. They measured 40 × 29 (39—41 × 29—31); L/W ratio: 1.4, had a M ~8 wide, that was pyramidal in shape, and an OR that was a large irregular mass ~14—15 wide, that showed diverticula extending between sporocysts; there was no PG. Sporocysts were ellipsoidal, 18 × 10 (ranges not given), with a small SB at the pointed end, but SSB, PSB, and SR were absent. Aoutil et al. (2005) named and described this species from only two oocysts. Given the normal variability of oocysts from all eimerian species discharged by mammals in general, and rabbits in particular, we believe there is not enough quantitative and qualitative morphological information available to name this form a valid species. Although Aoutil et al. (2005) provided a line drawing and a good photomicrograph of this form in their description, they did not provide all of the quantitative and qualitative information needed and expected in a modern, standardized species description (e.g., age of the oocysts when studied, sporulation time and temperature, sporocyst L, W ranges; measurements of SZ, RB, etc.). This is disappointing and below acceptable standards; for these reasons we must conclude that this form be relegated as a *species inquirenda* until sufficient morphological data are available to characterize it accurately enough to evaluate its status vis-à-vis other established species. Finally, in presenting their description of this new species in their published paper (Aoutil et al., 2005) the authors list the name as "*Eimeria mazierae* n. sp. Aoutil & Landau"; this is not, and cannot be, the correct authority for this species, since there is no separate published paper by Aoutil and Landau describing and naming it. Thus, the full scientific name must be as we have listed it above.

Eimeria sp. Golemanski, 1975

Original host: *Lepus europaeus* Pallas, 1778, European hare.

Remarks: Golemanski (1975) found this form in 32% of an unstated number of hares in Bulgaria. The ovoidal oocysts were 25 × 18 (18—30 × 15—20) with a M and a MC; sporocysts were 11—16 × 7—8. The oocysts sporulated in 28—56 hr at 23°C in 3% $K_2Cr_2O_7$ solution. Nothing else is known about this species.

Eimeria sp. No. 1 Gvozdev, 1948

Original host: *Lepus tolai* Pallas, 1778, Tolai hare.

Remarks: Levine (pers. comm., see Acknowledgments) found a paper by Svanbaev (1979) that is unavailable to us. From that paper, Levine cited another paper by Gvozdev (1948) that was unavailable to him (and also to us). According to Levine, Svanbaev (1979) said that Gvozdev (1948) found oocysts representing two forms of *Eimeria* in the feces of the Tolai hare in Kazakhstan (former USSR). The first group of oocysts were ovoidal, yellow, 32—36 × 23—25 with a "double-contoured" wall and a well-formed M, but lacked an OR. Nothing else is known about this species.

Eimeria sp. No. 2 Gvozdev, 1948

Original host: *Lepus tolai* Pallas, 1778, Tolai hare.

Remarks: Levine (pers. comm., see Acknowledgements) found a paper by Svanbaev (1979) that is unavailable to us. From that paper Levine cited another paper by Gvozdev (1948) that was unavailable to him (and also to us). According to Levine, Svanbaev (1979) said that Gvozdev (1948) found oocysts representing two forms of *Eimeria* in the feces of the Tolai hare in Kazakhstan (former USSR). This second group consisted of oocysts that were spheroidal 14—15 wide, or ovoidal 14—18 × 10—13, with a well-defined M, and the sporocysts had a distinct SR. Nothing else is known about this species.

DISCUSSION AND SUMMARY

Coccidiosis in Hares

Pellérdy (1974) said that clinical disease and/or even lethal coccidiosis develops in hares only under special circumstances such as their transport and/or confinement to limited space for breeding or other purposes that could favor massive infection with sporulated oocysts. He also felt that younger animals were more susceptible than older ones. The main symptom of the disease in hares and jack rabbits is diarrhea, which can become lethal.

There are isolated reports that at least five *Eimeria* species found in *Lepus* species are associated with moderate to severe pathology including *E. europaea*, *E. hungarica*, *E. leporis*, *E. robertsoni*, and *E. townsendi*.

Eimeria europaea

Levine and Ivens (1972) said that *E. europaea* is "possibly" pathogenic, while Bouvier (1967) said that *E. europaea* may cause serious lesions and hemorrhage in the cecum of young hares, but more discrete lesions in adults. Tacconi et al. (1995) found this species in the intestinal contents of four dead juvenile *L. europaeus* in Italy. The oocysts were always found with oocysts of *E. hungarica*, *E. robertsoni*, and *E. townsendi* in these dead hares, which all showed severe enteritis, but there was no way to determine which species caused the pathology and death or whether it was the synergistic or overwhelming presence of all four species.

Eimeria hungarica

Bouvier (1967) recovered *E. hungarica* from a European hare in Switzerland and said that it had severe enteritis. Post-mortem examination of the "**cachectic** carcass" revealed extensive epithelial desquamation and hemorrhagic inflammation in the intestine. Numerous masses of (unidentified) endogenous stages were seen microscopically and the cause of the enteritis was purported to be a mixed infection with *E. leporis* and *E. robertsoni*, but chiefly, *E. hungarica*. Tacconi et al. (1995) found this species in the intestinal contents of four dead juvenile *L. europaeus* in Italy. The oocysts were always found with oocysts of *E. europaea*, *E. robertsoni*, and *E. townsendi* in these dead hares, which all showed severe enteritis, but there was no way to determine which species caused the pathology and death or whether it was the synergistic or overwhelming presence of all four species.

Eimeria leporis

Bouvier (1967) said that *E. leporis* may cause severe enteritis, sometimes hemorrhagic, especially in young hares. Pellérdy et al. (1974) suggested that *E. leporis* was responsible for the death of 14/227 (6%) hares in Hungary that died spontaneously of coccidiosis. The main gross lesions consisted of grayish-white foci 1–2 mm wide along the entire length of the small intestine, especially the ileum. The villi were enlarged and the focal lesions were often confluent. The intestinal capillaries also were noted to be dilated and congested. Pellérdy et al. (1974) killed two young hares, which already had what they said was a "slight" infection with *E. leporis*, by administering 2 million sporulated oocysts to each of them. They concluded that the hare's immune response to this eimerian is very weak, and that this low immunogenicity may account for the high prevalence of this species in hares.

Eimeria robertsoni

Madsen (1938) said that one *L. arcticus* fed an unspecified number of oocysts of *E. robertsoni* died one week later with emaciation, suggestions of diarrhea (whatever that means), a mucosanguine exudate in the posterior small intestine, and an enormously swollen urinary bladder.

Eimeria townsendi

Bouvier (1967) said that pathology due to this species was "null," while Tacconi et al. (1995) found this species in the intestinal contents of four dead juvenile *L. europaeus* in Italy. The oocysts of this species were always found with oocysts of *E. europaea*, *E. hungarica*, and *E. robertsoni* in these dead hares, which all showed severe enteritis as the cause of death, but there was no way to determine which species caused the pathology and death or whether it was the synergistic or overwhelming presence of all four species.

Finally, Gološin and Tešić (1963) and Gološin et al. (1963) reported "considerable" losses from coccidiosis (species not identified) among hares exported from Yugoslavia to Italy and France.

Host Specificity In *Lepus* Eimerians

The degree of **host specificity** varies from host group to host group; it has been studied best in mammals, and to a lesser degree in birds, especially domesticated stock/flock animals. There currently are two lines of evidence used to help determine the host specificity of a coccidian species: well-controlled cross-transmission studies and finding the parasite in one or more host species different from those in which it was originally described. Experimental cross-transmission studies can prove conclusively whether or not a coccidium can live and reproduce in another host species. Finding oocysts (which then must be sporulated before identification can be attempted) in the feces of a host different from the one in which it was originally recorded can suggest a lack of host specificity, but such evidence has its drawbacks when working with coccidia. There are many reasons for this: sporulated oocysts have only a small suite of structural characters that can help

distinguish one species from another; oocysts from different host species can have very similar morphometrics even though they are different species (Upton et al., 1992); sporulated oocysts of some species may naturally vary greatly in size (up to 40%) when discharged over the length of the patent period and, thus, appear to be different (Duszynski, 1971; Gardner and Duszynski, 1990; Duszynski et al., 1992); oocysts produced by multiple gamonts developing in a single host cell tend to be smaller than oocysts that develop individually in one host cell (Cheissin, 1967); the size of the inoculating dose, the immune and nutritional status of the host, and numerous other biotic interactions, particularly the genome of both parasite and host, must all additionally contribute to the host specificity, or lack thereof, attributed to each *Eimeria* species. Thus, some coccidia can be exceedingly host specific, being limited only to one host species (as far as we currently know); some may cross species, but not genus boundaries; some can be transmitted to and/or be found in sister genera in the same host family; and one species has even been reported to cross familial lines, but this seems rare (De Vos, 1970).

Ten eimerians for which a *Lepus* species is the type host have been attempted to be experimentally transmitted to the domestic rabbit, *O. cuniculus*, and six of these same species have been attempted to be transmitted to the Iowa cottontail, *S. floridanus* (see text above and Table 11.2). In every cross-transmission attempt the results were negative, indicating that *Lepus* eimerians are very host specific for members of their host genus, at least based on the evidence available to us now (Nieschulz, 1923; Carvalho, 1943; Pellérdy, 1954a, b, 1956; Lucus et al., 1959; Samoil and Samuel, 1977b).

Reiterating these studies, Carvalho (1943) was not able to transmit *E. americana*, *E. campania*, *E. robertsoni*, *E. sculpta*, *E. semisculpta*, *E. septentrionalis*, or *E. townsendii* either to the domestic rabbit or to the Iowa cottontail with oocysts isolated from *L. townsendii*. Pellérdy (1954a, b, 1956) could not infect the domestic rabbit with *E. campania, E. europaea, E. hungarica, E. leporis, E. robertsoni, E. semisculpta,* or *E. townsendi* isolated from *L. europaeus*. Nieschulz (1923) and Lucas et al. (1959) were unable to infect domestic rabbits with *E. leporis* isolated from *L. europaeus*. Samoil and Samuel (1977b) were not able to infect *O. cuniculus* with various numbers of sporulated oocysts (100, 30,600, or 50,000) of *E. robertsoni* isolated originally from a naturally infected *L. americanus*.

Four eimerians for which *Sylvilagus* species are the type host have been reported in various surveys of *Lepus* species. Gill and Ray (1960) reported *E. minima* in *L. nigricollis*, but we and others (Levine and Ivens, 1972) think they likely were dealing with *E. hungarica*. Aoutil et al. (2005) in France recorded what they called *E. auduboni* in *L. granatensis*, but admitted it was doubtful that the form they saw was identical to that found in an American *Sylvilagus*; we believe it is likely a new species from *Lepus*, and agree it is not *E. audubonii* from *Sylvilagus*. Mandal (1976) said he recovered oocysts of *E. neoleporis* in a *Lepus* sp. from Punjab, India, and oocysts of *E. sylvilagi* in *L. nigricollis* from Kashipur, India, but their validity as a true parasites of any *Lepus* species seems unlikely to us and needs to be verified by experimental and/or molecular work.

Nine eimerians for which *O. cuniculus* is the type host have been reported in various surveys of *Lepus* species. Ryšavý (1954) said he found *E. exigua* in *L. europaeus* in the Czech Republic. Mandal (1976) reported oocysts of *E. intestinalis* in the intestine of a *Lepus* sp. from Kashipur, India. *Eimeria irresidua* was reported in both *L. nigricollis* and the domestic rabbit by Gill and Ray (1960), but they only provided measurements for sporulated oocysts from domestic rabbits, and Ryšavý (1954) said he found it in the small intestine of *L. europaeus* in the Czech Republic. *Eimeria magna* was reported in three species of *Lepus*: *L. californicus* by Henry (1932), *L. europaeus* by Ryšavý (1954), and *L. nigricollis*

by Gill and Ray (1960), but the latter only gave measurements for sporulated oocysts in the domestic rabbit. Mandal (1976) reported oocysts of *E. matsubayashii* in *Lepus* sp. from Kashipur, India, but its validity as a parasite of *Lepus* species needs to be verified. *Eimeria media* was reported by Henry (1932) from one *L. californicus* in California, and Gill and Ray (1960) reported finding it in both the domestic rabbit and in *L. nigricollis* in India, saying that oocysts collected from the rectal contents of wild hares were "morphologically indistinguishable" from oocysts taken from *O. cuniculus*. Matschoulsky (1941) said he found oocysts that resembled *E. media* in *L. timidus* from Buryat-Mongol (former USSR). All of these reports supply few or no data to support these identifications. Boughton (1932) said he found oocysts of *E. perforans* in *L. americanus* in Western Canada, Henry (1932) said she saw *E. perforans* oocysts in one *L. californicus* from California, Robertson (1933) reported *E. perforans* in one *L. europaeus* from England, and Ryšavý (1954) reported its oocysts both in *O. cuniculus* and in *L. europaeus* in the Czech Republic, all with no supporting evidence. Gill and Ray (1960) reported finding *E. perforans* both in the domestic rabbit and in *L. nigricollis* at Kashipur, India, but, once again, the measurements they gave for sporulated oocysts were only for oocysts collected from domestic rabbits. Ryšavý (1954) said he found *E. piriformis* both in *O. cuniculus* and in *L. europaeus* in the Czech Republic. Finally, Boughton (1932, p. 535) said he found oocysts of *E. stiedai* in the American hare, *L. americanus*, in Western Canada, but gave no evidence to support his identification, and Henry (1932, p. 282) said that she saw *E. stiedai* in one Black-tailed jackrabbit from California, while Ryšavý (1954) said he found it in the liver of both *O. cuniculus* and in *L. europaeus* in the Czech Republic, where it was abundant in the domestic rabbits he sampled. Of these nine species for which *O. cuniculus* is the type host, only *E. stiedai* seems to be a valid parasite of *Lepus* species, having been found in the liver hepatocytes of *Lepus* species in surveys and having been transmitted both to *Lepus* and *Sylvilagus* species experimentally (above and Chapters 6, 7).

Parasite Discovery

Of the ten *Lepus* species from which eimerians have been named, *L. californicus* does not have any valid species known from it; this species is common and widely distributed throughout northern Mexico and the western USA, and needs to be examined over its range to determine what coccidia species it harbors. The remaining nine *Lepus* species that have been examined for coccidia have from three to 17 *Eimeria* species known from them as valid parasites (Table 5.1). As expected, *L. tolai*, which has been examined the fewest number of times, has only three eimerians described from it, while *L. europaeus*, which has been surveyed the most extensively, has 17 eimerians known from it. Of the 31 valid eimerians that infect

Lepus species, 15 are known from only a single host species, six are known to infect two *Lepus* species, four infect three *Lepus* species, three infect four *Lepus* species, one infects six, one infects seven, and one infects eight *Lepus* species (Table 5.1). These results are dependent, of course, on the number of individuals, the number of times, and the number of localities from which each host species has been sampled. Nonetheless, we think from these data and our experience it is fair to say that most eimerians in *Lepus* may travel freely between species within the genus (especially *E. robertsoni*, *E. stiedai*, and *E. leporis*). Finally, 22 of 32 (69%) of the extant *Lepus* species have never been examined for coccidia and six of the 10 species that have been examined have been sampled three or fewer times (Table 5.1). If we cautiously assume that each species of *Lepus* has at least four eimerians that are unique to it, then there still may be about 100 new eimerians yet to be discovered once all *Lepus* species have been systematically examined. There is still a lot of work to do!

TABLE 5.1 All Known Coccidia Species[1] (Apicomplexa: Eimeriidae) Reported from Jackrabbits/Hares in the Genus *Lepus* L., 1758 (Family Leporidae)

Lepus spp.	*Eimeria* spp.	Natural (N) or experimental (E)	References
americanus	*athabascensis*	N	Samoil & Samuel, 1977a
	holmesi	N	Samoil & Samuel, 1977a
	keithi	N	Samoil & Samuel, 1977a
	leporis	N	Samoil & Samuel, 1977a
	perforans	N (?)	Boughton, 1932
	robertsoni	N	Samoil & Samuel, 1977a, b
	rochesterensis	N	Samoil & Samuel, 1977a
	rowani	N	Samoil & Samuel, 1977a
	ruficaudati	N	Samoil & Samuel, 1977a
	stiedai	N	Boughton, 1932
	townsendi	N	Samoil & Samuel, 1977a

(Continued)

TABLE 5.1 All Known Coccidia Species[1] (Apicomplexa: Eimeriidae) Reported from Jackrabbits/Hares in the Genus *Lepus* L., 1758 (Family Leporidae) (*cont'd*)

Lepus spp.	*Eimeria* spp.	Natural (N) or experimental (E)	References
arcticus	groenlandica	N	Madsen, 1938; Levine & Ivens, 1972
	leporis	N	Madsen, 1938; Levine & Ivens, 1972
	robertsoni	N	Madsen, 1938; Carvalho, 1943
	sculpta	N	Madsen, 1938
	semisculpta	N	Madsen, 1938; Pellérdy, 1956
	septentrionalis	N	Madsen, 1938
californicus	magna	N (?)	Henry, 1932
	media	N (?)	Henry, 1932
	perforans	N (?)	Henry, 1932
	stiedai	N	Henry, 1932
capensis	europaea	N	Moreno Montañez et al., 1979
	exigua	N (?)	Ogedengbe, 1991
	hungarica	N	Moreno Montañez et al., 1979
	leporis	N	Moreno Montañez et al., 1979; Romero-Rodriguez, 1976
	magna	N (?)	Ogedengbe, 1991; Vila-Viçosa & Caeiro, 1997
	media	N (?)	Ogedengbe, 1991; Vila-Viçosa & Caeiro, 1997
	perforans	N (?)	Ogedengbe, 1991; Vila-Viçosa & Caeiro, 1997
	piriformis	N (?)	Vila-Viçosa & Caeiro, 1997
	robertsoni	N	Vila-Viçosa & Caeiro, 1997
	stiedai	N	Ogedengbe, 1991; Vila-Viçosa & Caeiro, 1997
europaeus	cabareti	N	Aoutil et al., 2005
	campania	N	Pastuszko, 1961a
	coquelinae	N	Aoutil et al., 2005
	europaea	N	Pellérdy, 1956; Aoutil et al., 2005
	exigua	N (?)	Ryšavý, 1954
	hungarica	N	Pellérdy, 1956; Litvenkova, 1969 (?)
	irresidua	N (?)	Ryšavý, 1954
	lapierrei	N	Aoutil et al., 2005
	leporis	N	Nieschulz, 1923; Ryšavý, 1954; Aoutil et al., 2005

(Continued)

TABLE 5.1   All Known Coccidia Species[1] (Apicomplexa: Eimeriidae) Reported from Jackrabbits/Hares in the Genus *Lepus* L., 1758 (Family Leporidae) (*cont'd*)

Lepus spp.	*Eimeria* spp.	Natural (N) or experimental (E)	References
	macrosculpta	N	Sugár, 1979
	magna	N (?)	Ryšavý, 1954
	perforans	N (?)	Robertson, 1933; Ryšavý, 1954
	piriformis	N (?)	Ryšavý, 1954
	reniai	N	Aoutil et al., 2005
	robertsoni	N	Pastuszko, 1961a; Bouvier, 1967; Gräfner et al., 1967; Arnastauskene & Kazlauskas, 1970; Kutzer & Frey, 1976; Sugár et al., 1978; Jirouš, 1979; Terracciano et al., 1988; Tasan & Özer, 1989; Pakandl, 1990; Tacconi et al., 1995
	sculpta	N	Sugár et al., 1978; Pakandl, 1990
	semisculpta	N	Pastuszko, 1961a; Gräfner et al., 1967; Arnastauskene & Kazlauskas, 1970; Sugár et al., 1978; Jirouš, 1979; Berg, 1981; Chroust, 1984; Terracciano et al., 1988; Pakandl, 1990
	septentrionalis	N	Tasan & Özer, 1989; Pakandl, 1990;
	stefanskii	N	Pastuszko, 1961a
	stiedai	N/E	Ryšavý, 1954; Bouvier, 1967; Litvenkova, 1969; Entzeroth & Scholtyseck, 1977; Aoutil et al., 2005
	tailliezi	N	Aoutil et al., 2005
	townsendi	N	Pastuszko, 1961a; Bouvier, 1967; Gräfner et al., 1967; Kutzer & Frey, 1976; Sugár et al., 1978; Jirouš, 1979; Berg, 1981; Terracciano et al., 1988; Pakandl, 1990
granatensis	*auduboni*	N (?)	Aoutil et al., 2005
	bainae	N	Aoutil et al., 2005
	cabareti	N	Aoutil et al., 2005
	coquelinae	N	Aoutil et al., 2005
	gantieri	N	Aoutil et al., 2005
	hungarica	N	Aoutil et al., 2005
	lapierrei	N	Aoutil et al., 2005
	leporis	N	Aoutil et al., 2005

(*Continued*)

TABLE 5.1 All Known Coccidia Species[1] (Apicomplexa: Eimeriidae) Reported from Jackrabbits/Hares in the Genus *Lepus* L., 1758 (Family Leporidae) (*cont'd*)

Lepus spp.	*Eimeria* spp.	Natural (N) or experimental (E)	References
	macrosculpta	N	Aoutil et al., 2005
	nicolegerae	N	Aoutil et al., 2005
	pierrecouderti	N	Aoutil et al., 2005
	reniai	N	Aoutil et al., 2005
	ruficaudati	N	Aoutil et al., 2005
	stiedai	N	Aoutil et al., 2005
nigricollis	*campania*	N	Gill and Ray, 1960
	hungarica	N	Gill and Ray, 1960
	irresidua	N (?)	Gill and Ray, 1960
	leporis	N	Gill and Ray, 1960
	magna	N (?)	Gill and Ray, 1960
	media	N (?)	Gill and Ray, 1960
	minima	N (?)	Gill and Ray, 1960
	perforans	N (?)	Gill and Ray, 1960; Mandel, 1976
	punjabensis	N	Gill and Ray, 1960
	robertsoni	N	Mandal, 1976
	ruficaudati	N	Gill and Ray, 1960
	sylvilagi	N (?)	Mandal, 1976
spp.	*intestinalis*	N (?)	Mandal, 1976
	irresidua	N (?)	Mandal, 1976
	matsubayashii	N (?)	Mandal, 1976
	minima	N (?)	Mandal, 1976
	neoleporis	N (?)	Mandal, 1976
	stiedai	N	Tyzzer, 1902; Von Braunscheweig, 1969; Mandal, 1976
timidus	*europaea*	N	Litvenkova, 1969 (?)
	exigua	N (?)	Matschoulsky, 1941
	hungarica	N	Litvenkova, 1969 (?)
	leporis	N	Soveri and Valtonen, 1983
	magna	N	Vila-Viçosa & Caeiro, 1997

(*Continued*)

TABLE 5.1   All Known Coccidia Species[1] (Apicomplexa: Eimeriidae) Reported from Jackrabbits/Hares in the Genus *Lepus* L., 1758 (Family Leporidae) (*cont'd*)

Lepus spp.	*Eimeria* spp.	Natural (N) or experimental (E)	References
	media	N (?)	Matschoulsky, 1941
	perforans	N (?)	Matschoulsky, 1941
	robertsoni	N	Pastuszko, 1961a
	septentrionalis	N	Yakimoff et al., 1936
	stiedai	N	Litvenkova, 1969
	townsendi	N	Burgaz, 1973; Soveri & Valtonen, 1983
tolai	*gobiensis*	N	Gardner et al., 2009
	leporis	N	Yakimoff et al., 1931
	stiedai	N	Svanbaev, 1977
towensendii	*americana*	N	Carvalho, 1943
	campania	N	Carvalho, 1943; Levine and Ivens, 1972
	robertsoni	N	Madsen, 1938; Carvalho, 1943; Pellérdy, 1956
	sculpta	N	Madsen, 1938
	semisculpta	N	Carvalho, 1943
	septentrionalis	N	Carvalho, 1943
	townsendi	N	Carvalho, 1943; Pellérdy, 1956
Ten host species	43 *Eimeria* (31 valid)		

[1] *Forty-three Eimeria species reported and/or described from the lagomorph genus Lepus. We believe that 31 of them likely are valid species, but that the following are not: E. auduboni, E. exigua, E. intestinalis, E. irresidua, E. magna, E. matsubayashii, E. media, E. minima, E. neoleporis, E. perforans, E. piriformis, and E. sylvilagi (see text).*
(?) not considered a valid parasite of *Lepus* (see text).

Coccidia (Eimeriidae) of the Family Leporidae: Genus *Oryctolagus*

FIGURE 6.1 The European or Old World rabbit, *Oryctolagus cuniculus* L., 1758.

INTRODUCTION

As noted in Chapter 5, this Family is composed of 11 genera with 60 species. One of the eight monotypic genera, consisting only of the wild Old World rabbit, *Oryctolagus cuniculus* and its domesticated sibling, sometimes referred to as *O. c. domesticus*, has had a long and interesting history with humans, and because of this is of great economic, medical, and ecological importance worldwide. This wild rabbit digs burrows, or warrens, where it spends most of its time when not canvassing the immediate environment to forage. Unlike hares and jackrabbits (*Lepus* species), *O. cuniculus*, like *Sylvilagus* species (cottontails) are **alitricial**. Warrens may have two to 10 individuals. Mature males and females are better at fighting off predators than are juveniles, and females are more territorial than males. Wild rabbits can be highly aggressive, and competition between males often leads to injury or death. In addition to hostile displays, males squirt urine on challengers as a form of territorial marking, while the most common response to a challenge is immediate attack. Their powerful back legs are used as weapons, kicking at an opponent's underside, as well as biting and scratching with the front paws.

Because of its importance to humans, especially in biomedical research worldwide, *O. cuniculus* has been better studied for its intestinal parasites than any other lagomorph species. They have been reported to harbor17 *Eimeria* species (Table 6.1); however, we believe that one of these, *E. leporis* from *L. europaeus*, is not a valid parasite of this rabbit. Species of both *Lepus* (hares, Chapter 5) and *Sylvilagus* (cottontails, Chapter 7) have been reported to be infected with eimerians known to parasitize *O. cuniculus*, either naturally or experimentally; most of these reports from surveys are probably erroneous, and for the moment should be considered as such until supported by molecular markers and/or controlled experimental cross infections. It is also of interest to note that oocysts from other, related coccidian genera (e.g., *Isospora, Caryospora, Cyclospora*) have never been observed in *O. cuniculus* despite the many surveys that have been done on this species over the last century.

HOST GENUS *ORYCTOLAGUS* LILLJEBORG, 1874

EIMERIA COECICOLA CHEISSIN (KHEYSIN), 1947

Synonym: *Eimeria oryctolagi* Ray and Banik, 1965.

Type host: *Oryctolagus cuniculus* (L., 1758) (syn. *Lepus cuniculus*), European (domestic) rabbit.

Type locality: ASIA: Russia (former USSR).

Other hosts: None reported to date.

Geographic distribution: AFRICA: Egypt; ASIA: Azerbaijan (former USSR), India, Russia (former USSR), Taiwan; AUSTRALIA: Western

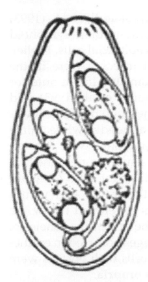
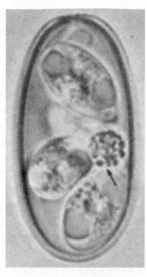

FIGURES 6.2, 6.3 Line drawing of the sporulated oocyst of *Eimeria coecicola* from Gill and Ray, 1960, with permission from Springer Science and Business Media, copyright holders for the *Proceedings of the Zoological Society* (supersedes *Proceedings of the Zoological Society of Calcutta*). Photomicrograph of a sporulated oocyst of *E. coecicola* from Norton et al., 1977, with permission from *Parasitology*.

Australia; EUROPE: France, Hungary, Poland; MIDDLE/NEAR EAST: Iran, Saudia Arabia, Syria, Turkey; SOUTH AMERICA: Brazil.

Description of sporulated oocyst: Oocyst shape: elongate-ellipsoidal or cylindroidal with end opposite M rounded; number of walls: 1; wall characteristics: smooth, light yellow to light brown, somewhat thickened, forming a small ridge at M; L × W: 35.5 × 19.5 (23−40 × 15−21); L/W ratio: 1.9 (1.5−2.3); M: present; M characteristics: well-pronounced, surrounded by slightly thickened wall; OR: present; OR characteristics: round, 4.5 (1−7) wide; PG: absent. Distinctive features of oocyst: elongate-ellipsoidal shape with wide M at one end; also, oocysts discharged on the first day of patency are smaller (33 × 17) than those discharged at the end of patency (35.5 × 19.5; Cheissin, 1947c).

Description of sporocyst and sporozoites: Sporocyst shape: elongate-ovoidal to ellipsoidal to spindle-shaped (line drawing); L × W: 16−17 × 8−9 (12 × 5.5 according to Gill and Ray, 1960); L/W ratio: 2.1; SB: present (line drawing); SSB, PSB: both absent; SR: present, ~2−4 wide; SZ: elongate, lying head-to-tail in sporocyst, each with one RB at wider end. Distinctive features of sporocyst: spindle-shaped with pointed end.

Prevalence: Levine (1973b) said this species apparently is rare in domestic lab rabbits, but is common in wild ones. Veisov (1982) examined 3,150 rabbits from three different areas/districts of the Cuba-Khachmazskoy zone in Azerbaijan and found 2,905 (81%) infected with nine species of *Eimeria*, including *E. coecicola*; however, we were not able to determine how many of the rabbits examined harbored this species. Santos and Lima (1987) first documented this species in Brazil and Santos-Mundin and Barbon (1990) reported it in 118/375 (31.5%) rabbits in Minas Gerais, Brazil. Darwish and Golemansky (1991) found this species in 27/75 (36%) domestic rabbits from four localities in Syria. Polozowski (1993) examined feces from 246 rabbits in six farm rabbitries in the Wroclaw District of Poland and found 51 (21%) infected with this species; 49/51 (96%) infected rabbits were < 3-mo-old while the other 22/51 (4%) were 4−12-mo-old. None of 48 rabbits 13 to > 24-mo-old that were examined harbored this species. Hobbs and Twigg (1998) examined fresh fecal samples from wild *O. cuniculus* in Western Australia and found 89/1200 (7%) to harbor this species; 50 of their sporulated, ellipsoidal oocysts measured 34.5 × 18.5 (29.5−42 × 16−20) with sporocysts 14.5 × 7.7 (13−16 × 7−9). Grès et al. (2003) found *E. coecicola* in 5−32% of 254 wild rabbits sampled from six different localities in France. They identified 10 eimerians from these 254 rabbits and, apparently, 100% were infected with one or more of these 10. Unfortunately, they did not specifically mention how many individuals were infected with multiple species vs. those with single species infections. Razavi et al. (2010) found this species in 2/71 (3%) wild rabbits in southern Iran; their oocysts were 34.1 × 18.7 (25−28 × 16−22), L/W 1.8, with a M and OR 5.3 (4−6). Li et al. (2010) reported *E. coecicola* in 46/642 (7%) *O. cuniculus* from pet shops (22/642) and farms (24/342) in 11 districts on the island of Taiwan. Oncel et al. (2011) reported this species in 10/10 (100%) 2−3-mo-old dead kids from a commercial rabbitry in Kocaeli, Turkey; ellipsoidal oocysts had a M and OR and were 34.3 × 17.4 (33−38 × 15−18). Prevalence rates of from 13−80% were reported from domesticated rabbits in various other regions in Turkey (Merdivenci, 1963; Tasan and Özer, 1989; Çetindağ and Biyikoğlu, 1997; Karaer, 2001).

Sporulation: Exogenous. Oocysts sporulated by the end of day 3 at 22°C (Cheissin, 1947c); Mandal (1976) said that oocysts (originally identified as *E. oryctolagi*) sporulated in 32−48 hr or in 3 days (as *E. coecicola*, see *Remarks* below); Kasin and Al-Shawa (1987) said sporulation occurred in 56 hr in 2.5% potassium dichromate ($K_2Cr_2O_7$) at 25−28°C; Razavi et al. (2010) said that sporulation occurred in 62 hr in 2.5% $K_2Cr_2O_7$ at 25−28°C; Li et al. (2010) said that sporulation occurred in 36−48 hr in 2.5% $K_2Cr_2O_7$ at 25°C.

Prepatent and patent periods: Prepatent period is 8−9 days (Pakandl, 1989) and patent period is 7−9 days (Cheissin, 1947c, 1968).

Site of infection: The life cycle of this species was studied by Cheissin (1947c, 1967, 1968) and by Pakandl (1989) in experimentally infected rabbits by light, scanning, and transmission electron microscopy. Later, Pakandl et al. (1993) added information on the migration of SZ and the early merogony in gut-associated lymphoid tissue. They (1993) observed SZ in the duodenum and jejunum until 32 hr PI, first in the villous epithelial cells and in cells they said resembled intraepithelial lymphocytes. The SZ then moved into cells of lymphatic follicles of gut-associated lymphoid tissue (Peyer's patches, sacculus rotundus, vermiform

appendix). At 64 hr PI, Pakandl et al. (1993) observed the first merogony in these lymphoid cells and in membranous epithelial cells. Earlier, Pakandl (1989) said that meronts were in the epithelial cells of the ileum, 10−15 cm anterior to the cecum, while Pakandl et al. (1993) said that first- and second-generation meronts developed in the vermiform appendix while third- and fourth-generations meronts were located in the epithelium of the ileum, as per Cheissin (1947c, 1967, 1968), until about day 7 PI, but gamogony occurred in the cecum and vermiform appendix. Gamonts developed below the N of epithelial cells in the appendix (Cheissin, 1967). All endogenous stages occurred in epithelial cells, except that the cells with oocysts were submerged in the **tunica propria**.

Endogenous stages: Cheissin (1947c, 1967, 1968) first described the endogenous stages; he said that meronts were 12−15 wide and had 8−12 merozoites, each 10 × 0.8, but did not know the exact number of generations. SZ were seen in lymphatic follicles of the vermiform appendix and other lymphoid tissues in the gut as early as 64 hr PI (Pakandl et al., 1993) and up to day 4 PI (Pakandl, 1989) by transmission electron microscopy. Pakandl (1989) said these were first-generation meronts that developed by endodyogeny, but their later work (Pakandl et al., 1993) suggested these were second-generation meronts; some of these second-generation merozoites had 2−3 N, and continued to be present on days 5−6 PI, increasing in size during that time. Pakandl et al. (1993) did not give measurements of meronts found in gut-associated lymphoid tissues, but stated that they contained "up to 12 merozoites." Pakandl's (1989) first (probably second) generation meronts were 6.6 × 6.4 (5−9.5 × 4.5−8), each with two merozoites that were 6.5 × 2.6 (4.5−8 × 1.5−3.5) on day 4 PI, but by day 6 PI meronts were 11.0 × 8.3 (8−13.5 × 6−10), with 2−8 merozoites that were 9.8 × 2.9 (8−13 × 2−4). Second (third?) generation meronts appeared at day 6 PI and were 16.0 × 13.3

(11.5—22.5 × 7—17), each with 8—47 merozoites that were 11.2 × 1.6 (8.5—13 × 1—2). Third (fourth?) generation meronts also were first seen on day 6 PI and these measured 10.1 × 8.1 (7—14 × 4—10), each with 4—25 merozoites that were 8.4 × 1.8 (6—11 × 1.5—3). Fourth (fifth?) generation meronts appeared on day 7 PI and they were 16.1 × 10.1 (8—28 × 7—17), each with 12—80 merozoites that measured 5.1 × 1.0 (4.5—6 × 0.5—1). Pakandl's (1989) second- and fourth-generation merozoites were only uninucleate, while third-generation merozoites were only multinucleate. Pakandl (1989) said that both mature macro- and microgamonts were seen on day 8 PI. Macrogamonts were 24.9 × 14.9 (20—28.5 × 11—18) and microgamonts were 22.6 × 15.3 (15—31 × 11—25). Cheissin (1967) said that microgametocytes were 20—22 × 12—17 and contained biflagellate microgametes, 2.5—3 × 1.5, and that macrogamonts were not larger than the microgamonts.

Cross-transmission: None to date.

Pathology: Cheissin (1947c) said that in heavy infections the crypts became considerably enlarged with many white spots at the surface of the cecum (corresponding to the accumulation of hematocytes in the crypts). Pellérdy (1954a) found that *E. coecicola* (which he called *E. neoleporis*) was highly pathogenic for domestic rabbits. It caused characteristic changes in the vermiform process, ileocecal valve, and throughout the large intestine, and most of the infected young rabbits died. Ten days after artificial infection, the small intestine mucosa was somewhat inflamed and catarrhal, but no oocysts were found. The walls of the large intestine, ileocecal valve and vermiform process were thickened, **hyperemic**, and covered internally by a layer of **purulent** material; the blood vessels on their serosae were engorged and enlarged. Large numbers of gamonts and meronts were present in the large intestine, but only a few in the posterior part of the ileum. Tissue eosinophilia was not present. In severely affected areas, the normal epithelium was destroyed and sloughed off so that only patches remained. Vitovec and Pakandl (1989) also reported on the pathogenicity in rabbits with prolonged infection of at least 20 days. Pathological changes apparently began to develop 4 days PI, with the beginning of merogony, and were characterized by an inflammatory infiltration and abundant pyogenic component in the lamina propria, swelling, and coalescence of upper parts of the appendix mucosa above **atrophied** domes, where spaces filled with stagnating inflammatory exudate, endogenous stages of the parasite, and desquamated epithelium. At day 8 PI, the epithelium with late endogenous stages became hyperplastic, proliferated into the lamina propria, and became necrotic. This alteration of the epithelium and exposure of the appendix lamina propria occurred during gamogony, about 10 days PI. Groups of immature oocysts and pieces of oocysts remained in the lamina propria, at least for 10 days, and were resorbed by granulomatous inflammatory structures with abundant multinucleated cells (Vitovec and Pakandl, 1989). Oncel et al. (2011) reported that 10 dead kids from a commercial rabbitry in Kocaeli, Turkey, were all thin with reduced fat stores and muscle wasting, had rough hair coats, and the small intestines were distended and filled with gray-green ingesta; the intestinal mucosa was hyperemic and oedematose. However, they could not attribute the pathology they observed to *E. coecicola* since all 10 kids also were infected with *E. intestinalis* and *E. perforans*, as determined by oocysts in their feces (Oncel et al., 2011).

Material deposited: None.

Remarks: Cheissin (1947b) was studying variation between two other rabbit eimerians, *E. magna* and *E. media*, when he found a single oocyst that he said resembled *E. media*. He infected a young rabbit with that one oocyst and produced a patent infection that gave him, "in due course a multitude of oocysts which have been found to differ however from any species so far known, *viz.* from *E. stiedai, E. magna, E. demia, E. irresidua, E. perforans* and *E. piriformis."* He was able to infect other domestic rabbits with the progeny from his initial infection with a single oocyst, and was able to study the endogenous stages, which led him to conclude that this was an independent species, *E. coecicola.* Description of sporulated oocysts (above) is based on Cheissin (1947c, 1967, 1968). Oocysts of *E. coecicola* are similar in shape and size to those of *E. magna* and *E. media,* but differ by having endogenous stages that occur in the epithelium of both the small and large intestine, while those of *E. magna* are sometimes found in the large intestine, but in the connective tissue only, and those of *E. media* are never found in the large intestine. In addition, *E. coecicola* oocysts are slightly longer than those of *E. media* and smaller in width than those of *E. magna.* During the last days of the patent period, many of the oocysts lack an OR. Pellérdy (1954a, 1965, 1974) believed that this species was a synonym of *E. neoleporis,* but Cheissin (1968) gave reasons to the contrary. The greatest differences are in the location and size of the meronts and merozoites. The prepatent and patent periods and pathogenicity also are different. Mandal (1976) reported on oocysts he called *E. oryctolagi* Ray and Banik, in domestic rabbits from Calcutta, India, but distinguished them from oocysts he thought were *E. coecicola,* from Ludhiana, Punjab, India. He (1976) said the former oocysts had a two-layered wall and were ellipsoidal, 38.7 × 10.1 (sic) (28.5−47 × 12.5−28.5) with L/W 2.0, and had both a M and OR; the piriform sporocysts were 10 × 6 (8.5−14.5 × 4.5−8.5) with L/W 1.7, and had both a SB, SR, and SZs that were 10.5 long. Mandal (1987) corrected his earlier mistake on the mean width of *E. oryctolagi* to be 19.1 and argued that it should be treated as a distinct species (see *E. oryctolagi,* below). The oocysts he described as *E. coecicola* were ovoidal, 27.5−33 × 14−19.5, with an OR and had ovoidal sporocysts, 12 × 5.5, with a SR. Kasim and Al-Shawa (1987)

examined 263 adult domestic rabbits from three regions in Saudia Arabia from 1984−1985 and found 95 (36%) to be infected with this species. Their (1987) oocysts were cylindroidal, 33.4 × 19.3 (28−40 × 15.5−22), L/W ratio of 1.7 (1.4−2.2), with a M and OR, and sporocysts 15.1 × 7.2 (12−19 × 6−9), L/W 2.1 (1.8−2.4). Pakandl (1989) studied all life-cycle stages of *E. coecicola* and found ovoidal to cylindroidal oocysts that were 35.2 × 20.2 (33−38 × 18−22), L/W ratio of 1.7, with a M 3.8 (2.5−5) wide, and an OR 6.5 (5−8) wide; these contained ovoidal sporocysts, 16.9 × 7.6 (15−19 × 7−8), L/W ratio of 2.2, with a SR 7.0 × 3.7 (5−9.5 × 3.5−5) and a small SB at the tip of the sporocyst. Pakandl (1989) also noted two unusual observations in his study of the life cycle of this species, development by endodyogeny (a process typical of heteroxenous coccidia like *Toxoplasma* and *Sarcocystis* spp.) and the occurrence of multinucleate merozoites, which he mentions as "an exceptional phenomenon in the life cycle of coccidia." Licois et al. (1992b) found that immunity, as manifested by excretion of oocysts after a challenge infection, was absolute after only one immunizing dose of 1,000 oocysts, and they observed no cross-protection when rabbits were challenged with *E. magna, E. flavescens,* or *E. intestinalis.* Grés et al. (2003) examined 254 wild rabbits from six localities in France, and this is one of 10 species they identified. Oocysts measured by Li et al. (2010) from *O. cuniculus* in Taiwan were 33.4 × 21.8 (33−36 × 20−24) with sporocysts 12.7 × 7.4 (9.5−16 × 6−9.5).

EIMERIA EXIGUA YAKIMOFF, 1934

Synonym: Eimeria exigua Type I of Yakimoff, 1934; *Eimeria exigua* var. *septentrionalis* Madsen, 1938, *pro parte; Eimeria hungarica* Pellérdy, 1956, *pro parte.*

Type host: Oryctolagus cuniculus (L., 1758) (syn. *Lepus cuniculus*), European (domestic) rabbit.

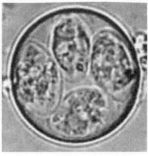

FIGURES 6.4, 6.5 Line drawing of the sporulated oocyst of *Eimeria exigua* from Carvlaho, 1943, with permission from John Wiley & Sons Ltd. (current rights owner of the *Iowa State College Journal of Science*). Photomicrograph of a sporulated oocyst of *E. exigua* from Jelinková et al., 2008, with permission from *Veterinary Parasitology*.

Type locality: ASIA: Russia, St. Petersburg (Leningrad).

Other hosts: *Lepus arcticus* Ross, 1819, Arctic hare (?); *Lepus capensis* L., 1758, Cape hare (?); *Lepus europaeus* Pallas, 1778, European hare (?); *Lepus timidus* L., 1758, Mountain hare (?); *Sylvilagus floridanus* (J.A. Allen, 1890), Eastern cottontail (?).

Geographic distribution: AFRICA (?): Nigeria (?): ASIA: India, Russia, Taiwan; AUSTRALIA: New South Wales, Western Australia; EUROPE: France (including a small French island off the coast of Brittany, Île Beniguet), Hungary, Italy, Poland, Spain; MIDDLE/NEAR EAST: Saudia Arabia, Syria.

Description of sporulated oocyst: Oocyst shape: mostly subspheroidal; number of walls: not given, but appears as 1 (line drawing); wall characteristics: outer is smooth, colorless; L × W: 14 × 13 (12−21 × 9−18); L/W ratio: 1.4 (1.0−1.9); M, OR, PG: all absent. Distinctive features of oocyst: spheroidal to subspheroidal shape, with lack of M, OR, and PG.

Description of sporocyst and sporozoites: Sporocyst shape: not stated, but appears to be ellipsoidal (line drawing); L × W: 9 × 5 (8−10 × 4−6); L/W ratio: 1.8; SB, SSB, PSB: all absent (in original line drawing, Yakimoff, 1934); SR:

present (?) (not pictured in original line drawing, Yakimoff, 1934); SZ: not described. Distinctive features of sporocyst: lack of a SB.

Prevalence: Francalancia and Manfredini (1967) said *E. exigua* was rare in rabbits in Italy, but Stodart (1971) found it in 216/656 (33%) wild rabbits in eastern Australia (New South Wales). Pellérdy (1974) also said the occurrence of *E. exigua* is "relatively rare, although it has been observed in rabbits all over the world." In an unpublished Master's thesis, Ogedengbe (1991) said he found *E. exigua* in 19% of the rabbits surveyed in Kaduna State, Nigeria. However, he (1991) neglected to state how many rabbits were surveyed, nor did he mention the host species; Anonymous (2012) listed the Cape hare, *L. capensis*, as the only lagomorph species found in Nigeria. Darwish and Golemansky (1991) found this species in 19/75 (25%) domestic rabbits from four localities in Syria. Hobbs and Twigg (1998) examined fresh fecal samples from wild *O. cuniculus* in Western Australia and found 750/1200 (62.5%) to harbor *E. exigua*; 50 sporulated, subspheroidal oocysts measured 16 × 13 (12−19 × 11−15) with sporocysts 8.5 × 5.1 (6−10 × 4.5−6). Grès et al. (2003) found *E. exigua* in 32 to 72% of 254 wild rabbits sampled from six different localities in France. They identified 10 eimerians from these 254 rabbits and, apparently, 100% were infected with one or more of these 10. Unfortunately, they did not specifically mention how many individuals were infected with multiple species vs. those with single species infections. Li et al. (2010) reported *E. exigua* in 9/642 (1%) *O. cuniculus* in pet shops (3/642) and on farms (6/342) in 11 districts on the island of Taiwan.

Sporulation: Exogenous, 36−48 hr (Pastuzko, 1961a; Pastuzko, 1963), 2 days at 20°C (Lizcano and Romero, 1969), or 17 hr at 26°C (Coudert et al., 1995); Li et al. (2010) said that sporulation occurred in 16−36 hr in 2.5% $K_2Cr_2O_7$ at 25°C.

Prepatent and patent periods: Prepatent period is 7 days (Coudert et al., 1995).

Site of infection: All endogenous stages localized along the whole length of the small intestine, with individual meronts and gamonts concentrated in specific parts of the gut. As development progressed, endogenous stages successively moved from the duodenum toward the ileum. All stages were in epithelial cells of the middle and upper part of the villi (Jelínková et al., 2008). No parasites were found in the large intestine.

Endogenous stages: Jelínková et al. (2008) were the first to document the endogenous development of *E. exigua* by both light and electron microscopy. Four asexual stages precede gamogony and all stages are smaller than comparable stages of other rabbit coccidia known to date. First-generation meronts were observed at 72—90 hr PI in the duodenum-jejunum; they were 4.6 × 3.5, each with two merozoites that were 4.1 × 2.0. Second-generation meronts were of two types (as reported in several other rabbit coccidia life cycles), both in the jejunum-ileum: Type A were small, 5.7 × 4.3, and each produced, via endomerogony, only two merozoites, each with two N, and were 5.2 × 2.4; their Type B second-generation meronts were 7.7 × 5.0, and produced, via ectomerogony, 3—7 merozoites that had one N each and were 5.5 × 1.9. Third-generation Type A meronts appeared at 126 hr PI in the jejunum-ileum region and were 6.6 × 4.8, with 2—4 merozoites that measured 6.4 x2.2, each with two N, while the Type B third-generation meronts were 11.5 × 6.7, with 4—16 mononuclear merozoites, that were 5.8 × 1.7. Fourth-generation Type A meronts appeared 144 hr PI in the ileum and were 6.8 × 5.4, with two binucleate merozoites that measured 6.1 × 3.3; Type B fourth-generation meronts were 9.6 × 6.7, with 4—13 mononuclear merozoites that were 6.1 × 1.9. Gamonts were seen 144—188 hr PI in the ileum; macrogamonts, which had relatively small numbers of wall-forming bodies in the cytoplasm, were 9.4 × 12.1, and microgamonts were 8.8 × 10.9 (Jelínková et al., 2008).

Cross-transmission: Pellérdy (1956) was unable to infect two hares, *Lepus europaeus*, about 3—4 wk old, with a mixture of coccidia from the domestic rabbit that presumably contained *E. exigua*.

Pathology: Coudert et al. (1995) grouped *E. exigua* together with *E. perforans* and *E. vejdovskyi* as slightly pathogenic, a condition they associated to the localization of these species in more distal positions in the epithelial cells of the intestinal villi (vs. those that localize in cells deep within the crypts).

Material deposited: None.

Remarks: Yakimoff (1934), who named this species, described three oocyst types in his rabbit from the (former) USSR, of which his Type I was *E. exigua*; the others were ovoidal or ellipsoidal. The oocysts of his Type I were spheroidal to subspheroidal, 14.5 × 12.7, with a smooth wall, but lacked a M and OR. Description of the sporulated oocyst, above, is an amalgam taken from Yakimoff (1934), Pastuszko (1961a), and Lizcano Herrera and Romero Rodriguez (1969). Morgan and Waller (1940) said they found spheroidal oocysts of *E. exigua*, ~14 wide, in 8/210 (4%) *S. floridanus* in Iowa, but they were likely seeing *E. minima*. Matschoulsky (1941) said he found oocysts that resembled this species in *Lepus timidus* from Buryat-Mongol (former USSR) and that the spheroidal oocysts were 15.7 × 13.6 (15—18 × 13—15) with sporocysts 7—8 × 4—5, but he was likely looking at *E. hungarica*, which hadn't been described and named yet by Pellérdy (1956). Cheissin (1947a, 1967) and others considered this species to be a synonym of *E. perforans*. Ryšavý (1954) said he found this species in *L. europaeus* in the Czech Republic with oocysts 16.5—20.8 × 9.9—17.5 and sporocysts 6.5 × 3.0, but it was certainly not *E. exigua*; he (1954) also said he found it in the duodenum and small intestine, but did not explain how he determined this. Pellérdy (1954a) differentiated this form in the domestic rabbit from one which he (Pellérdy, 1956) later named *E. hungarica* from the European hare, *L. europaea*. He was unable to transmit *E. hungarica* to the rabbit from the hare. Pastuszko (1963) examined 9,845 rabbits from 17 districts in Poland; these included 502 rabbits by necropsy and 9,343 fecal samples tested. She (Pastuszko, 1963) said that this species caused coccidiosis in rabbits in Poland; her oocysts were 14.5 × 12.7 (12–21 × 10–17.5; her Table 1). Later, Pellérdy (1974) said his oocysts were 14.5 × 12.7 (10–18 × 9–16). Coudert et al. (1995) isolated and characterized the first pure strain of *E. exigua* from oocysts found by a colleague in tame rabbits in Italy; these oocysts were 10–18 × 11–16. They (1995) did controlled experimental infections in SPF laboratory rabbits that gave a more accurate picture of the endogenous development; that is, the endogenous stages of *E. exigua* do not invade host cell nuclei (see *Endogenous stages*, above). Grès et al. (2000) sampled 254 wild rabbits (*O. cuniculus*) from five localities in France and on a small French island off the coast of Brittany (Île Beniguet). They found intensity (number of oocysts/g feces) heavier in young than in adult rabbits and said that it increased in the autumn and winter and declined in the spring, while prevalences increased progressively during autumn and winter, were at a maximum during spring, and then declined in the summer (Grès et al., 2000). The oocysts they found in the feces were spheroidal to subspheroidal, lacked both a M and OR, and were thus called *E. exigua*. However, in rabbits they necropsied, they reported endogenous stages within the nuclei of intestinal mucosal cells in stained smears they made, and then they made a leap of faith, saying that these stages produced the oocysts they had identified as *E. exigua*. This was in contrast to the earlier work of Coudert et al. (1995), who said endogenous stages do not enter the N of intestinal cells. Thus, the intranuclear stages reported by Grès et al. (2000) probably belonged to a currently unknown new species of wild hares in France. Grés et al. (2003) examined 254 wild rabbits from six localities in France and this is one of 10 species they identified. Jelínková et al. (2008) described the

endogenous development in detail and firmly established that *E. exigua* is, in fact, a valid species of *O. cuniculus*, although one that occurs only rarely in field samples. Oocysts measured by Li et al. (2010) from *O. cuniculus* in Taiwan were 17.1 × 13.6 (13–21 × 11–19) with sporocysts 7.1 × 4.7 (5–8 × 3–5). Its validity as a parasite of *Lepus* or *Sylvilagus* species needs to be verified by cross-transmission and/or molecular studies.

EIMERIA FLAVESCENS MAROTEL & GUILHON, 1941

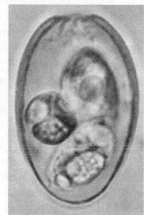

FIGURES 6.6, 6.7 Line drawing of the sporulated oocyst of *Eimeria flavescens* from Catchpole and Norton, 1979, with permission from *Parasitology*. Photomicrograph of a sporulated oocyst of *E. flavescens* from Norton et al., 1979, with permission from *Parasitology*.

Synonym: *Eimeria pellerdyi* Coudert, 1977a, b; *Eimeria hakei* Coudert; *Eimeria irresidua* Kessel and Jankiewicz, 1931 of Francalani and Manfredini, 1967.

Type host: *Oryctolagus cuniculus* (L., 1758) (syn. *Lepus cuniculus*), European (domestic) rabbit.

Type locality: EUROPE: Belgium.

Other hosts: None to date.

Geographic distribution: AUSTRALIA: Western Australia, Tasmania/Macquarie Island;

EUROPE: Belgium, England, France, Poland; MIDDLE/NEAR EAST: Iran, Saudia Arabia; SOUTH AMERICA: Brazil.

Description of sporulated oocyst: Oocyst shape: ovoidal; number of walls: 2; wall characteristics: outer is smooth, yellow, ~1.4 thick; inner is a dark line, ~0.4 thick, with the wall slightly thickener at the broad end, around the M, than on the sides; L × W: 31.7 × 21.4 (25—37 × 14—24); L/W ratio: 1.5 (1.2—2.1); M: present; M characteristics: prominent, well-defined, 4.3 (2—7) wide; OR, PG: both absent. Distinctive features of oocyst: thick two-layered wall and prominent M.

Description of sporocyst and sporozoites: Sporocyst shape: elongate-ovoidal; L × W: 15.3 × 8.6 (13—17 × 7—10); L/W ratio: 1.75 (1.2—2.1); SB: small, present at pointed end; SSB, PSB: both absent; SR: present as a sub-spheroidal cluster of granules, ~6.3; SZ: elongate, 17 × 5 after excystation and have one clear RB, ~4 wide, at more rounded end. Distinctive features of sporocyst: SZ longer than sporocyst with one clear RB.

Prevalence: Catchpole and Norton (1979) found *E. flavescens* in 95/596 (16%) rabbits reared for meat production. Peeters et al. (1981) found it in 108/770 (14%) in the type host from commercial rabbitries and in 124/282 (44%) sick rabbits in traditional rabbitries in Belgium. Santos and Lima (1987) first documented *E. flavescens* in Brazil and Santos-Mundin and Barbon (1990) reported it in 22/375 (6%) rabbits in Minas Gerais, Brazil. Polozowski (1993) examined feces from 246 rabbits in six farm rabbitries in the Wroclaw District of Poland and found 163 (66%) infected with *E. flavescens*; 131/163 (72%) infected rabbits were < 3-mo-old, 8/163 (5%) were 4—12-mo-old, 18/163 (11%) were 13—24-mo-old, while only 6/163 (< 4%) were > 24-mo-old. Hobbs and Twigg (1998) examined fresh fecal samples from wild *O. cuniculus* in Western Australia and found 715/1200 (60%) to harbor *E. flavescens*; 50 sporulated, ovoidal oocysts measured

32.1 × 21.4 (22—37 × 20—23) with sporocysts 14.6 × 8.4 (12—17 × 9—9). Grès et al. (2003) found *E. flavescens* in 90—100% of 254 wild rabbits sampled from six different localities in France. They identified 10 eimerians from these 254 rabbits and this was the only eimerian found in all 254 rabbits. Unfortunately, they did not specifically mention how many individuals were infected with multiple species vs. those with single species infections. Razavi et al. (2010) found *E. flavescens* in 8/71 (11%) wild rabbits in southern Iran; their oocysts were 31.6 × 18.9 (26—36 × 14—23), L/W 1.7, with a M, but lacking an OR, and sporocysts were 15.1 × 9.3 (13—17 × 7—11).

Sporulation: 38 hr or less at 27°C in 2% (w/v) $K_2Cr_2O_7$ solution (Norton et al., 1979; Kasim and Al-Shawa, 1987); Razavi et al. (2010) said that sporulation occurred in 48 hr in 2.5% (w/v) $K_2Cr_2O_7$ at 25—28°C.

Prepatent and patent periods: Prepatent period is 9 (8—11) days, and both prepatency and the patent period are dose and weight dependent (Norton et al., 1979).

Site of infection: Epithelial cells of the lower small intestine, cecum, and colon.

Endogenous stages: First-generation meronts were in the epithelial cells at the base of the villi or in the glands of the lower small intestine, while second- to fifth-generation meronts are in those of the cecum and colon. Second-, third- and fourth-generation meronts are in the superficial epithelium, while fifth-generation meronts are in the glands. First-generation meronts were 15 × 9 at 72 hr, and contained 16 first-generation merozoites, 9 × 2.5. At 88—115 hr PI, mature second-generation meronts were 16.1 × 13.1 (12.5—18 × 8—16) and contained 16 (12—20) second-generation merozoites that measured ~10 × 2. Third-generation meronts were 15.9 × 12.3 (12.5—19 × 9—16) at 120 hr PI and produced, on average, 15.5 (12—20) third-generation merozoites that measured 11 × 2. Fourth-generation meronts were 14.5 × 9.6 (10—19 × 7—13) at 120—156 hr PI with 16

(12–24) fourth-generation merozoites, 17 × 2.5. Fifth-generation meronts were in the deepest layers of the cecal mucosa, but were few in number; they measured 15.8 × 13.0 (14–18 × 11–16) at 144–156 hr and contained 50 fifth-generation merozoites, ~5 × 2. Both gamonts and gametes are in the glands of the cecum and colon at about 163 hr PI. Microgamonts were 30.4 × 23.0 and produced many flagellated microgametes and a residuum while macrogametes measured 29.9 × 19.8 and had many wall-forming bodies (Norton et al., 1979).

Cross-transmission: None to date.

Pathology: Norton et al. (1979) said that this species is very pathogenic for young (6-wk-old) Dutch rabbits, with even low doses of sporulated oocysts (10^4) producing severe enteritis with high mortality. Clinical signs included anorexia, weakness, and diarrhea, which caused soiling of the hind quarters. Heavily infected rabbits produced mucoid or watery feces that contained strips of intestinal epithelium and variable amounts of blood (Norton et al., 1979).

Material deposited: None.

Remarks: The original description of *E. flavescens* by Marotel and Guilhon (1941) noted that it produced smaller ovoidal oocysts than did *E. irresidua*, that these lacked an OR, and that it developed in the large intestine. However, the name was largely overlooked and/or ignored because Levine (1973b), Pellérdy (1965, 1974), and Levine and Ivens (1972) regarded it as a synonym of *E. media*. Later, Coudert (1977a, b) recognized the same unique characters (lack of an OR, development in the large intestine) in an eimerian from rabbits and, perhaps, unaware of the description by Marotel and Guilhon (1941), named a distinct species, *E. pellerdyi*, a name already in use for a camel eimerian. According to Norton et al. (1979), Coudert later emended that name to *E. hakei* (we can find no reference to this emendation), but the rules of priority of the International Code of Zoological Nomenclature (Ride et al., 2000) require that the name be suppressed as a junior synonym

of *E. flavescens*. However, it wasn't until Coudert and Norton (1979, in an abstract) and Norton et al. (1979) redescribed *E. flavescens* in detail and studied its endogenous development and other aspects of its biology that it has become recognized as a valid parasite species of *Oryctolagus*. The strain Norton et al. (1979) studied was isolated in 1960 when a coccidia-free Dutch domestic rabbit received three sporulated oocysts from a fecal sample submitted for diagnosis; the oocysts collected were passaged six times in coccidia-free rabbits and then sporocysts from the last passage were frozen in liquid nitrogen (N_2) to form a stabilate in 1969; oocysts of this stabilate were used for their work in 1979. They pointed out that *E. flavescens* differs from *E. media* in the greater width of the sporulated oocyst, in the absence of an OR, and its location in the gut where endogenous development occurs; mostly in the large intestine for *E. flavescens*, but only in the small intestine for *E. media* (Norton et al., 1979). Kasim and Al-Shawa (1987) examined 263 adult domestic rabbits from three regions in Saudia Arabia from 1984–1985 and found 45 (17%) to be infected with this species; their (1987) oocysts were ovoidal, 32.0 × 21.2 (27–37.5 × 17–25), L/W 1.5 (1.4–2.0), with a M, but without an OR, and sporocysts 15.0 × 8.4 (12.5–17 × 7–10), L/W 1.8 (1.7–2.0). Interestingly, *E. flavescens* has been reported from *O. cuniculus* on Macquarie Island, a Tasmanian State Reserve since 1978. Grés et al. (2003) examined 254 wild rabbits from six localities in France, and this is one of 10 species they identified.

EIMERIA INTESTINALIS CHEISSIN, 1948

Synonyms: *Eimeria agnosta* Pellérdy, 1954b; *Eimeria piriformis* Gvéléssiani and Nadiradze, 1945; *Eimeria piriformis* Cheissin, 1948 (Pellérdy, 1953); non *Eimeria piriformis* Kotlan and Pospesch, 1934; non *Eimeria piriformis* Marotel and

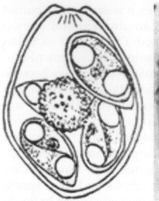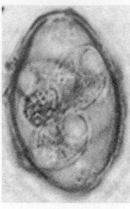

FIGURES 6.8, 6.9 Line drawing of the sporulated oocyst of *Eimeria intestinalis* from Gill and Ray, 1960, with permission from Springer Science and Business Media, copyright holders for the *Proceedings of the Zoological Society* (supersedes *Proceedings of the Zoological Society of Calcutta*). Photomicrograph of a sporulated oocyst of *E. intestinalis* from Hobbs and Twigg, 1998, with permission from the *Australian Veterinary Journal*.

Guilhon, 1941; non *Eimeria piriformis* Lubimov, 1934.

Type host: *Oryctolagus cuniculus* (L., 1758) (syn. *Lepus cuniculus*), European (domestic) rabbit.

Type locality: ASIA: Russia (former USSR).

Other hosts: *Lepus* sp. (?) (Mandal, 1976).

Geographic distribution: AFRICA: Egypt; ASIA: Azerbaijan (former USSR); India, Russia (former USSR); AUSTRALIA: Western Australia; EUROPE: Belgium, France, Hungary, Italy, Portugal; MIDDLE/NEAR EAST: Saudia Arabia, Syria, Turkey.

Description of sporulated oocyst: Oocyst shape: broadly pear-shaped to ovoidal; number of walls: 1; wall characteristics: smooth, yellow to light brown, somewhat thickened around M; L × W: 27 × 18 (21–36 × 15–21); L/W ratio: 1.5; M: present at narrow end of oocyst; M characteristics: well-defined with wall somewhat thickened around it; OR: present; OR characteristics: conspicuous, granular, ~5 (3–10) wide; PG: absent. Distinctive features of oocyst: pear-shape with a distinct M and OR.

Description of sporocyst and sporozoites: Sporocyst shape: oblong-ovoidal; L × W: 15 × 10; L/W ratio: 1.5; SB: small, present at pointed end (line drawing); SSB, PSB: both absent; SR: small or a round, compact mass, 1–6 wide; SZ: elongate, lying lengthwise head-to-tail in sporocyst, with one clear RB at larger end. Distinctive features of sporocyst: none.

Prevalence: Peeters et al. (1981) found *E. intestinalis* in 92/770 (12%) commercial rabbitries and in 118/282 (42%) sick rabbits in traditional rabbitries in Belgium. Veisov (1982) examined 3,150 rabbits from three different areas/districts of the Cuba-Khachmazskoy zone in Azerbaijan and found 2,905 (81%) infected with nine species of *Eimeria*, including *E. intestinalis*; however, we were not able to determine how many of the infected rabbits harbored this species. Darwish and Golemansky (1991) found this species in 9/75 (12%) domestic rabbits from four localities in Syria. Polozowski (1993) examined feces from 246 rabbits in six farm rabbitries in the Wroclaw District of Poland and found 81 (33%) infected with *E. intestinalis*; 76/81 (94%) infected rabbits were < 3-mo-old, 3/81 (4%) were 4–12-mo-old, 2/81 (2%) were 13–24-mo-old, while no rabbits > 24-mo-old were infected. Hobbs and Twigg (1998) examined fresh fecal samples from wild *O. cuniculus* in Western Australia and found 230/1200 (19%) to harbor *E. intestinalis*; 50 sporulated, piriform oocysts measured 29.1 × 18.2 (26–34 × 16–20) with sporocysts 12.9 × 7.2 (11–14 × 6–8). Grès et al. (2003) found *E. intestinalis* in 0–16% of 254 wild rabbits sampled from six different localities in France. They identified 10 eimerians from these 254 rabbits and, apparently, 100% were infected with one or more of these 10. Unfortunately, they did not specifically mention how many individuals were infected with multiple species vs. those with single species infections. In most surveys, this species seems relatively uncommon when compared to other eimerians infecting *O. cuniculus*. However, Oncel et al. (2011) reported *E. intestinalis* in 10/10 (100%) dead kids (2–3-mo-old) from a commercial rabbitry in Kocaeli, Turkey; their piriform oocysts had a M and OR and were 29.4 × 17.5 (28—33 × 15—20). Prevalence rates of from 13% to 80% were reported from domesticated rabbits in various other regions in Turkey (Merdivenci, 1963; Tasan and Özer, 1989; Çetindağ and Biyikoğlu, 1997; Karaer, 2001).

Sporulation: Exogenous. Oocysts sporulated within 1—2 days at room temperature (Pellérdy, 1965), in 6 days at 15°C (Cheissin, 1948) or in 48 hr at 27°C in 2% (w/v) $K_2Cr_2O_7$ solution (Kasim and Al-Shawa, 1987).

Prepatent and patent periods: Prepatent period is 9—10 days, while the patent period is 6—8 days (Cheissin, 1948) and no longer than 10 days (Cheissin, 1967). However, Licois et al. (1990) selected for early development of oocysts in rabbits and produced a precocious line of *E. intestinalis* in which the prepatent period was reduced from 215 hr to < 144 hr PI.

Site of infection: Epithelium of the villi and crypts, usually the lower part of the small intestine (ileum) below the duodenum, but the gamonts also develop in the epithelial cells of the cecum and appendix (Cheissin, 1948, 1958a, 1967); thus, the oocysts form in both the small and large intestines. The first- and second-generation meronts occupy 15% the length of the small intestine of 1-mo-old rabbits, while third-generation meronts and gamonts occupy about 40% the length of the lower small intestine and also are found in the large intestine (Cheissin, 1948, 1958a, 1967).

Endogenous stages: Cheissin (1948, 1958a, 1967) and Pellérdy (1953, 1965, 1974) differ somewhat in their accounts of the endogenous stages, and Pellérdy (1953) described these under the name *E. piriformis*. There are either three or four merogenous stages, depending on how they are numbered. First-generation meronts developed in the ileum at the base of the villi and in the crypts, 10—50 cm anterior to the ileocecal valve, and matured by 4—6 or 7—8 days PI; they measured 15—25 wide and contained 20—60 (Cheissin, 1948, 1958a, 1967)

or 70—150 (Pellérdy, 1974) spindle-shaped merozoites, each 7—8 × 2. Beginning on day 5 PI, second (or second and third?) generation meronts were of two types, both in the distal part of the villus. Type A second-generation meronts were seen on day 5 PI and matured on day 9 PI. They were large, 13—30 wide, with 70—150 slender second-generation merozoites, each ~10—12 × 0.5—1, arranged within the meront like "bunches of bananas." Type B second-generation meronts appeared 7—8 days PI, and were 13—16 wide, with 35—80 merozoites, ~9—11 × 1 (Cheissin, 1948). However, in later accounts by Cheissin (1958a, 1967), second-generation meronts were 10—12 wide with five or 10—20 short, broad merozoites, each 8—10 × 1—1.5; these two stages/types are present until about day 9 PI. Between days 7 and 10 PI, small, 13—15 wide, third (or fourth?) generation meronts appeared in the distal part of the villi with 15—35 narrow merozoites (Cheissin, 1967), while Pellérdy (1974) said they were 10—16 wide with 15—18 slender merozoites each. On day 8 PI, the merozoites migrated from the small intestine into the cecum and appendix and spread for a long distance along the intestine (Cheissin, 1967). Cheissin (1948) said that all meronts had a residuum, while Pellérdy (1953, 1974) said that there was none in the Type B second-generation meronts. Few, if any, meronts can be found beyond 11 days PI. Cheissin (1958a) described the cytological and histochemical reactions of the endogenous stages of *E. intestinalis*. He found different quantities of glycogen in larger meronts, in merozoites, in RBs of microgametocytes, in growing macrogametes, in unsporulated oocysts, in SZs, and in the sporoplasm of unsporulated oocysts. He attributed its accumulation to the parasite sequestering existing carbohydrate from the tissues of the host epithelium and said that such energy stores were necessary for the movement and penetration of each of the motile stages (merozoites, microgametes, sporozoites) and the long-term survival of the non-motile (oocyst) stages outside the host. However, more recent accounts on the invasion and endogenous development of this species (Licois et al., 1992b; Drouet-Viard et al., 1994) provide a more detailed and complex picture. First, in experimental infections with 3×10^7 sporocysts, excystation occurred in < 10 min in the duodenum and proximal jejunum and SZs began penetrating the duodenal epithelium as early as 10 min PI (Drouet-Viard et al., 1994). From there they migrated to the ileum, by an "unknown non-lumenal tissue route," where they entered epithelial cells along the entire length of the villi and their development was studied from 24 hr onward by Licois et al. (1992b).

Licois et al. (1992b) reported four merogonous (asexual) generations, with the first three, but not the fourth, each having two types of meronts; their first, which they regarded as female, had mononuclear merozoites and their second, regarded as male, had polynuclear merozoites. Mononuclear first-generation meronts were 10.3×8.6 with 15—25 merozoites that were 2.6 long, while the polynuclear first-generation meronts were 7.0×5.9 with 3—5 merozoites that were 1.9×1.8; both meronts peaked in numbers at 80 (36—144) hr PI. Mononuclear second-generation meronts were 8.2×7.2 with 10—20 merozoites that were 4.9×1.8, while the polynuclear meronts were 7.7×5.5 with 5—8 merozoites that were 5.3×1.7; both second-generation meronts peaked in numbers at 96 (64—168) hr PI. Mononuclear third-generation meronts were 20.7×13.8 with 60—120 merozoites that were 6.6×0.7, while the polynuclear meronts were 14.1×11.3 with < 30 merozoites that were 7.9×1.7; both third-generation meronts peaked in numbers at 144 (96—192) hr PI. All fourth-generation meronts were mononuclear, 12.4×9.5, with < 30 merozoites that were 6.4×0.5, and peaked in numbers at 210 (168—240) hr PI.

According to Cheissin (1958a), gametogony began about 185—192 hr PI and gamonts developed in the small intestine, cecum, and ascending colon, generally above the host cell nucleus (HCN); in the small intestine, these stages were in the epithelial cells of the villi, while gametogony that occurred in the large intestine always occurred in cells of the crypts (Cheissin, 1958a). Licois et al. (1992b), however, said they first recognized young gamonts at 144 hr PI and by 168 hr PI both macrogametocytes and zygotes were plentiful. Beyer and Ovchinnikova (1964) were the first to quantify the RNA content in growing macrogametes of *E. intestinalis* using the **microspectrophotometer** to scan fixed, stained macrogametes. Their results indicated that although the RNA concentration decreased as the macrogamont grew, the total amount of RNA in the cytoplasm of the macrogamete continuously increased, with the highest value being measured just before fertilization and formation of the oocyst wall. For *E. intestinalis* microgamonts, the quantity of RNA rose evenly throughout its development, which was somewhat different than the increase they saw of RNA in the microgamonts of *E. magna* (see below).

Microgametogony was studied by Cheissin (1958a, 1964, 1965, 1967) and Snigirevskaya (1969a, b), the latter with both the light and electron microscope. Microgamonts are 5 wide on day 12 PI and contained 2—3 N. In the first phase of microgametogony, the microgamont enlarges and the number of N increases. When it was 10—12 wide, it had 16—24 N. It remained ovoidal and attained a size of $20—27 \times 18—20$ (Cheissin, 1967). The N at first were scattered in the cytoplasm, but later they came to lie on the periphery. The microgamont was rarely divided into cytomeres. The N decreased in diameter as their numbers increased. A large mitochondrion formed, presumably as the result of fusion of several smaller ones, and was associated with each N. The **anlagen** of the **basal bodies** of the future microgamete flagella appeared when the microgamonts were 12—16 wide and they lie near the N. The two flagella form during the first phase and come to lie on the surface of the microgamont. During the second phase, the microgametes differentiated. The N did not increase in number, but they elongated, eventually becoming comma-shaped and occupied the posterior $^2/_3$ of the body. Large protuberances formed on the microgamont; each contained a N, a single mitochondrion, and the two flagellar anlagen. As the N elongated, so did the flagella. The microgametes remained attached to the microgamont by their anterior end until the pointed **perforatorium** was formed, after which they separated, leaving a large residuum of cytoplasm. About 250—500 microgametes were formed by each microgamont. They were about 3—4 × 0.5, with flagella 5—7 long. Macrogametes were piriform when mature. The same host cell may harbor both macrogametes and microgametes. Mosevich and Cheissin (1961) and Snigirevskya (1968, 1969a, b, c, 1972) used the EM to investigate the various endogenous stages of this species and said that meronts, merozoites, and young micro- and macrogametocytes all possessed **micropores**, but the mature microgametes themselves did not. Snigirevskya (1972) also noted that all endogenous stages were limited to one plasma membrane, 70—80 Å thick, but in the pellicle of budding meronts, microgametocytes, and microgametes, an auxiliary layer appeared, underlying the plasmic membrane. Macrogametes had numerous micropores which acted as **cytostomes** and Snigarevskaya (1969c) said that nutrition was via these micropores in young macrogametes and by **pinocytosis** in mature ones.

Cross-transmission: None reported to date.

Pathology: Although relatively rare, E. intestinalis can be pathogenic, especially in young domesticated rabbits; in fact, some regard it as one of the most pathogenic coccidia of rabbits (Courdet, 1976; Licois et al., 1978a, b; Catchpole and Norton, 1979; Peeters et al., 1984). According to Cheissin (1948), this species is non-pathogenic for adult rabbits, even in large doses; however, it is pathogenic for 1-mo-old rabbits. He said that they lost their appetites on day 7 PI, sat motionless, and that their stomachs swelled; on days 9—10 PI, diarrhea appeared, which lasted 3—4 days. Heavily infected rabbits died on days 12—13 PI, while those with light infections recovered. Cheissin (1948) found pathogenic changes in the intestine 7—8 days PI. The lower part of the small intestine was significantly enlarged and its wall was thickened, with white spots (containing meronts and gamonts) on its surface. Epithelial cells infected with meronts or gamonts were greatly enlarged, as were the villi that housed these infected cells. Nine to 10 days PI, a caseous exudate containing epithelial cells with parasites, leukocytes, free merozoites, and oocysts accumulated in the intestinal lumen. Intense desquamation of the epithelium occurred with massive formation of oocysts 10—11 days PI. Pellérdy (1953, 1954b) said that experimental infections caused more or less severe intestinal **catarrh** and diarrhea, and might kill young rabbits. At necropsy, edema and grayish-white foci, which can coalesce to form a homogeneous, sticky, purulent layer, might be found in the intestine. Coudert (1976, 1979a, b) said that E. intestinalis is extremely pathogenic and Licois et al. (1978a, b) said that in young rabbits infected 10 days earlier with either E. intestinalis or with what they called E. pellerdyi (= E. flavescens?) the pathogenesis of diarrhea resulting from these coccidial infections was as much the result of nutritional deficiency, especially potassium, as of a malfunctioning of the digestive tract. Coudert et al. (1978) infected 2-mo-old rabbits with $3 × 10^4$ oocysts of E. intestinalis and E. flavescens (= E. pellerdyi) and found that by 10 days PI there were big changes (+/−) in the relative values of different plasma proteins, including albumin, and α-, β-, and γ-globulins, accompanied by high increases of **lipidemia** and **uremia**. Licois and Mongin (1980) looked at the correlation of sodium (Na^+) and potassium (K^+) related to water content in rabbits infected with E. intestinalis and E. flavescens and reported that (1) the defect in water absorption was due to a lack of Na^+ reabsorption at the site of the coccidial infections and (2) in the colon, Na^+ was reabsorbed against K^+ secretion. Later, Coudert et al. (1993) concluded that the LD_{50} for E. intestinalis is $0.5-1.0 \times 10^4$ sporulated oocysts, but that diarrhea is erratic. Oncel et al. (2011) noted that 10/10 dead rabbit kids in Turkey were thin with reduced fat stores and muscle wasting, had rough hair coats, and the small intestines were distended and filled with gray-green ingesta; the intestinal mucosa was hyperemic and edematose. However, they could not attribute the pathology noted only to E. intestinalis because all 10 kids also were infected with E. coecicola and E. perforans, as determined by oocysts in their feces.

Material deposited: None.

Remarks: Pellérdy (1954b) was working with oocysts from a "wild rabbit (O. cuniculus)" that resembled those of E. piriformis (Kotlán and Pospesch, 1934) except that they possessed an OR, which those of E. piriformis lack. He was apparently unaware of the description of E. intestinalis by Cheissin (1948) at the time. To unravel this mystery, he infected a "domestic rabbit" (likely a lab-reared O. cuniculus) with these oocysts. The oocysts produced from this experimental infection were found to have an OR after they sporulated. This led him to name a new species, E. agnosta, which he distinguished from E. piriformis because its oocysts had an OR, which those of E. piriformis lacked; however, he provided no line drawing or photomicrograph with his description, which made it a species inquirenda. Once Pellérdy (1954a) discovered Cheissin's account of E. intestinalis (1949), he synonymized E. agnosta with it. Pellérdy (1954b) reported the oocysts of E. intestinalis to be 27×18 ($23-30 \times 15-20$) and later (1974) as $27-32 \times 17-20$. According to Gill and Ray (1960), who reported E. intestinalis in India,

a few oocysts have concave surface plaques which give the wall a wavy appearance. Kasim and Al-Shawa (1987) examined 263 adult domestic rabbits from three regions in Saudia Arabia from 1984−1985, and found seven (3%) to be infected with E. intestinalis. Their (1987) oocysts were pear-shaped, 29.2×19.3 ($25-34 \times 17-22$), L/W 1.5 (1.5−1.6), with a M and an OR, and sporocysts 13.1×7.3 ($10-16.5 \times 6-8$), L/W 1.8 (1.5−2.0). Vila-Viçosa and Caeiro (1997) reported E. intestinalis in O. cuniculus in Portugal, with piriform oocysts that measured $33-37.5 \times 18-20.6$. Grés et al. (2003) examined 254 wild rabbits from six localities in France and E. intestinalis as one of 10 species they identified.

When meronts and gamonts of E. intestinalis developed in epithelial cells of the villi of the small intestine, the activity level of **alkaline phosphatase** decreased sharply compared to that of cells of non-infected parts of the intestine. Alkaline phosphatase is localized in the striated border of the epithelial cells of the intestinal mucosa. The more intense the infection, the more clearly the disappearance of alkaline phosphatase from the striated border is expressed. The processes of digestion and absorption of material in the small intestine were disturbed in the same way (work of Beyer, 1960 [paper unavailable], as cited by Cheissin, 1967). Cheissin (1965, microgametogenesis), Cheissin and Snigirevskaya (1965, merozoites), Mosevich and Cheissin (1961, merozoites), Snigirevskaya (1969a, b, microgametogenesis) and Snigirevskaya (1969c, macrogametogenesis) all studied the ultrastructure of various developmental stages of E. intestinalis. Snigirevskaya (1968) found that the meronts have several micropores, including one at the level of the nucleus, that are used for ingestion of nutrients. The merozoites have an **apical complex** consisting of a **conoid**, two **rhoptries**, **micronemes**, 24−30 subpellicular **microtubules**, and several mitochondria (Mosevich and Cheissin, 1961; Cheissin and Snigirevskaya, 1965). Schrecke and Dürr (1970) found that 1-day-old rabbits could be infected if enough oocysts were used. Susceptibility to infections rose slowly with age until weaning, when there was a marked increase. They thought this was due mainly to diet-induced physiological and biochemical properties of the intestine. Licois and Coudert (1980b) attempted to suppress immunity in rabbits immunized against *E. intestinalis*. They inoculated lab rabbits with five progressively larger doses of oocysts (800, 3200, 12,800, 51,200, 204,800) every 2 weeks up to the third infection, then weekly with the final two infection doses. They found that a single infection with *E. intestinalis* developed an immunity in rabbits, which strongly reduced the excretion of oocysts and the clinical signs of illness following subsequent infections; in hyperimmunized rabbits it was not possible for them to break immunity with immunodepressors, or by provoking diarrhea with an antibiotic, or by infecting the rabbits with another species of coccidia (*E. piriformis*).

Licois et al. (1990) developed a precocious line of *E. intestinalis* (EiP). In addition to shortening the prepatent period, they reported the following observations for EiP: (1) oocyst sporulation time was the same for EiP as for the wild strain from which it was developed; (2) excreted unsporulated oocysts and the morphology and measurements of sporulated oocysts were similar between EiP and the wild strain; (3) in EiP sporocysts, a huge SR was located in two of four sporocysts, while no SR was seen in the other two; (4) the EiP line had a reproductive potential 1,000 times lower than that of its parent strain; and (5) as judged by weight gain, mortality, and lesions in the jejunum and ileum, pathogenicity of the EiP strain was substantially reduced when compared to those characteristics in rabbits infected with the wild strain. Coudert et al. (1993) looked at the immunizing potential of *E. intestinalis* and said that as few as six oocysts was sufficient to minimize the clinical expression of disease following challenge with 3×10^3 oocysts 14 days later, and that oocyst output

was reduced by 60%. Moreover, inoculation with 600 oocysts was sufficient for total suppression of the effects of challenge infection.

EIMERIA IRRESIDUA KESSEL & JANKIEWICZ, 1931

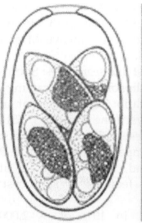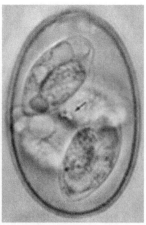

FIGURES 6.10, 6.11 Line drawing of the sporulated oocyst of *Eimeria irresidua* from Catchpole and Norton, 1979, with permission from *Parasitology.* Photomicrograph of a sporulated oocyst of *E. irresidua* from Norton et al., 1979, with permission from *Parasitology.*

Synonym: *Eimeria elongata* Marotel & Guilhon, 1941.

Type host: *Oryctolagus cuniculus* (L., 1758) (syn. *Lepus cuniculus*), European (domestic) rabbit.

Type locality: NORTH AMERICA: USA: California, Los Angeles.

Other hosts: *Sylvilagus floridanus* (J.A. Allen, 1890), Eastern cottontail; *Lepus nigricollis* F. Cuvier, 1823 (syn. *Lepus ruficaudatus* Geoffroy, 1826), Indian hare (?); see *Remarks* below.

Geographic distribution: ASIA: Azerbaijan (former USSR); India; AUSTRALIA: New South Wales, Western Australia; EUROPE: Belgium, Czech Republic, France, Italy, Poland, Portugal; MIDDLE/NEAR EAST: Iran, Saudia Arabia,

Syria, Turkey; NORTH AMERICA: USA: California; SOUTH AMERICA: Brazil.

Description of sporulated oocyst: Oocyst shape: ellipsoidal to slightly ovoidal, blunt at the end with the M, such that the side walls appear parallel; number of walls: 1; wall characteristics: thicker on end with M, but tapering slightly at opposite end, and usually light yellow or dark brown; L × W: 38.3 × 23.8 (25—49 × 16—28); L/W ratio: 1.6; M: present; M characteristics: concave, conspicuous; OR, PG: both absent. Distinctive features of oocyst: concave M and lack of both OR and PG.

Description of sporocyst and sporozoites: Sporocyst shape: ovoidal to ellipsoidal (line drawing); L × W: 20 × 10; L/W ratio: 2.0; SB: present, at pointed end of sporocyst (line drawing); SSB, PSB: both absent; SR: present; SR characteristics: irregularly elongate, compact and granular, 4—9 wide; SZ: lie end-to-end, with a large RB at wider end. Distinctive features of sporocyst: distinctive, pointed SB.

Prevalence: Reported in 10% of > 2,000 domestic rabbits from southern California studied by Kessel and Jankiewicz (1931). Lund (1950) found *E. irresidua* in 48/1200 (4%) domestic rabbits from the same area. On the basis of a 35-mo study in a free-living population of wild rabbits on an experimental enclosure in Australia, Mykytowycz (1956, 1962) found that, once established within the population, *E. irresidua* was highly persistent, tended to maintain a high level of infection regardless of the season, was fairly common in rabbits of all ages, and when it occurred its oocysts were usually present in moderately high numbers. From 1954—1958, Pastuszko (1963) examined 9,845 rabbits from 17 districts in Poland; these included 502 rabbits by necropsy and 9,343 fecal samples tested. She said (1963) that *E. irresidua* caused coccidiosis in rabbits in Poland; the oocysts were 38.3 × 25.6 (23—43 × 14—27; her Table 1). Niak (1967) found it in 16/84 (19%) rabbits from four rabbitries in Iran. Stodart (1971) said she found it in 505/656 (77%) wild

rabbits in eastern Australia (New South Wales). Peeters et al. (1981) found it in 39/533 (7%) fecal samples from commercial rabbitries and in 21/282 (7%) sick rabbits in traditional rabbitries in Belgium. Veisov (1982) examined 3,150 rabbits from three different areas/districts of the Cuba-Khachmazskoy zone in Azerbaijan and found 2,905 (81%) infected with nine species of *Eimeria*, including *E. irresidua*; however, we were not able to determine how many of the rabbits examined harbored *E. irresidua*. Kasim and Al-Shawa (1987) examined 263 adult domestic rabbits from three regions in Saudia Arabia from 1984—1985, and found 79 (30%) to be infected with *E. irresidua*. Their (1987) oocysts were ellipsoidal, 37.1 × 23.8 (33—43 × 21—26), L/W 1.6 (1.4—2.0), with a M, but without an OR, and sporocysts were 18.5 × 9.3 (16—22 × 8—11), L/W 2.0 (1.6—2.2). Santos and Lima (1987) first documented *E. irresidua* in Brazil and Santos-Mundin and Barbon (1990) reported it in 123/375 (33%) rabbits in Minas Gerais, Brazil. Darwish and Golemansky (1991) found *E. irresidua* in 18/75 (24%) domestic rabbits from four localities in Syria. Polozowski (1993) examined feces from 246 rabbits in six farm rabbitries in the Wroclaw District of Poland and found 166 (67.5%) infected with *E. irresidua*; 141/166 (85%) infected rabbits were < 3-mo-old, 6/166 (4%) were 4—12-mo-old, 14/166 (8%) were 13—24-mo-old, while only 5/166 (3%) were > 24-mo-old. Hobbs and Twigg (1998) examined fresh fecal samples from wild *O. cuniculus* in Western Australia and found 107/1200 (9%) to harbor this species; 50 of their sporulated, ellipsoidal oocysts measured 39.1 × 24.3 (35—45 × 21—27) with sporocysts 18.7 × 9.7 (16—21 × 9—11). Yakhchali and Tehrani (2007) found *E. irresidua* in 122/436 (21%) New Zealand white rabbits from two rabbitries in northwestern Iran. Razavi et al. (2010) found *E. irresidua* in 3/71 (4%) wild rabbits in southern Iran; their oocysts were 37.2 × 23.8 (35—41 × 19—26), L/W 1.6, with a M, but lacking an OR, and sporocyst were 17.9 × 8.7 (14—22 × 7—11).

Sporulation: Exogenous. Oocysts sporulated in 65 (50–70) hr in 2% (w/v) $K_2Cr_2O_7$ solution (Kessel and Janciewicz, 1931; Carvalho, 1943); 50 hr at 20°C (Lizcano Herrera and Romero Rodriguez, 1969); 46 hr or less at 27°C (Norton et al., 1979) in 2% (w/v) $K_2Cr_2O_7$ solution; 50–90 hr at 18–20°C (Pastuszko, 1963) or 48 hr at 27°C in 2% (w/v) $K_2Cr_2O_7$ solution (Kasim and Al-Shawa, 1987); Razavi et al. (2010) said that sporulation occurred in 56 hr in 2.5% $K_2Cr_2O_7$ at 25–28°C.

Prepatent and patent periods: Prepatent period is 8–9 days (Carvalho, 1943; Norton et al., 1979), while the patent period is 17–21 days (Carvalho, 1943) or 10–13 days, but can last 16 days (Cheissin, 1940, 1967).

Site of infection: Endogenous development occurs throughout the small intestine, and can be found within epithelial cells of the villi from the duodenum to the jejunum to the lower ileum (Kessel and Jankiewicz, 1931; Rutherford, 1943). The heaviest infection is located in the upper 45–46 cm of the upper third of the small intestine (Rutherford, 1943). Only merogony occurs in the epithelium, while gamogony, fertilization, and oocyst formation take place in the subepithelium. The endogenous stages develop along the entire villus and in the crypts (Cheissin, 1967). The meronts and gamonts of this species also are reported to localize above the host cell nucleus (HCN) in the epithelial cells they occupy.

Endogenous stages: There are at least four generations of meronts. Six hr after inoculation with sporulated oocysts, SZ are free in the intestinal lumen and, after penetration, young meronts began developing in the epithelium, especially in the first 45 cm below the stomach. Rutherford (1943) first described the life-cycle stages of this species. He said that each of the first two merogenous stages had a Type A meront and a Type B meront. Type A meronts developed distal to the HCN, while Type B meronts developed proximal to the HCN. In the first-generation meronts, both Types A and B became mature about day 6 PI. Type A first-generation meronts were 8.5 × 7.2 and contained ~4 (2–10) banana-shaped merozoites, pointed at both ends, that measured 6.5–7.2 × 1.5–1.8, while Type B first-generation meronts were 17 × 13 and contained 36–48 banana-shaped merozoites that were 11.5 × 1.8. The Type B first-generation meronts pushed past the basement membrane with their host cells bulging into the subepithelial layers. Both types (A and B) of first-generation meronts liberated their merozoites at about the same time by gradual necrosis of the tips of the villi. Second-generation meronts, also with Types A and B, matured on days 9–10 PI. The second-generation Type A meronts produced 8–10 merozoites that were 7.2 × 2.0 and were pointed at both ends, while the Type B meronts produced 48–75 merozoites that were 7.2 × 1.3. Norton et al. (1979) redescribed merogony in this species, which they said took place in the jejunum and ileum. Their first-generation meronts were found in the glands, the second-generation meronts in the lamina propria, and their third and fourth-generation meronts in the villous epithelium. They saw first-generation meronts as early as 96 hr, but there were not enough for them to get good measurements. According to Norton et al. (1979), first-generation merozoites penetrated neighboring cells below the epithelium and appeared to migrate through the lamina propria towards the base of the villi. Colonies of second-generation meronts were present throughout the entire small intestine, except the first and last 20 cm (Norton et al., 1979). Norton et al. (1979) said that at ~120 hr PI, mature second-generation meronts were 35.7 × 30.7 (29–39 × 26–35) with ~50 merozoites that were 15.8 × 2.4. Their third- and fourth-generation meronts could not be distinguished morphologically, but they occurred at slightly different times PI, with third-generation meronts at 144 hr PI and fourth-generation meronts at 168 hr PI. Meronts from these two generations, combined, were

16.4 × 25.0 (20−29 × 16−25), with 40−50 merozoites, 15.8 × 2.5, tightly packed around a small
RB. Cheissin (1967) also looked at endogenous
development of this species and said that at
the same time as the second-generation meronts, "even larger meronts are found" that
measured 7−10, each with 1−20 merozoites,
10−12.5 × 2−4 (among the largest known in
the rabbit coccidia). Could this be a third-
generation meront or is it just a larger version
of the second generation? Also according to
Cheissin (1967), on day 5 PI, small meronts,
5−7, each with 2−4 merozoites were found; he
called these the third-generation meronts, but
for unexplained reasons. Then on day 7 PI, meronts 7−12 long were found which contained
12−25 merozoites, each 5−8 long; Cheissin
(1967) called these fourth-generation meronts.
Also, unlike many species of coccidia from
non-lagomorphs, which have only one N per
merozoite, sickle-shaped polynuclear merozoites were observed in *E. irresidua* and a few
others, also from rabbits (*E. magna*, *E. media*)
(Cheissin, 1967). Such merozoites have from
two to eight N situated along their length, and
they are larger than the mononuclear merozoites of the same generation.

Beginning on about days 5−6 PI, young gamonts appear. Rutherford (1943) thought that
Type B second-generation merozoites turned
into macrogametes. Young macrogametes
occurred at the base of the epithelial cells in
close association with connective tissue
elements, much like the Type B first-generation
merozoites. At day 6 PI, the gametes and merozoites did not appear to be limited to any particular region of the mucosa. Young macrogametes
were spheroidal, 3.5−4.5 wide, and had a central
karyosome in the N and an indefinite nuclear
border with little cytoplasm. At day 8 PI, they
were 14−16 wide with a N, ~3 wide, and a large
karyosome. By day 9 PI, those that had been
fertilized had formed an oocyst wall with a M,
and were about 36 × 24 (Rutherford, 1943). Norton et al. (1979) found macrogamonts in cells of

the epithelium and lamina propria that were
33.9 × 20.5. Microgamonts were mature at day
9 PI and were 28.5 × 20.0, with a large number
of biflagellate microgametes (Rutherford,
1943). Norton et al. (1979) first saw young gamonts at 192 hr PI, which matured over the next
24 hr. These measured 37.2 × 32.6 and were
found in the villous epithelium of the jejunum
and ileum. Each microgamont produced many
flagellated microgametes with a central N.
Cheissin (1967) said microgametocytes were
30−60 × 30−50 and formed many
microgametes.

Cross-transmission: Kessel and Jankiewicz
(1931) were not able to transmit this species to
domestic chickens or guinea pigs. Carvalho
(1943) transmitted it experimentally to the
cottontail rabbit, *S. floridanus mearnsi*. Pellérdy
(1954a) was not able to infect the hare, *L. europaeus*, with this species isolated from the
domestic rabbit.

Pathology: This is one of the more pathogenic
of the intestinal coccidia of rabbits according to
Kessel and Jankiewicz (1931) and Levine and
Ivens (1972). Most endogenous stages of *E. irresidua* lay below the HCN, causing the host cell to
deteriorate, and, depending on the size of the
inoculating dose, the virulence of the strain,
and the susceptibility of the host, cell destruction results in inflammation of varying severity
(Pellérdy, 1974). Kessel and Jankiewicz (1931)
said that *E. irresidua* destroyed many epithelial
cells and caused inflammation and hyperemia.
There may be **extravasation** of blood from capillaries and the epithelium may slough and
become denuded. However, Norton et al.
(1979) said that it is not pathogenic, that heavy
infections produce only a transient pause in
weight gain, while Pellérdy (1974) said it is,
and should be regarded as a pathogenic
coccidium of the tame rabbit, even though it is
relatively rare.

Material deposited: None.

Remarks: Most of the description of the sporulated oocyst, above, is from the original

description (Kessel and Jankiewicz, 1931) and from Carvalho (1943); however, Norton et al. (1979) redescribed the sporulated oocysts of this species as follows: broadly ellipsoidal, with M, ~4 (3—7) wide; wall two-layered; outer is smooth, yellow, 1.4 thick; inner as a dark line, 0.4 thick; L × W 38.4 × 23.2 (35—42 × 19—28); L/W ratio 1.65; OR presence variable, being absent in 20/170 (12%) oocysts they studied; of the remaining oocysts, 57% had an OR composed of only a few granules, while 31% had a spheroidal OR ~2 (1.5—5) wide. Sporocysts were elongate ovoidal, 19 × 9 (15—22 × 7—11), with a distinct SB and a prominent SR; SZ were elongate, 22 × 4 after excystation in bile and trypsin, with 1—3 clear RBs each. Oocysts of *E. irresidua* resemble those of *E. magna* and *E. stiedai*, but differed by being larger, by having pointed sporocysts, and by having a conspicuous M that was concave instead of being convex like those of the other two species. Oocysts of *E. irresidua* also did not exhibit the marginal thickening of the wall around the M, which was seen in *E. magna*. Ryšavý (1954) said he found *E. irresidua* in *L. europaeus* in the Czech Republic with oocysts that were 28.9—42.8 × 13.6—26.7, with sporocysts 11 × 7, and that these oocysts sporulated in 50—60 hr; he (1954) also said the parasite was in the small intestine (but didn't say how he determined this) and that it was abundant in the hares and rabbits he sampled. Morgan and Waller (1940) said they found *E. irresidua* in the majority of 210 cottontails, *S. floridanus*, in Iowa, and that it had a large, conspicuous M on oocysts that were 36.5 × 18.4 μm, but Carvalho (1943) was unable to find it in *S. floridanus* from the same area. Carvalho (1943, p. 110) said, "it has also been reported from *L. californicus* and *S. floridanus* ...," but he did not provide a literature citation for his statement and we are unable to find any such reference. However, Carvalho (1943) described a similar form from *L. campestris townsendii* as *E. irresidua* form *campanius*; this is now *E. campania* and is most likely the species

that occurs in hares and the species seen by Ryšavý (1954). The form described as *E. irresidua* from the Indian field hare, *L. nigricollis*, by Gill and Ray (1960), also is most probably *E. campania*. Vila-Viçosa and Caeiro (1997) reported *E. irresidua* in Portugal, with ovoidal oocysts from *O. cuniculus* being 35.6—37.5 × 22.5—24.3. It has been successfully transmitted to *Sylvilagus*, but the validity of *E. irresidua* as a parasite of *Lepus* species is doubtful and will need to be verified by cross-transmission and/or molecular studies.

EIMERIA LEPORIS NIESCHULZ, 1923 (FIGURES 5.26, 5.27)

Synonyms: non *Eimeria leporis* Nieschulz of Morgan and Waller, 1940 and of Waller and Morgan, 1941; *Eimeria leporine* Nieschulz, 1923 of Sugár et al. (1978).

Type host: *Lepus europaeus* Pallas, 1778, European hare.

Remarks: Ryšavý (1954) said he found abundant numbers of oocysts of *E. leporis* in both *O. cuniculus* and in *L. europaeus* in the Czech Republic, with oocysts that were 25.5—36.3 × 13.6—17.5, sporocysts 12 × 6.7, and that these oocysts sporulated in 48—60 hr. However, Nieschulz (1923), Pellérdy (1956), and Lucas et al. (1959) were unable to infect domestic rabbits, *O. cuniculus*, with this species isolated from *L. europaeus*. We believe that molecular and/or cross-transmission confirmation is needed before we can accept *E. leporis* as a valid species of *O. cuniculus*. See Chapter 5 for the complete description of *E. leporis*.

EIMERIA MAGNA PÉRARD, 1925b

Synonym: *Eimeria perforans* var. *magna* Pérard, 1925b.

Type host: *Oryctolagus cuniculus* (L., 1758) (syn. *Lepus cuniculus*), European (domestic) rabbit.

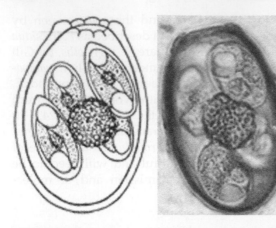

FIGURES 6.12, 6.13 Line drawing of the sporulated oocyst of *Eimeria magna* from Carvalho, 1943, with permission from John Wiley & Sons Ltd. (current rights owner of the *Iowa State College Journal of Science*). Photomicrograph of a sporulated oocyst of *E. magna* from Hobbs and Twigg, 1998, with permission from the *Australian Veterinary Journal*.

Type locality: EUROPE: France.

Other hosts: *Lepus californicus* Gray, 1837, the Black-tailed jackrabbit (?); *Lepus capensis* L., 1758, Cape hare (?); *Lepus europaeus* Pallas, 1778, European hare (?); *Lepus nigricollis* F. Cuvier, 1823 (syn. *L. ruficaudatus*), the Indian hare (?); *Lepus timidus* L., 1758, the Mountain hare (?); and *Sylvilagus floridanus* (J.A. Allen, 1890), Eastern cottontail (experimental, Carvalho, 1943). Its validity as a parasite of *Lepus* species needs to be verified by cross-transmission and/or molecular studies.

Geographic distribution: AFRICA (?): Nigeria (?); ASIA: Azerbaidzhan (former USSR), India, Japan, Taiwan; AUSTRALIA: New South Wales, Western Australia; EUROPE: Belgium, Czech Republic, France, Germany, Italy, Poland, Portugal; MIDDLE/NEAR EAST: Iran, Iraq, Saudi Arabia, Syria; NORTH AMERICA: USA: California, Iowa; SOUTH AMERICA: Brazil.

Description of sporulated oocyst: Oocyst shape: ovoidal to ellipsoidal; number of walls: 2, of which the outer layer is easily detached; wall characteristics: dark yellow to orange to brownish and appears truncated at the end

with the M; L × W: 35 × 24 (27–41 × 17–29); L/W ratio: 1.5; M: present; M characteristics: a collar-like thickening of outer layer around M, but this may not be seen if the outer layer has been detached; OR: present; OR characteristics: large, 4–12 wide, although in exceptional cases, Cheissin (1947b) said there may be none; PG: absent. Distinctive features of oocyst: collar-like thickening of outer wall layer around M and a very large OR.

Description of sporocyst and sporozoites: Sporocyst shape: ovoidal to ellipsoidal (line drawing); L × W: 15 × 7.8 (11–16 × 6–9); L/W ratio: 1.9; SB: present, small (Pellérdy, 1974) or absent (Pérard, 1925b; Levin and Ivens, 1972); SSB, PSB: both absent; SR: present; SR characteristics: coarsely granular and appears as a small cluster (1–3 or up to 4–5 wide) in the middle of sporocysts between SZs; SZs: lie head-to-tail around the SR, each with a central N and one large RB at their broad end. Excysted SZ were 19.7 × 3.3 (Ryley and Robinson, 1976). Distinctive features of sporocyst: distinct SR and the very small (or absence of) SB.

Prevalence: This species is fairly common. Kessel and Jankiewicz (1931) found it in about 200 of > 2,000 (19%) domestic rabbits in California. Lund (1950) reported *E. magna* in almost 100/1,200 (8%) domestic rabbits in southern California. Mykytowycz (1956, 1962) did a 35-mo study of a free-living population of wild rabbits in an experimental enclosure in Australia. He found that the average level of infection fluctuated markedly with the season, but was unable to find a satisfactory explanation for the fluctuation. *Eimeria magna* was one of the least common species that he found in rabbits of all ages, and it occurred most frequently in 1-mo-old rabbits. Niak (1967) found it in 25/84 (30%) rabbits from four rabbitries in Iran. Mirza (1970) found it in 11/31 (35%) domestic rabbits in Iraq. Stodart (1971) found *E. magna* in 154/656 (23.5%) wild rabbits in New South Wales, eastern Australia. Peeters et al. (1981) found it in 393/770 (51%) fecal samples from commercial, domestic rabbitries and in 104/282 (37%) sick rabbits in traditional rabbitries in Belgium. Veisov (1982) examined 3,150 rabbits from three different areas/districts of the Cuba-Khachmazskoy zone in Azerbaijan and found 2,905 (81%) infected with nine species of *Eimeria*, including *E. magna*; however, we were not able to determine how many of the rabbits examined harbored this species. Kasim and Al-Shawa (1987) examined 263 adult domestic rabbits from three regions in Saudia Arabia from 1984–1985, and found 109 (41%) to be infected with *E. magna*. Santos and Lima (1987) first documented *E. magna* in Brazil and Santos-Mundin and Barbon (1990) reported it in 146/375 (39%) rabbits in Minas Gerais, Brazil. Darwish and Golemansky (1991) found *E. magna* in 25/75 (33%) domestic rabbits from four localities in Syria. In an unpublished Master's thesis, Ogedengbe (1991) said he found *E. magna* in 19.5% of the rabbits surveyed in Kaduna State, Nigeria. However, he (1991) neglected to state how many rabbits were surveyed, nor did he mention the host species; Anonymous (2012) listed the Cape hare, *L. capensis*, as the only lagomorph species found in Nigeria. Polozowski (1993) examined feces from 246 rabbits in six farm rabbitries in the Wroclaw District of Poland and found 163 (66%) infected with *E. magna*; 149/163 (91%) infected rabbits were < 3-mo-old, 3/163 (2%) were 4–12-mo-old, 9/163 (5.5%) were 13–24-mo-old, while only 2/163 (1%) were > 24-mo-old. Hobbs and Twigg (1998) examined fresh fecal samples from wild *O. cuniculus* in Western Australia and found 391/1,200 (33%) with oocysts of *E. magna* and 50 of their sporulated, ovoidal oocysts measured 35.2 × 21.8 (30–42 × 18–26), with sporocysts 15.3 × 8.5 (14–18 × 7–10). Grès et al. (2003) found *E. magna* in 17–42% of 254 wild rabbits sampled from six different localities in France. They identified 10 eimerians from these 254 rabbits and, apparently, 100% were infected with one or more of these 10. Unfortunately, they did not specifically mention how many individuals were infected with multiple species vs. those with single species infections. Yakhchali and Tehrani (2007) found *E. magna* in 152/436 (35%) New Zealand white rabbits from two rabbitries in northwestern Iran. Razavi et al. (2010) found *E. magna* in 12/71 (17%) wild rabbits in southern Iran; their oocysts were 35.1 × 23.9 (29–38 × 17–29), L/W 1.5, with a M and with an OR, 11.6 (10–14) wide, and sporocyst 15.3 × 8.6 (12–17 × 6–10). Li et al. (2010) reported *E. magna* in 101/642 (16%) *O. cuniculus* from pet shops (63/642) and farms (38/342) in 11 districts on the island of Taiwan.

Sporulation: Exogenous. Oocysts sporulated in a minimum of 2 days (Pérard, 1925b); 3–5 days at room temperature (Cheissin, 1967); 48–52 hr for small oocysts and 62–72 hr for large ones (Cheissin, 1947b); 50 (48–55) hr (Carvalho, 1943; Kasim and Al-Shawa, 1987); 50–60 hr (Matsubayashi, 1934); 2–3 days at 18–20°C (Pastuszko, 1961b, 1963); or 4 days (Mandal, 1976). Ryley and Robinson (1976) studied samples from fresh rabbit droppings placed at a range of temperatures and examined them at 6-hr intervals for percentage sporulation. They (1976) found that sporulation proceeded rapidly at temperatures from 18 to 30°C, being completed by about 48 hr, and that oocysts held at 0–4°C did not sporulate, but did so normally when transferred to a higher temperature, reaching 90+% sporulation in 3 days at 25°C; Razavi et al. (2010) saw sporulation in 52 hr in 2.5% $K_2Cr_2O_7$ at 25–28°C; Li et al. (2010) said that sporulation occurred in 32–44 hr in 2.5% $K_2Cr_2O_7$ at 25°C.

Prepatent and patent periods: Prepatent period is 7–9 days (Cheissin, 1940) or 6–7 days PI (Ryley and Robinson, 1976). The patent period is 15–19 days or 180–186 hr PI when infected with 150–200 oocysts and 216 hr PI when infected with one oocyst (Cheissin, 1940, 1967) or 16 (12–21) days (Carvalho, 1943).

Site of infection: Endogenous developmental stages were found in columnar epithelial cells of the villi in the middle to lower part of the small intestine (Pérard, 1925b; Shazly et al., 2005), but Carvalho (1943) said that cells of the cecum and large intestine were infected as well. Matsubayashi (1934) said it located in the "tunica propria of mucous membrane of the intestine." Rutherford (1943), who first described the life cycle, said it developed in the villar epithelial cells either above or below the HCN from the middle jejunum to the posterior end of the ileum, but not in the duodenum. Cheissin (1967) said that in an average-intensity infection, the first-generation meronts localized about 15 (10–20) cm from the cecum and the second-generation meronts were found about 30 cm along the length of the colon. In an intense infection, endogenous stages spread over the length of the lower half of the small intestine, with gamonts sometimes found in the appendix and cecum (Cheissin, 1947a, 1967). Development of all stages began in epithelial cells, but infected cells sink into the tunica propria where mature meronts, gamonts, and oocysts were localized.

Endogenous stages: The SZ of *E. magna* enter their host cells within 12 hr after oral ingestion of sporulated oocysts. Pérard (1925b) and Rutherford (1943) were the first to study the life cycle of this species within intestinal epithelial cells of the rabbit; unfortunately, Rutherford's description of the asexual stages is quite confusing: "On the 4th day young and adult merozoites will be seen which belong to two distinct types previously mentioned as occurring in *E. irresidua*." This leads into a description of Type A and Type B merozoites, each with a first and second generation. These are so intertwined in his description as to create sufficient confusion that, from our point of view, his description of these early stages is not of much use. Cheissin (1940, 1965, 1967) believed that Rutherford's Type A and Type B meronts and merozoites belonged to different generations. He described the process as follows: the SZ infected epithelial cells, but then sink into the tunica propria, where

mature meronts, gamonts, and oocysts develop. Young, first-generation meronts developed as early as 48 hr PI, and measured 11 × 7, with 8–24 merozoites (Cheissin, 1967, p. 32). By 92–96 hr (or days 5–6) PI, larger, second-generation meronts appeared; they were 14–20 wide and formed 6–40 second-generation merozoites that were 8.9 × 2.1 (Cheissin, 1967, p. 29) or they were 14–36 wide and contained as many as 40 merozoites (Cheissin, 1967, p. 32); he attributed size differences seen in second-generation meronts and merozoites to sexual dimorphism. At 120–168 hr, or days 6–7 PI, even larger third-generation meronts appeared that were 25–40 wide and had 30–80 merozoites, each 5–10 long (Cheissin, 1967). On days 7–8 PI, another generation of meronts (and perhaps two) appeared. They were 25–45 long and each had 30–60 or, perhaps, up to 125 merozoites, which were 6 × 2. Matsubayashi (1934) saw one meront with 60 merozoites.

From day 4 PI, it was often possible to see sickle-shaped polynuclear merozoites (vs. those which have only one N per merozoite; Cheissin, 1967, pp. 41–42); multinuclear merozoites also have been observed in other *Eimeria* species from *O. cuniculus* (e.g., *E. coecicola*, *E. irresidua*, others). These merozoites have from two to eight N situated along their length, and they are larger than the mononuclear merozoites of the same generation. The binuclear second-generation merozoites of *E. magna* measured 8 × 2, trinuclear merozoites were 9 × 3, and hexanuclear merozoites were 10 × 3. The binuclear third-generation merozoites measured 10 × 3, the tetranuclear third-generation merozoites were 13 × 4, and the octanuclear third-generation merozoites were 15 × 5 (Cheissin, 1967). The polynuclear merozoites formed in small numbers in one meront. When this happened, the merozoites formed always had the same number of N. Cheissin (1940, 1947a, 1967) reported that the cytoplasm of meronts contained round or bacillus-shaped mitochondria with numerous short internal tubes in *E.*

magna and in those of *E. irresidua* (Cheissin, 1967, pp. 42—45). There seemed to be no synchrony in the development of asexual generations and, in some cases, meronts of different generations were present in cells of the intestinal mucosa at the same time. Danforth and Hammond (1972) studied merogony of these multinucleate forms with the electron microscope. In rabbits killed on day 4 PI they found that some of the meronts had uninucleate merozoites, while others had multinucleate ones; they interpreted the multinucleate forms to be merozoite-shaped meronts which contained **rough endoplasmic reticulum**, mitochondria, **micronemes**, clear globules, and, occasionally, lipid droplets, and some N contained a **nucleolus**. Danforth and Hammond (1972) thought that the presence of multinucleate merozoites indicated that there might be another merogonic generation in the same **parasitophorous vacuole** (PV). Merogony (with **conoid** formation, etc.) was taking place within these meronts; that is, **endopolygeny** was occurring. Senaud and Černá (1969) described the ultrastructure of merozoites and their merogonous process in this species.

Ryley and Robinson (1976) studied the endogenous development of *E. magna* at 6-hr intervals from both fresh intestinal scrapings squashed in saline and in fixed and sectioned intestinal stages stained with hematoxylin and eosin. Their (1976) work began to clear some of the confusion generated by the complex and seemingly conflicting observations of earlier colleagues. They (1976) found at least five cycles of merogony before gametogony; these later were confirmed by Shazly et al. (2005), but they then added another layer of confusing terminology. According to Shazley et al. (2005) each generation of meronts produced both Type A (micromeront) and Type B (macromeront) stages with micromeronts producing a small number of large macromerozoites, while macromeronts produced a large number of smaller micromerozoites. The first-generation meronts appeared 24—30

hr PI, only in the glands, and were 15 × 11. When mature, these meronts produced Type A micromeronts, 9.4 × 8.2, with two to eight to 12 multinucleate macromerozoites and Type B macromeronts, 12.2 × 10.7, with 20 uninucleate micromerozoites. All other stages following the first-generation meronts took place in the upper parts of the small intestinal villi. The second-generation meronts appeared ~50 hr PI, again with two types: second-generation micromeronts (Type A) were 10.7 × 9.6, and produced 2—9 merozoites, while second-generation Type B meronts were 11.3 × 11.7, and produced 20—30 merozoites by ~66 hr PI. Third-generation meronts were reported at 70 hr PI. Third-generation Type A micromeronts were 11.2 × 10.1 and contained 4—10 macromerozoites and Type B macromeronts were 14.2 × 12.8 and produced 20—50 micromerozoites. Fourth-generation micromeronts were 11.8 × 10.7, with 4—14 macromerozoites, and fourth-generation macromeronts were 14.9 × 13.1 with 20—50 micromerozoites, ~102 hr PI. Fifth-generation micromeronts were 12.1 × 11.2, with 4—16 macromerozoites and fifth-generation macromeronts were 15.6 × 14.2, with 20—60 micromerozoites, after 120 hr PI. Shazly et al. (2005), who also examined the ultrastructure of these stages, said that both Type A and Type B merozoites of all five generations have all of the typical Apicomplexa characters including pellicle, micropores, conoid, rhoptries, micronemes, polar rings, and subpellicular microtubules and, further, said that all five generations were formed by ectomerogony. This contradicted what others who worked on the developmental stages of rabbit eimerians (e.g., see *E. coecicola*, *E. exigua*, *E. intestinalis*, others) had reported— that smaller (Type A) meronts have fewer merozoites that are polynucleate and develop by endomerogony. Shazly et al. (2005) did not study the development or ultrastructure of the sexual stages, but Al-Ghamdy et al. (2005) did.

Sexual stages were first seen at 96 hr PI, but maximal sexual activity was noted at 120—192 hr PI (Rutherford, 1943; Cheissin, 1940, 1965, 1967; Speer and Danforth, 1976; Speer et al., 1973b). Al-Ghamdy et al. (2005) said that sexually differentiated fourth- and fifth-generation merozoites transformed into micro- or macrogamonts at 125 hr PI. Rutherford's (1943) rendition of gamogony was sketchy. He found it impossible to distinguish early gametes from one another. He did say that "mature microgametocytes show terminal flagella, and what appears to be a nucleus located in the mid-region." By days 5—6 PI, he was able to recognize spheroidal macrogametes that had a N with a prominent **karyosome** (nucleolus); larger macrogametes were 18—20 wide, with large plastic globules in their cytoplasm. Beyer and Ovchinnikova (1964) also were the first to quantify RNA content in growing macrogametes of *E. magna* using the microspectrophotometer to scan fixed, stained macrogametes. Their results indicated that although the RNA concentration decreased as the gamont grew, the total amount of RNA in the cytoplasm of the macrogamete continuously increased, with the highest value being measured just before fertilization and formation of the oocyst wall. For *E. magna* microgamonts, however, they saw a delay in RNA increase between their gamont classes II and III. Then after the delay, the quantity of RNA rose steadily during the remainder of development. This differed somewhat from the steady increase of RNA seen in the microgamonts of *E. intestinalis* (above). They explained this difference arguing as follows: "soon after its establishment within the host's cell in the intestinal epithelium, [*E. magna*] changes its location for that in the tunica propria. This change, followed by the adaptation to the new life conditions in tunica propria, as compared with those in the intestinal epithelium, might account for the delay in the increase of the RNA amount in *E. magna* which occurs at the first half of the macrogamete development."

Cheissin (1940, 1965, 1967) studied gametogony and the descriptions that follow mostly are

from his work. Microgamonts were recognizable as early as day 5 PI and were 24 × 14. When mature, at days 6—7 PI, they measured 40—17, with biflagellated microgametes that were 3—5 × 0.6—0.8, and the flagella were 15 long (Cheissin, 1967). Cheissin (1965) described microgamete formation in detail, using both light and electron microscopes, and divided the process into two phases. In his phase I, the microgamont increased in size, its N multiplied, and when it had 16—24 N, it was about 10—12 wide. It remained ovoidal, and attained a size of 25—40 × 20—30. At first, the N were scattered in the cytoplasm, but later they came to lie on the periphery, forming several **cytomeres**. The N decreased in diameter as their numbers increased. In the early stages they were about 8 wide, but by the time there were 150—200 of them, each was about 1.7 wide. A large mitochondrion formed in association with each N, presumably as the result of fusion of several smaller ones. The **anlagen** of the **basal bodies** of the future flagella of microgametes appeared when they were 12—16 wide. They, too, were seen near the N. The flagella formed during phase I and came to lie on the surface of the microgamonts. During phase II the microgametes differentiated. The N did not increase in number, but elongated, eventually becoming comma-shaped. Large protuberances formed on the microgamont; each contained a N, mitochondrion, and flagellar anlagen. As the N elongated, the flagella did also. The microgametes remained attached to the microgamont by their anterior end and remained attached until the **perforatorium** was formed, after which they separated, leaving a large residuum of cytoplasm. About 750—1,000 microgametes were formed by each microgamont, each ~3—4 × 0.5 with two flagella, 5—7 long. Speer and Danforth (1976) and Al-Ghamdy et al. (2005) also studied the fine structure of microgametogensis. The former authors (1976) confirmed that microgametes have two flagella plus 8—10 **microtubules**. Al-Ghamdy et al. (2005) added that microgamonts were recognizable by the presence of peripherally arranged N and the presence of one or two centrioles between the N and limiting membrane of the gamont. Their (2005) mature microgamonts were 52.3 × 50.8 and produced 150–250 microgametes, each with two flagella.

Macrogametes are initially spheroidal and could be recognized on days 5–6 PI; their N had a large karyosome and contained plastic granules (wall-forming bodies), which later formed the oocyst wall. After fertilization they had a typical wall with a distinct M and measured 30–35 × 15–20. Speer et al. (1973b) studied macrogametogenesis in rabbits killed 5.5 days PI. **Micropores** were present in all stages. Macrogamonts were recognized by the presence of wall-forming bodies of types I and II. Type II wall-forming bodies and lipid globules were the first cytoplasmic inclusions to appear early in the development of the macrogamont, while type I wall-forming bodies appeared later, were smaller, and were distributed peripheral to the type II wall-forming bodies. Both gamete types had two phases: growth and differentiation.

Cross-transmission: Both Becker (1933) and Carvalho (1943) transmitted *E. magna* experimentally from the domestic rabbit to the cottontail rabbit, *S. floridanus mearnsii*, but Pellérdy (1954a) was not able to infect the hare, *Lepus europaeus*, with this species isolated from the domestic rabbit. Despite the fact that Pellérdy (1965, 1974) listed both *Lepus* and *Sylvilagus* species as hosts, these animals are not natural hosts. Burgaz (1973) said she was able to transmit *E. magna* oocysts from *O. cuniculus* to *L. timidus*, but her paper must be viewed with caution because she did not give specific information regarding how she identified the eimerian species, nor did she detail the procedures regarding whether or not the hosts were infected prior to inoculation. We think it likely that she was dealing either with *E. robertsoni* or *E. townsendi*. Morgan and Waller (1940) reported *E. magna* oocysts in *S. floridanus* from Iowa, but

Carvalho (1943) found no *E. magna* oocysts in this same cottontail species, also from Iowa. Kessel and Jankiewicz (1931) were unable to infect the domestic chicken or guinea pig with oocysts of *E. magna*, Yakimoff and Iwanoff-Gobzem (1931) were unable to infect the dog, chicken, pigeon, white rat, or mouse, and Becker (1933) could not infect the Norway rat.

Pathology: This may be one of the most pathogenic of the intestinal coccidia of the rabbit (Levine, 1973b; Coudert, 1989; Pakandl et al., 1996c). Only a few hundred oocysts of some strains may produce signs of disease, and inoculation with 3×10^5 sporulated oocysts may cause death (Lund, 1949), while other strains seem to be less pathogenic, with doses of 1 million oocysts not causing death. Ryley and Robinson (1976) inoculated lab rabbits with 1×10^5 or 2×10^5 sporulated oocysts and said that pathogenic effects were mostly mild, amounting to growth depression over days 4–7 PI with softening of the feces and only occasional diarrhea and/or death. The principal signs of disease were loss of weight, loss of appetite, and diarrhea. A good deal of mucus may be passed in the feces of infected rabbits, while the animals lose their appetites and grow thin. The intestinal mucosa became hyperemic and inflamed, and epithelial sloughing occurred. Francalanci and Manfredini (1967) reported severe hemorrhagic-catarrhal enteritis of the ileum and jejunum. On the other hand, Coudert (1976) found relatively mild effects in rabbits inoculated with 5×10^5 oocysts, with weight gains and feed intakes reduced on days 2–7 PI, and diarrhea on days 4–9 PI, but there were no deaths. Fioramonti et al. (1981) looked at intestinal motility as it affected the transit of food materials in rabbits infected with 1×10^5 *E. magna* oocysts and found that intestinal hypomotility and increased flow of digesta seemed to be the two primary disturbances, the consequences of which were changes in the gastric and cecal motility contributing to an increased transit time of digesta.

Material deposited: None.

In vitro cultivation: Speer and Hammond (1971) cultivated this species in Madin-Darby bovine kidney (MDBK) cells and epithelioid embryonic bovine liver (BEL-1) cells, and obtained mature first- and second-generation meronts, but the intracellular SZ did not develop in bovine fibroblastic embryonic liver cells, lamb trachea cells, or aggregates of primary embryonic bovine kidney cells. The PV containing the SZ was formed by the host cell (Jensen and Hammond, 1975). Speer and Hammond (1971) found in the MDBK and BEL-1 cultures that first-generation meronts looked like "sporozoite-shaped schizonts" with 2–8 N, and later became spheroidal. First-generation merozoites were apparently formed by radial budding. Only four (2–8) first-generation merozoites were formed, in 3 days, as compared with six (2–10) merozoites in 3–4 days in rabbits. At 3–4 days in cell cultures there were ~13 (4–48) merozoites. These first-generation merozoites entered new host cells, and some of them turned into meronts, which formed 2–4 second-generation merozoites. Some of these entered new host cells, but they did not develop further. Speer et al. (1973a) inoculated MDBK, embryonic bovine trachea, and primary rabbit kidney cell cultures, starting with uni- and multinucleate merozoites obtained by scraping the intestinal mucosa of rabbits inoculated 3–5.5 days earlier. Merozoites formed meronts containing 2–80 merozoites and most of these remained where they were as "merozoite-shaped" meronts and formed 2–48 uninucleate merozoites of a further generation; thus, the resultant composite meront contained 40–500 merozoites. Some of these entered other cells at 60–120 hr and formed a third tissue-culture generation of small and medium-sized meronts or gamonts. The macrogamonts moved by means of pseudopods in MDBK cell cultures, even going from one host cell to another (Speer and Hammond, 1972a, b). In order to obtain further development,

Speer and Hammond (1972a, b) inoculated MDBK cultures with merozoites from mucosal scrapings of a rabbit inoculated 5 days and 6 hr earlier. They observed mature microgamonts 32 wide after 72 and 80 hr, and mature oocysts after 72 hr. Mature macrogamonts, presumably at 60 hr, were 28.5×21 in size. Speer and Danforth (1976) studied the fine structure of microgametogensis in rabbit kidney cell cultures and it was the same as in the rabbit intestine. Speer (1979) studied the development of gamonts and oocysts in monolayer, cell-line cultures of embryonic bovine trachea (EBT), MDBK, and primary rabbit kidney cells, starting with merozoites. Merozoites from rabbits inoculated 3 days earlier entered host cells and produced another generation of merozoites only. Merozoites from rabbits inoculated 3.5–5.5 days earlier formed both gamonts and meronts. First- and second-generation merozoites produced both gamonts and meronts, but third-generation merozoites produced only gamonts. Mature microgamonts were 34 wide, unsporulated oocysts were 31×22, and sporulated oocysts were 32×23. Oocysts obtained from cell cultures were sporulated and then inoculated into rabbits by gavage and rabbits passed oocysts 6–10 days later. Sporozoites developed to first- or second-generation meronts in cell cultures. Jensen and Edgar (1976) studied the penetration of *E. magna* SZs into EBT cells and showed that the host cell membrane was not broken during entry of the parasite, but it did undergo alterations characterized by blebbing of vesicles, thickening, and eventual disorganization once penetration was completed. They (1976) proposed that rhoptry secretions aided penetration by changing the cell's surface characteristics, which produced an eventual breakdown of the invaginated portion of the host cell membrane.

Remarks: Prior to Pérard's (1925b) description and naming *E. magna*, other workers had considered oocysts of this species to be a larger form of *E. perforans* or had confused them with those of *E. stiedai* (Rutherford, 1943). The description of the sporulated oocyst is based on Pérard (1925b), Carvalho (1943), Pellérdy (1965, 1974) and Cheissin (1947b, 1967). Oocysts of *E. magna* produced large multinucleated merozoites, which is similar to the endogenous development seen in *E. piriformis*, *E. stiedai*, and others (Pellérdy, 1953, Pellérdy & Dürr, 1970). Pérard (1925b) originally described this species as a variety of *E. perforans*, but *E. magna* oocysts are much larger and possess a M. Tyzzer (1929) hypothesized that when the epithelial cells of the cecal mucosa of chickens were infected with coccidia, the cells actively migrated into the underlying connective tissue. This opinion was confirmed by Cheissin (1940, 1947a, 1967) using the endogenous stages of *E. magna* (and other) species of rabbit eimerians. He said that infected epithelial cells are pushed out of the epithelial layer by the new cells growing in the crypts and submerge into the tunica propria. Gousseff (1931) measured oocysts of *E. magna* that were 31.6 × 18.8 (24–39 × 13–28), with sporocysts 9.0 × 5.6 (6–15 × 4–8). Henry (1932) said that she saw this eimerian, and three others reported from the domestic rabbit, in one Black-tailed jackrabbit from San Andreas, California, but we view her report as totally unreliable based on what we know today. Matsubayashi (1934) said this species infected mostly young domestic and wild rabbits in Japan. Morgan and Waller (1940) said they found *E. magna* in *S. floridanus* in Iowa, that it had ovoidal oocysts with a conspicuous M, and that the oocysts measured 34.8 × 24. Ryšavý (1954) said he found this species in both *O. cuniculus* and in *L. europaeus* in the Czech Republic with oocysts that were 30.6–45.6 × 19.8–26.7, sporocysts 12 × 7.5, and that these oocysts sporulated in 48–65 hr. He said (1954) the parasite was found in the "blind gut" (cecum?), but only rarely in the hares and rabbits they sampled. Gill and Ray (1960) reported finding *E. magna* both in the domestic rabbit and in *L. nigricollis* (syn. *L.*

ruficaudatus) collected in Punjab, India. The measurements they gave for sporulated oocysts were only for oocysts collected from domestic rabbits, but they did include one line drawing of an oocyst from *L. nigricollis* (syn. *L. ruficaudatus*), with three line drawings of oocysts from *O. cuniculus*, and they are remarkably similar. From 1954–1958, Pastuszko (1963) examined 9,845 rabbits from 17 districts in Poland; these included 502 rabbits by necropsy and 9,343 fecal samples tested. She said (1963) she found *E. magna* causing coccidiosis in rabbits in Poland; these oocysts were 35 × 24 (28–40 × 20–27) (her Table 1). Ryley and Robinson (1976) measured oocysts from domestic rabbits after experimental infections designed to study the endogenous development; their oocysts were 35.9 × 24.1. Mandal (1976) said he recovered oocysts of this species in *L. nigricollis* (syn. *L. ruficaudatus*) from Punjab, India, with ellipsoidal to ovoidal oocysts that were 20.3–31.5 × 18–25.5, with an OR, and had ovoidal sporocysts, 15.3–18 × 7–8.8, with a SR. Kasim and Al-Shawa (1987) measured oocysts that were ovoidal, 35.4 × 24.2 (29–40 × 21–26.5), L/W 1.5 (1.4–1.5), with a M and an OR, and sporocysts 17.2 × 9.0 (12–20 × 7–11.5), L/W 1.9 (1.8–2.3). Vila-Viçosa and Caeiro (1997) reported *E. magna* in both *O. cuniculus* and *L. capensis* in Portugal, with the ovoidal oocysts from *O. cuniculus* being 35.6–37.5 × 22.5–24.3. Oocysts measured by Li et al. (2010) from *O. cuniculus* in Taiwan were 36.6 × 27.6 (32–40 × 23–35), with sporocysts 15.2 × 8.8 (11–17 × 8–11).

Cheissin (1947b) said that oocysts appearing early after experimental infection were smaller and paler than those discharged later during patency. Cheissin (1940) said that infections with one sporulated oocyst produced 6–8 × 10^5 oocysts. Schrecke and Dürr (1970) found that 1-day-old rabbits could be infected if enough oocysts were used. Susceptibility to infection rose slowly with age until weaning, when there was a marked increase. They thought that this was due mainly to diet-induced physiological and biochemical properties of the intestine. Ryley and Robinson (1976) documented the excystation of sporozoites from oocysts exposed to CO_2 at body temperature followed by exposure to trypsin and bile. Leysen et al. (1989) used monoclonal antibodies (Mabs) to examine surface antigens on SZs of *E. stiedai* and *E. magna*. Five of 106 Mabs raised against *E. magna* reacted with the surface of live SZs and some were able to agglutinate SZs in suspension. The surface-active Mabs were specific for the species against which they were raised and they did not react with SZs of *E. intestinalis*, *E. flavescens*, *E. piriformis*, or *E. irresidua*.

EIMERIA MATSUBAYASHII TSUNODA, 1952

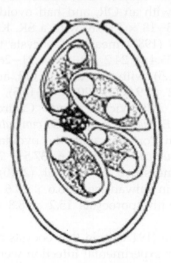

FIGURE 6.14 Line drawing of the sporulated oocyst of *Eimeria matsubayashii* from Levine and Ivens, 1972, with permission from the *Journal of Protozoology*.

Type host: *Oryctolagus cuniculus* (L., 1758) (syn. *Lepus cuniculus*), European (domestic) rabbit.
Type locality: ASIA: Japan.
Other hosts: *Lepus* sp. (?) (Mandal, 1976).

Geographic distribution: ASIA: India; Japan; MIDDLE/NEAR EAST: Syria.

Description of sporulated oocyst: Oocyst shape: broadly ovoidal to ellipsoidal; number of walls: unknown, but probably 2; wall characteristics: light-yellow, smooth, of even thickness throughout except around M, where it increases in thickness; L × W: 24.8 × 18.3 (22−30 × 14−22); L/W ratio: 1.4; M: present; M characteristics: described as "prominent" and 2−4 wide (Gill and Ray, 1960); OR: present; OR characteristics: 6.2 wide; PG: absent. Distinctive features of oocyst: large size with prominent M and large OR.

Description of sporocyst and sporozoites: Sporocyst shape: ovoidal; L × W: 7 × 6; L/W ratio: 1.2; SB: present (according to Gill and Ray, 1960, but not shown in their line drawing); SSB, PSB: both absent; SR: present (according to Gill and Ray, 1960, but not shown in their line drawing); SZ: elongate, broader at one end than the other, lying head-to-tail in sporocyst, with one clear RB at the wider end. Distinctive features of sporocyst: very small size, ovoidal shape, and very thin walled.

Prevalence: 1/1 (100%) in the type host; Darwish and Golemansky (1991) reported this species in 9/75 (12%) domestic rabbits from four localities in Syria.

Sporulation: Exogenous. Oocysts sporulated within 32−40 hr at 28° C or in 2−3 days (Mandal, 1976).

Prepatent and patent periods: Prepatent period is 7 days.

Site of infection: Upper small intestine, primarily the ileum, but in a heavy infection endogenous stages occasionally were found in the duodenum and also in the posterior half of the jejunum, but rarely in the cecum, and colon (Tsunoda, 1952).

Endogenous stages: Both merogony and gamogony were described by Tsunoda (1952), and Pellérdy (1974) confirmed that the endogenous development "scarcely differed from the corresponding stages of *E. media* and *E. magna*."

Cross-transmission: None to date.

Pathology: According to Tsunoda (1952), heavy infections caused inflammation of the ileum characterized by **diphtheritic enteritis** that resulted in diarrhea in infected hosts.

Material deposited: None.

Remarks: Matsubayashi (1934) said oocysts he saw (now *E. matsubayashii*) might be a variant of *E. media*, but the shape, smaller size, and presence of an OR indicated it is a separate species (Tsunoda, 1952). Cheissin (1967), on the other hand, thought it was "highly probable" that this is not a valid species, since the oocysts are very similar to those of *E. intestinalis*, while the endogenous stages resemble those of *E. media*. However, Tsunoda (1952) said that cross-immunity tests he did showed that this species was different from both *E. media* and *E. magna*. Gill and Ray (1960) said the oocysts they measured in Indian domestic rabbits were 26.3 × 16.5 (23.5−29.5 × 14.5−19), L/W 1.7. Mandal (1976) said he recovered oocysts of this species in *Lepus* sp. from Kashipur, India, with ellipsoidal to ovoidal oocysts that were 23.5−29.5 × 14.5−19.3 with an OR and had ovoidal sporocysts, 7 × 6 with a SR. Its validity as a parasite of *Lepus* species seems unlikely and needs verification by cross-transmission and/or molecular studies.

EIMERIA MEDIA KESSEL & JANKIEWICZ, 1931

Synonym: *Eimeria media* Kessel, 1929.

Type host: *Oryctolagus cuniculus* (L., 1758) (syn. *Lepus cuniculus*), European (domestic) rabbit.

Type locality: NORTH AMERICA: USA: California.

Other hosts: *Lepus capensis* L., 1758, Cape hare (?); *Lepus nigricollis* F. Cuvier, 1823 (syn. *L. ruficaudatus*), Indian hare (?); *Lepus timidus* L., 1758, Mountain hare (?); *Sylvilagus floridanus* (J.A. Allen, 1890), Eastern cottontail (experimental, Carvalho, 1943).

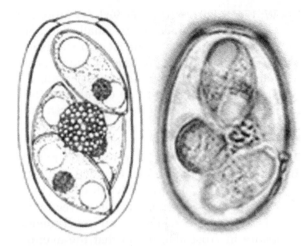

FIGURES 6.15, 6.16 Line drawing of the sporulated oocyst of *Eimeria media* from Catchpole and Norton, 1979, with permission from *Parasitology*. Photomicrograph of a sporulated oocyst of *E. media* from Hobbs and Twigg, 1998, with permission from the *Australian Veterinary Journal*.

Geographic distribution: AFRICA (?): Nigeria (?); ASIA: Azerbaijan (former USSR), China, India, Japan, Philippines, Taiwan; AUSTRALIA: New South Wales, Western Australia; EUROPE: Belgium, Finland, France, Italy, Poland, Portugal, Sweden (?); MIDDLE/NEAR EAST: Iran, Iraq, Saudi Arabia, Syria; NORTH AMERICA: USA: California, Illinois (?), Iowa; SOUTH AMERICA: Brazil.

Description of sporulated oocyst: Oocyst shape: ovoidal to ellipsoidal; number of walls: described as 1 (Kessel, 1929), but Matsubayashi (1934) said there were 2; wall characteristics: smooth, delicate pink to a moderately dark orange-pink, or yellowish to light brown, slightly convex in region of M; L × W: 31.2 × 18.5 (27−36 × 15−22); L/W ratio: 1.7 (Kessel and Jankiewicz, 1931); M: present; M characteristics: convex, ~3.7 (3−4.5) wide, and protruding with thickening of wall surrounding it; OR: present; OR characteristics: spheroidal, 5.2 (3−8) wide; PG: none. Distinctive features of oocyst: convex, protruding M and large spheroidal OR.

Description of sporocyst and sporozoites: Sporocyst shape: fusiform or elongate-ovoidal;

L × W: 15.7 × 8.0 (7–16.5 × 6–9); L/W ratio: 2.0; SB: present; SB characteristics: small, flat; SSB, PSB: both absent; SR: present; SR characteristics: compact, spheroidal, composed of rough granules and usually near the sporocyst wall, 2.7 (1–3) wide; SZ: appear to lie end-to-end (line drawing). Distinctive features of sporocyst: fusiform shape with small, flat SB, and distinct SR.

Prevalence: Kessel and Jankiewicz (1931) reported *E. media* in ~12% of > 2,000 rabbits and Lund (1950) found it in 48/1,200 domestic rabbits, both in California. Mykytowycz (1956, 1962), who did a 35-mo study in a free-living population of wild rabbits on an experimental enclosure in Australia, found that the average level of infection fluctuated markedly with the season, but he was unable to find a satisfactory explanation for the fluctuation; *E. media* was one of the least common species that he found in rabbits of all ages. Niak (1967) found it in 6/84 (7%) rabbits from four rabbitries in Iran. Mirza (1970) found it in 14/31 (45%) domestic rabbits in Iraq. Stodart (1971) examined 656 wild *O. cuniculus* in New South Wales, eastern Australia, and found 625 (95.3%) to be infected with seven eimerians based on their sporulated oocysts, but she lumped *E. media* and *E. stiedai* together and said that 387/656 (59%) were infected with this mixture (also see Stodart, 1968a, b). Peeters et al. (1981) found it in 393/770 (51%) fecal samples from commercial rabbitries and in ~200/282 (71%) sick rabbits in traditional rabbitries in Belgium. Veisov (1982) examined 3,150 rabbits from three different areas/districts of the Cuba-Khachmazskoy zone in Azerbaijan and found 2,905 (81%) infected with nine species of *Eimeria*, including *E. media*; however, we were not able to determine how many of the rabbits examined harbored this species. Ryan et al. (1986) reported *E. media* in 1/1 (100%) *S. floridanus* in Illinois. Santos and Lima (1987) first documented *E. media* in Brazil, and Santos-Mundin and Barbon (1990) reported it in 125/375 (33%) rabbits in Minas Gerais, Brazil. Ogedengbe (1991), in an unpublished Master's thesis, said he found this species in 59% of the rabbits surveyed in Kaduna State, Nigeria. However, he (1991) neglected to state how many rabbits were surveyed, nor did he mention the host species; Anonymous (2012) listed the Cape hare, *L. capensis*, as the only lagomorph species found in Nigeria. Darwish and Golemansky (1991) found *E. media* in 5/75 (7%) domestic rabbits from four localities in Syria. Polozowski (1993) examined feces from 246 rabbits in six farm rabbitries in the Wroclaw District of Poland and found 154 (63%) infected with *E. media*; 139/154 (90%) infected rabbits were < 3-mo-old, 9/154 (6%) were 4–12-mo-old, 5/154 (3%) were 13–24-mo-old, while only 1/154 (<1%) was > 24-mo-old. Hobbs and Twigg (1998) examined fresh fecal samples from wild *O. cuniculus* in Western Australia and found 536/1,200 (48%) to harbor *E. media*; 50 sporulated, ellipsoidal oocysts they measured were 29.1 × 17.2 (23–34 × 14–20), with sporocysts 13.9 × 7.1 (11–16 × 6–9). Grès et al. (2003) found *E. media* in 29–69% of 254 wild rabbits sampled from six different localities in France. They identified 10 eimerians from these 254 rabbits and, apparently, 100% were infected with one or more of these 10 eimerians. Unfortunately, they did not specifically mention how many individuals were infected with which multiple species vs. those with single species infections. Yakhchali and Tehrani (2007) found this species in 36/436 (8%) New Zealand white rabbits from two rabbitries in northwestern Iran. Razavi et al. (2010) found *E. media* in 10/71 (14%) wild rabbits in southern Iran; their oocysts were 29.7 × 18.3 (19–36 × 14–22), L/W 1.6, with a M and with an OR, 6.2 (5–7) wide, and sporocyst were 13.1 × 6.9 (9.5–17 × 5–7). Li et al. (2010) reported *E. media* in 158/642 (25%) *O. cuniculus* from pet shops (98/642) and farms (60/342) in 11 districts on the island of Taiwan.

Sporulation: Exogenous. Oocysts sporulated within 40–52 hr (Kessel and Jankiewicz, 1931); 44 hr (Gill and Ray, 1960); 52 (42–58) hr

(Carvalho, 1943); 2–3 days (Cheissin, 1967) in 2% (w/v) $K_2Cr_2O_7$ solution; 44 hr (Mandal, 1976); 36–60 hr at 18–20°C (Pastuszko, 1963); or 48 hr at 27°C in 2% (w/v) $K_2Cr_2O_7$ solution (Kasim and Al-Shawa, 1987); Razavi et al. (2010) saw sporulation in 36 hr in 2.5% (w/v) $K_2Cr_2O_7$ at 25–28°C; Li et al. (2010) said that sporulation occurred in 10–36 hr in 2.5% (w/v) $K_2Cr_2O_7$ at 25°C.

Prepatent and patent periods: The prepatent period is 5–6 days (Pellérdy and Babos, 1953), or 6–7 days (Carvalho, 1943), while the patent period is 6–10 days (Cheissin, 1940, 1967) or 15–18 days (Carvalho, 1943).

Site of infection: Most endogenous development occurs in the duodenum and the upper part of the small intestine, but sometimes the endogenous stages are found as far as the middle of the small intestine. Rutherford (1943) said endogenous stages could be found "throughout the entire small intestine." Pakandl et al. (1996b) said the entire development occurred in the duodenum and jejunum, with a few stages seen in the ileum. Within infected epithelial cells the developing stages are found either above or below the HCN in the epithelial cells of the intestinal villi (Cheissin, 1967). Pellérdy and Babos (1953) said they found gamonts in the epithelium of the large intestine and appendix (but this likely was a double infection with *E. media* and *E. coecicola* because stages of *E. media* have never been confirmed in the large intestine by other authors: Cheissin, 1967; Pakandl, 1988; Coudert, 1989; Coudert et al., 1995; Pakandl et al., 1996b). Only merogony occurs in the epithelium, while gamogony, fertilization, and oocyst formation take place in the subepithelium with all stages in the distal region of the villi according to Cheissin (1967), while Pakandl et al. (1996c) found all developmental stages in endothelial cells of the wall and tips of the villi, but none in the crypts or subepithelium. The meronts and gamonts of this species also are reported to localize above the HCN in the epithelial cells they occupy.

Endogenous stages: Rutherford (1943), Pellérdy and Babos (1953), Cheissin (1967), and Pakandl et al. (1996c) described the life cycle of this species. There are three merogonous stages according to both Cheissin (1967) and Pakandl et al. (1996c), with the first and second-generation meronts situated along the entire villus, while third-generation meronts are found at the distal end of the villus. Mature first-generation meronts (presumably Type A) appeared on days 2–3 PI, were 12–16 wide, and contained 6–11 merozoites (Cheissin, 1967); Pakandl et al. (1996c) found Type A meronts, but did not measure them or count merozoites. Pakandl et al. (1996c) found first-generation Type B meronts at 24 hr PI to contain 8–20 merozoites. On day 4 PI, ovoidal Type A first-generation meronts formed, were 9–12 long, and contained 2–18 merozoites, 7–10 long (Cheissin, 1967). Pakandl et al. (1996c) found second-generation meronts at 40 hr PI and said that Type A had 2–6 merozoites and Type B had 10–20 merozoites. According to Cheissin (1967), third-generation meronts on days 4–6 PI were 18–25 in size, and formed 35–130 merozoites, each 6–7 × 1–1.5; Pakandl et al. (1996c) reported third-generation meronts at 76 hr PI and said Type A meronts had 2–6 merozoites and Type B had 10–40 merozoites. Rutherford (1943) said that nuclear division of the first-generation meronts began 2 days PI and merozoites appeared at day 4 PI. He said there were two types of first-generation meronts, Type A, which produced 2–10 merozoites, 6 × 1.5, and Type B, which produced 12–36 merozoites, 5 × 1; he (1943) also identified two types of second-generation meronts that he found by day 6 PI: Type A, ~7 × 6, that produced 2–8 merozoites, each 8 × 2, and Type B, 17 × 10, that produced an unspecified number of merozoites that were ~4 × 1. At the time of his work (Rutherford, 1943), there were only four eimerians known to infect rabbits, and he may have assumed that they all had similar life cycles; thus, he missed the third asexual generation. Later, Pakandl et al. (1996c) confirmed that *E. media* had three asexual generations and that "two types of meronts were observed in each generation." Their Type A meronts gave rise to large polynucleated merozoites that multiplied by endomerogony; these were always present in small numbers and their Type B meronts, which were more slender and numerous than Type A, produced multiple merozoites via ectomerogony. These authors (1996c), along with Rutherford (1943) and Streun et al. (1979), suggested that Type A meronts seem to be male and Type B meronts were female.

Gamonts were found at the distal end of the villus and they also penetrated into the connective tissue of the villi. Cheissin (1967) said gamogony began on days 4–6 PI and microgametocytes measured 15–25 × 9–17. Rutherford (1943) said that both macro- and microgamonts appeared 5–6 days PI; the latter were spheroidal, ~17 wide, and produced microgametes, ~2 long, each with two flagella and macrogametes that were about 25 × 14 before fertilization. Pakandl et al. (1996c) found the first gamonts at day 4 PI (96 hr).

Cross-transmission: Kessel and Jankiewicz (1931) were unable to infect the domestic chicken or guinea pig. Carvalho (1943) transmitted this species from the domestic rabbit to the cottontail, *S. floridanus*. Pellérdy (1954a) was not able to infect the hare, *L. europaeus*, with this species isolated from the domestic rabbit. Burgaz (1973) said she was able to transmit *E. media* oocysts from *L. timidus* to *O. cuniculus*, but her paper must be viewed with caution because she did not give specific information regarding how she identified the coccidian species, nor did she detail the procedures regarding whether or not the hosts were infected prior to inoculation. We think it likely that she was dealing with another eimerian from *Lepus* (either *E. robertsoni* or *E. europaea*).

Pathology: This species is moderately to very pathogenic. Pellérdy and Babos (1953) found that feeding 50,000 oocysts to young, susceptible rabbits generally caused fatal coccidiosis characterized by severe enteritis with destruction and sloughing of the intestinal epithelium. In fatal cases, the cecum, and especially the vermiform process, was markedly swollen and thickened, with masses of coccidia stages in it; the cecum became grayish-white, with conspicuous dark red blood vessels. The rest of the cecum and large intestine contain grayish white nodules on the mucosa and these nodules tended to coalesce. If enough nodules were present, the mucosa of both the large and small intestines became swollen, catarrhal, and inflamed, with an adhesive yellowish mucous exudate; occasionally there were patches of petechial hemorrhages in the hyperemic mucosa. Sanyal and Srivastava (1988) found no clinical symptoms evident in rabbits experimentally infected with 10^2, 10^3, and 10^4 *E. media* oocysts, whereas doses of 10^5 and 10^6 oocysts resulted in anorexia, depression, weight loss, and diarrhea, and mortality reached 25% and 50% in these two groups, respectively. On the other hand, Coudert (1976) found that weight gains were reduced or there was weight loss, in rabbits inoculated with 500,000 oocysts, and food intake was reduced on days 2–5 and 7–9 PI; five of 24 rabbits had slight diarrhea on day 5 PI, but there were (apparently) no deaths.

Material deposited: None.

Remarks: Kessel (1929) first named this species as new, said it was limited to the intestine, and gave only oocyst length and width measurements in an abstract of a paper he presented at the 5th Annual Meeting of the American Society of Parasitologists, Des Moines, Iowa. Unfortunately, the "description" was woefully inadequate by any measure, it was not accompanied by an illustration, and new species cannot be named in abstracts according to the *International Code of Zoological Nomenclature* (see Ride et al., 2000), which must be applied retroactively. Thus, the name became a *species inquirenda* and the authority for this name used by subsequent authors (e.g.,

Pellérdy, 1974 and others, who cite *Eimeria magna* Kessel, 1929) is incorrect. It was not until Kessel and Jankiewicz (1931) (re)named it as new, described the sporulated oocyst (but not the sporocysts), and added a line drawing that the name should be considered valid (as used above). In their description they said that the oocysts of this species were intermediate in size between those of *E. perforans* and *E. magna* and so named it *E. media* (Kessel and Jankiewicz, 1931). Oocysts of *E. media* also resemble those of *E. stiedai* in shape and color, but differ by being smaller, by having a M that is more protruded (convex), by developing in the intestines instead of the liver, and by possessing an OR. Henry (1932) said that she found oocysts of what she thought was probably *E. media* in a jack rabbit, *L. californicus*, in California, but she did not describe the oocysts and provided no data to support her identification. Hegner and Chu (1930), in the Philippines, found oocysts (their Type 3, Figs. 1–3; also see Madsen, 1938) that likely were *E. media*; their oocysts were 31 × 19 (L/W, 1.6). Matsubayashi (1934) said this species infected mostly young domestic and wild rabbits in Japan. Chang (1935) mentioned various coccidia from laboratory rabbits, one of which was probably *E. media* (Chang's *Eimeria* sp., p. 155, Fig. 8; also see Madsen, 1938), the oocysts of which were 34.9 × 20.4 with L/W 1.7. Matschoulsky (1941) said he found oocysts that resembled *E. media* in *L. timidus* from Buryat-Mongol (former USSR) and that the ovoidal to ellipsoidal oocysts had a M ~5–6 wide, and were 28.5 × 17.5 (28–29 × 16.5–18.5), with sporocysts 10 × 7. Lampio (1946) reported this species in Finland from the European hare, but gave no other information. Ecke and Yeatter (1956) reported what they called *E. media* from 3/32 (9%) cottontails, *S. floridanus*, in Illinois, but we think it is much more likely they were probably observing *E. honessi*. Gill and Ray (1960) reported oocysts in the rectal contents of an unspecified number of Indian field hares, *L. nigricollis* (syn. *L.*

ruficaudatus), which they said were structurally indistinguishable from those of *E. media*; the oocysts they measured were 30.5 × 18.0 (25–37 × 16–20), but they carried out no cross-transmission experiments, and we feel that both they and Henry (1932) were dealing with another eimerian species limited to *Lepus*. Pastuszko (1963) examined 9,845 rabbits from 17 districts in Poland from 1954–1958; these included 502 rabbits by necropsy and 9,343 fecal samples tested. She said (1963) that she found this species causing coccidiosis in rabbits in Poland, and the oocysts she measured were 31.2 × 18.8 (27–36 × 15–22) (her Table 1). Schrecke and Dürr (1970) found that 1-day-old rabbits could be infected if "enough oocysts" were used. Susceptibility to infection rose slowly with age until weaning, when there was a marked increase in susceptibility. They thought this was due mainly to diet-induced physiological and biochemical properties of the intestine. Mandal (1976) said he recovered oocysts of *E. media* in *Lepus* sp. from Punjab, India, with ovoidal oocysts that were 25–37 × 16–20, with an OR, and ovoidal sporocysts, 14–15 × 9–10, with a SR. Pakandl (1986b) described two morphological types of oocysts from the domestic rabbit that he called *E. media*, but later (Pakandl, 1988) described the second form as *E. vejdovskyi* (see species description below). Still later, Pakandl et al. (1996b) said, "the description of the life cycle of *E. media* given by Pakandl (i.e., *E. vejdovskyi* according to Pakandl, 1988) corresponds to the last two generations [of *E. media*] ...." Kasim and Al-Shawa (1987) examined 263 adult domestic rabbits from three regions in Saudia Arabia from 1984–1985, and found 133 (51%) to be infected with *E. media*. Their (1987) oocysts were ovoidal to ellipsoidal, 30.0 × 18.7 (25.5–34 × 15–22), L/W 1.6 (1.4–1.8), with a M and an OR, and sporocysts 12.9 × 6.6 (9–16 × 5–8.5), L/W 2.0 (1.8–2.3). Vila-Viçosa and Caeiro (1997) reported this species in both *O. cuniculus* and *L. capensis* in Portugal, with the ellipsoid oocysts from *O. cuniculus* being 30–31.8 × 15.7–16.8. Grés et al. (2003) examined 254 wild rabbits from six localities in France and this is one of 10 species they identified. Oocysts measured by Li et al. (2010) from *O. cuniculus* in Taiwan were 20.3 × 18.3 (22–35 × 16–21) with sporocysts 13.8 × 7.6 (8–17 × 5–11).

Licois et al. (1992a) developed a precocious strain of *E. media* by selecting the earliest oocysts discharged after an infection, sporulating them, and reinoculating into "clean" rabbits for 12 such passages. This reduced the prepatent period from about 144 hr (6 days) to 72 hr (3 days). The precocious line was less pathogenic than the parent strain, a mixture of *E. media* oocysts from five wild strains isolated in France, Guadeloupe, Balearic Isles, Ivory Coast and Poland by the authors (Licois et al., 1992a), and its "multiplication rate" (oocysts inoculated vs. oocysts discharged) was lower. Rabbits infected with oocysts of the precocious line were immunized to challenge with the wild strain, as measured by weight gain, although some oocysts were discharged, showing they were not totally protected. Minor morphological changes were noted in the sporulated oocysts of the precocious strain in that SZ contained one large RB rather than the two smaller RBs in the SZ from the parent strain. Pakandl et al. (1996b) also developed a precocious strain of *E. media* and said it was characterized by lacking the third merogonus stage. They also found that sporulated oocysts produced by the precocious line had a "complete absence of the refractile body," presumably within the SZ. Its validity as a parasite of *Lepus* species is doubtful and needs to be verified by cross-transmission and/or molecular studies.

EIMERIA NAGPURENSIS GILL & RAY, 1960

Type host: *Oryctolagus cuniculus* (L., 1758) (syn. *Lepus cuniculus*), European (domestic) rabbit.

FIGURE 6.17   Line drawing of the sporulated oocyst of *Eimeria nagpurensis* from Gill and Ray, 1960, with permission from Springer Science and Business Media, copyright holders for the *Proceedings of the Zoological Society* (supersedes *Proceedings of the Zoological Society of Calcutta*).

Type locality: ASIA: India: Nagpur, Mukteswar.

Other hosts: *Lepus* sp. (?) (Mandal, 1976).

Geographic distribution: ASIA: Azerbaijan (former USSR), India; MIDDLE/NEAR EAST: Iran.

Description of sporulated oocyst: Oocyst shape: barrel-shaped, longitudinal sides parallel at least in the middle third; number of walls: 1; wall characteristics: thin, but prominent and of equal thickness throughout, and colorless or with slight yellow tint; L × W: 23–13 (20–27 × 10–15); L/W ratio: 1.8; M, OR, PG: all absent. Distinctive features of oocyst: parallel sides of oocyst and absence of M, OR and PG.

Description of sporocyst and sporozoites: Sporocyst shape: "oat-shaped" (= elongate spindle-shape), with one end sharply pointed while the other end is rounded (line drawing); L × W: 15 × 5; L/W ratio: 3.0; SB: present; SB characteristics: sharply pointed; SSB, PSB: both absent; SR: present; SR characteristics: granular; SZ: elongate; 12.5 × 2, small RB, 1.75 wide at wider end, with centrally located N. Distinctive features of sporocyst: spindle-shape, high L/W ratio, sharply pointed SB.

Prevalence: Niak (1967) found this species in 2/84 (2%) rabbits from four rabbitries in Iran. Veisov (1982) examined 3,150 rabbits from three different areas/districts of the Cuba-Khachmazskoy zone in Azerbaijan and found 2,905 (81%) infected with nine species of *Eimeria*, including *E. nagpurensis*; however, we were not able to determine how many of the rabbits examined harbored this species.

Sporulation: Exogenous. Oocysts sporulated in 2.5% (w/v) potassium dichromate solution at 28°C or in 2–3 days (Mandal, 1976).

Prepatent and patent periods: Unknown.

Site of infection: Unknown. Oocysts collected from feces.

Endogenous stages: Unknown.

Cross-transmission: None to date.

Pathology: Unknown.

Material deposited: None.

Remarks: Gill and Ray (1960) reported finding this species in domestic rabbits, but they did not mention how many rabbits were examined or what the prevalence was. Martínez Fernández et al. (1969) thought that the oocysts of this species were morphologically identical to those of *E. perforans*, yet they agreed to regard *E. nagpurensis* as a valid species until unequivocal proof for, or against, its independence emerged from future studies. We think the oocysts, and especially the sporocysts, of *E. nagpurensis* are sufficiently different in shape and size from other rabbit coccidia to warrant its status as a separate species. Mandal (1976) said he recovered oocysts of this species in a *Lepus* sp. from Maharashtra, India, but he did not add any new descriptive information and just repeated the measurements given in the original description by Gill and Ray (1960). Later, Mandal (1987) again repeated only the descriptive parameters of Gill and Ray (1960), listed the type host as *O. cuniculus*, and stated that there were no other hosts known for *E. nagpurensis* (contradicting his 1976 report of finding it in a *Lepus* sp.).

EIMERIA NEOLEPORIS CARVALHO, 1942 (FIGURE 7.14)

Synonyms: *E. leporis* Nieschulz of Morgan and Waller, 1940, and Waller and Morgan, 1941; *E. neoleporis* Carvalho, 1943 of Gill and Ray, 1960 *(lapsus calami)*; non *E. neoleporis* Nieschulz, 1923; non *E. neoleporis* Carvalho of Pellérdy, 1954a, 1965.

Type host: *Sylvilagus floridanus* (J.A. Allen, 1890), Eastern cottontail.

Remarks: *Eimeria neoleporis* was originally described from the Eastern cottontail rabbit in Iowa, USA (Carvalho, 1942), but it has been transmitted many times to *O. cuniculus* (see Carvalho, 1943). Mandal (1976) said he recovered oocysts of this species in *Lepus* sp. from Punjab, India, with ellipsoidal oocysts that were 30–54 × 16–22, without an OR, and had ovoidal sporocysts, 16.5 × 8, with a SR. Later, Mandal (1987) copied Carvalho's (1942) line drawing, gave the same size ranges for oocysts and sporocysts, listed the type host as *O. cuniculus*, and listed no other hosts known for *E. neoleporis* (contradicting his 1976 report of finding it in a *Lepus* sp.). Vila-Viçosa and Caeiro (1997) reported this species only in *O. cuniculus* (but not in *L. capensis*; see their Table XIII) in Portugal, with the ellipsoid oocysts from *O. cuniculus* being 37.5–43.1 × 14.2–16.8. See Chapter 7 for complete description of *E. neoleporis*.

EIMERIA ORYCTOLAGI RAY & BANIK, 1965b

Type host: *Oryctolagus cuniculus* (L., 1758) (syn. *Lepus cuniculus*), European (domestic) rabbit.

Type locality: ASIA: India: Calcutta.

Other hosts: None reported to date.

FIGURE 6.18 Line drawing of the sporulated oocyst of *Eimeria oryctolagi* Ray and Banik, 1965b, from the *Indian Journal of Health*.

Geographic distribution: ASIA: India.

Description of sporulated oocyst: Oocyst shape: ovoidal to ellipsoidal; number of walls: described as 2, but only 1 in the original line drawing; wall characteristics: outer is slightly greenish and inner is pale orange; L × W (N = 100): 38.7 × 19.1 (28.5–47 × 12.5–28.5); L/W ratio: 2.0; M: present; M characteristics: thin, 4.5 wide; OR: present; OR characteristics: granular, ~6.0 wide; PG: absent. Distinctive features of oocyst: high L/W ratio, with distinct M and large OR.

Description of sporocyst and sporozoites: Sporocyst shape: piriform to ellipsoidal; L × W: 10 × 6 (8–14 × 4–8); L/W ratio: 1.7; SB: present at narrow end; SSB, PSB: both absent; SR: present; SR characteristics: compact mass of granules, 3–4.5; SZ: elongated, tapering at one end, ~10.5 long. Distinctive features of sporocyst: piriform shape with SB and compact OR of many granules.

Prevalence: Unknown.

Sporulation: Exogenous. Oocysts sporulated within 32–48 hr at 31°C.

Prepatent and patent periods: Unknown.
Site of infection: Presumed to be the intestine.
Endogenous stages: Unknown.
Cross-transmission: None to date.
Pathology: Unknown.
Material deposited: None.

Remarks: This species was first named and partially described by Ray and Banik (1965a) in an abstract at a meeting in India, without photomicrographs or line drawings; thus, it was initially a *species inquirenda*. Later that year, it was described, minimally, in the *Indian Journal of Animal Health*, with a line drawing (Ray and Banik, 1965b), but it has not been observed since its original description(s). In the paper by Ray and Banik (1965b), their description of the oocysts said, "it measures 28–46 microns in length, (average 8.5 microns)" (sic). Mandal (1976) included this species in his account of the coccidia of Indian vertebrates and said, "Type specimen was examined and described. No noticeable differences are found," though it is impossible to tell how he examined and measured oocysts from the "type" sample 11 years later. He (1976, p. 97) said the oocysts were 38.7 × 10.1 (sic) (29–47 × 13–29). Later, Mandal (1987, p. 300) corrected the oocyst width to 19.1. Oocysts of *E. oryctolagi* are quite similar to those of *E. coecicola* (see above). It also resembles *E. neoleporis*, described from the Eastern cottontail, *S. floridanus*, except that it possesses a large OR, which is absent in sporulated oocysts of *E. neoleporis* (see Chapter 7). Pellérdy (1974) said, "not only does the validity of this *E. neoleporis*-like species seem questionable, but also its being a parasite of the domestic rabbit at all," but Mandal (1987) argued, "*E. oryctolagi* can be treated as a distinct species of domestic rabbit till further studies" (sic). Since it has not been seen in nearly 50 years, and since no cross-infections have been performed, the question of its validity remains until it is found again and its molecular signature is compared to other closely related species to which it is so similar.

EIMERIA PERFORANS (LEUCKART, 1879) SLUITER & SWELLENGREBEL, 1912

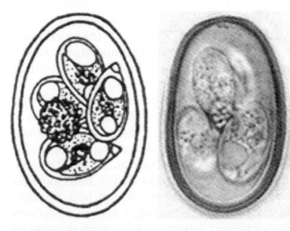

FIGURES 6.19, 6.20 Line drawing of the sporulated oocyst of *Eimeria perforans* from Carvalho, 1943, with permission from John Wiley & Sons Ltd. (current rights owner of the *Iowa State College Journal of Science*). Photomicrograph of a sporulated oocyst of *E. perforans* from Hobbs and Twigg, 1998, with permission from the *Australian Veterinary Journal*.

Synonym: *Coccidium cuniculi* Rivolta, 1878; *Coccidium perforans* Leuckart, 1879; *Eimeria nana* Marotel and Guilhon, 1941; *Eimeria lugdunumensis* Marotel and Guilhon, 1942.

Type host: *Oryctolagus cuniculus* (L., 1758) (syn. *Lepus cuniculus*), European (domestic) rabbit.

Type locality: EUROPE: Germany (?).

Other hosts: *Lepus californicus* Gray, 1837, Blacktailed jack rabbit (?); *Lepus capensis* L., 1758, Cape hare (?); *Lepus europaeus* Pallas, 1778, European hare (?); *Lepus timidus* L., 1758, Mountain hare (?); *Sylvilagus brasiliensis* (L., 1758), Tapeti; *Sylvilagus floridanus* (J.A. Allen, 1890), Eastern cottontail (experimentally).

Geographic distribution: AFRICA (?): Nigeria (?); ASIA: Azerbaijan (former USSR), China, India, Japan, Philippines, Taiwan; AUSTRALIA: New South Wales, Western Australia; EUROPE: Belgium, Czech Republic, England, Finland, France, Germany, Italy, Poland, Portugal; MIDDLE/NEAR EAST: Iran, Iraq, Saudia Arabia, Syria, Turkey; NORTH AMERICA: USA: California, Iowa; SOUTH AMERICA: Brazil.

Description of sporulated oocyst: Oocyst shape: highly polymorphic, being described as ovoidal, ellipsoidal, cylindroidal, or subspheroidal by different authors, or ellipsoidal, but not ovoidal, according to Kessel and Jankiewicz (1931); number of walls: 2; wall characteristics: smooth, delicate pink to a moderately dark orange-pink, or yellowish to light-brown; the larger the oocyst, the more clearly one can see the thickening in the external wall around the M, which is less conspicuous in this species than in any other rabbit species (Kessel and Jankiewicz, 1931); L × W: 23 × 14 (15–31 × 11–20); L/W ratio: 1.6; M: sometimes present (?) or absent; M characteristics: cryptic, rarely visible in smaller oocysts during routine examination; OR: present; OR characteristics: spheroidal, 3.0 (1–5) wide; PG: absent. Distinctive features of oocyst: highly variable shape and cryptic M being visible only in larger oocysts.

Description of sporocyst and sporozoites: Sporocyst shape: fusiform to elongate-ellipsoidal, slightly pointed at one end; L × W: 8 × 4 (5.5–9 × 3.5–5); L/W ratio: 2; SB: present at pointed end; SB characteristics: small; SSB, PSB: both absent; SR: present; SR characteristics: small, granular, ~1–1.5; SZ: appear to lie end-to-end (line drawing), with one end broader than the other and with one clear RB at large end. Distinctive features of sporocyst: none.

Prevalence: Kessel and Jankiewicz (1931) found this species in ~600/2,000 (30%) rabbits in California. Lund (1950) found it in 24/1,200 (2%) domestic rabbits in southern California. Mykytowycz (1956, 1962) conducted a 35-mo study in a free-living population of wild rabbits on an experimental enclosure in Australia, and found that, once established within the population, *E. perforans* was highly persistent and tended to maintain a high level of infection regardless of the season. It was one of the most common species that he encountered and he found it just as often in older rabbits as in young ones. Francalancia and Manfredini (1967) found it in 88/100 (88%) outbreaks of coccidiosis on different farms in the district of Venice, Italy; they also found *E. exigua, E. intestinalis, E. irresidua, E. magna, E. media,* and *E. stiedai,* but failed to say what percentage of their rabbits were infected with the other species; in the same year, Niak (1967) reported *E. perforans* in 4/84 (5%) rabbits from four rabbitries in Iran. Mirza (1970) found it in 17/31 (55%) rabbits in Iraq. Stodart (1971) found it in 505/656 (77%) rabbits in New South Wales, Australia (also see Stodart 1968a, b). Peeters et al. (1981) found it in 470/770 (61%) fecal samples from commercial rabbitries and 231/282 (79%) sick rabbits in traditional rabbitries in Belgium. Veisov (1982) examined 3,150 rabbits from three different areas/districts of the Cuba-Khachmazskoy zone in Azerbaijan and found 2,905 (81%) infected with nine species of *Eimeria,* including *E. perforans;* however, we were not able to determine how many of the infected rabbits harbored this species. Kasim and Al-Shawa (1987) examined 263 adult domestic rabbits from three regions in Saudia Arabia from 1984—1985, and reported 162 (62%) to be infected with *E. perforans.* Santos and Lima (1987) first documented this species in Brazil, and Santos-Mundin and Barbon (1990) reported it in 100/375 (27%) rabbits in Minas Gerais, Brazil. Ogedengbe (1991), in an unpublished Master's thesis, said he found this species in 30% of the rabbits surveyed in Kaduna State, Nigeria. However, he (1991) neglected to state how many rabbits were surveyed, nor did he mention the host species; Anonymous (2012) listed the Cape hare, *L. capensis,* as the only lagomorph species found in Nigeria. Darwish and Golemansky (1991) found this species in 45/75 (60%) domestic rabbits from four localities in Syria. Polozowski (1993) examined feces from 246 rabbits in six farm rabbitries in the Wroclaw District of Poland and found 208 (84.5%)

infected with this species; 174/208 (84%) infected rabbits were < 3-mo-old, 10/208 (5%) were 4—12-mo-old, 20/208 (10%) were 13—24-mo-old, while only 4/208 (2%) were > 24-mo-old. Hobbs and Twigg (1998) examined fresh fecal samples from wild *O. cuniculus* in Western Australia and found 1010/1,200 (84%) to harbor *E. perforans;* 50 of their sporulated, ellipsoidal oocysts were 22.3 × 14.1 (17—38 × 11.5—17), with sporocysts 10.5 × 5.7 (8—14 × 5—7). Grès et al. (2003) found *E. perforans* in all 254 wild rabbits sampled from six different localities in France. They identified 10 eimerians from these 254 rabbits and *E. perforans,* along with *E. flavescens,* apparently was found in all 254. Unfortunately, they did not specifically mention how many individuals were infected with multiple species vs. those with single species infections. Yakhchali and Tehrani (2007) found this species in 28/436 (6%) New Zealand white rabbits from two rabbitries in northwestern Iran. Razavi et al. (2010) found *E. perforans* in 13/71 (18%) wild rabbits in southern Iran; their oocysts were 21.3 × 14.1 (14—29 × 12—22), L/W 1.5, without a M, but with an OR, 4.2 (4—5) wide, and sporocysts were 8.5 × 4.7 (5—10 × 4—6). Li et al. (2010) reported *E. perforans* in 58/642 (9%) *O. cuniculus* from pet shops (31/642) and farms (27/342) in 11 districts on the island of Taiwan. Oncel et al. (2011) reported this species in 10/10 (100%) dead kids (2—3-mo-old) from a commercial rabbitry in Kocaeli, Turkey; their ellipsoidal oocysts lacked a M, but had an OR and were 23.2 × 15.8 (18—28 × 13—18). Prevalence rates from 13% to 80% were reported from domesticated rabbits in various other regions in Turkey (Merdivenci, 1963; Tasan and Özer, 1989; Çetindağ and Biyikoğlu, 1997; Karaer, 2001).

Sporulation: At least 30 hr, but up to 46 hr (Kessel and Jankiewicz, 1931); 40 hr (Matsubayashi, 1934); 30—55 hr (Carvalho, 1943); 1—2 days (Cheissin, 1967); 46 hr at 18°C, 30 hr at 22°C, and 21 hr at 26°C (Coudert et al., 1979); 24 hr at 27°C in 2% (w/v) $K_2Cr_2O_7$ solution (Kasim and Al-Shawa, 1987); Razavi et al.

(2010) saw sporulation in 36 hr in 2.5% (w/v) $K_2Cr_2O_7$ at 25−28°C; Li et al. (2010) said sporulation occurred in 12−36 hr in 2.5% (w/v) $K_2Cr_2O_7$ at 25°C.

Prepatent and patent periods: Prepatent period is 5−6 days (Rutherford, 1943; Martínez Fernández et al., 1969, 1970) or 4−5 days (Carvalho, 1943), while the patent period is 12−24 days (Carvalho, 1943), 6−7 days (Cheissin, 1940, 1967), or 29−32 days (Martínez Fernández et al., 1969, 1970). The latter said that 95% of the oocysts were eliminated during the first 7 days in primary infections and 84% during the first 7 days when reinfected.

Site of infection: Endogenous stages were reported in the epithelial cells of the villi and crypts in the middle part of the small intestine, but some stages also were found from the duodenum to the lower part of the small intestine. However, Cheissin (1967, p. 115) said, "The endogenous stages (of *E. perforans*) develop near the bottom of the crypts of the large intestine." The meronts and gamonts of this species also were reported to localize either above or below the HCN in the epithelial cells of the intestine. Streun et al. (1979) described the endogenous stages in the life cycle of *E. perforans* in SPF rabbits starting with a pure strain of *E. perforans* derived from a single oocyst. They found the endogenous stages in epithelial cells of the villi of the duodenum and also in the jejunum and ileum, but not in the large intestine.

Endogenous stages: Matsubayashi (1934) found meronts in the epithelium of the intestine and said that each had > 100 merozoites. The actual number of meronts has not been determined exactly, but there may be two or four or some combination thereof. The life cycle was first described by Rutherford (1943), but his description may be more confusing than helpful. He said that the SZs rounded up and entered the host cells within 12 hr after inoculation. The SZs became first-generation meronts, which formed merozoites 3−4 days PI. These first-generation meronts (Rutherford, 1943) were designated Type A and Type B. Type A first-generation meronts (he called them "pockets") were 6−10 wide, with 4−8 merozoites that were 7 × 1.5 when formed, blunt at both ends, and had a delicate N surrounded by a clear area and were nearer to one end than the other. These merozoites entered other epithelial cells to form second-generation meronts which, in turn, produced 4−8 Type A merozoites that were crescent-shaped, with a large, central N surrounded by a clear area. These merozoites, he said, continued the endogenous cycle, forming new meronts. Rutherford (1943) first saw Type B meronts on day 3 PI. By day 4 PI, 8−24 merozoites were present, each ~4 × 1, in a meront "pocket," 7−8 × 4−6; most of these merozoites were pointed at both ends, with a central N. Rutherford (1943) believed that Type A first-generation merozoites gave rise to Type A second-generation meronts and merozoites and then microgamonts, while first-generation Type B merozoites gave rise to second-generation Type B meronts and merozoites and then to macrogamonts, but he gave no evidence for this idea. He said it took 4 days PI for Type A merozoites to mature and 1 day for Type B merozoites to mature. He also said that it took 5 days for *E. perforans* to complete its life cycle. His interpretations do not agree with those of Scholtyseck and his co-workers (below), or with those of Cheissin (1967), or with those of Streun et al. (1979), none of whom mentioned Types A and B meronts/merozoites in *E. perforans*. According to Cheissin (1967) first-generation meronts appeared on days 3−4 PI; they were 13−14 long and formed 50−120 narrow merozoites, which were 7−9 × 0.5. What may be small second-generation meronts measured 5−10 wide and had 4−8 merozoites that were 4−5 × 1. Streun et al. (1979) found first-generation meronts 12−60 hr PI and second-generation meronts 72−84 hr PI. At each time interval they reported two types of meronts: (1) meronts with 16−36 small, mononucleated merozoites produced by

ectomerogony; and (2) meronts with 2—12 large multinucleated merozoites produced by endomerogony. They considered the first group of meronts to be female and the second group to be male, but gave no convincing evidence for that view.

Scholtyseck and Spiecker (1964) described the fine structure of all stages of *E. perforans* and provided a diagram of its life cycle. Scholtyseck (1965a) described the fine structure of the meronts, then the microgametes (1965b), and Scholtyseck and Piekarski (1965) that of the merozoites. The merozoites were 8 × 2—3 and bounded by two narrow double membranes, of which the outer was thin and enclosed the whole cell, and the inner was thicker and did not extend across the anterior and posterior ends. Intracellular organelles included a typical conoid, a polar ring, two rhoptries, a variable number of micronemes, 24 subpellicular microtubules, a N in the posterior half bounded by a double membrane containing many pores, a Golgi apparatus just anterior to the N, and an opening at the posterior end, ~0.2 wide. The meronts were ~20 × 6 and had a highly developed endoplasmic reticulum, which formed a concentric system of fissures by means of which the merozoites split off from the meronts. This process of merozoite formation by splintering the entire meront into individual pieces may be a general characteristic feature of merogony. Scholtyseck et al. (1969), using *E. perforans* and *E. stiedai*, were among the first to describe the formation of the two major layers of the oocyst wall, the so-called dark bodies that form the outer layer, and the "wall-forming bodies" that form the inner layer, during endogenous development.

Scholtyseck (1962) was the first to study the fine structure of the macrogametocytes, microgametocytes, and oocysts of *E. perforans*. Microgametocytes measured 12—15 (Cheissin, 1967) and were not larger than the macrogametes. Microgametes were 3 long, and the flagella were 10 long (Cheissin, 1967). Hammond et al.

(1967), Scholtyseck (1965a, b), and Scholtyseck et al. (1972) described the fine structure of the microgamonts and microgametes. They said the microgamonts were 10—14 wide and had numerous micropores. Each produced relatively few microgametes. These contained a variable number of spheroidal to broadly ellipsoidal N, numerous mitochondria, numerous fine granules, small and large vacuoles near the center of the cell, osmophilic lipid inclusions, a well-defined endoplasmic reticulum, and what were probably glycogen granules; they had no tube-like extrusions. Each N formed a slender microgamete, about 6 long. They were covered by a unit membrane and had a perforatorium, lacked micropores, and had a mitochondrion, ~2.5 long, and three flagella (while the microgametes of most other coccidia are known to have only two, e.g., *E. magna*, *E. intestinalis*).

Scholtyseck et al. (1966, 1970a, b, 1971) and Mehlhorn and Scholtyseck (1974) described the fine structure of the macrogametes. The macrogamete surface had a single layer of tube-like extrusions (microtubules) at least 1.3 long and ~65 nm wide; the tubule walls had transverse striations, ~9 nm wide every 16.5 nm, and some of them extended across the PV, making direct contact with the host cell. The macrogametes contained glycogen granules, mitochondria, an endoplasmic reticulum, vacuoles, and lipid granules. After fertilization, two kinds of wall-forming bodies in the macrogametes formed the two layers of the oocyst wall.

Cross-transmission: Kessel and Jankiewicz (1931) were unable to infect the domestic chicken or guinea pig, and Yakimoff and Iwanoff-Gobzem (1931) could not infect the dog, chicken, pigeon, white rat, or mouse with *E. perforans*. Lee (1934) was not able to infect the domestic dog with it. Carvalho (1943) transmitted *E. perforans* from the domestic rabbit to the cottontail, *Sylvilagus floridanus mearnsii*. Pellérdy (1954a) was not able to infect the hare, *Lepus europaeus*, with *E. perforans* oocysts isolated from the domestic rabbit. Burgaz (1973) was unable to transmit this species to *L. timidus*, although her identifications and methods are questionable.

Pathology: *Eimeria perforans* may be one of the less pathogenic intestinal coccidia of rabbits, but it may nevertheless cause mild to moderate signs of pathology in rabbits if the infection is heavy enough. The duodenum can become enlarged and edematous, sometimes chalky white; the jejunum and ileum may contain white spots and streaks, and there may be petechiae in the cecum. Pellérdy (1965, 1974) said that this species is not pathogenic. Weisbroth and Scher (1975) reported a fatal case of **intussusception** associated with an *E. perforans* infection in a 14-wk-old pet male New Zealand rabbit in New York, USA. The intussusception was ileo-ileal in a region that was heavily parasitized by *E. perforans* upon histological examination of the tissues. Intussusception is believed to occur in all mammals, but is most common in humans, dogs, horses, cows, and sheep, and in all species it occurs most frequently in the young. The infected rabbit exhibited early symptoms of shock-like depression and lethargy for 2 days prior to death. The authors (Weisbroth and Scher, 1975) found it difficult to evaluate the significance of the infection with *E. perforans* as a causal factor of the intussusception; however, **hyperperistalsis** is known to be commonly observed in severe coccidial infections, and they hypothesized that in this case the hyperperistalsis was of such a degree as to progress to "irreducible pathologic intussusception." Coudert et al. (1979) isolated a pure strain of *E. perforans* in SPF rabbits from a single oocyst. They then inoculated SPF rabbits that were 5, 7, 9 or 17 wks old with either 8×10^4, 4×10^5, or 2×10^6 sporulated oocysts; they said this parasite was only mildly pathogenic; causing a moderate decrease in fecal consistency and a decrease in body weight and food consumption within 2 days PI, which disappeared by day 4 PI. Li and Ooi (2009) said they found occult blood only in the feces of rabbits infected with *E. perforans*, but not those infected

with five other rabbit eimerian (*E. coecicola, E. exigua, E. magna, E. media, E. piriformis*) and concluded that "concurrent infection with other *Eimeria* species could exacerbate the pathogenicity of *E. perforans*" and that "*E. perforans* can be considered as comparatively the most virulent species examined." However, Licois (2009) wrote a rebuttal letter highly critical of every aspect of the paper by Li and Ooi (2009) including, but not limited to: the number of eimerian species that infect rabbits, discrepancies between methods described and the results reported, the way data were quantified, various incompatibility statements made by the authors vs. the data in the table and figure they cited, and their opinion that "the presence of blood in the feces is not sufficient to sustain the pathogenicity of a micro-organism," and, in summary, "For all this (sic) reasons I disagree totally with the conclusions of the authors." Oncel et al. (2011) noted that 10/10 dead kids were thin with reduced fat stores and had muscle wasting, rough hair coats, and their small intestines were distended and filled with gray-green ingesta; the intestinal mucosa was severely hyperemic and **edematous**. However, they could not attribute the pathology noted to *E. perforans*, since all 10 kids also were infected with *E. intestinalis*, as determined by oocysts in their feces.

Material deposited: None.

Remarks: Leuckart (1879) was the first to suggest that the intestinal coccidium of rabbits (*E. perforans*) differed from the liver form, *E. stiedai*. It wasn't until Railliet and Lucet (1891a, b) inoculated healthy rabbits with oocysts of this species (intestinal origin), and produced only intestinal coccidiosis, that the early views of Leuckart (1879) to distinguish the oocysts of *E. perforans* from those of *E. stiedai* were conclusively supported. Sluiter and Swellengrebel (1912) clearly distinguish the oocysts of *E. perforans* from those of *E. stiedai*, but then for a time all *Eimeria* species found in the lagomorph intestine were thought to be *E. perforans*. As a result, many earlier reports of *E. perforans* in cottontails and hares must be discounted. For example, Henry (1932) reported it from *L. californicus* without describing it; she may have been dealing with *E. groenlandicus*. Robertson (1933) said he found *E. perforans* in one *L. europaeus* from England that had oocysts 18—31 × 13—20, with a M and an OR, and sporocysts 12.5 (10—14) long with a SB and SR, which he said, "disappeared fairly rapidly in the majority"; he likely was looking at oocysts of *E. groenlandicus*. Hegner and Chu (1930), in the Philippines, found coccidia in laboratory rabbits, including oocysts of *E. perforans* (also see Madsen, 1938), which measured 24 × 16, L/W 1.25. Correa (1931), in South America, recorded oocysts that were 26.4 × 16.5, L/W 1.25, from the small intestine that were almost certainly those of *E. perforans*. Chang (1935) mentioned various coccidia from laboratory rabbits, one of which was probably *E. perforans*, which had oocysts that were 22.5 × 15, L/W 1.5. Matsubayashi (1934) said this species infected mostly young domestic and wild rabbits in Japan. Madsen (1938) described *E. perforans* var. *groenlandicus* from the Greenland hare, *L. arcticus*, but this species was actually *E. groenlandicus*. Carini (1940) reported it from *S. brasiliensis* in Brazil, but the sporocysts that he described were much larger than those of *E. perforans*. Morgan and Waller (1940) said they found ellipsoidal oocysts of *E. perforans* in the majority of 210 *S. floridanus* in Iowa, with an average size of 26.4 × 16.2, but Carvalho (1943) thought they probably mistook it for *E. environ*. He never found *E. perforans* in wild *S. floridanus* in Iowa, although he reported that he transmitted it experimentally to this cottontail. Matschoulsky (1941) said he found oocysts that resembled *E. perforans* in *L. timidus* from Buryat-Mongol (former USSR), and that the ovoidal oocysts were 18.0 × 15.1 (16.5—20 × 15—16.5) with ovoidal sporocysts 7—9 × 4—7. Lampio (1946) said he found *E. perforans* in Finland from the European hare, but gave no other information. Ryšavý (1954) said he found *E. perforans* in both *O. cuniculus* and

L. europaeus in the Czech Republic, with oocysts that were 10—23 × 14—29 and sporocysts 8.5 × 4.5, and that these oocysts sporulated in 30—60 hr; it was found in the small intestine and was abundant in the hares and rabbits they sampled (Ryšavý, 1954), but he didn't mention from which rabbit species the oocysts he measured were collected. Ecke and Yeatter (1956) reported *E. perforans* on the basis of <u>un</u>sporulated oocysts from 6/32 (19%) cottontails, *S. floridanus*, in Illinois, but again it is likely that they were dealing with *E. environ*. Lucas et al. (1959) said that *E. perforans* was very rare in the hare in France, but they were probably finding oocysts of *E. groenlandicus*. Gill and Ray (1960) said that they found *E. perforans* in *L. nigricollis* in India, but it was probably *E. groenlandicus* as well. Pastuszko (1963) examined 9,845 rabbits from 17 districts in Poland from 1954—1958; these included 502 rabbits by necropsy and 9,343 fecal samples tested. She said (1963) that she found *E. perforans* causing coccidiosis in rabbits in Poland; the oocysts were 25.5 × 15.5 (15—29 × 11—17) (her Table 1). Seddon (1966) said that *E. perforans* commonly infected both domestic rabbits and hares in Australia and has been the cause of death of captured hares, but offered no empirical evidence. Martínez Fernández et al. (1970) carried out a biometric study of 1,000 oocysts of *E. perforans* on the third and fourth passages from a single oocyst. They noted that the oocysts were extremely pleomorphic, especially as to the structure of the OR. They concluded that *E. lugdunumensis* (from Marotel and Guilhon, 1942) and *E. nagpurensis* (see above) could not be distinguished from *E. perforans*, and that their validity should be determined by histological and immunological studies. Pellérdy (1965, 1974) added *E. nana* (from Marotel and Guilhon, 1941) to that mix. Vila-Viçosa and Caeiro (1997) reported this species in both *O. cuniculus* and *L. capensis* in Portugal, but gave only measurements of the ellipsoid oocysts from *O. cuniculus* as being 20.6—22.5 × 13.1—15; the oocysts they saw in the cape hare likely were a different species, but they didn't describe it. We would add that all of the forms thought to be *E. perforans* found in *Lepus* and *Sylvilagus* species must one day be sorted out with molecular studies.

The description of the sporulated oocyst (above) is based mainly on that of Kessel and Jankiewicz (1931), because earlier descriptions were based on a mixture of species and are, therefore, not trustworthy. Gousseff (1931) measured oocysts that were 19.9 × 12.9 (15—34.5 × 9—24) with sporocysts 5.6 × 3.8 (5—9 × 3—6), which likely were oocysts of *E. perforans*. Carvalho (1943) said oocysts were 21.5 × 15.5 (15—30 × 11—20). Lucas et al. (1959) found oocysts that were 23 × 16 (24—30 × 14—20), and Cheissin (1967) said the oocysts were 21.5 × 15.5 (13—31 × 11—17). Kasim and Al-Shawa (1987) measured oocysts that were ellipsoidal, 20.9 × 13.7 (14—27 × 11—19.5), L/W 1.5 (1.2—1.8), without a M, but with an OR, and sporocysts were 9.7 × 5.1 (8—11 × 4—7), L/W 1.9 (1.7—2.3). Coudert et al. (1979) isolated a pure strain of *E. perforans* in **SPF** rabbits from a single oocyst. Its oocysts varied markedly in shape from subspheroidal to elongate-ellipsoidal to ovoidal, but the larger ones were more "characteristic" than the others; 96% of the total oocyst population were 22 × 14 (17—28 × 12—16), with L/W 1.6. They (1979) added that there was a M and an OR, but gave no further structural details. The cytoplasm of epithelial cells infected with gamonts of *E. perforans* shows an increase in the number and concentration of mitochondria between the N of the infected epithelial cell and the developing parasite. Electron microscope studies have shown that many mitochondria vacuolize and degenerate. Numerous fibrils appeared in the cytoplasm of the host cell surrounding macrogametes of *E. perforans* (Scholtyseck, 1963, 1964). Oocysts measured by Li et al. (2010) from *O. cuniculus* in Taiwan were 20.9 × 13.0 (16—29 × 8—20), with sporocysts 8.6 × 4.1 (5—12 × 3—7). Mandal (1976, 1987) said he recovered oocysts

of *E. perforans* in *L. ruficaudatus* from Kashipur, India, with ellipsoidal oocysts that were 17—32 × 12.5—19.5, with an OR, and had ovoidal sporocysts, 7.8 × 3.8, with a SR, and a sporulation time of 48 hr. Schreckee and Dürr (1970) found that 1-day-old rabbits could be infected with *E. perforans* if enough oocysts were used. Susceptibility to infection rose slowly with age until weaning, when there was a marked increase. They thought that this was due mainly to diet-induced physiological and biochemical properties of the intestine.

EIMERIA PIRIFORMIS KOTLÁN & POSPESCH, 1934

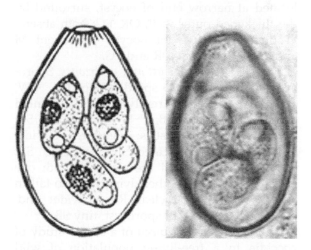

FIGURES 6.21, 6.22 Line drawing of the sporulated oocyst of *Eimeria piriformis* from Levine and Ivens, 1972, with permission from the *Journal of Protozoology*. Photomicrograph of a sporulated oocyst of *E. piriformis* from Hobbs and Twigg, 1998, with permission from the *Australian Veterinary Journal*.

Synonyms: Eimeria pyriformis Kotlán and Pospesch, 1934 *lapsus calami*; non *Eimeria piriformis* Lubimov, 1934; *Eimeria piriformis* Marotel and Guilhon, 1941; non *Eimeria pellerdyi* Prasad, 1960; non *Eimeria pellerdyi* Coudert, 1977a, b.

Type host: *Oryctolagus cuniculus* (L., 1758) (syn. *Lepus cuniculus*), European (domestic) rabbit.

Type locality: EUROPE: Hungary.

Other hosts: *Lepus capensis* L., 1758, Cape hare (?), but unlikely.

Geographic distribution: ASIA: Azerbaijan (former USSR), Taiwan; AUSTRALIA: New South Wales, Western Australia; EUROPE: Belgium, Czech Republic, France, Hungary, Poland, Portugal, Spain; MIDDLE/NEAR EAST: Iran; Saudia Arabia, Syria.

Description of sporulated oocyst: Oocyst shape: piriform, often asymmetrical; number of walls: 2; wall characteristics: smooth, light yellowish-brown; the outer layer is thickened around the M; L × W: 29 × 18 (26−33 × 17−21); L/W ratio: 1.6; M: present; M characteristics: prominent, located at narrow end of oocyst, surround by the thickened outer wall; OR, PG: both absent. Distinctive features of oocyst: prominent M and absence of both OR and PG.

Description of sporocyst and sporozoites: Sporocyst shape: ovoidal to spindle-shaped; L × W: 10.5−13 × 6; SB: apparently present at pointed end of sporocyst; SSB, PSB: both absent; SR: present; SR characteristics: granular, 3−5 wide, usually centrally located in sporocyst; SZ: surround the SR, and are elongate, with one end wider that the other, lying head-to-tail in sporocyst with one clear RB at wider end. Distinctive features of sporocyst: tiny SB.

Prevalence: On the basis of a 35-mo study of coccidia in a free-living population of wild rabbits on an experimental enclosure in Australia, Mykytowycz (1956, 1962) found that, once established within the population, *E. piriformis* was highly persistent and tended to maintain a high level of infection regardless of the season. It was one of the most common species that he encountered, but occurred much more commonly in rabbits 8-mo-old and above than in rabbits 2-mo-old and below; when it did occur, however, its oocysts were usually present in high numbers. Niak (1967) found this species in 3/84 (4%) rabbits from

four rabbitries in Iran. Stodart (1971) found it in 459/656 (70%) wild rabbits in eastern Australia (also see Stodart 1968a, b). Peeters et al. (1981) found it in 23/770 (3%) fecal samples from commercial rabbitries and in 42/282 (15%) sick rabbits in traditional rabbitries in Belgium. Veisov (1982) examined 3,150 rabbits from three different areas/districts of the Cuba-Khachmazskoy zone in Azerbaijan and found 2,905 (81%) infected with nine species of *Eimeria*, including *E. piriformis*; however, we were not able to determine how many of the rabbits examined harbored this species. Darwish and Golemansky (1991) found this species in 8/75 (11%) domestic rabbits from four localities in Syria. Polozowski (1993) examined feces from 246 rabbits in six farm rabbitries in the Wroclaw District of Poland and found 95 (39%) infected with this species; 61/95 (64%) infected rabbits were < 3-mo-old, 4/95 (4%) were 4−12-mo-old, 27/95 (28%) were 13−24-mo-old, and 4/95 (4%) were > 24-mo-old. Hobbs and Twigg (1998) examined fresh fecal samples from wild *O. cuniculus* in Western Australia and found 604/1200 (50%) to harbor this species; 50 of their sporulated, piriform oocysts were 31.6 × 19.7 (24−36 × 18−21), with sporocysts 12.7 × 7.8 (11−14 × 7−9). Grès et al. (2003) found *E. piriformis* in 70−95% of 254 wild rabbits sampled from six different localities in France. They identified 10 eimerians from these 254 rabbits and, apparently, all were infected with one or more of these 10. Unfortunately, they did not specifically mention how many individuals were infected with multiple species vs. those with single species infections. Yakhchali and Tehrani (2007) found this species in 28/436 (6%) New Zealand white rabbits from two rabbitries in northwestern Iran. Li et al. (2010) reported *E. piriformis* in 16/642 (2.5%) *O. cuniculus* from pet shops (8/642) and farms (8/342) in 11 districts on the island of Taiwan.

Sporulation: Exogenous. Oocysts sporulated in 2 days (Kotlan and Pospesch, 1934; Lizcano Herrera and Romero Rodríguez, 1969) or in

3—4 days at 22°C; in 6 days at 15°C (Cheissin, 1948); in 48 hr at 18—20°C (Pastuszko, 1963); Li et al. (2010) said that sporulation occurred in 20—27 hr in 2.5% (w/v) $K_2Cr_2O_7$ at 25°C.

Prepatent and patent periods: Prepatent period is 9—10 days (Kotlán and Pospech, 1934; Cheissin, 1948), while the patent period is 10—12 days (Cheissin, 1948, 1967).

Site of infection: Kotlán and Pospech (1934) said that piriform oocysts could be found in the duodenum and that they increased in number toward the posterior part of the gut. Others have found endogenous development to occur in the large intestine, usually in the ascending colon, cecum, and appendix in the epithelial cells of the crypts and above the HCN (Cheissin, 1948, 1967; Pellérdy, 1953, 1965, 1974).

Endogenous stages: The life cycle was described by Cheissin (1948, 1967). The life cycle described for "*E. piriformis*" by Pellérdy (1953, 1965, 1974) was actually that of *E. intestinalis*. There are three merogonous stages. Mature first-generation meronts appeared by days 5—6 PI, reached a diameter of ~20, and contained 15—25 spindle-shaped merozoites, each ~15 × 2; second-generation meronts appeared 7—9 days PI and were smaller, 11—14 wide, with 25—55 short, narrow merozoites, 6—7 × 1. On days 9—10 PI, third-generation meronts appeared; these were 17—20 wide, and contained 15—50 merozoites that measured 11—12 × 1.5. Gamonts appeared on day 8 PI and oocysts started to accumulate in the lumen of the crypts 9—10 days PI. Microgametocytes were 15—18, monocentric, and were not larger than mature macrogametes, which were 20 wide (Cheissin, 1947a, 1948, 1967). Gamogony ended 14—15 days PI.

Cross-transmission: Pellérdy (1954a, 1956) was not able to infect the hare, *L. europaeus*, with this species isolated from the domestic rabbit. Burgaz (1973) was unable to transmit this species to *L. timidus*, although her identifications and methods are questionable.

Pathology: Not very much is known about whether or not this species causes any pathology in domestic rabbits. Cheissin (1948) found that inoculation of 1-mo-old rabbits with about 1,000 oocysts produced loss of appetite, sluggishness, rough hair coats, posterior **paresis**, and often diarrhea by 8 days PI; a dose of 10,000 oocysts produced death in 11—12 days PI.

Material deposited: None.

Remarks: Kotlán and Pospech (1934), in Hungary, found piriform oocysts in the feces of a doe and determined they had oocysts that differed from all other known coccidia from rabbits to that date. They (1934) also noted that oocysts passed on the first day of patency were smaller (23—24 × 16) than those passed later. Cheissin (1948, 1967) and Pellérdy (1953, 1965, 1974) differed as to where the endogenous stages of *E. piriformis* occur in the rabbit. Pellérdy said that they are in the small intestine, while Cheissin said that they occur in the large intestine. Levine and Ivens (1972) felt that the two authors were speaking of two different organisms; the form that Pellérdy called *E. piriformis* is really *E. intestinalis*. From 1954—1958, Pastuszko (1963) examined 9,845 rabbits from 17 districts in Poland; these included 502 rabbits by necropsy and 9,343 fecal samples tested. She (1963) reported that *E. piriformis* caused coccidiosis in rabbits in Poland; the oocysts she measured were 28 × 18 (26—32 × 15—25) (her Table 1). Ryšavý (1954) said he found *E. piriformis* in both *O. cuniculus* and in *L. europaeus* in the Czech Republic, with oocysts that were 29—33 × 18—25, but he did not measure sporocysts, nor did he determine sporulation time. The parasite was found in the small intestine of the hares and rabbits he sampled near Prague (Ryšavý, 1954) and he likely was dealing with different species with similar oocysts. Lizcano Herrera and Romero Rodríguez (1969) reported oocysts of *E. piriformis* that were 29 × 18. Vila-Viçosa and Caeiro (1997) reported this species in both *O. cuniculus* and *L. capensis* in Portugal, with the piriform oocysts from *O. cuniculus* being 33–37.5 × 18–20.6, but they did not provide measurements of the oocysts from *L. capensis*. These oocysts likely were another eimerian, but not enough information was given to guess which one. Oocysts measured by Li et al. (2010) from *O. cuniculus* in Taiwan were 29.3 × 20.0 (27–35 × 16–24) with sporocysts 14.0 × 7.3 (8–16 × 4–9).

EIMERIA ROOBROUCKI GRÉS, MARCHANDEAU & LANDAU, 2002

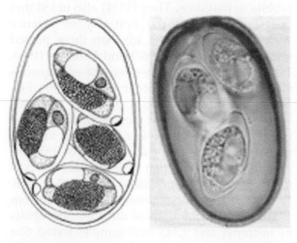

FIGURES 6.23, 6.24 Line drawing of the sporulated oocyst of *Eimeria roobroucki* from Grés et al., 2002, with permission from the authors and from *Zoosystema*. Photomicrograph of a sporulated oocyst of *E. roobroucki* from Grés et al., 2002, with permission from the authors and from *Zoosystema*.

Type host: *Oryctolagus cuniculus* (L., 1758) (syn. *Lepus cuniculus*), European (domestic) rabbit.

Type locality: EUROPE: France, exact locality not stated, only that this "species was found in five of six areas" examined: Arjuzanx (Landes), Donzère-Mondragon (Drôme), Massereau (Loire-Atlantique), Gerstheim (Bas-Rhin), island of Beniguet (Molène archipelago)."

Other hosts: None to date.

Geographic distribution: EUROPE: France.

Description of sporulated oocyst: Oocyst shape: ellipsoidal, generally symmetrical with a slight flattening on the sides around the M; number of walls: 2; wall characteristics: thick, smooth, brown, and very fragile, as it is frequently broken during handling; L × W (*N* = 30): 55 × 33.7 (ranges not given); L/W ratio: 1.6; M: present; M characteristics: prominent, ~8 wide, but the oocyst outer wall is not thickened around its periphery; OR, PG: both absent. Distinctive features of oocyst: the largest of all eimerian oocysts described from *O. cuniculus*, presence of a large M, and lacking both OR and PG.

Description of sporocyst and sporozoites: Sporocyst shape: elongate-ovoidal to spindle-shaped, pointed at both ends (line drawing and photomicrograph); L × W (not given; measurements are extrapolated from line drawing): 23.8 × 14.2; L/W ratio: 1.7; SB: described as "thin," present at one end of sporocyst; SSB: present, same width as SB, described as a clear half disc; PSB: absent; SR: present; SR characteristics: large mass of granules, centrally located in sporocyst, usually obscuring at least one SZ; SZ: elongate, with one end wider that the other, lying head-to-tail in sporocyst with one clear RB in the middle or at larger end (line drawing). Distinctive features of sporocyst: flat SB with a half disc SSB and a very large SR.

Prevalence: Grés et al. (2002) examined 254 rabbits from six areas in France; they said the prevalences were low, varying from 4% to 14% in the five localities in which this species was found, and that the maximum prevalence was during the winter. A year later, Grès et al. (2003) said they found *E. roobroucki* in 0–21% of 254 wild rabbits sampled from six different localities in France. In looking at both papers, these are the same six localities and the same 254 rabbits, yet they reported different prevalences. There are some other statements made in these papers that make the interpretation of their results uncertain. In the latter paper (2003), they identified 10 eimerians from these 254 rabbits and, apparently, 100% were infected with two of these 10 (*E. flavescens, E. perforans*). Unfortunately, they did not specifically mention how many individuals were infected with multiple species vs. those with single species infections.

Sporulation: Presumably exogenous, but unknown.

Prepatent and patent periods: Unknown, oocysts recovered from fecal samples

Site of infection: Unknown.

Endogenous stages: Unknown.

Cross-transmission: None to date.

Pathology: Unknown.

Material deposited: None.

Remarks: Grés et al. (2002) provided a line drawing and a photomicrograph in their original description, but they did not deposit either of these in an accredited museum, even though they said their photomicrograph (their Fig. 2) was a holotype. Nor did they provide all of the quantitative and qualitative information needed and expected in standard species descriptions (range of oocyst L, W measurements; sporocyst mean L, W measurements and their ranges; measurments of SR, SZ, RB, etc.). This is disappointing and does not meet best-practice standards.

EIMERIA STIEDAI (LINDEMANN, 1865) KISSKALT & HARTMANN, 1907

Synonyms: *Monocystis stiedai* Lindemann, 1865; *Psorospermium cuniculi* Rivolta, 1878; *Coccidium oviforme* Leuckart, 1879; *Coccidium cuniculi* (Rivolta, 1878) Labbé, 1899; *Eimeria cuniculi* (Rivolta, 1878) Wasielewski, 1904; *Eimeria oviformis* (Leuckart, 1879) Fantham, 1911; *Eimeria stiedai* var. *cuniculi* Graham, 1933.

Type host: *Oryctolagus cuniculus* (L., 1758) (syn. *Lepus cuniculus*), European (domestic) rabbit.

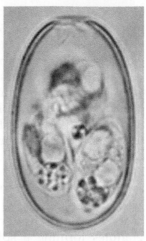

FIGURES 6.25, 6.26 Line drawing of the sporulated oocyst of *Eimeria stiedai* from Gill and Ray, 1960, with permission from Springer Science and Business Media, copyright holders for the *Proceedings of the Zoological Society* (supersedes *Proceedings of the Zoological Society of Calcutta*). Photomicrograph of a sporulated oocyst of *E. stiedai* from Norton et al., 1977, with permission from *Parasitology*.

Type locality: EUROPE: The Netherlands: Delft (?).

Other hosts: *Lepus americanus* Erxleben, 1777, Snowshoe hare (natural); *Lepus californicus* Gray, 1837, Black-tailed jackrabbit (natural); *Lepus capensis* L., 1758, Cape hare (?); *Lepus europaeus* Pallas, 1778, European hare (experimental); *Lepus timidus* L., 1758, Mountain hare (natural); *Sylvilagus audubonii* (Baird, 1858), Desert cottontail (experimental); *Sylvilagus floridanus* (J.A. Allen, 1890), Eastern cottontail (natural and experimental); *Sylvilagus nuttallii* (Bachman, 1837), Mountain cottontail (natural).

Geographic distribution: AFRICA: Nigeria; ASIA: Azerbaijan (former USSR), China, India, Kazakhstan, Philippines; AUSTRALIA: New South Wales, Western Australia; EUROPE: Belgium, Czech Republic, Finland, France, Germany, Hungary, Italy, The Netherlands, Poland, Portugal, Sweden, Switzerland; NEW ZEALAND; MIDDLE/NEAR EAST: Iran, Iraq, Saudia Arabia, Syria; NORTH AMERICA:

USA: California, Iowa, North Carolina, Washington DC, Wyoming; SOUTH AMERICA.

Description of sporulated oocyst: Oocyst shape: elongate-ovoidal to ellipsoidal; number of walls: 1−2; wall characteristics: smooth, thin, colorless, yellow-brown, or salmon-colored; generally considered to be composed of one layer, but Pellérdy (1965, 1974) said there is a thin outer layer that detaches readily and/or can be removed easily, and Scholtyseck et al. (1969) found two layers; L × W: 37 × 20 (31−42 × 17−25); L/W ratio: 1.8; M: present at narrow end of oocyst; M characteristics: 6−10 wide; OR: present; OR characteristics: a few small, rounded, refractile granules forming a spheroidal body, up to 3 wide, but generally smaller, located in the middle of the oocyst, partly obscured by sporocysts (Norton et al., 1977); PG: absent. Distinctive features of oocyst: large size, prominent M and absence of PG.

Description of sporocyst and sporozoites: Sporocyst shape: elongate ovoidal; L × W: 18 × 10 (17−18 × 8−10); L/W ratio: 1.8; SB: present; SSB, PSB: both absent; SR: present; SR characteristics: granular; SZ: elongate. Distinctive features of sporocyst: none.

Prevalence: This may be the most important coccidium of domestic rabbits. It is much less common in cottontails than some other species that infect *Oryctolagus* and it is extremely rare in hares. Kessel and Jankiewicz (1931) found it in ~180/> 2,000 (9%) domestic rabbits in California, while Lund (1950) found it in only 12/1,200 (1%) rabbits in southern California. Harkema (1936) reported it in the liver of 15/41 (37%) cottontails, *S. floridanus*, in the Duke Forest, Durham County, North Carolina. Honess (1939) said that he found it occasionally in *S. nuttallii* in Wyoming. Lampio (1946) reported *E. stiedai* in rabbits in Finland. Ryšavý (1954) said he found this species in both *O. cuniculus* and in *L. europaeus* in the Czech Republic. Gill and Ray (1960) noted that Ray (1945) already had reported *E. stiedai* in domestic rabbits in India and they gave a brief description of the sporulated oocyst and included a line drawing, but did not report its incidence in the rabbits they examined. Mykytowycz (1956, 1962), who did a 35-mo study in a free-living population of experimentally enclosed wild rabbits in Australia, found that the average level of infection with *E. stiedai* fluctuated considerably with season, but was unable to find a satisfactory explanation for the fluctuation. It tended to be especially common in 3-mo-old and younger rabbits, and to affect them more severely than older rabbits. Niak (1967) found *E. stiedai* in 18/84 (21%) rabbits from four rabbitries in Iran, and Bouvier (1967) found it in the liver of 20/> 2,000 (1%) gray hares, *L. europaeus*, in Switzerland. Mirza (1970) found it in 16/31 (52%) domestic rabbits in Iraq, while Burgaz (1970a, b) reported it in Sweden, from the European hare, but he gave no other information. Stodart (1971) examined 656 wild *O. cuniculus* in New South Wales, eastern Australia, found 625 (95.3%) to be infected, and identified seven eimerians based on their sporulated oocysts (also see Stodart 1968a, b). In her paper (1971), she lumped *E. stiedai* and *E. media* together and said 387/656 (59%) were infected with this mixture and that "11.6% of rabbits had lesions in the liver caused by *E. stiedai*." Flatt and Campbell (1974) found it in the livers of 1,562/17,354 (9%) domestic rabbits slaughtered for food in Iowa. Peeters et al. (1981) did not find *E. stiedai* in 770 fecal samples from commercial rabbitries, but did report it in 87/282 (7%) sick rabbits in traditional rabbitries in Belgium. Veisov (1982) examined 3,150 rabbits from three different areas/districts of the Cuba-Khachmazskoy zone in Azerbaijan and found 2,905 (81%) infected with nine species of *Eimeria*, including *E. stiedai*; however, we were not able to determine how many of the rabbits examined harbored this species. In an unpublished Master's thesis, Ogedengbe (1991) said he isolated this species in the gall bladders of three of the rabbits surveyed in Kaduna State, Nigeria. However, he (1991) neglected to state how many rabbits were surveyed, nor did he mention the host species; Anonymous (2012) listed the Cape hare, *L. capensis,* as the only lagomorph species found in Nigeria. Darwish and Golemansky (1991) found this species in 3/75 (4%) domestic rabbits from four localities in Syria; however, the authors believed this low prevalence was not a true representation of its real prevalence because they did not look at the livers of the 75 rabbits. On the other hand, Musongong and Fakae (1999) reported *E. stiedai* in the livers of 41/131 (37%) domesticated *O. cuniculus* in all five village markets in Nsukka Local Government Area of eastern Nigeria. Polozowski (1993) examined feces from 246 rabbits in six farm rabbitries in the Wroclaw District of Poland and found 180 (73%) infected with *E. stiedai;* 158/180 (88%) infected rabbits were < 3-mo-old, 7/180 (4%) were 4–12-mo-old, 15/180 (8%) were 13–24-mo-old, while 0/180 were > 24-mo-old. Hobbs and Twigg (1998) examined fresh fecal samples from wild *O. cuniculus* in Western Australia and found 310/1,200 (26%) to harbor *E. stiedai;* 50 of their sporulated, ellipsoid-ovoidal oocysts were 36.2 × 20.6 (34–40 × 18–22), with sporocysts 15.5 × 8.3 (14–17 × 7–9). Grès et al. (2003) found *E. stiedai* in 4–21% of 254 wild rabbits sampled from six different localities in France. They identified 10 eimerians from these 254 rabbits and, apparently, 100% were infected with one or more of these 10. Unfortunately, they did not specifically mention how many individuals were infected with multiple species vs. those with single species infections. Yakhchali and Tehrani (2007) found *E. stiedai* in 87/436 (20%) New Zealand white rabbits from two rabbitries in northwestern Iran, and Al-Mathal (2008) found it in the livers of 158/490 (32%) domestic rabbits examined for hepatic coccidiosis on three farms in the Eastern Province of Saudi Arabia.

Sporulation: Exogenous. Oocysts can sporulate between 10 and 35°C, but the optimum temperature is 20–27°C. The minimum time is 58 hr, but, in general, it takes about 3 days. Ayeni (1969) found that it took a minimum of 48 hr at 28°C, and if the oocysts were shaken continuously, most oocysts sporulated within 2–4 days. Ryšavý (1954) said that oocysts sporulated in 60–75 hr. In general, oocysts will sporulate in 2–4 days under most circumstances. A detailed structural picture of sporulating oocysts that includes the timing and duration of characteristic stages in this process has been described for only for a couple species of *Eimeria.* Rose (1958) reviewed much of the work on the sporogony of *E. stiedai,* and Wagenbach and Burns (1969) studied the structure and respiration of sporulating oocysts of *E. stiedai.* They pointed out that oocysts of this species go through the sporulation process with some degree of synchrony. An early increase in respiratory rate was followed by a depression in rate that correlated with the appearance of the early spindle stage. The rate again increased and then decreased toward a base rate during and after completion of sporulation.

Prepatent and patent periods: Smetana (1933a) reported that oocysts first appeared in the feces 3–4 wk PI, while Kotlán and Pellérdy (1949) said the prepatent period is 17 days, a time later confirmed by Cheissin (1967), or 14–16 days (Fitzgerald, 1970a, b; Sanyal and Sharma, 1990). The patent period is ~24 (21–30) days (Carvalho, 1943).

Site of infection: The route by which sporozoites of *E. stiedai* reach the liver from the intestinal tract has not been indisputably determined, but some clues are given by the work of Fitzgerald (1970a, b, 1972, 1974). He indicated (1970a, b) that SZs of *E. stiedai* were transported in monocytes and probably reached the liver by blood transportation. He noted that blood withdrawn from anesthetized rabbits by heart puncture still had SZs at least 7 days PI. Later Fitzgerald (1972) documented that *E. stiedai* was transfused by heart puncture to 59/76 (78%) coccidia-free uninfected recipient rabbits that received whole blood, erythrocytes, or leukocytes from donor rabbits orally inoculated up to 17 days earlier with *E. stiedai* oocysts. Pellérdy (1969a, b) and Fitzgerald (1970b) also reported the occurrence of liver infections with *E. stiedai* following ear vein inoculations with experimentally excysted SZ. Once in the liver, SZs located above the HCN in the bile duct epithelial cells, but the liver parenchyma cells were rarely invaded. Cheissin (1967) said that Yakimoff (1931, *Veterinary Protozoology*, unavailable to us; in Russian) reported finding oocysts of *E. stiedai* in the lymph nodes of a rabbit and in the mesenteric nodes, the bladder, and in the nasal passages. Numerous experimental infections of rabbits with this species have shown that the endogenous stages will not develop in the intestine (Cheissin, 1967). The question as to the routes by which SZs move from the intestine into the liver has not been entirely answered, although Pellérdy and Dürr (1970), who studied experimental infections in rabbits aged 18 days to 5 wk old, found SZs within 24 hr in the mesenteric lymph nodes, and from days 1 to 5 PI they detected the parasite in the livers by histological sections. The SZs of *E. stiedai* have been observed in the upper part of the intestine, but not endogenous stages. Smetana (1933b, c) said that the SZs penetrated the liver through the bloodstream, not the bile duct, with one line of evidence being that the endogenous stages of *E. stiedai* have developed in the livers of rabbits in which the bile duct was ligated. When SZs were introduced intravenously, the liver of rabbits became infected with the endogenous stages and SZs were found in the blood vessels of the liver after large doses of sporulated oocysts had been introduced into the intestine. However, no SZs were found in the lymph vessels. Apparently, SZs used the blood system to penetrate not only the liver, but also the kidneys (Cheissin, 1967). Owen (1970) infected donor rabbits with $2-4 \times 10^6$ sporulated oocysts and studied the route of migration of SZs, after initial penetration of the duodenal epithelium, by harvesting lymph node and bone marrow

samples at 12, 24 and 48 hr PI and liver, blood buffy layer, axillary lymph node, and washings from the ceolomic cavity at 48 hr PI; these tissues were homogenized and injected intravenously (IV) into 8-wk-old SPF recipient rabbits. Based on severity of liver involvement and oocyst production, she concluded there were viable SZs present in the mesenteric lymph node at 12 hr PI, in bone marrow at 24 hr PI, and in liver and buffy coat at 48 hr PI in donor rabbits. When excysted SZs were injected into the ear veins, they passed via the blood stream to the mesenteric lymph nodes and liver; some entered monocytes and passed within them to the liver, but most remained free of host cells during their passage through the blood stream (Slater et al., 1969). Dürr (1971) reported the presence of radioactivity in all organs of mice and rabbits being fed labeled SZs of *E. stiedai*, but found no direct evidence of actual SZs of *E. stiedai* in these organs. Later Dürr (1972) found SZs in the liver, mesenteric lymph nodes, and bone marrow of the femur and tibia of rabbits that he had infected with *E. stiedai* oocysts. He found them in the bone marrow only in macrophages in fresh smears, but also, on rare occasions, free between the cells in histological preparations. He concluded that *E. stiedai* SZs pass through the intestinal wall and go to the mesenteric lymph nodes via the lymph and were distributed through the body by the thoracic duct; he thought that the liver may be a filtration organ for SZs. Previously, it had been believed (Smetana, 1933b; Slater et al., 1969; Fitzgerald, 1970a, b; Pellérdy and Dürr, 1970) that the SZs were carried directly to the liver via portal blood. Dürr (1972) acknowledged that this also may occur, but that it "certainly plays a minor role"; he also concluded that the role of macrophages in their transportation was not clear. Horton (1967) found that *E. stiedai* migrated to the liver via the lymph vessels. After infected rabbits were killed, he found SZs in the lamina propria of the duodenum 5–8 hr PI, and then in the mesenteric lymph nodes, both free and within lymphatic monocytes, 12—84 hr PI. He found no SZs in the duodenal portal blood and believed that the SZs migrated from the duodenum to the mesenteric lymph nodes, probably in lymphatic monocytes. Compounding the issue of what happens to SZs of *E. stiedai* when they enter the rabbit gut and excyst, Fitzgerald (1974) transmitted *E. stiedai* to young, coccidia-free rabbits by injecting them by heart puncture with heart blood, erythrocytes or leukocytes harvested by heart puncture from donor rabbits that had been experimentally inoculated *per os* with sporulated oocysts up to 27 days earlier. Fifty-nine of 76 (77%) young rabbits injected in this manner developed *E. stiedai* infections; that is, infective forms were present in the blood that long.

Endogenous stages: von Wasielewski (1904) was among the first to study the excystation and early development of *E. stiedai* (syn. *E. cuniculi*), including penetration of epithelial cells by the SZs and formation of the early meronts and merozoites. Three years later, Kißkalt (= Kisskalt) and Hartmann (1907) described the sporulation process and rendered another line drawing of an early meront of *E. stiedai*. Smetana (1933b) detailed experimental infection of the liver by SZs of *E. stiedai*, but it was Horton (1967) who discovered SZ of *E. stiedai* in the mesenteric lymph nodes of rabbits to throw some light on the route of migration between the duodenum and the liver and he suggested the following sequence of events. After 5 hr in the duodenum most of the SZs had excysted from their sporocysts and began to penetrate the mucosa. Upon reaching the basement membrane, the SZs were phagocytosed and then transported within monocytes via the mesenteric lymph vessels to the lymph nodes. Then, 5—6 hr later, the SZs reached the sinuses of the lymph nodes within lymphatic monocytes. Later, the SZs left the monocytes and entered the portal blood, which carried them to the liver or, alternatively, they might have

been carried passively by the monocytes to the liver where they left the monocytes and entered biliary epithelial cells (Horton, 1967). The complete endogenous development of *E. stiedai* was studied in detail by a number of investigators, including Tyzzer (1902), Smetana (1933c), Kotlan and Pellérdy (1936, 1937), Pellérdy and Dürr (1970), Owen (1970) and others (see Pellérdy, 1965, 1974 for reviews), but the number of asexual stages is still uncertain. Once SZs enter epithelial cells of the bile ducts, some of them become elongate meronts with 2—3 N, but most round up and become meronts which produce 6—30 (usually 8—16) merozoites, ~8—10 × 1.5—2. Pellérdy and Dürr (1970) found five asexual generations within 12 days PI, after which gamonts were formed. However, parallel to gamogony, a sixth generation of merozoites was formed.

Pellérdy and Dürr (1970) detailed the merogenous stages as follows. Once inside bile duct epithelial cells, on about day 3 PI the SZ became wider at the anterior end and narrower at the posterior end, but did not round up. They were ~8—13 × 3—5, becoming first-generation meronts, which produced 5—7 merozoites that were 7.5 × 1.5. First-generation merozoites entered new bile duct epithelial cells, generally on the lumen side of the HCN, and became second-generation meronts, about the same size as the first-generation meronts, and produced 5—8 slender merozoites by day 4 PI. Third-generation meronts were in a PV, ~11 wide; they produced 10—20 slender merozoites that were slightly bent, measured 8 × 0.75, and lay side-by-side like a bunch of bananas. Third-generation merozoites then entered new cells to produce fourth-generation meronts that were 16 × 10, lying in a PV near the HCN, or on its luminal side. These were present on day 8 PI and produced 20—30 slender, slightly bent merozoites, 6—8 × 0.5—0.75, which lay in a circle around a dark-staining residuum. By day 11 PI, fifth-generation meronts were found in a PV, ~17—20 × 14, and had produced

30—40 merozoites, each ~8—9 × 1—1.5, and which lay randomly in their meronts around an ellipsoidal residual body. All five of these asexual generations were designated as Type B (Pellérdy and Dürr, 1970). They (1970) also designated Type A meronts and merozoites that apparently developed coincidentally with those they designated as Type B.

Type A first-generation meronts were similar to Type B's first-generation meronts. Type A second-generation meronts were seen at 4—6 days PI. They were in a PV, ~6—10 × 6—8, and contained only two hook-shaped merozoites, ~6 × 2. Third-generation Type A meronts were in vacuoles, ~12 × 9, with 5—7 merozoites, each ~8 × 1.5—2; these resembled Type B second-generation merozoites (present on day 4 PI), but were plumper, darker, and more homogeneous. The Type A fourth-generation meronts were found in vacuoles, ~13.5 × 9, on day 10 PI, and each had 3—4 plump merozoites, ~9 × 2. On day 12 PI, Pellérdy and Dürr (1970) found fifth-generation Type A meronts in vacuoles, ~12—13 × 8—9, which produced 5—8 plump merozoites that were 10 × 2.5—3 and had several small N. By day 15 PI, they saw sixth-generation meronts in vacuoles, ~20 × 12, each with ~15 merozoites, that looked like those of the previous Type A generation. Later on, and coincident with gamogony, they found stages of merogony that resembled those seen on day 10 PI. With time, merogony became progressively less abundant in bile ducts containing gamonts and oocysts, but development in different bile ducts was not synchronous, so that young gamonts or merogonic stages predominated in some bile ducts, while mature gamonts and free oocysts predominated in others.

Scholtyseck (1965a) described the ultrastructure of merogony and the cytochemical detection of glycogen (Scholtyseck (1964), and Scholtyseck and Piekarski (1965) detailed the fine structure of merozoites. Scholtyseck (1965a) said there are two sizes of meronts; one, ~25 wide, which was seen early in an infection, and the other which is 8—10 wide. Merozoites were 8 × 2—3, with a cell surface composed of two narrow double membranes separated by a zone of less electron-dense material (Scholtyseck and Piekarski, 1965). The outer membrane was relatively thin and continuous over the entire cell, while the inner membrane was relatively thick and was not present at the anterior and posterior ends. The anterior had a typical conoid and 24 subpellicular microtubules extended posteriorly from this end. A variable number of rhoptries or micronemes extended longitudinally through the cytoplasm, and there was an opening, ~0.2 wide, at the posterior end of each merozoite. The N was in the posterior half and it had a double membrane containing many pores. There was a Golgi apparatus in the cytoplasm just above the anterior end of the N. Merozoites, like sporozoites, contained clear globules (Scholtyseck and Piekarski, 1965). Heller (1971) studied the fine structure of merogony in the so-called small meronts. In these, the merozoites developed in two phases. First, four large merozoite-like meronts were formed. Their N divided repeatedly, forming several small merozoites within each merozoite-like meront, resulting in the presence of many merozoites within a single PV. Heller and Scholtyseck (1971) studied the fine structure of the formation of the large meronts and merozoites. The latter develop both at the surface of, and in the interior, of their meronts; in the latter case they were associated with deep invaginations of the cell membrane. The first indication of merozoite formation was the appearance of a conoid near the N. Each conoid was accompanied by 24 microtubules, which become the subpellicular microtubules. The rhoptries grew posteriorly as osmophilic pouch-like structures within the conoid. The merozoite's outer membrane was formed when the meront outer membrane folded down around the conoid, and the inner layer (actually a double membrane) appeared within it. The polar ring was at the anterior end of the inner membrane.

Micronemes became visible when the rhoptries were formed. The process of merozoite formation resembled internal merogony as well as endodyogeny. Černá and Sénaud (1971) noticed some peculiarities in the fine structure of the merozoites. On day 13 PI they (1971) found some meronts with merozoites that had two or more N, a phenomenon seen in the development of several other rabbit eimerians. Scholtyseck (1973) confirmed that both endodyogeny and merogony occurred in *E. stiedai*, and proposed that merogony is a secondary process of reproduction, which developed from endodyogeny. Scholtyseck and Ratanavichien (1976) found that endodyogeny took place in the epithelial cells of the bile ducts of the rabbit. They said that this confirmed the theory that endodyogeny is the primary and fundamental process of asexual reproduction in the coccidia, and that merogony is a secondary process that has developed from endodyogeny. Finally, in Scholtyseck's (1965a) study of the fine structure of meronts, he said the cytoplasm of this stage had a large number of membranes in the endoplasmic reticulum, which were situated concentrically around the nuclei. Apparently, the formation of fissures occurred along these membranes in the cytoplasm of the meronts, leads to a separation of the merozoites during segmentation, and this process of merozoite formation by splintering the entire meront into individual pieces may be a general characteristic feature of merogony, according to Cheissin (1967).

Pellérdy and Dürr (1970) first saw sexual stages 10 days PI, but they could not differentiate macro- from microgamonts until day 15 PI and by day 16 PI there were large numbers of both gamonts present. Scholtyseck et al. (1966) described the fine structure of the macrogametes. Macrogametocytes were 20–25 wide and had a large residual body. The parasite cell surface was a unit membrane with **osmophilic** granules lying underneath it in some places. The cytoplasm had a fine, dense granulation, which distinguished it clearly from the cytoplasm of the host cell. Macrogamont cytoplasm contained glycogen granules, a well-defined endoplasmic reticulum, "dark bodies" (wall-forming bodies I), wall-forming bodies (II), dumbbell-shaped mitochondria with internal tubules, and lipid inclusions and had several micropores about 110 nm deep and 110 nm wide. These were larger than occur in other species of *Eimeria* and they lacked microtubules at their surface, which other *Eimeria* are reported to have. Hammond et al. (1967) and Ball et al. (1988) described the fine structure of the microgamonts. They were about 15 wide and lay in a PV. In the early stages, they contained unusually numerous, and widely distributed, vesicles and also canals of the endoplasmic reticulum in the finely granulated cytoplasm. Microgamonts also had numerous mitochondria containing tubular structures near the N; the mitochondria were typical of *Eimeria*. The N had a compact nucleolus, densely granular nucleoplasm, and a membrane with numerous pores. The microgamont cell membrane was extremely thin, with an underlying osmophilic layer in some places.

Cross-transmission: Jankiewicz (1941) and Herman and Jankiewicz (1943) transmitted *E. stiedai* from the domestic rabbit to the cottontail, *S. audubonii valicola* from California. Hsu (1970) infected *S. floridanus* and found that it was highly pathogenic for this cottontail. Burgaz (1973) successfully transmitted *E. stiedai* from *O. cuniculus* to *L. timidus*; although her identification and labotatory methods have been questioned by us and others (e.g., Samoil and Samuel, 1981), we give more credence to this result based on the success of others in transmitting this eimerian to *Lepus*. Varga (1976) infected nine hares (*L. europaeus*) with *E. stiedai* oocysts from *O. cuniculus* and found that each of them, killed 18–20 days after experimental infection, had macroscopic liver lesions, but that they were less numerous than those in the domestic rabbit. Entzeroth and Scholtyseck (1977) transmitted *E. stiedai* from the domestic rabbit to *L. europaeus* by feeding the hares sporulated oocysts; then from 1970 to 1976, Scholtyseck et al. (1979) transferred sporulated oocysts of *E. stiedai* from *O. cuniculus* to six young, coccidia-free *L. europaeus* and were able to detect all stages of the endogenous cycle from sporozoites to oocysts by both light and electron microscopy. Pérard (1924a, b) was unable to infect the Norway rat, house mouse, dog, lamb, or kid. Kessel and Jankiewicz (1931) could not infect the domestic chicken or guinea pig. Yakimoff and Iwanoff-Gobzem (1931) were unable to infect the dog, chicken, pigeon, white rat, or mouse. Lee (1934) could not infect a dog. Schrecke and Dürr (1970) conducted excystation and infection experiments with *E. stiedai* and the oocysts of intestinal coccidia of rabbits (*E. intestinalis*, *E. magna*, *E. media*, *E. perforans*) by infecting newborn albino rats (*Rattus norvegicus*), albino mice (*Mus musculus*), multimammate mice (*Mastomys natalensis*), and chicks (*Gallus gallus*) during their first day of life. In the mammals, the rate of excystation was minimal in very young animals and rose gradually with age; in the baby chicks, only small numbers of SZ excysted during the first hr PI. Aly (1993) established infection of *E. stiedai* in 20/25 (80%) immunocompromised mice given 0.03 mg dexamethasone for 10 days before inoculation with 250,000 sporulated oocysts from rabbits and for 2 wk after infection. She (1993) reported that the prepatent period was 30—35 days (vs. 17—18 days in rabbits) and that patency continued for 2 mo after inoculation when the experiment was terminated. The infected mice showed enlargement of the liver and distension of their abdomens with ascites.

Pathology: It is now well known that change in enzyme activity, particularly in the serum and liver, is an important parameter in the diagnosis and early detection of many diseases. Dunlap et al. (1959) were among the first to document that infection of rabbits with *E. stiedai* significantly changed their blood and liver enzymes and other parameters. They (1959) found increased serum β- and γ-**globulin** and β-**lipoprotein**, but decreased α-**lipoprotein** in their infected rabbits; in addition, bile flow, liver weight, and **bromsulfophthalein metabolites** in the bile increased, while bromsulfophthalein excretion decreased. Hoenig et al. (1974) found that during *E. stiedai* infection in rabbits, serum **glutamic**, **pyruvic** and **oxaloacetic transaminases**, bilirubinemia and **lipemia** were increased, and total serum proteins decreased; the greater the infective dose of oocysts, the higher the mortality. Abdel-Ghaffar et al. (1990) infected rabbits with 10^5 sporulated oocysts and sampled blood and liver enzymes for 23 days PI, looking specifically at levels of **serum glutamic oxaloacetic transaminase (SGOT)**, **serum glutamic pyruvic transaminase (SGPT)**, **alkaline phosphatase (ALP)**, and **acid phosphatase (AcP)** activity. They found serum SGOT activity in both male and female infected rabbits was highly significantly increased vs. uninfected controls. The SGOT levels in the livers also increased in both sexes, but not significantly. Other non-significant increases included SGPT in the livers of both sexes. On the other hand, both AlkP and AcP activity increased in the serum of both sexes, but decreased in the livers of both sexes. Sanyal and Sharma (1990) also looked at body and liver weight, and several blood parameters (SGOT, SGPT, hypoglycemia, hypoproteinemia, bilirubinemia) during *E. stiedai* infections, but they used graded experimental infections of 10^2, 10^3, 10^4, and 10^5 sporulated oocysts. Four of six rabbits infected with 10^5 oocysts and two of six infected with 10^4 oocysts died, but different grades of infection did not affect body weights between experimental groups and the control group; however, the weight of livers increased up to 10-fold in infected hosts. In addition, SGOT levels showed a steady increase on day 14 PI with a secondary increase on day 28 PI. SGPT levels increased up to day 14 PI and declined thereafter. There was continuous bilirubinemia and transient hypo-proteinemia and hypoglycemia on day 21 PI. Necrotic changes in the livers accompanied by subsequent release of enzymes into the serum might be expected to cause the decreased tissue levels and increased serum levels for these enzymes. Thus, increased levels in these blood serum enzymes may be a good indicator, not only of pathological lesions, but also a degenerative destruction of the tissues involved.

Many workers have described the physical and physiological effects of *E. stiedai* on the liver (Smetana, 1933b; Kotaln and Pellérdy, 1937; Pellérdy and Dürr, 1970; Stingl, 1974; Barriga and Arnont, 1979, 1981; Al-Mathal, 2008; Freitas et al., 2009, 2010; others), and indeed the lesions are so characteristic as to be easily recognized with the naked eye. Seddon (1966) said that *E. stiedai* is present in about 75% of wild and domesticated rabbits in all districts of New South Wales, Australia, showing marked lesions of hepatic coccidiosis in rabbits in apparently good condition. In mild cases of liver coccidiosis there may be no signs, but in more severe cases rabbits lose their appetites and grow thin, diarrhea results, and mucous membranes become **icteric**. The disease is more severe in young animals than in the old (Levine, 1973b; Gomez-Bautista et al., 1987), and it may be chronic or acute, with death occurring 21–30 days PI. In experimentally infected rabbits, the first reaction of the host tissue can be seen histologically on day 2 PI. The **reticuloendothelial** elements in the liver are noticeably increased, and there is some leukocytic infiltration in the vicinity of the bile ducts. These changes were more pronounced on days 3 and 4 PI, but decreased thereafter. From day 6 PI and beyond, the leukocytes were replaced around the bile ducts by connective tissue. Some of the signs of pathology were due to interference with liver function. The liver may become markedly enlarged and white circular nodules or elongated yellowish cords may appear in it. At first these are sharply circumscribed, but later they

tended to coalesce. There were enormously enlarged bile ducts filled with developing parasites and tremendous hyperplasia of the bile duct epithelial cells. Instead of forming a simple, narrow tube, the epithelium was thrown into great, **arborescent** folds, and each cell contained a parasite. Stingl (1974) studied the differences in the bile duct systems between domestic rabbits infected with *E. stiedai* and uninfected rabbits. He found: (1) length and diameter of the bile ducts dilated significantly with increasing infective doses due to "stenosis of the papilla duodeni;" (2) both intra- and extra-hepatic bile ducts experienced "cholangitis catarrhalis chronica coccidiosa;" (3) the gall bladder also became infected, but usually later and much less than the bile ducts, and only merogony, but no gamogony occurred there; (4) both epithelial proliferation and mucous production increased in infected rabbits, and the increased mucous seemed to have a protective effect against infection on neighboring cells; and (5) **eosinophilia** and **histiocytosis** both increased in the tissue surrounding all parts of the infected bile duct system. Barriga and Arnoni (1981) studied the pathophysiology of hepatic coccidiosis in New Zealand rabbits infected with from 10^2 to 10^4 oocysts for 50 days PI. They identified four specific pathophysiological events during their infections: (1) a phase of indirect damage to the hepatocytes that took place during the first 2 wk of infection and was characterized by increased transaminases; (2) a cholestatic period consequent to the production of oocysts that began suddenly in the third week, diminished gradually towards the seventh week, and was characterized by a rise of bilirubinemia and lipemia; (3) a stage of metabolic dysfunction that began in the third or fourth week, intensified for the next 3 wk, started to recover during the seventh week PI, and was characterized by hypoproteinemia and hypoglycemia; and (4) a period of immunodepression characterized by the inability of the heavily infected host to inhibit oocyst production. Deaths seemed to be related to depression of the host's immunity (Barriga and Arnoni, 1979).

Material deposited: None.

In vitro cultivation: Fitzgerald (1970c) inoculated 150,000 sporozoites, from oocysts of *E. stiedai* harvested from rabbit livers, into the chorioallantoic cavity of 10-day-old chick embryos; later, he found early gametocytes in the chorioallantoic membrane (CAM) of 15-day-old embryos 5 days PI. These stages localized in the epithelial cells of the CAM and were "nearly identical" to comparable stages of *E. stiedai* in the bile duct epithelium of the rabbit. His photomicrograph showed microgamonts, but he said his results were not consistent. Krieg (1971) studied the in vitro excystation of sporozoites of *E. stiedai*, and their eventual viability, when exposed to various chemical agents and physical (temperature) parameters. Coudert and Provôt (1974) took merozoites from the bile of rabbits infected with *E. stiedai* 120–132 hr PI and inoculated these into newborn rabbit kidney cell cultures in which different generations of meronts developed. John et al. (1999) examined the carbohydrates present on the surface of *E. stiedai* SZs, detected lectin-binding sites, and tried to determine their functional role in the invasion process of primary rabbit liver biliary epithelial cells in vitro.

Remarks: *Eimeria stiedai* was originally described from the domestic rabbit, *O. cuniculus*, by Lindemann (1865). Since then it has been reported from *Lepus* species by Tyzzer (1902), Nieschulz (1923), Henry (1932), Robertson (1933), Mandal (1976) and others, with most succeeding writers accepting previous reports unquestioningly. At least some of these reports are probably erroneous. For example, Nieschulz (1923) said nothing about the liver in his paper, when he wrote that, "*E. stiedai*" caused diffuse lesions in a large part of the small intestine. Henry (1932, p. 282) said merely that she recognized *E. stiedai* oocysts in the feces of a jack rabbit, *L. californicus*, in California, USA, without providing any empirical evidence. According to Levine (pers. comm. in an unpublished ms), "there is something in the literature about R.V. Boughton (1932) having found *E. stiedai* in the snowshoe hare, *L. americanus* in Manitoba, but all he reported was that he found ellipsoidal or ovoidal oocysts, mostly flattened at the micropyle end and 36–52 × 24–27 in the intestine, and that they were 'possibly' *E. stiedai*." Boughton's (1932, p. 535) data do not support his identification. Robertson (1933) said that he found *E. stiedai* in hares in England, but his illustrations were not of this species (Pellérdy, 1965). Litvenkova (1969) said she found this species in *L. europaeus* and *L. timidus* in Byelorussia, but gave no description. Svanbaev (1979) reported *E. stiedai* oocysts in the feces of *L. europaeus* and *L. tolai* in Kazakhstan, but again, without much empirical evidence to support that identification.

Other, mostly later, authors provided more compelling evidence of *E. stiedai* in *Lepus* species when they began to document tissue stages in the liver of wild rabbits they surveyed. Ryšavý (1954) said he found this species in both *O. cuniculus* and in *L. europaeus* in the Czech Republic with oocysts that were 34.4 × 19.3 (30–40 × 15–21), sporocysts 13–18 × 7.7; the parasite was found in the liver and was abundant in the domestic rabbits they sampled. According to Levine (unpublished ms), von Braunschweig (1965) found *E. stiedai* "in the bile ducts of a hare (presumably *L. europaeus*) in Germany, and mentioned that it had been found in hares 11 times between 1908 and 1947 at the Institut Galli-Valerio in Switzerland." Bouvier (1967) found it in the livers of 20 out of > 20,000 gray hares, *L. europaeus*, in Switzerland. Mandal (1976) said he recovered *E. stiedai* in *Lepus* sp. from Kashipur, India, with ovoidal to ellipsoidal oocysts that were 26–40 × 16–24, without an OR, and had ovoidal sporocysts, 17 × 9, with a SR, and that oocysts sporulated in 72 hr. Vila-Viçosa and Caeiro (1997) reported this species in both *O. cuniculus* and *L. capensis* in Portugal, with the ellipsoid oocysts from *O. cuniculus* being 33—37.5 × 16.8—19.5. Grés et al. (2003) examined 254 wild rabbits from six localities in France and this is one of 10 species they identified. Aoutil et al. (2005) reported *E. stiedai* in *L. europaeus* in France; their oocysts were 37 × 22 (33—40 × 19—24) with sporocysts, 18 × 10. Based on these latter studies and on reliable cross-transmission experiments, we conclude that *E. stiedai* can and does infect *Lepus* species, but that it is relatively rare in hares and jackrabbits.

Levine and Ivens (1972) began using the spelling "*stiedai*" even though the original description used "*stiedae*," a convention that most other authors have now willingly followed. They argued, correctly, that Lindemann (1865) named it for Ludwig Stieda, and according to Article 27 of the International Rules of Zoological Nomenclature, the genitive ending must be used; thus, "*ae*" would indicate that the species was named for a woman, and "*ai*" for a man (Levine and Ivens, 1972).

Eimeria stiedai has special significance in that it dates to Leeuwenhoek (1674), who saw oocysts of what were almost certainly this species in rabbit bile and partially described them in his unpublished seventh letter to the Royal Society (Dobell, 1932; Wenyon, 1926). The history of this species also is interesting in that Bosc (1898) in France was possibly the first to see and document intracellular development of endogenous stages, which he thought were tumors produced by the parasite. Later, Wenyon (1926) said that the first published description of pathology in the duodenum and liver by *E. stiedai* was by Hake in 1839, who suggested that unsporulated oocysts were pus globules in rabbit livers (Levine, 1973a). Since then it has been found and described by many workers, but the earlier ones were of the opinion that all coccidia of the rabbit belonged to the same species, so their descriptions cannot be used because they often were confusing oocysts of intestinal forms

with *E. stiedai* (see above for some examples). The description of the sporulated oocyst (above) is based on those of Pérard (1924a, b), Kessel and Jankiewicz (1931) and Pellérdy (1965, 1974). Hegner and Chu (1930), in the Philippines, found coccidia in laboratory rabbits, including oocysts of *E. stiedai* (also see Madsen, 1938), which measured 35 × 23, L/W 1.5. Correa (1931), in South America, recorded oocysts that were 36 × 23, L/W 1.6, from the liver that were certainly those of *E. stiedai*. Gousseff (1931) measured oocysts that were 31.4 × 19.2 (24—43 × 15—28) with sporocysts 9.0 × 6.0 (6—17 × 4—9). Oocysts measured by Yakimoff (1933a, b) were 34 × 22 (24—41 × 17—31). Smetana (1933a) did a detailed experimental study on the excystation of oocysts of *E. stiedai in vitro*. Chang (1935) mentioned various coccidia from laboratory rabbits in China, including oocysts of *E. stiedai* (also see Madsen, 1938), which measured 37.5 × 22.1, L/W, 1.7. Pastuszko (1963) examined 9,845 rabbits from 17 districts in Poland, from 1954—1958; these included 502 rabbits by necropsy and 9,343 fecal samples tested. She (1963) found that *E. stiedai* caused coccidiosis in rabbits in Poland; her oocysts measured 37.2 × 21.5 (20—40 × 15—25) (her Table 1).

Even suckling rabbits are susceptible to experimental infection (Dürr and Pellérdy, 1969; Schrecke, 1969), although they are more resistant than older ones. Bull (1958) studied about 5,000 wild *O. cuniculus* at Hawk's Bay, New Zealand, between 1950 and 1958. He used liver lesions as an index of infection, but also made microscopic examinations of the bile of 960 of these rabbits. Liver lesions were the better index of infection than the presence of oocysts; they healed slowly and were sometimes present after oocyst production had ceased. On the other hand, in light infections oocysts were present in the absence of gross lesions. Sex of the host had little effect on the infection rate, but age was extremely important. The infection rate was highest, 91% in 1950—1951, in rabbits 6—11-wk-old, and much lower in older and younger age groups. In young rabbits the infection rate was 49% in 1950–1951, rose to 60% in 1951–1952, and then began a steady decline to 6% by 1955–1956. At the same time, the rabbit population declined markedly. The location where the rabbits were captured also had a marked effect on the infection rate. Rates were high on a grassy paddock, whereas they were low on an open riverbed with little forage or cover. Bull (1958) thought that E. stiedai might help control rabbit populations. He estimated that in 1951–1952 about half of the rabbits born died of coccidiosis before they reached the age of 120 days. However, the population was still fairly high at the end of the breeding season, and poisoning had to be resorted to. Thus, coccidiosis, by itself, did not keep the rabbit population to a satisfactorily low level, but it reduced the amount of poisoning that was required and provided some economic advantage. On the other hand, Mykytowycz (1956, 1962), who studied E. stiedai and the intestinal coccidia of a free-living population of wild rabbits on an experimental enclosure in Australia, concluded that coccidiosis was not responsible for the general mortality pattern he saw, as most of his rabbits died of myxomatosis. He found no important difference in the number of oocysts discharged between the sexes, but he did find lower oocyst counts in the progeny of the dominant doe in the pecking order than in the progeny of the other does. He concluded that social behavior and seasonal pasture changes were among the most significant factors—aside from myxomatosis—influencing population numbers. Predation by birds was much more important as a cause of death than was coccidiosis.

Dürr et al. studied both sporogony (1971) and the excystation of sporozoites (1972) in oocysts of E. stiedai and Dürr and Reiser (1972) looked at the resistance of oocysts to both X-rays and gamma irradiation. Coudert et al. (1972) studied the resistance of oocysts to ultrasonic waves and found that waves (20 kHz, 23.9 W/cm^2,

1.74 cm^2) applied for 2 min can destroy 99.5% of unsporulated oocysts in a suspension. With other colleagues (Coudert et al., 1973) they looked at the influence of temperature on the respiration and duration of segmentation of E. stiedai oocysts. Sénaud et al. (1976) used the electron microscope to study the development of the micropore in the endogenous stages of E. stiedai in the liver, and of several other sporozoan species, including Plasmodium cathemerium; the formation of the micropore was similar in all species studied with the outer unit membrane of the pellicle forming the continuous lining of the micropore cavity and the inner pellicular complex forming the concentric osmophilic ring, which surrounds the neck of the micropore invagination.

Klesius et al. (1976) demonstrated the development of delayed hypersensitivity (DH) in rabbits infected with a particulate antigen fraction prepared from nonsporulated oocysts; their antigen produced dermal indurations similar to that seen in tuberculin reactions. This skin reactivity (DH) was detected in 11/28 (39%) infected rabbits at 10 days PI and in 53/55 (96%) rabbits by 20–30 days PI, and skin reactivity was passively transferred to non-infected rabbits with lymphocyte suspensions and cell-free transfer factor, but not with serum from infected skin-reactive animals. Leysen et al. (1989) used monoclonal antibodies (Mabs) to examine surface antigens on the SZ of both E. stiedai and E. magna. Ten of 220 Mabs raised against E. stiedai reacted with the surface of live SZs and some were able to agglutinate SZs in suspension. The surface-active Mabs were specific for the species against which they were raised and they did not react with SZ of E. flavescens, E. intestinalis, E. irresidua, or E. piriformis. Eimeria stiedai, thus, seems to be unique among rabbit coccidia: its 18S rDNA sequence (1345 bp) formed a relatively long branch with about a 100 bp deletion (see Kvičerová et al., 2008), it is the only eimerian known from rabbits that develops extraintestinally, i.e., in the bile ducts, its OR is composed of only a few small refractile granules, and it is able, at least experimentally, to infect both hares (*Lepus* spp.) and cottontails (*Sylvilagus* spp.) (Jankiewicz, 1941; Herman and Jankiewicz, 1943; Hsu, 1970; Pellérdy and Dürr, 1970; Pellérdy, 1974; Varga, 1976; Entzeroth and Scholtyseck, 1977; Scholtyseck et al., 1979; Eckert et al., 1995).

EIMERIA VEJDOVSKYI (PAKANDL, 1988) PAKANDL & COUDERT, 1999

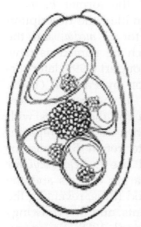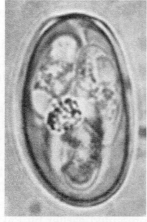

FIGURES 6.27, 6.28 Line drawing of the sporulated oocyst of *Eimeria vejdovskyi* from Pakandl, 1988, with permission from *Folia Parasitologica* (Praha). Photomicrograph of a sporulated oocyst of *E. vejdovskyi* from Pakandl, 1988, with permission from *Folia Parasitologica* (Praha).

Synonym: *Eimeria media* Kessel, 1929 (see Pakandl and Coudert, 1999 and *Remarks*, below).

Type host: *Oryctolagus cuniculus* (L., 1758) (syn. *Lepus cuniculus*), European (domestic) rabbit.

Type locality: EUROPE: Czechoslovakia, Klec near Lomnice nad Lužnicí (South Bohemia).

Other hosts: None reported to date.

Geographic distribution: EUROPE: Czechoslovakia.

Description of sporulated oocyst: Oocyst shape: ellipsoidal, slightly asymmetrical (line drawing,

Pakandl, 1988) with the greatest width usually behind the midlength of oocyst; number of walls: 2; wall characteristics: outer is smooth, yellowish or light brown, interrupted by the M, while inner is flattened or slightly concave; L × W: 32.9 × 19.2 (30—37 × 18—21) (see *Remarks*, below); L/W ratio: 1.7; M: present; M characteristics: 3.7 (3—4.5) wide; OR: present; OR characteristics: spheroidal, composed of many granules, 5.7 (4—7) wide; PG: absent. Distinctive features of oocyst: slightly asymmetrical shape with M and large OR composed of many granules.

Description of sporocyst and sporozoites: Sporocyst shape: fusiform; L × W: 15.7 × 8.0 (14—16.5 × 7—9); L/W ratio: 2.0; SB: present; SB characteristics: small, flat; SSB, PSB: both absent; SR: present; SR characteristics: compact, composed of rough granules, usually near sporocyst wall; SZ: not described, but it appeared to lie end-to-end with a large RB at the more rounded end. Distinctive features of sporocyst: markedly pointed at the end with the SB.

Prevalence: Unknown.

Sporulation: Exogenous. Oocysts sporulated in 2—3 days at room temperature (Pakandl, 1988, for *E. media*, now considered to be the mensural and biological data for *E. vejdovskyi*).

Prepatent and patent periods: Prepatent period is 9—10 days (Pakandl, 1988; Pakandl and Coudert, 1999). Patent period unknown.

Site of infection: Small intestine. All endogenous stages developed exclusively in the epithelium of the ileum (Pakandl and Coudert, 1999). No parasite stages were found at 18 hr PI, but by 36 hr PI, SZs were seen at the bottom of crypts, often in Paneth cells (Pankadl and Coudert, 1999).

Endogenous stages: Five generations of meronts developed at different intervals. All five generations had two types of meronts: Type A, which formed a small number of larger merozoites with several nuclei via endodyogeny, and Type B, which formed a larger number of uninucleate merozoites via ectomerogony. First-generation meronts were seen 54—72 hr PI in epithelial cells at the bottom of the crypts; first-generation Type A meronts were SZ-shaped, but with several N inside, while Type B first-generation meronts produced 20—50 merozoites via ectomerogony. Second-generation meronts were found 72—108 hr PI in epithelial cells of the crypts; meronts of both types were nearly spheroidal and formed two to four or up to eight nearly spheroidal merozoites; Type A second-generation merozoites had two N, whereas Type B merozoites were uninucleate. The third-generation meronts (first generation of Pakandl, 1988; see *Remarks*), more elongated than those in the second generation, were seen 108—126 hr PI, also in the crypts, but not at the bottom. Third-generation Type A meronts were 6.3 × 6.0 (5—7.5 × 5—7) and produced 2—4 fat merozoites with two, three, or four N, while Type B meronts were 8.1 × 6.9 (5.5—15 × 5—12; see Pakandl, 1988) and formed 4—200 smaller uninucleate merozoites. Fourth-generation meronts were seen in the wall epithelial cells of intestinal villi, between 126 and 162 hr PI. These were the largest meronts in endogenous development of this species. Type A were 19.8 × 12.9 (13.5—24 × 10—14.5) and produced 25—75 binucleated merozoites, and Type B were 24.7 × 13.5 (17—35 × 9—18.5) and produced 50—200 uninucleated merozoites. The fifth-generation meronts occurred 162—180 hr PI, but some were present as late as 198 hr PI; these were located in the upper part of the villi. Fifth-generation Type A meronts were 6.1 × 6.2 (5—8.5 × 4—7) and formed 2—4 merozoites, with 2—3 N, while Type B fifth-generation meronts were 6.8 × 6.2 (4.5—11 × 5—8) and produced 4—20 merozoites, with one N (Pakandl, 1988; Pakandl and Coudert, 1999). Gamogony began at 198 hr PI in epithelial cells in the upper parts of the villi. Macrogamonts were 27.1 × 14.9 (19—30.5 × 11.5—19) and mature microgamonts were 29.9 × 16.1 (21—37 × 12—22.5).

Cross-transmission: None to date.

Pathology: Unknown.

Material deposited: Institute of Parasitology, Czechoslovak Academy of Sciences, České Budějovice, Czechoslovakia, Protozool. Coll. No. H-PA-003a-c.

Etymology: The specific epithet is given in honor of the Czech zoologist František Vejdovský.

Remarks: Pakandl (1988) originally described this species, but he confused it with *E. media*. He (1988) identified two species with similar oocyst morphology, one with larger oocysts and a 9—10 day prepatent period, and the second with smaller oocysts and a 5—6 day prepatent period; he called the former *E. media* because oocyst size resembled the description given by Kessel (1929) for *E. media*, and the latter, *E. vejdovskyi*, which he named as new. However, Pakandl and Coudert (1999) later corrected the earlier decision by Pakandl (1988). They argued that since all other researchers used the name *E. media* for the species with smaller oocysts and the shorter prepatent period, they proposed to retain *E. media* for this smaller species and use *E. vejdovskyi* for the other species with larger oocysts and the 9—10 day prepatent period (thus, reversing the descriptions, measurements, and line drawings presented earlier by Pakandl, 1988). So, sporulated oocysts of *E. vejdovskyi* clearly resemble those of *E. media* in size (32.9 × 19.2 [30—37 × 18—21], L/W ratio 1.7 vs. 29.1 × 18.2 [24—33 × 15—20], L/W ratio 1.6, respectively). However, the prepatent periods, number and size of endogenous stages, and their timing during development differ significantly (see Table 1, Pakandl, 1988, but transpose the names), and starch gel electrophoresis showed marked differences in mobility of both **lactate dehydrogenase (LDH)** and **glucosophosphate isomerase (GPI)** between the two species.

Pakandl and Coudert (1999) were unable to locate SZs at 18 hr PI in the ileum, duodenum, or jejunum, but they saw SZs in the ileal crypts at 36 hr PI, before they started merogony. Although the site of entry of the SZs is unknown, they (1999) speculated it seemed probable that "as in other rabbit species, the SZ enter the intestine in the duodenum and upper jejunum and then migrate by a route which remains unclear."

Polynucleated merozoites have been described in many rabbit coccidia (Pellérdy and Dürr, 1970; Ryley and Robinson, 1976; Norton et al., 1979; Streun et al., 1979; Pakandl, 1988; Licois et al., 1992a, b; Pakandl et al., 1996a, b, c; Pakandl and Coudert, 1999), although this does not seem to be common in coccidian species from other hosts. Streun et al. (1979) first articulated the theory that the two types of meronts and merozoites precede macro- and microgamonts, respectively. In *E. vejdovskyi* Type A numbers decreased from generations 1 through 5, apparently due to the smaller numbers of merozoites in each Type A meront. Pakandl and Coudert (1999) suggested that this corresponded to a smaller number of microgamonts compared with the number of macrogamonts and that "it seems probable that Type A and B meronts of *E. vejdovskyi* and other rabbit *Eimeria* species are sexually determined and, unlike coccidian species from other hosts, this is also expressed in their morphology."

SPECIES INQUIRENDAE (1)

Eimeria pellerdi Coudert, 1977a, b

Homonym: *Eimeria pellerdyi* Prasad, 1960, from the camel.

Original host: *Oryctolagus cuniculus* (L., 1758) (syn. *Lepus cuniculus*), European (domestic) rabbit.

Remarks: Coudert (1977a, b) described this form as a new species from the domestic rabbit in France. This name is a homonym of *E. pellerdyi* Prasad, 1960, from *Camelus bactrianus* (see Levine, 1974; Pellérdy, 1974). Coudert's organism appears to be *E. flavescens* Marotel and Guilhon, 1941.

DISCUSSION AND SUMMARY

Historic Relationship Between *O. cuniculus* and humans

Oryctolagus cuniculus is one of the rare mammals originally domesticated in Western Europe. From about the late Pleistocene until Classic Antiquity, rabbits only occupied the Iberian Peninsula and a narrow area in southern France (Monnerot et al., 1994). Little substantial evidence of humanity's relationship with the European rabbit is documented until the medieval period, but it can be surmised that transportation by humans throughout antiquity, including the Phoenicians, Greeks, and Romans, resulted in this species becoming globally distributed throughout most of Europe, into South America and Australia and on to many oceanic islands. The European rabbit is the only species of rabbit to be domesticated, probably beginning in the sixteenth century, when warrens began to be conserved (Rougeot, 1981), and all pet breeds of rabbits are of this species. After this period, three kinds of rabbits were living in Western Europe: wild rabbits, rabbits kept in warrens for hunting as a food source, and domestic stocks (Audoin, 1986).

Coccidia and Coccidiosis in European Wild and Domestic Rabbits

Coccidiosis refers to the disease state caused by infection with coccidia. The beginning of a disease condition in an infected animal can be indicated by anorexia with reduced food intake, sluggish behavior, weakness, rough hair coat, and either a pause in weight gain or noticeable thinning and weight loss. Such visible symptoms are caused by the parasite's development in the gut (and/or liver), which affects many physical and physiological parameters including, but not limited to: denuding and sloughing of gut epithelium, intestinal hypomotility (constipation) or hypermotility (diarrhea), inflammation, thickening and enlargement of infected sections, extravasation of blood capillaries, intestinal intussusception, hemorrhage in a hyperemic mucosa, malabsorption, changes in plasma proteins and electrolytes, and death. Infection and (sometimes) disease caused by the 16 valid eimerians for which O. cuniculus is the type host occur in two forms, intestinal and/or hepatic, and are dependent primarily upon the species involved, the size of the inoculating dose, and the age, nutritional, and immune status of the rabbit, with juveniles more susceptible than adults. The latter can act as carriers and distribute the oocysts widely in their feces. We know nothing, or almost nothing, about the pathology of five of these 16 (31%), all intestinal species: E. nagpurensis, E. oryctolagi, E. piriformis, E. roobroucki, and E. vejdovskyi. Four species (25%) can be considered slightly-to-mildly pathogenic; these are E. exigua, E. matsubayashii, E. neoleporis, and E. perforans. It needs mention that E. neoleporis was originally described and studied in S. floridanus (Carvalho, 1942) and later transmitted to O. cuniculus (Carvalho, 1943, 1944). It is reported to live in the area of the ileocecal valve and anterior cecum in both host species, but is more pathogenic in S. floridanus than in O. cuniculus. Eimeria flavescens, E. media, and E. stiedai (liver only) can be considered pathogenic, especially in young rabbits and in high inoculation/infective doses. And from the information available to us at present, the most pathogenic intestinal forms are E. coecicola, E. intestinalis, E. irresidua, and E. magna.

Site and Host Specificity in O. cuniculus

The evolution of host and site specificity of parasites, in general, and Eimeria species in particular, has been called one of the most intriguing questions of all parasitological research (Kvičerová et al., 2008). These specificities differ in degrees.

Site Specificity

Most coccidian species for which endogenous development is known undergo asexual (merogony) and sexual (gamogony) development within specific cells of the gastrointestinal tract, but not all species are found in the gut. Eimeria stiedai undergoes its development in epithelial cells of the bile duct and (sometimes) parenchymal cells of the liver of rabbits, while other mammalian eimerians have been found to develop in cells of the gall-bladder (goat), placenta (hippopotamus), epididymis (elk), uterus (impala), genitalia of both sexes (hamsters), bile duct (chamois), and pyloric antrum (kangaroo) (see Duszynski and Upton, 2001; Duszynski, 2011). Once eimerians have localized within their specific organ system of choice, they seem to be limited to specific zones within that system, specific cells within that zone, and specific locations within those cells. Thus, one species may be found only in cells of the ileum (e.g., E. vejdovskyi), another only in the epithelial cells of the lower small intestine, cecum and colon (e.g., E. flavescens), and a third in the epithelial cells of the bile ducts (e.g., E. stiedai). Within their specific regions, one may be found only in cells at the base of the crypts of Lieberkühn, another in cells along the lengths of the villi, a third in endothelial cells of the lacteals of the villi, or a fourth in the epithelial cells of bile ducts, but usually not in the hepatocysts of the liver. And the final degree of specificity is within their host cells. Some develop below the striated border (of microvilli) of epithelial cells and above the HCN, some below the HCN and a few within the HCN (Duszynski and Upton, 2001; Duszynski, 2011). Of the 16 eimerian species which we consider to be valid, 15 are intestinal parasites, while only E. stiedai is found in the liver's bile ducts, where its development occurs in their epithelial cells. Of the intestinal species, nothing is known about the locations of endogenous development for four (27%) of them. The location of development in the rabbit is completely unknown for *E. nagpurensis* and *E. roobroucki*, while for *E. oryctolagi* and *E. vejdovskyi* we know only that they are reported to live in "the small intestine"; the actual location of their endogenous stages has not been studied. Five species (33%) live in the upper intestines, distributed from the duodenum to the lower ileum, dependent upon the size of the inoculating dose; these include *E. exigua*, *E. irresidua*, *E. matsubayashii*, *E. media*, and *E. perforans*, and their more specific locations vary from superficial epithelial cells on the villi to more proximal locations within the lamina propria and in cells within or at the base of the crypts (see text). The remaining six (40%) intestinal eimerians seem to be less parochial in the choice of location within the rabbit gut and live in the ileum and cecum (*E. coecicola*), the ileum through the colon (*E. flavescens*, *E. magna*, *E. neoleporis*) or even, during heavy infecting doses, from the duodenum and jejunum through the cecum, appendix, and ascending colon (*E. intestinalis*, *E. piriformis*).

Host Specificity

In reality, strict host specificity probably rarely, if ever, exists in nature because it would not be to the parasite's advantage to limit its reproductive opportunities to a single host, especially if that host is reasonably rare or solitary or territorial. Thus, it's our contention that most coccidia may be infective to several host species, at least within the same host genus, and likely infective to hosts within closely related (sister genera) taxa. Among the Eimeriidae, *Eimeria* species are considered to exhibit the highest level of specificity for their definitive host (Marquardt, 1973; Hnida and Duszynski, 1999a), while *Sarcocystis*, with its cyst-forming heteroxenous species (Sarcocystidae), are generally more specific for their intermediate host than for their definitive host(s) (Dubey et al., 1989; Votýpka et al., 1998; Doležel et al., 1999; Šlapeta et al., 2003). The degree of specificity in mammals seems to vary from host group to

host group. For example, *Eimeria* from goats cannot be transmitted from goats to sheep or vice versa (Lindsay and Todd, 1993), but *Eimeria* from cattle (*Bos*) often are found in American bison (*Bison*) (Ryff and Bergstrom, 1975; Penzhorn et al., 1994), while *Eimeria* species that infect certain rodents (Sciuridae) seem to easily cross host genus boundaries (Todd and Hammond, 1968a, b; Wilber et al., 1998). At least one rodent eimerian, *E. chinchilla*, originally isolated from the chinchilla, was experimentally transmitted to seven genera of wild rodents in two families (de Vos, 1970), and *E. separata* from the rat (*Rattus*) will infect certain genetic strains of mice, *Mus* species (Mayberry and Marquardt, 1973; Mayberry et al., 1982). We also know that genetically altered (Rose and Millard, 1985) and immunosuppressed mammals (Todd et al., 1971; Todd and Lepp, 1972; Nowell and Higgs, 1989; Aly, 1993) are susceptible to infection with *Eimeria* species to which they otherwise might be naturally resistant. These studies help emphasize that numerous biotic interactions between host and parasite must contribute, in concert, to any host specificity (or lack thereof) in the coccidia, especially the genetic constitution of both participants (Duszynski and Upton, 2001). Host specificity in *O. cuniculus* seems fairly strict. Based on the evidence available at this time, the majority, 10/15 (67%), of valid intestinal species are reported only from *O. cuniculus* with a few isolated, and questionable, reports from *Lepus* and/or *Sylvilagus* species, but without supporting cross-transmission or molecular work; these include *E. coecicola*, *E. exigua*, *E. flavescens*, *E. intestinalis*, *E. matsubayashii* *E. nagpurensis*, *E. oryctolagi*, *E. piriformis*, *E. roobroucki*, and *E. vejdovskyi*. Pellérdy (1954a, 1956) was unable to cross-transmit several of these to *L. europaeus*, but most species have no experimental work to refute or support their specificity for *O. cuniculus* (see text). Of the remaining five intestinal species, four (*E. irresidua*, *E. magna*, *E. media*, *E. perforans*) were discovered first in

O. cuniculus, but have been transmitted to at least one species of Sylvilagus, and the fifth (E. neoleporis) was discovered originally in Sylvilagus and was cross-transmitted to O. cuniculus; none of these has yet been transmitted to Lepus species (see text). The only species to infect the liver, E. stiedai, is the least specific of the eimerians of O. cuniculus, as it is known to also infect both Lepus and Sylvilagus species, both naturally and experimentally.

TABLE 6.1 All Reported Coccidia Species[1] (Apicomplexa: Eimeriidae) from the Old World Wild European Rabbit, *Oryctolagus cuniculus*, and its Domesticated Subspecies, *O. c. domesticus* (Family Leporidae)

Domestic/farmed (D)/ European (wild) (W)	*Eimeria* spp.	Natural (N) &/or experimental (E)	References
D/W	coecicola	N/E	Cheissin, 1947, 1967; Gill & Ray, 1960
D/W	exigua	N	Yakimoff, 1934; Pastusko, 1961a, 1963; Lizcano Herrera & Romero Rodríguez, 1969; Stodart, 1971; Coudert et al., 1995; Hobbs & Twigg, 1998; Jelinková et al., 2008; Li et al., 2010
D/W	flavescens	N	Marotel & Guilhon, 1941; Catchpole & Norton, 1979; Norton et al., 1979; Peeters et al., 1981; Kasim & Al-Shawa. 1987; Polozowski, 1993; Hobbs & Twigg, 1998
D/W	intestinalis	N	Cheissin, 1948, 1967; Pellérdy, 1953, 1954a, b, 1965, 1974; Peeters et al., 1981; Veisov, 1982; Darwish & Golemansky. 1991; Hobbs & Twigg, 1998; Oncel et al., 2011
D/W	irresidua	N	Kessel & Jankiewicz, 1931; Cheissin, 1940, 1967; Carvalho, 1943; Lund, 1950; Mykytowycz, 1956,1962; Niak, 1967; Stodart, 1971; Norton & Joyner, 1979; Peeters et al., 1981
W	leporis	N (?)	Nieschulz, 1923; Ryšavý, 1954; Pellérdy, 1956; Lucas et al., 1959
D/W	magna	N/E	Pérard, 1925b; Kessel & Jankiewicz, 1931; Becker, 1933; Carvalho, 1934; Cheissin, 1947b, 1967; Lund, 1950; Niak, 1967; Mirza, 1970; Stodart, 1971; Pellérdy, 1974; Peeters et al., 1981; Veisov, 1982
D	matsubayashii	N/E	Matsubayashi, 1934; Tsunoda, 1952; Gill & Ray, 1960; Cheissin, 1967
D/W	media	N/E	Kessel & Jankiewicz, 1931; Matsubayashi, H. 1934; Carvalho, 1943; Pellérdy, 1954b; Pakandl et al., 1966c; Cheissin, 1967; Hobbs & Twigg, 1998; Grès et al., 2003; Razavi et al., 2010
D	nagpurensis	N	Gill & Ray, 1960; Martínez Fernández et al., 1969; Mandal, 1976

TABLE 6.1 All Reported Coccidia Species[1] (Apicomplexa: Eimeriidae) from the Old World Wild European Rabbit, *Oryctolagus cuniculus*, and its Domesticated Subspecies, *O. c. domesticus* (Family Leporidae) (*cont'd*)

Domestic/farmed (D)/ European (wild) (W)	*Eimeria* spp.	Natural (N) &/or experimental (E)	References
D	*neoleporis*[1]	E	Carvalho, 1943; Mandal, 1976, 1987
D	*oryctolagi*	N	Ray & Banik, 1965b; Mandal, 1976, 1987
D/W	*perforans*	N/E	Leuckart, 1879; Sluiter & Swellengrebel, 1912; Carvalho, 1943; Kessel & Jankiewicz, 1931; Lund, 1950; Francalancia & Manfredini, 1967; Peeters et al.,1981; Veisov, 1982; Mirza, 1970; Niak, 1967; Santos & Lima, 1987; and many other authors (see text)
D/W	*piriformis*	N	Kotlán & Pospech, 1934; Cheissin, 1948, 1967; Pellérdy, 1953, 1965, 1974; Levine & Ivens, 1972; Pastuszko, 1963; Peeters et al., 1981; Veisov, 1982; Niak, 1967; Stodart, 1971; Darwish & Golemansky, 1991; and many other authors (see text)
W	*roobroucki*	N	Grés et al., 2002, 2003
D/W	*stiedai*[1, 2]	N/E	Lindemann, 1865; Kisskalt & Hartmann, 1907; Kessel and Jankiewicz, 1931; Harkema, 1936; Lampio (1946); Lund, 1950; Ryšavý, 1954; Gill & Ray, 1960; Mykytowycz, 1962; Bouvier, 1967;Niak, 1967; Litvenkoa, 1969; Von Braunschweig, 1965; Bouvier, 1967; Burgaz, 1970a, b; Mirza, 1970; Stodart, 1971; Flatt & Campbell, 1974; Entzeroth & Scholtyseck, 1977; Svanbaev, 1979; Peeters et al., 1981; Veisov, 1982; Ogedengbe, 1991; Darwish & Golemansky, 1991; and many other authors (see text)
D	*vejdovskyi*	N/E	Pakandl, 1988; Pakandl & Coudert, 1999
	17 *Eimeria* (16 valid)		

[1] *Seventeen* Eimeria *species reported/described from the lagomorph monotypic genus* Oryctolagus. *We believe all these eimerians are valid parasites of* O. cuniculus *and/or* O. c. domesticus, *except* E. leporis *(see text). Several of these species have been reported to occur naturally in various* Lepus *and* Sylvilagus *species; however, most of those reports are likely misidentifications except for* E. stiedai. *Jankiewicz (1941), Herman and Jankiewicz (1943), Entzeroth and Scholtyseck (1977), and Scholtyseck et al. (1979) all experimentally transmitted* E. stiedai *from* O. cuniculus *to* L. europaeus, *and Carvalho (1943) successfully transmitted* E. neoleporis *from* S. floridanus *to* O. cuniculus.

[2] Eimeria stiedai *has been reported to occur naturally in several species of* Sylvilagus, *including* S. audubonii, S. braziliensis, S. floridanus, *and* S. nuttalli, *but only Jankiewicz (1941) and Herman and Jankiewicz (1943) transmitted it from* O. cuniculus *to* S. audubonii *and Hsu (1970) transmitted it to* S. floridanus.

(?) Not considered a valid parasite of *Oryctolagus*.

Coccidia (Eimeriidae) of the Family Leporidae: Genus *Sylvilagus*

The Biology and Identification of the Coccidia (Apicomplexa) of Rabbits of the World
http://dx.doi.org/10.1016/B978-0-12-397899-8.00007-X

189

INTRODUCTION

The "cottontail" rabbits include 17 species that are distributed throughout North (including Canada), Central, and South America, as far south as northern Argentina. In appearance, the majority somewhat resemble the wild European rabbit, *Oryctolagus cuniculus* (Chapter 6) and, at one time, several other rabbit genera (e.g., *Brachylagus*, Chapter 4) were synonymized by Hall (1981) and treated as subgenera of *Sylvilagus*. However, most mammalogists now accept the taxonomic scheme of Hoffmann and Smith (2005), which we use here. Many *Sylvilagus* species are so similar in appearance that even experts have difficulty distinguishing them in the field, and only careful examination of preserved specimens, especially the cleaned skulls, is essential for correct identification (Findley, 1987). Most members of this genus have a short stub tail, called a scut, that is brown above, but with a white underside that shows prominently when they are fleeing, thus giving them the name "cottontails." Most, but not all, species in the genus raise **altricial** young in fur-lined nests called forms.

Although similar in appearance, the rabbits in this genus show great diversity in habitat selection. Some, like *S. floridanus*, are **euryoecious** because they can be found in deserts, farmlands, prairies, swamps, woodlands, and in hardwood,

FIGURE 7.1 The Mountain Cottontail rabbit, *Sylvilagus nuttallii* (Bachman, 1837).

rain, and boreal forests, while other species in the genus, for example *S. palustris*, must be considered **stenoecious**, because they are narrowly restricted to marshes and are limited by the availability of water. The largest ranges are those of males during the breeding season. All species are terrestrial, although a few, like *S. nuttallii*, are sometimes said to be semiarboreal because they climb into juniper trees to feed. Most species move by hopping a few cm up to 1 m at a time, but, when startled or frightened, some can cover 3—5 m in the first several leaps, and then hop shorter distances in a zigzag manner at maximum speed of up to about 35 km/hr. At least one species, *S. palustris*, walks instead of hopping and swims extensively; it is believed that all species probably are capable of swimming (Nowak, 1991).

Cottontails do not dig burrows, as far as is known, but some species use burrows made by other animals, while most seek shelter in brush piles and shallow depressions in the soil or vegetation and these may be connected by regular trails. Females of most species dig holes for nests, which the female lines with hair plucked from her underside, and in which their young are reared (Findley, 1987). The female does not reside in the nest, but crouches above it and the baby rabbits climb to the top of it to nurse. Cottontails are usually nocturnal or **crepuscular**, but sometimes are seen at all times of the day and are active throughout the year (and the coyotes are thankful for that). During the winter they feed on the bark and twigs of woody vegetation, but otherwise they eat a wide variety of herbivorous vegetation. From these diets, cottontails excrete two kinds of feces: hard, brown fecal pellets, from which their digestion has extracted almost all of the nutrients, and soft greenish pellets, which are re-ingested and provide them a supplement of vitamin B (Nowak, 1991).

Most *Sylvilagus* species are solitary, but males may congregate in small groups during the breeding season to pursue a female in estrous. The males of some species (e.g., *S. floridanus*) will fight each other to establish dominance hierarchies that determine mating priorities, while the females exhibit far less rigid hierarchies. In general, breeding varies with latitude, beginning later in higher latitudes and elevations and earlier in lower ones. Some species also breed all year long, especially in lower latitudes and altitudes, and most species produce from two to seven litters each year. Gestation varies from 27 to 42 days depending upon the species, and litter size can vary from one to a dozen babies. Babies open their eyes in 4 to 7 days, are out of the nest in about 14 days, and are weaned and independent by 5 weeks of age. Individuals of some species may be sexually mature by 11—12 weeks and are capable of reproducing during their first year of life, but most do not reproduce until their second year.

Only four of 17 (23.5%) species in this genus have been examined for coccidia and, combined, they are reported to be host to 19 *Eimeria* species (Table 7.1), but not species from other, related genera (e.g., *Isospora*, *Cyclospora*). However, at least some of these survey reports must certainly be incorrect as discussed below.

HOST GENUS *SYLVILAGUS* GRAY, 1867

EIMERIA AUDUBONII DUSZYNSKI AND MARQUARDT, 1969

Type host: *Sylvilagus audubonii* (Baird, 1858), Desert cottontail.

Type locality: NORTH AMERICA: USA: Colorado, Larimer County, near Ft. Collins.

Other hosts: *Sylvilagus floridanus* (J.A. Allen, 1890), Eastern cottontail.

Geographic distribution: NORTH AMERICA: USA: Colorado, Pennsylvania.

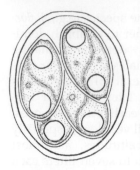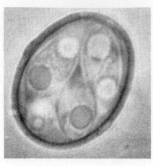

FIGURES 7.2, 7.3 Line drawing of the sporulated oocyst of *Eimeria audubonii* from Duszynski and Marquardt, 1969, with permission of the *Journal of Protozoology*. Photomicrograph of a sporulated oocyst of *E. audubonii* from Duszynski and Marquardt, 1969, with permission of the *Journal of Protozoology*.

Description of sporulated oocyst: Oocyst shape: spheroidal to slightly ovoidal; number of walls: 2; wall thickness: ~1; wall characteristics: outer is smooth, thin, yellow, of uniform thickness; inner is distinctly greenish, becoming thin or absent at the anterior end giving the appearance of a M; L × W: 21.2 × 17.1 (15−25 × 13−20); L/W ratio: 1.2 (1.0−1.5); M: indistinct if present, but probably absent; OR, PG: both absent. Distinctive features of oocyst: inner wall a distinct greenish color and thins at the slightly ovoidal end.

Description of sporocyst and sporozoites: Sporocyst shape: ellipsoidal; L × W: 12.9 × 5.8 (10−15 × 4−8); L/W ratio: 2.2 (1.6−2.8); SB: present; SSB, PSB, SR: all absent; SZ: comma-shaped, with a small central N and large, posterior RB. Distinctive features of sporocyst: absence of SR.

Prevalence: In 38/100 (38%) in the type host; in 20/139 (14%) *S. floridanus*.

Sporulation: Exogenous. Oocysts were sporulated after 5 days in 2.5% aqueous potassium dichromate ($K_2Cr_2O_7$) solution at 20°C.

Prepatent and patent periods: Unknown.

Site of infection: Unknown. Oocysts recovered from fecal contents.

Endogenous stages: Unknown.

Cross-transmission: None to date.

Pathology: Unknown.

Material deposited: None.

Remarks: Duszynski and Marquardt (1969) said that sporulated oocysts of *E. audubonii* somewhat resembled those of *E. exigua* Yakimoff, 1934, of the European wild or domestic rabbit, *O. cuniculus*, from the former USSR, *E. hungarica* Pellérdy, 1956, of the European hare, *L. europaeus*, from Hungary, and *E. punjabensis* Gill and Ray, 1960, of the Indian hare, *L. nigricollis*, from India. However, their hosts and geographic distributions are all spatially isolated from each other, so it is very likely they are all different species. The sporulated oocysts of *E. audubonii* differ from those of *E. punjabensis* by lacking a prominent OR; from those of *E. hungarica* by being larger (21.2 × 17.1 vs. 14 × 13) and lacking one or two OR granules; and from those of *E. exigua* by having larger sporocysts with a SB (10−15 × 4−8 vs. 8−10 × 4−6). Wiggins et al. (1980) found this species in *S. floridanus* in Pennsylvania, and recorded its measurements as follows: sporulated oocyst, 20.9 × 16.1 (17.5−24.5 × 12−21), lacking a M, PG and OR and the sporocysts, 11 × 6 (10.5−13 × 5.5−7), lacking a SR, having a RB in each SZ, ~3.5 wide, but they did not mention the presence/absence of a SB, SSB, PSB or N.

EIMERIA AZUL WIGGINS AND ROTHENBACHER, 1979

Type host: *Sylvilagus floridanus* (J.A. Allen, 1890), Eastern cottontail.

Type locality: NORTH AMERICA: USA: Pennsylvania, no specific locality given in the original description.

Other hosts: None to date.

Geographic distribution: NORTH AMERICA: USA: Pennsylvania.

Description of sporulated oocyst: Oocyst shape: ellipsoidal; number of walls: 1; wall thickness:

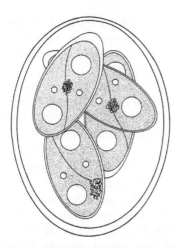

FIGURE 7.4 Line drawing of the sporulated oocyst of *Eimeria azul* from Wiggins and Rothenbacher, 1979, with permission of the *Journal of Parasitology.*

~1; wall characteristics: smooth, thinner at slightly pointed end (line drawing); L × W: 22.9 × 16.7 (19.5–27 × 15–19); L/W ratio: 1.3; M, OR, PG: all absent. Distinctive features of oocyst: thin, one-layered, smooth wall without a M.

Description of sporocyst and sporozoites: Sporocyst shape: fusiform; L × W: 11.8 × 5.8 (8–14 × 3–7); SB: present at pointed end; SSB, PSB: both absent; SR: present; SR characteristics: scattered or compact granules in midsection of sporocyst or at one end; SZ: comma-shaped, dark, lying lengthwise in sporocysts, with central N and large RB at rounded end. Distinctive features of sporocyst: fusiform shape.

Prevalence: In 8/70 (17%) of the type host (Wiggins and Rothenbacher, 1979); 15/139 (11%) from the same host in 1980 (Wiggins et al., 1980).

Sporulation: Exogenous. Oocysts sporulated within 1 wk in 2.5% aqueous $K_2Cr_2O_7$ solution.

Prepatent and patent periods: Unknown.

Site of infection: Unknown. Oocysts recovered from feces and intestinal contents.

Endogenous stages: Unknown.

Cross-transmission: None to date.

Pathology: Unknown.

Material deposited: None.

Remarks: Wiggins and Rothenbacher (1979) compared the sporulated oocysts of *E. azul* to those of other eimerians from *Sylvilagus, Oryctolagus,* and *Lepus* species and said it differed from most by lacking a M. Its sporocysts differ from those of *E. audubonii* by having a SR, which those of *E. audubonii* lack, from oocysts of *E. minima* by being larger (22.9 × 16.7 vs. 13.4 × 10.8), from *E. paulistana* by being smaller (22.9 × 16.7 vs. 43 × 24), from *E. perforans, E. hungarica, E. leporis,* and *E. punjabensis* by lacking an OR, from *E. exigua* by being larger (22.9 × 16.7 vs. 15.7 × 13.0), and from *E. nagpurensis* in size and shape of sporocysts and by lacking a PG. Wiggins et al. (1980) reported *E. azul* a second time from *S. floridanus* in Pennsylvania and these had sporulated oocysts, 21.9 × 16.6 (18–27 × 14.5–19), lacking an OR, PG, and M, with sporocysts, 11.9 × 5.6 (10.5–13 × 5.5–7), lacking a SR, having SZ with a distinct RB, ~3.5 wide, but with no mention by Wiggins et al. (1980) of the presence/absence of a SB, SSB, PSB or N.

EIMERIA ENVIRON HONESS, 1939

Synonym: Eimeria perforans of Morgan and Waller, 1940.

Type host: Sylvilagus nuttallii (Bachman, 1837), Mountain cottontail.

Type locality: NORTH AMERICA: USA: Wyoming, Albany County, near Agriculture Experiment Station of the University of Wyoming.

Other hosts: Sylvilagus audubonii (Baird, 1858), Desert cottontail; *Sylvilagus floridanus* (J.A. Allen, 1890), Eastern cottontail.

Geographic distribution: NORTH AMERICA: USA: Colorado (Duszynski and Marquardt, 1969); Illinois (Ecke and Yeatter, 1956); Iowa (Carvalho, 1943); Pennsylvania (Wiggins et al., 1980); Wisconsin (Dorney, 1962); Wyoming (Honess, 1939; Honess and Winter, 1956).

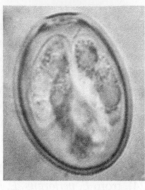

FIGURES 7.5, 7.6 Line drawing of the sporulated oocyst of *Eimeria environ* from Duszynski and Marquardt, 1969, with permission of the *Journal of Protozoology*. Photomicrograph of a sporulated oocyst of *E. environ* from Duszynski and Marquardt, 1969, with permission of the *Journal of Protozoology*.

Description of sporulated oocyst: Oocyst shape: elongate-ovoidal to broadly ellipsoidal; number of walls: 2; wall thickness: ~1; wall characteristics: smooth, of irregular thickness, being thinnest at end opposite M and thickest on sides and around margin of M; L × W: 26.8 × 20.5 (22–30 × 16–23); L/W ratio: 1.3; M: present; M characteristics: ~5–6 wide, distinct and variable in appearance, sometimes appearing convex and not disrupting the smooth contour of the wall, and sometimes sunken into the oocyst wall giving a flattened appearance; OR, PG: both absent. Distinctive features of the oocyst: distinct M with variable appearance.

Description of sporocyst and sporozoites: Sporocyst shape: ellipsoidal to elongate-ovoidal, bluntly pointed at one end; L × W: 15.6 × 8.3 (12–18 × 7–9.5); L/W ratio: 1.9; SB: present; SSB, PSB, SR: all absent; SZ: banana-shaped with small central N and posterior RB. Distinctive features of the sporocyst: no SR and sporocysts bluntly pointed at one end.

Prevalence: Unknown in type host, *S. nuttallii* (Honess, 1939); in 29/45 (64%) *S. floridanus* from Iowa (Carvalho, 1943); 16/32 (50%) *S. floridanus* from Illinois (Ecke and Yeatter, 1956); 150/161 (93%) from *S. floridanus* in Wisconsin (Dorney, 1962); 52/58 (90%) *S. audubonii* from Colorado (Duszynski and Marquardt, 1969); 102/139 (73%) *S. floridanus* in Pennsylvania (Wiggins et al., 1980).

Sporulation: Exogenous. Minimum sporulation time was 50 hr and the majority of oocysts sporulated in 68–70 hr in 2% (potassium?) dichromate solution (Honess, 1939), probably at room temperature. Carvalho (1943) said sporulation ranged from 48 to 75 hr and averaged 60 hr in 3% (w/v) $K_2Cr_2O_7$ maintained at 30°C, and Duszynski and Marquardt (1969) reported that most oocysts were sporulated after 5 days in 2.5% (w/v) $K_2Cr_2O_7$ solution at 20°C.

Prepatent and patent periods: Prepatency is 6 days; patency is 15–18 days.

Site of infection: Small intestine (Carvalho, 1943).

Endogenous stages: Unknown.

Cross-transmission: Carvalho (1943) inoculated five young and two adult domestic rabbits (*O. cuniculus*) with 100,000 oocysts of *E. environ*, but he was unable to show that infection took place. Unfortunately, Carvalho (1943) did not state the length of time that feces were examined post-inoculation.

Pathology: Mildly pathogenic with loss of weight and appetite, but no diarrhea; experimental inoculations with 80,000 sporulated oocysts were not lethal (Carvalho, 1943) for five young *S. floridanus*.

Material deposited: None.

Remarks: The original description of this species by Honess (1939) included a photomicrograph of a sporulated oocyst, but no line drawing, and some of the descriptive features were minimal. Carvalho (1943) reported sporulated oocysts from Iowa cottontails (*S. floridanus*) as ovoidal to broadly ellipsoidal, 25.7 × 18.5 (20–32.5 × 14–23), with a M, ~5–6 wide, but lacking an OR and PG; the sporocysts were ovoidal, 13–18.5 × 7–9.5, lacking a SB, but the SZs had a large posterior RB. Duszynski and Marquardt (1969) described sporulated oocysts from *S. audubonii* from Colorado as elongate ovoidal, 25.4 × 18.3 (20—30 × 17—21), L/W ratio 1.4 (1.1—1.7), with a M, but lacking both OR and PG; their sporocysts were ellipsoidal, 14.2 × 6.4 (12—17 × 6—8) with a L/W ratio 2.2 (1.9—2.8), and possessed a small SB and SZs with one posterior RB. Wiggins et al. (1980) examined *S. floridanus* from Pennsylvania and said the oocysts were 25.8 × 17.4 (19—35 × 11—24.5) with a M, ~4.9 wide, but lacked an OR and PG; their sporocysts were 13.1 × 6.4 (10—17.5 × 4.5—10), lacked a SR, and their SZs had a posterior RB, ~4 wide, but no mention was made about the presence/absence of SB, SSB, PSB.

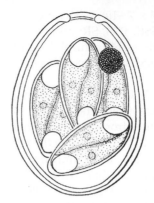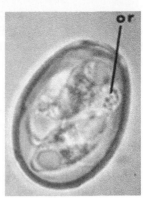

FIGURES 7.7, 7.8 Line drawing of the sporulated oocyst of *Eimeria honessi* from Duszynski and Marquardt, 1969, with permission of the *Journal of Protozoology*. Photomicrograph of a sporulated oocyst of *E. honessi* from Duszynski and Marquardt, 1969, with permission of the *Journal of Protozoology*.

EIMERIA EXIGUA YAKIMOFF, 1934 (FIGURES 6.4, 6.5)

Synonym: *Eimeria exigua* Type I of Yakimoff, 1934.

Type host: *Oryctolagus cuniculus* (L., 1758) (syn. *Lepus cuniculus*), European (domestic) rabbit.

Remarks: Morgan and Waller (1940) said they found spheroidal oocysts of *E. exigua*, about 14 wide, in 8/210 (4%) *S. floridanus* from Iowa, but Carvalho (1943) thought these likely were oocysts of *E. minima*, which he described as new (1943), stating that *E. exigua* likely does not exist in wild *Sylvilagus* species. Although five eimerians from *O. cuniculus* have been successfully cross-transmitted to *S. floridanus* (Becker, 1933; Jankiewicz, 1941; Carvalho, 1943; Herman and Jankiewicz, 1943), no one yet has tried to transmit *E. exigua* to *Sylvilagus* species. Thus, it is unknown whether or not *E. exigua* can infect *Sylvilagus* species. See Chapter 6 for the complete description of *E. exigua*.

EIMERIA HONESSI (CARVALHO, 1943) EMEND. LEVINE AND IVENS, 1972 AND PELLÉRDY, 1974

Synonyms: *Eimeria media* form *honessi* Carvalho, 1943; variety of *Eimeria media* Kessel, 1929 of Honess, 1939; non *Eimeria honessi* Landers, 1952.

Type host: *Sylvilagus floridanus* (J.A. Allen, 1890), Eastern cottontail.

Type locality: NORTH AMERICA: USA: Iowa, Story County, near Ames.

Other hosts: *Sylvilagus audubonii* (Baird, 1858), Desert cottontail; *Sylvilagus nuttallii* (Bachman, 1837), Mountain cottontail.

Geographic distribution: NORTH AMERICA: USA: Colorado, Illinois, Iowa, Pennsylvania, Wisconsin, Wyoming.

Description of sporulated oocyst: Oocyst shape: ellipsoidal to ovoidal; number of walls: 1 or 2 (?); wall thickness: not stated in original description, but given as ~1 by other authors; wall characteristics: circular thickening surrounding the M; L × W: 28.2 × 19.6 (24—33 × 17—24); L/W ratio: 1.4; M: present; M characteristics: lateral circular thickening of the oocyst wall imparts a truncated appearance to the oocyst; OR: present; OR characteristics: ovoidal, well-defined, 5—14 wide; PG: absent. Distinctive features of oocyst: truncated appearance at area of M and large OR.

Description of sporocyst and sporozoites: Sporocyst shape: ellipsoidal to elongate-ovoidal;

L × W: 13.7 × 7.6 (10—17 × 7—10); L/W ratio: 2.3 (1.8—3.0); SB: present; SSB, PSB, SR: all absent; SZ: banana-shaped with small central N and one posterior RB. Distinctive features of sporocyst: absence of SR.

Prevalence: Found in 3/45 (7%) *S. floridanus*, the type host, from Iowa (Carvalho, 1943); in 24/139 (17%) *S. floridanus* from Pennsylvania (Wiggins et al., 1980); in 47/58 (81%) *S. audubonii* from Colorado (Duszynski and Marquardt, 1969); in 150/161 (93%) *S. floridanus* from Wisconsin (Dorney, 1962); and in 3/32 (9%) *S. floridanus* from Illinois (Ecke and Yeatter, 1956, although they called it *E. media*).

Sporulation: Exogenous. Minimum of about 45 hr, but most oocysts required at least 70 hr (Honess, 1939). Oocysts sporulated after 5 days in 2.5% (w/v) aqueous $K_2Cr_2O_7$ solution at 20°C according to Duszynski and Marquardt (1969).

Prepatent and patent periods: Unknown.

Site of infection: Unknown. Oocysts recovered from feces and intestinal contents.

Endogenous stages: Unknown.

Cross-transmission: Carvalho (1943) gave a "light dose" of sporulated oocysts to one domestic rabbit, *O. cuniculus*, and no infection resulted; unfortunately, he did not mention any history of previous infection in this rabbit.

Pathology: Unknown.

Material deposited: None.

Remarks: This species was originally described by Honess (1939) as a new variety of *E. media*, but his description was minimal. Carvalho (1943) called this eimerian "*E. media* form *honessi*" and said, "Dimensions and characteristics are the same as those given by Honess (1939)" for *E. media*. Both Levine and Ivens (1972) and Pellérdy (1974) proposed the current name because protozoologists "do not agree with the creation of lower taxons for coccidia." Thus, Carvalho's (1943) eimerian becomes the type species and its host, *S. floridanus*, is the type host from Iowa. We refer to the descriptive characters given by Duszynski and Marquardt

(1969), who found this form in *S. audubonii* from Colorado. They described ovoidal oocysts with a smooth outer wall of irregular thickness, thinnest at the end opposite the M and thickest on the sides and around the margin of the M, which was variable in appearance, sometimes appearing convex and not disrupting the smooth contour of the oocyst wall, and sometimes sunken into the oocyst wall giving it a flattened, truncated appearance. Their oocysts were 25.5 × 17.8 (19—33 × 14—21), L/W 1.4 (1.1—1.8), with two wall layers, ~1 thick. The PB was absent, but a spheroidal, well-defined OR was 2.6 wide. Sporocysts were ellipsoidal, 13.8 × 6.1 (10—18 × 4—8), L/W 2.3 (1.8—3.0), with a small SB, but SSB, PSB, SR were all absent. SZs were banana-shaped with a small central N and a posterior RB. This species differs from *E. media* in lacking a SR and in having a SB and has an oocyst wall of two distinct layers. Wiggins et al. (1980) examined *S. floridanus* from Pennsylvania and said the oocysts were 27.1 × 17.6 (21—35 × 14—25.5) with a M, ~4.8 wide, and an OR, ~3.7 wide, but lacked a PG; their sporocysts were 13.4 × 6.4 (11.5—17 × 4—8.5), lacked a SR, and their SZs had a posterior RB, ~4.5 wide, but no mention was made by Wiggins et al. (1980) about the presence/absence of a SB, SSB, or PSB.

EIMERIA IRRESIDUA KESSEL AND JANKIEWICZ, 1931 (FIGURES 6.10, 6.11)

Synonym: *Eimeria elongata* Marotel & Guilhon, 1941.

Type host: *Oryctolagus cuniculus* (L., 1758) (syn. *Lepus cuniculus*), European (domestic) rabbit.

Remarks: This species was originally described from the domestic rabbit, *O. cuniculus*, by Kessel and Jankiewicz (1931) from southern California, USA, and since that time it has been reported in wild and domesticated rabbits worldwide. Carvalho (1943) said he successfully transmitted *E. irresidua* experimentally to *S. floridanus* in Iowa, with oocysts he isolated from the tame rabbit; the prepatent period was 9 days and the patent period varied from 8 to 17 days. Morgan and Waller (1940) said they found *E. irresidua* in the majority of 210 cottontails, *S. floridanus*, in Iowa, and that it had a large, conspicuous M on oocysts that were 36.5 × 18.4 μm. See Chapter 6 for the complete description of *E. irresidua*.

EIMERIA LEPORIS NIESCHULZ, 1923 (FIGURES 5.26, 5.27)

Synonyms: non *Eimeria leporis* Nieschulz of Morgan and Waller, 1940 and of Waller and Morgan, 1941; *Eimeria leporine* Nieschulz, 1923 of Sugár et al. (1978).

Type host: *Lepus europaeus* Pallas, 1778, European hare.

Remarks: Morgan and Waller (1940) and Waller and Morgan (1941) reported oocysts from the cottontail, *Sylvilagus floridanus*, in Iowa and Wisconsin, under the name *E. leporis*; they said the infection localized in the cecum, and that oocysts averaged 36.6 × 18.8. However, Carvalho (1943) argued, and we agree, that they were actually seeing oocysts of *E. neoleporis*. Nieschulz (1923), Pellérdy (1956), and Lucas et al. (1959) were unable to infect the domestic rabbit, *O. cuniculus*, with *E. leporis* isolated from *L. europaeus*. Although Carvalho (1943) did not have access to *E. leporis* oocysts, he stated that, "all species from the jack rabbit were fed to young tame rabbits and cottontails, but results have shown that they are unable to grow in either host." Burgaz (1973) was unable to transmit *E. leporis* oocysts from *L. timidus* to *O. cuniculus*, but her paper must be viewed with caution (Chapter 6). Thus, the validity of *E. leporis* being able to infect *Sylvilagus* species seems unlikely to us and needs to be verified either by cross-transmission studies or by gene sequencing or both. Until then, we maintain the opinion that it is not a valid parasite of *Sylvilagus* species. See Chapter 5 for the complete description of *E. leporis*.

EIMERIA MAGNA PÉRARD, 1925b (FIGURES 6.12, 6.13)

Type host: *Oryctolagus cuniculus* (L., 1758) (syn. *Lepus cuniculus*), European (domestic) rabbit.

Remarks: This species was originally described from the domestic rabbit, *O. cuniculus*, by Pérard (1925b) in France, but it has been found in wild and domestic rabbits worldwide. Becker (1933) said he was able to infect and produce clinical coccidiosis (and death) with this species in *S. floridanus* in Iowa, using oocysts isolated from the tame rabbit. Morgan and Waller (1940) said they found it in *S. floridanus* in Iowa and that the oocysts were ovoidal, 34.8 × 24, with a conspicuous M. Carvalho (1943) confirmed Becker's work by transmitting this species from tame rabbits to *S. floridanus*, and said that *E. magna* produced clinical coccidiosis in these rabbits. From 3–6 days PI, his cottontails showed rough coats, loss of appetite, and softened fecal pellets. The prepatent period was 4 days with a patent period lasting 14 days. See Chapter 6 for the complete description of *E. magna*.

EIMERIA MAIOR HONESS, 1939

Type host: *Sylvilagus nuttallii* (Bachman, 1837), Mountain cottontail.

Type locality: NORTH AMERICA: USA: Wyoming, Albany County, near the Agriculture Experiment Station of the University of Wyoming.

Other hosts: *Sylvilagus audubonii* (Baird, 1858), Desert cottontail; *Sylvilagus floridanus* (J.A. Allen, 1890), Eastern cottontail.

Geographic distribution: NORTH AMERICA: USA: California, Colorado, Iowa, Pennsylvania, Wisconsin, Wyoming.

 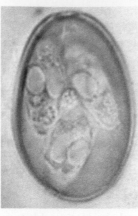

FIGURES 7.9, 7.10 Line drawing of the sporulated oocyst of *Eimeria maior* from Duszynski and Marquardt, 1969, with permission of the *Journal of Protozoology*. Photomicrograph of a sporulated oocyst of *E. maior* from Duszynski and Marquardt, 1969, with permission of the *Journal of Protozoology*.

Description of sporulated oocyst: Oocyst shape: ovoidal, tapering slightly toward M; number of walls: 2; wall characteristics: outer is yellow-brown, rough, striated, and of uneven thickness, being thinnest opposite M and thickest around it; L × W: 47 × 31 (44—51 × 26—37); L/W ratio: 1.5; M: present; M characteristics: a distinct thinning with a slight indentation of wall, ~8.5 wide; OR: present; OR characteristics: variable size and shape, commonly spheroidal, 2—8.5; PG: absent. Distinctive features of oocyst: large size, distinct M, and two-layered wall.

Description of sporocyst and sporozoites: Sporocyst shape: ovoidal; L × W: 21 × 10 (18—25 × 8—12); L/W ratio: 2.1; SB: present; SSB, PSB: both absent; SR: present: SR characteristics: well-defined, varying in shape from diffuse granules to spheroidal to ovoidal; SZ: banana-shaped with small central N and large, posterior RB, and lie lengthwise head-to-tail in sporocyst. Distinctive features of sporocyst: large size of sporocyst and large posterior RB of SZ.

Prevalence: Unknown for type host; in 6/45 (13%) *S. floridanus* from Iowa (Carvalho, 1943); in 36/100 (36%) *S. audubonii* from Colorado

(Duszynski and Marquardt, 1969); in 10/139 (7%) *S. floridanus* from Pennsylvania (Wiggins et al., 1980); and in 13/161 (8%) *S. floridanus* from Wisconsin (Dorney, 1962).

Sporulation: Exogenous. Minimum sporulation time was 50 hr and the majority of oocysts sporulated in 80 hr in 2% (potassium?) dichromate solution (Honess, 1939; Herman and Jankiewicz, 1943), probably at room (?) temperature. Carvalho (1943) said sporulation ranged from 70 to 95 hr and averaged 85 hr in 3% (w/v) $K_2Cr_2O_7$ solution maintained at 30°C, and Duszynski and Marquardt (1969) reported that most oocysts sporulated after 5 days in 2.5% (w/v) $K_2Cr_2O_7$ solution at 20°C.

Prepatent and patent periods: Prepatency is 6—7 days and patency is 9—14 days with an average of 11 days (Carvalho, 1943).

Site of infection: Unknown. Oocysts recovered from intestinal contents.

Endogenous stages: Unknown.

Cross-transmission: Carvalho (1943) fed 50,000 sporulated oocysts of *E. maior* to three young and two adult domestic rabbits (*O. cuniculus*), but none developed infections.

Pathology: Non-pathogenic (Carvalho, 1943).

Material deposited: None.

Remarks: The original description of this species by Honess (1939) included a photomicrograph of a sporulated oocyst, but no line drawing and some of the descriptive features were minimal. Carvalho (1943) reported on sporulated oocysts from Iowa cottontails (*S. floridanus*) as ovoidal, slightly tapered toward the M, 48.4 × 29.5 (44—57 × 25—35), L/W 1.6, with a M, ~8.5 wide, with an OR varying in size and shape, ~4.2—7.1, but sometimes dispersed, and lacking a PG; his sporocysts were ovoidal, 13—18 × 7—10, with a SB at the more pointed end of the sporocyst; the SZs were banana-shaped, with a central N and a small posterior RB. Herman and Jankiewicz (1943) said they found *E. maior* in *S. audubonii vallicola* in California; their oocysts were 41.8 × 26.5 (38—46 × 24—29) with a dark amber oocyst wall and a conspicuous M. Duszynski and Marquardt (1969) described sporulated oocysts from *S. audubonii* from Colorado as broadly ovoidal, 41.5 × 27.0 (37–51 × 24–35), L/W 1.5 (1.3–1.8), with a M, 3.8 (2–6) wide, and a well-defined spheroidal OR, 4.4 (2–6) wide, but no PG; their sporocysts were ovoidal, 22.1 × 8.5 (19–25 × 8–10) with a L/W ratio 2.6 (2.2–3.3), with a small SB and a well-defined ovoidal SR, ~2–9, and banana-shaped SZs with a central N and a posterior RB. Wiggins et al. (1980) examined *S. floridanus* from Pennsylvania and said the oocysts were 42.6 × 24.8 (27–48 × 17.5–30) with a M, ~6.2 wide, and an OR, ~4.6 wide, but lacking a PG; their sporocysts were 18.1 × 8.0 (14–26 × 5.5–11) with a SR, ~4.2 wide, and their SZs had a posterior RB, ~5.5 wide, but no mention was made by Wiggins et al. (1980) about the presence/absence of a SB, SSB, PSB. According to Levine (pers. comm.), the form found by Morgan and Waller (1940) in *S. floridanus* in Iowa, and called *E. paulistana* by them, was probably *E. maior* (and Carvalho, 1943).

EIMERIA MEDIA KESSEL & JANKIEWICZ, 1931 (FIGURES 6.15, 6.16)

Type host: *Oryctolagus cuniculus* (L., 1758), European (domestic) rabbit.

Remarks: This species was originally described from the domestic rabbit, *O. cuniculus*, by Kessel (1929) and Kessel and Jankiewicz (1931). Based on oocyst morphology, Carvalho (1943) reported finding it naturally in one Eastern cottontail, *S. floridanus*, in Iowa. He also said that he was able to cross-transmit *E. media* from the domestic rabbit to *S. floridanus* and that it produced an infection (prepatency 5–6 days, patent period 14–16 days), but demonstrated no "symptomatology." Ecke and Yeatter (1956) reported finding this species in 3/32 (9%) *S. floridanus* from Illinois, and Dorney

(1962) reported finding it in 150/161 (93%) *S. floridanus* from Wisconsin. See Chapter 6 for the complete description of *E. media*.

EIMERIA MINIMA CARVALHO, 1943

FIGURE 7.11 Line drawing of the sporulated oocyst of *Eimeria minima* from Levine and Ivens, 1972, with permission of the *Journal of Protozoology*.

Synonym: *Eimeria exigua* Yakimoff, 1934 of Morgan and Waller, 1940.

Type host: *Sylvilagus floridanus* (J.A. Allen, 1890), Eastern cottontail.

Type locality: NORTH AMERICA: USA: Iowa, Story County, Ames.

Other hosts: None reported to date.

Geographic distribution: NORTH AMERICA: USA: Iowa, Pennsylvania.

Description of sporulated oocyst: Oocyst shape: subspheroidal; number of walls: 1; wall characteristics: of uniform thickness, colorless to straw-tinged; L × W: 13.4 × 10.8 (11–15 × 9–14); L/W ratio: 1.2; M, OR: both absent; PG: present. Distinctive features of oocyst: small size, lack of OR and M, and presence of PG.

Description of sporocyst and sporozoites: Sporocyst shape: ovoidal; L × W: 6 × 2.5 (ranges not given); L/W ratio: 2.4; SB: present; SSB, PSB: both absent; SR: present; SR characteristics:

distinct and granular; SZ: as usual, with very small refractive granules (RB?). Distinctive features of sporocyst: small size.

Prevalence: In 6/45 (13%) of the type host; in 8/210 (4%) *S. floridanus* from Iowa (under the name *E. exigua*; see Morgan and Waller, 1940); and in 3/139 (2%) *S. floridanus* from Pennsylvania (Wiggins et al., 1980).

Sporulation: Exogenous. Oocysts sporulated in 140–160 hr in both 3% (w/v) potassium dichromate and/or in 2% formic acid (Carvalho, 1943).

Prepatent and patent periods: Prepatency 6 days, patent period is 12–16 days.

Site of infection: Small intestine.

Endogenous stages: Unknown.

Cross-transmission: Oocysts of this species were not infective to two young domestic rabbits, *O. cuniculus* (Carvalho, 1943).

Pathology: Unknown.

Material deposited: None.

Remarks: Gill and Ray (1960) reported *E. minima* in the Indian field hare (*Lepus ruficaudatus*), but they were likely dealing with *E. hungarica* (from Levine and Ivens, 1972). Mandal (1976) also said he recovered oocysts of this species in *L. ruficaudatus* from Punjab, India, with subspheroidal oocysts that were 10–15.5 × 9–15, without an OR, and with ovoidal sporocysts, 5 × 2.8, with a SR. There is no way to know which species Mandal (1976) actually was dealing with, but we do not believe it was *E. minima*. In addition, due to the failure to cross-infect *E. minima* into the domestic rabbits, *O. cuniculus*, Carvalho (1943) felt that past reports on *E. exigua* (from *Oryctolagus*) in *S. floridanus* were probably referring to *E. minima*. Wiggins et al. (1980) examined *S. floridanus* from Pennsylvania, and said they found oocysts of *E. minima* that were 13 × 12 (12–14 × 11–12.5), lacking both a M and OR, but with a PG; their sporocysts were 6.1 × 4.0 (4–11 × 3–5), with a SR, and their SZs had a posterior RB, but no mention was made by them about the presence/absence of SB, SSB, PSB.

EIMERIA NEOIRRESIDUA DUSZYNSKI AND MARQUARDT, 1969

Type host: *Sylvilagus audubonii* (Baird, 1858), Desert cottontail.

Type locality: NORTH AMERICA: USA: Colorado, Larimer County, near Ft. Collins.

Other hosts: *Sylvilagus floridanus* (J.A. Allen, 1890), Eastern cottontail.

Geographic distribution: NORTH AMERICA: USA: Colorado, Pennsylvania.

Description of sporulated oocyst: Oocyst shape: ovoidal to ellipsoidal; number of walls: 2; wall thickness: ~1; wall characteristics: outer is smooth, irregular in thickness, being thinnest at end opposite M and thickest on sides and around margin of M; L × W: 25.7 × 17.9 (19–31 × 15–20); L/W ratio: 1.4 (1.2–1.75); M: present; M characteristics: distinct and variable in appearance, sometimes appearing convex and not disrupting the smooth contour of the oocyst wall, and sometimes sunken into the oocyst wall giving the wall a flattened appearance; OR, PG: both absent. Distinctive features of oocyst: variable appearance of the M and absence of OR and PG.

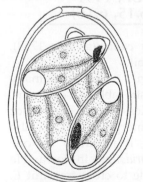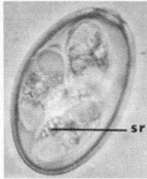

FIGURES 7.12, 7.13 Line drawing of the sporulated oocyst of *Eimeria neoirresidua* from Duszynski and Marquardt, 1969, with permission of the *Journal of Protozoology*. Photomicrograph of a sporulated oocyst of *E. neoirresidua* from Duszynski and Marquardt, 1969, with permission of the *Journal of Protozoology*; sr - sporocyst residuum.

Description of sporocyst and sporozoites: Sporocyst shape: ovoidal to ellipsoidal with bluntly pointed end that contains the SB; L × W: 14.5 × 6.4 (11−18 × 6−8); L/W ratio: 2.3 (1.7−2.8); SB: present as a thickening at pointed end; SSB, PSB: both absent; SR: present; SR characteristics: granular, diffuse to spheroidal or ellipsoidal, up to 6 wide, but varied by being present in 1−4 sporocysts in any single oocyst; SZ: banana-shaped with small central N and a large posterior RB. Distinctive features of sporocyst: bluntly-pointed end of sporocyst with SB and SR present in only 1−4 sporocysts in any single oocyst.

Prevalence: In 20/58 (34%) of the type host and in 46/139 (33%) *S. floridanus* from Pennsylvania (Wiggins et al., 1980).

Sporulation: Exogenous. Oocysts were sporulated after 5 days in 2.5% aqueous $K_2Cr_2O_7$ solution at 20°C.

Prepatent and patent periods: Unknown.

Site of infection: Unknown. Oocysts recovered from cecal contents.

Endogenous stages: Unknown.

Cross-transmission: None to date.

Pathology: Unknown.

Material deposited: None.

Remarks: Duszynski and Marquardt (1969) compared the sporulated oocysts of *E. neoirresidua* to those of two eimerians to which they were structurally similar: *E. irresidua* Kessel and Jankiewicz, 1931, in the domestic rabbit, *O. cuniculus*, from southern California, USA, and from which it was described, and to those of *E. pintoensis* da Fonseca, 1932, from *S. brasiliensis* from Brazil. They found that *E. neoirresidua* oocysts differed from the former by being smaller (25.7 × 17.9 vs. 38.3 × 25.6), by the shape of the sporocysts, the presence of a SB, and the irregularity in the occurrence of the SR. They differed from those of the latter by having slightly smaller sporocysts (12−14 × 5−7 vs. 11−18 × 6−8) and by the presence of a SB, which the sporocysts of *E. pintoensis* apparently lack. Wiggins et al. (1980) examined *S. floridanus*

from Pennsylvania, and said they found oocysts of *E. neoirresidua* that were 25.2 × 16.8 (20−35 × 12.5−23) with a M, ~4.7 wide, but lacking both OR and PG; their sporocysts were 12.6 × 6.4 (10−17 × 5−7.5) with a SR, ~2.8 wide, and their SZ had a posterior RB, ~4 wide, but no mention was made by them about the presence/absence of a SB, SSB, or PSB.

EIMERIA NEOLEPORIS CARVALHO, 1942

Synonyms: *Eimeria leporis* Nieschulz of Morgan and Waller, 1940, and Waller and Morgan, 1941; *Eimeria neoleporis* Carvalho, 1943 of Gill and Ray, 1960 (*lapsus calami*); [non] *Eimeria neoleporis* Nieschulz, 1923; [non] *Eimeria neoleporis* Carvalho of Pellérdy, 1954a, 1965, 1974.

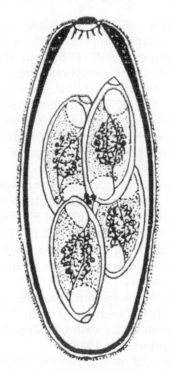

FIGURE 7.14 Line drawing of the sporulated oocyst of *Eimeria neoleporis* from Levine and Ivens, 1972, with permission of the *Journal of Protozoology*.

Type host: *Sylvilagus floridanus* (J.A. Allen, 1890), Eastern cottontail.

Type locality: NORTH AMERICA: USA: Iowa, Story County, near Ames.

Other hosts: *Oryctolagus cuniculus* (L., 1758) (syn. *Lepus cuniculus*), European (domestic) rabbit, both natural and experimental; *Sylvilagus audubonii* (Baird, 1858), Desert cottontail.

Geographic distribution: ASIA: India (?); NORTH AMERICA: USA: California, Iowa, Illinois, Pennsylvania, Wisconsin.

Description of sporulated oocyst: elongate-ellipsoidal to slightly cylindroidal, tapering somewhat toward M; number of walls: 2; wall characteristics: outer is pinkish-yellow, smooth, and the same thickness throughout, but enlarges noticeably toward M; L × W: 38.8 × 19.8 (33—44 × 16—23); L/W ratio: 2.0; M: present; OR, PG: both absent. Distinctive features of oocyst: distinct M.

Description of sporocyst and sporozoites: Sporocyst shape: ellipsoidal; L × W: 17 × 8—9; SB: present: SSB, PSB: both absent; SR: present; SR characteristics: large, occupying $^1/_3$ of sporocyst; SZ: banana-shaped, 14—15.5 × 4—5, lying lengthwise head-to-tail in the sporocysts, with a central N and one clear RB at each end of SZ. Distinctive features of sporocyst: length, presence of SB, and large SR.

Prevalence: In 4/15 (27%) of the type host; in 12/45 (26%) in same host species a year later (Carvalho, 1943); in 43/139 (24.5%) from Pennsylvania (Wiggins et al., 1980), in 10/32 (31%) from Illinois (Ecke and Yeatter, 1956), and in 35/161 (22%) from Wisconsin (Dorney, 1962), all in *S. floridanus*.

Sporulation: Exogenous. Oocysts sporulated in 50 to 75 h, on average 60 hr, in 3% aqueous $K_2Cr_2O_7$ solution (Carvalho, 1942) or 48 hr (Herman and Jankiewicz, 1943).

Prepatent and patent periods: Prepatency is 11—14 days and the patent period is 8—16 days (Carvalho, 1942) or 12 and 10 days, respectively (Carvalho, 1944).

Site of infection: Localized in the ileo-cecal valve and apical process of the cecum in experimentally infected domestic rabbits.

Endogenous stages: Carvalho (1944) said he was able to grow this species easily in over 65 tame/domestic rabbits, and from these he described and drew, but did not measure, most of the endogenous stages. Doses up to 600,000 oocysts were given to 18-day-old coccidia-free *O. cuniculus*, which were then killed every 48 hr until completion of the life cycle. The first generation of young meronts localized in the epithelial cells of the apical process, mainly those at the bottom of the crypts, but Carvalho (1944) reported that mature meronts were principally in the tunica propria of the mucous membrane, stating, "This migration of infected cells agrees with that described by Matsubayashi (1934) and Cheissin (1940) for certain other species in rabbits." Mature first-generation meronts contained 43—48 first-generation merozoites around a central residual mass, and these merozoites averaged 20.5 × 3, and were found in smears until day 8 PI. Second-generation meronts were completely mature by day 7 PI with 59—72 merozoites that measured 25.7 × 1.5 each, with one N and 4—5 "siderophilic" granules, and were found in the intestinal lumen until 9 days PI. Carvalho (1944) reported that third-generation meronts were fully mature by 9 days PI and produced two types of merozoites in the same meront, 14 smaller ones, 18 × 3.5, and 60—86 larger ones, 31.5 × 1.5. His suggestion was that the smaller ones continued into a fourth generation of merogony, while the larger merozoites give rise to gametocytes. Having made that suggestion, Carvalho (1944) made no further mention of the fourth-generation meronts, nor about the number of merozoites it produced.

Gametocytes began their development about 10 days PI and both macro- and microgametocytes completed their full development in 2 days. The microgametocyte N divided several times to form spheroidal black bodies, which later elongated to become comma-shaped. By day 12 PI, the microgametes were free in the tissues and in the intestinal lumen; each had two flagella at one end, ~0.6 long, and measured 3.5 × 0.5. Macrogametocytes were in the apical process of the cecum and in the ileo-cecal valve, where they occurred in large numbers forming what Carvalho (1944) called oocyst nests. Carvalho (1944) also noted that individual epithelial cells harbored as many as 18 young macrogametocytes, but that the average number in each infected cell was three to five; Carvalho (1944) suggested the name "oocystic nest-cell" to describe epithelial cells with multiple infections. Young oocysts were seen by day 11 PI, when they passed into the cecal lumen and usually remained there or in the large intestine for 24–36 hr before being passed in the feces.

Cross-transmission: Carvalho (1942) infected 26/26 tame rabbits, *O. cuniculus*, and said that he obtained five consecutive passages through this host. A year later he (Carvalho, 1943) stated, "several infections were obtained with this species when oocysts were fed to young tame rabbits." Finally, Carvalho (1944) said that 54 young domestic rabbits, all free from coccidia, were infected with *E. neoleporis* oocysts and that, "with the exception of one naturally immune rabbit and a few old does and bucks, all were susceptible." To further confirm that this species easily passes between genera, Carvalho (1944) infected three *S. floridanus*, each about 34 days old, with 150,000 oocysts from the same culture as that administered to and infective for the domestic rabbits, and all became infected.

Pathology: Carvalho (1944) described the pathogenic effects of this species. After day 5 PI in the domestic rabbit (*O. cuniculus*), the apical process of the cecum became whitish, enlarged, and showed hyperemic blood vessels. The ileo-cecal valve was attacked more during the later stages of infection, and Carvalho (1944) saw some infections when it became so enlarged "as almost to obstruct the passage of

intestinal contents to the large intestine and cecum." During the first 3 days of patency, the cecal contents, large intestine, and fecal pellets were hemorrhagic and the lumen was filled with caseous material including merozoites and leukocytes. Although Carvalho (1944) said that the epithelial layer of the mucosa was badly damaged during oocyst discharge, accompanied by hemorrhage and massive number of lymphocytes and eosinophils migrating into the submucosa, he concluded that, at least in the domestic rabbit, "In no case was loss of weight noted, even with dosages as high as 150,000 oocysts, that diarrhea was also absent, and that the only apparent symptom was partial loss of appetite during the 2 or 3 first days."

Upon completing his cross-infection and pathology studies, Carvalho (1944) concluded that *S. floridanus*, the natural host for *E. neoleporis*, showed a more marked susceptibility to the parasite than did domestic rabbits. In support of this conclusion, he pointed out that 2 days before elimination of the first oocysts (from three *S. floridanus*), they "became quiet, developed a rough coat and showed a marked loss of appetite;" one had diarrhea 2 days after oocysts were first seen in the feces, while the other two had softened pellets for 4 days. In the cottontail, the most evident symptom was weight loss, in which one cottontail lost 50% of its body weight. The other two cottontails lost less weight from day 8 PI up to day 3 after patency ended. In addition to weight loss in the cottontails, the numbers of oocysts they discharged were much greater, by at least 60%, than from the tame rabbits. And, given the same dosages, cottontails lost weight during infection, while domestic rabbits continued to gain.

Material deposited: None.

Remarks: Carvalho (1942, 1943) compared *E. neoleporis* to *E. leporis* and found that they differ in their intestinal location and that their sporulated oocysts differed by *E. neoleporis* having a distinct M, which *E. leporis* lacked. In addition, he was able to infect the domestic rabbit, *O. cuniculus*, with *E. neoleporis*. Carvalho (1944) said that the localization of endogenous stages he found in the rabbit intestine was the same as that reported by Waller and Morgan (1941) for *E. leporis* in cottontails and presumed they were likely dealing with *E. neoleporis*.

Carvalho (1944) also looked at immunity that developed in the domestic rabbit to infection with *E. neoleporis*. His infections demonstrated that domestic rabbits acquired total immunity to *E. neoleporis* following an inoculation dose of 6,000 or more oocysts, while lighter doses only led to partial immunity in the host. He also looked at age resistance against *E. neoleporis*. He infected four adult does and one buck, each with 150,000 oocysts, and was unable to demonstrate that they became infected. However, similar infections given to young domestic rabbits < 120 days old were positive. In the natural cottontail host, there was no evidence of age resistance to *E. neoleporis*, since adult as well as young cottontails were readily susceptible to infection, and Carvalho (1944) found positive infections in field-captured cottontails of all ages.

Pellérdy (1954a, 1965, 1974) found a coccidium which he identified as *E. neoleporis* in domestic rabbits in Hungary; he thought that *E. coecicola* was a synonym of *E. neoleporis*, but Cheissin (1968) showed that they were different (although the arguments he presented may not be convincing to everyone). Thus, Pellérdy (1954a), according to Cheissin (1968), was dealing with *E. coecicola* and not *E. neoleporis*. Cheissin's (1968) arguments were that these are separate species that differ in oocyst length, presence (*E. coecicola*) or absence (*E. neoleporis*) of an OR, localization of the meronts in the gut, the timing of endogenous development, the length of the prepatent and patent periods, and in their adaptation to different hosts (also see Levine and Ivens, 1972).

Herman and Jankiewicz (1943) reported *E. neoleporis* in *S. a. vallicola* in California; their

(Type III) cylindroidal oocysts were 31.5 × 16.7 (29–35 × 15–19), L/W 1.9, had an OR, ~2.5–3 wide, and had a delicate M. Mandal (1976) said he recovered oocysts of this species in *Lepus* sp. from Punjab, India, with ellipsoidal oocysts that were 30–54 × 16–22, without an OR, and with ovoidal sporocysts, 16.5 × 8, with a SR. Finally, Wiggins et al. (1980) examined *S. floridanus* from Pennsylvania, and said they found oocysts of *E. neoleporis* that were 38.1 × 18.4 (28–46 × 12–26.5) with a M, ~5.5 wide, and with an OR, ~1.6, but lacking a PG; their sporocysts were 15.7 × 7.2 (13.5–20 × 6–10) with a SR, ~5.1 wide, and their SZs had a posterior RB, ~4.5 wide, but no mention was made by Wiggins et al. (1980) about the presence/absence of a SB, SSB, or PSB.

EIMERIA PAULISTANA DA FONSECA, 1933

Type host: *Sylvilagus brasiliensis* (L., 1758) (syn. *Sylvilagus minensis* Thomas, 1913), Tapeti or Brazilian cottontail.

FIGURE 7.15 Line drawing of the sporulated oocyst of *Eimeria paulistana* from Levine and Ivens, 1972, with permission of the *Journal of Protozoology*.

Type locality: SOUTH AMERICA: Brazil: São Paulo.

Other hosts: None to date.

Geographic distribution: SOUTH AMERICA: Brazil.

Description of sporulated oocyst: Oocyst shape: elongate-ellipsoidal, flattened at M end; number of walls: 3; wall characteristics: outer is transversely striated and thicker near M, middle is "relatively thick," and inner layer is thin and adherent to middle layer; L × W: 43 × 23 (40–43 × 23–24); L/W ratio: 1.9; M: present; M characteristics: distinct, broad, at end of oocyst where the wall thins dramatically; OR, PG: both absent. Distinctive features of oocyst: oocyst wall with three distinct layers, with outer being striated; wall thins to a membrane-like structure at end with the M.

Description of sporocyst and sporozoites: Sporocyst shape: ovoidal; L × W: 15. × 7.5 (ranges not given); L/W ratio: 2.1; SB: likely is present, but small; SSB, PSB: both absent; SR: present; SR characteristics: scattered granules lying between the SZs (line drawing); SZ: elongate, comma-shaped, lying lengthwise head-to-tail in sporocyst with 1–2 clear RB in each SZ. Distinctive features of sporocyst: tiny SB.

Prevalence: Not given for definitive host.

Sporulation: Exogenous. Oocysts sporulated in ~5 days at 19–21°C.

Prepatent and patent periods: Unknown.

Site of infection: Unknown. Oocysts recovered from feces.

Endogenous stages: Unknown.

Cross-transmission: Da Fonseca (1933) was unable to infect the domestic rabbit with this species.

Pathology: Unknown.

Material deposited: None.

Remarks: The oocysts of this species are close in size to those of *E. maior* (43 × 23 [40–43 × 23–24] vs. 47 × 31 [44–51 × 26–37]), but lack an OR, which those of *E. maior* clearly possess; few other oocyst or sporocyst structures are very similar. The species recorded under this name from 4/210 (2%) *S. floridanus* in Iowa, by Morgan and Waller (1940), was probably *E. maior* (see Carvalho, 1943).

EIMERIA PERFORANS (LEUCKART, 1879) SLUITER AND SWELLENGREBEL, 1912 (FIGURES 6.19, 6.20)

Synonyms: *Coccidium cuniculi* Rivolta, 1878; *Coccidium perforans* Leuckart, 1879; *Eimeria nana* Marotel and Guilhon, 1941; *Eimeria lugdunumensis* Marotel and Guilhon, 1942.

Type host: *Oryctolagus cuniculus* (L., 1758) (syn. *Lepus cuniculus*), European (domestic) rabbit.

Remarks: This species was originally described from the domestic rabbit, *O. cuniculus*, in Europe. Carini (1940) reported finding *E. perforans* in *S. brasiliensis* from Brazil, and Morgan and Waller (1940) said they found ellipsoidal oocysts of this species in the majority of 210 *S. floridanus* in Iowa, with an average size of 26.4 × 16.2 μm. Carvalho (1943) successfully transmitted this species from the domestic rabbit to three young cottontails, *S. floridanus*. Herman and Jankiewicz (1943) reported this species in *S. a. vallicola* collected on the San Joaquin Experimental Range in California. Ecke and Yeatter (1956) reported finding this species in 6/32 (19%) *S. floridanus* from Illinois, but their tentative identification was based on unsporulated oocysts! See Chapter 6 for the complete description of *E. perforans*.

EIMERIA PINTOENSIS DA FONSECA, 1932

Type host: *Sylvilagus brasiliensis* (L., 1758) (syn. *Sylvilagus minensis* Thomas, 1913), Tapeti or Brazilian cottontail.

Type locality: SOUTH AMERICA: Brazil: Butantan, São Paulo.

Other hosts: *Sylvilagus floridanus* (J.A. Allen, 1890), Eastern cottontail.

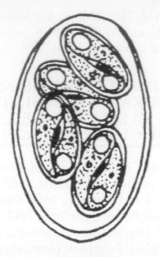

FIGURE 7.16 Line drawing of the sporulated oocyst of *Eimeria pintoensis* from Levine and Ivens, 1972, with permission of the *Journal of Protozoology*.

Geographic distribution: SOUTH AMERICA: Brazil; NORTH AMERICA: USA: Illinois.

Description of sporulated oocyst: Oocyst shape: ovoidal; number of walls: 1; wall characteristics: smooth, narrowing toward M end, yellowish-green; L × W: 23.5 × 15.5 (23–25.5 × 15–16); L/W ratio: 1.5; M: present; M characteristics: indistinct, especially in unsporulated oocysts; OR, PG: both absent. Distinctive features of oocyst: indistinct M.

Description of sporocyst and sporozoites: Sporocyst shape: ovoidal; L × W: 12–14 × 5–7; SB, SSB, PSB: all absent; SR: present; SR characteristics: only a few granules between SZs; SZ: elongate, lying head-to-tail in sporocyst and having one RB at large end. Distinctive features of sporocyst: absence of SB.

Prevalence: Unknown in type host; in 4/32 (12.5%) *S. floridanus* (?) from Illinois (Ecke and Yeatter, 1956).

Sporulation: Exogenous. Oocysts sporulated in 2 days.

Prepatent and patent periods: Unknown.

Site of infection: Unknown. Oocysts recovered from feces.

Endogenous stages: Unknown.

Cross-transmission: Da Fonseca (1932) was unable to infect the domestic rabbit with this species.

Pathology: Unknown.

Material deposited: None.

Remarks: Da Fonseca (1932) noticed that the oocysts increased in length slightly during sporulation from 21.5–25.5 to 23–26.5. As far as we can tell, this species has not been described from any other hosts since its original description. The evidence for this species occurring in *S. floridanus* (from Illinois, or elsewhere) is questionable because the report of its occurrence in North American cottontails was based on <u>un</u>sporulated oocysts (Ecke and Yeatter, 1956)!

EIMERIA POUDREI DUSZYNSKI AND MARQUARDT, 1969

Type host: *Sylvilagus audubonii* (Baird 1858), Desert cottontail.

Type locality: NORTH AMERICA: USA: Colorado, Larimer County, near Ft. Collins.

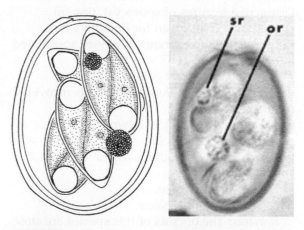

FIGURES 7.17, 7.18 Line drawing of the sporulated oocyst of *Eimeria poudrei* from Duszynski and Marquardt, 1969, with permission of the *Journal of Protozoology*. Photomicrograph of the sporulated oocyst of *Eimeria poudrei* from Duszynski and Marquardt, 1969, with permission of the *Journal of Protozoology*; sr - sporocyst residuum; or - oocyst residuum.

Other hosts: None reported to date.

Geographic distribution: NORTH AMERICA: USA: Colorado.

Description of sporulated oocyst: Oocyst shape: ovoidal to ellipsoidal; number of walls: 2; wall thickness: ~1; wall characteristics: smooth outer wall of irregular thickness, thinnest at end opposite M and thickest on sides and around margin of M; L × W: 26.0 × 18.1 (20−31 × 15−21); L/W ratio: 1.4 (1.2−1.8); M: present; M characteristics: distinct and variable in appearance, sometimes appearing convex, not disrupting the smooth contour of the oocyst wall and sometimes sunken into oocyst wall giving a flattened appearance; OR: present; OR characteristics: well-defined, granular, diffuse to spheroidal, 2−6 wide; PG: absent. Distinctive features of oocyst: presence of distinct OR and M.

Description of sporocyst and sporozoites: Sporocyst shape: ellipsoidal to ovoidal with bluntly pointed end that contains the SB; L × W: 14.4 × 6.4 (10−18 × 6−8); L/W ratio: 2.3 (1.7−2.8); SB: present; SSB, PSB: both absent; SR: present; SR characteristics: well-defined, granular, round to ellipsoidal, 1−5 wide, and varies by being present in 1−4 sporocysts of a single oocyst; SZ: banana-shaped with small central N and a posterior RB. Distinctive features of sporocyst: irregularity in the presence of a SR.

Prevalence: In 13/58 (22%) in the type host.

Sporulation: Exogenous. Oocysts were sporulated after 5 days in 2.5% (w/v) aqueous $K_2Cr_2O_7$ solution at 20°C.

Prepatent and patent periods: Unknown.

Site of infection: Unknown. Oocysts recovered from cecal contents.

Endogenous stages: Unknown.

Cross-transmission: None to date.

Pathology: Unknown.

Material deposited: None.

Remarks: Sporulated oocysts of *E. poudrei* resemble, in shape and size, those of *E. matsubayashii* Tsunoda, 1952, described from the domestic rabbit, *O. cuniculus*, in Japan. However, they differ from those of *E. matsubayashii* by the irregular occurrence of the SR in only some of the sporocysts, and by the size of the sporocysts (14.4 × 6.4 vs. 7 × 6; see Levine and Ivens, 1972). Also, the two species are isolated from each other by host and continent.

EIMERIA STIEDAI (LINDEMANN, 1865) KISSKALT AND HARTMANN, 1907 (FIGURES 6.25, 6.26)

Synonyms: *Psorospermium oviforme* Remack, 1854; *Monocystis stiedae* Lindemann, 1865; *Psorospermium cuniculi* Rivolta, 1878; *Coccidium oviforme* Leuckart, 1879; *Eimeria oviformis* (Leuckart, 1879) Fantham, 1911.

Type host: *Oryctolagus cuniculus* (L., 1758) (syn. *Lepus cuniculus*), European (domestic) rabbit.

Remarks: This species was originally described from the domestic rabbit, *O. cuniculus*, by Lindemann (1865). Species of both *Sylvilagus* (cottontails) and *Lepus* (hares) have been reported to be infected with *E. stiedai* either naturally or experimentally or both. Thus, it seems that this is the only eimerian that can infect all three genera, although in *Lepus* and *Sylvilagus* species it seems to be rare in nature. It should be noted that at least some of these early reports are probably erroneous. For example, Harkema (1936) reported finding *E. stiedai* in "white cysts in the liver" of 15/41 (37%) *S. floridanus* from the Duke Forest, Durham County, North Carolina, but provided no measurements, line drawing, or photographic evidence of the structure and composition of the sporulated oocyst. Honess and Winter (1956) reported finding *E. stiedai* in *S. nuttallii* in Carbon County, Wyoming, but gave little descriptive information. Jankiewicz (1941) and Herman and Jankiewicz (1943) reported experimental infection of *S. audubonii* with *E. stiedai*, and Hsu (1970) infected four *S. floridanus* from Iowa with oocysts of *E. stiedai* recovered from *O. cuniculus*; he said the prepatent period in the cottontails was 19−20 days with a patent period of 12–14 days. Hsu (1970) found that it was highly pathogenic for this cottontail, and that sporulated oocysts from cottontails were similar in size and structure to those from domestic rabbits. He then took sporulated oocysts from *S. floridanus* and reported they were infective to *O. cuniculus*, thus completing Koch's postulates. See Chapter 6 for the complete description of *E. stiedai*.

EIMERIA SYLVILAGI CARINI, 1940

Type host: *Sylvilagus brasiliensis* (L., 1758), Tapeti or Brazilian cottontail.

Type locality: SOUTH AMERICA: Brazil: São Paulo.

Other hosts: *Sylvilagus floridanus* (J.A. Allen, 1890), Eastern cottontail; *Lepus nigricollis* F. Cuvier, 1823 (syn. *L. ruficaudatus* Geoffroy, 1826) (?), but this seems unlikely.

Geographic distribution: ASIA: India (?); NORTH AMERICA: USA: Illinois, Iowa, Pennsylvania, Wisconsin; SOUTH AMERICA: Brazil: São Paulo.

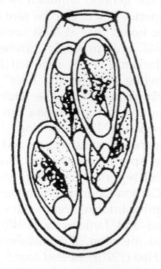

FIGURE 7.19 Line drawing of the sporulated oocyst of *Eimeria sylvilagi* from Levine and Ivens, 1972, with permission of the *Journal of Protozoology*.

Description of sporulated oocyst: Oocyst shape: ovoidal to ellipsoidal; number of walls: 2; wall characteristics: outer is smooth, with transverse striae (Carini, 1940) or not (Carvalho, 1943), inner is yellow-brown, thickened around end with M; L × W: 26–28 × 15–16 in original description; M: present; M characteristics: ~6 wide; OR, PG: both absent. Distinctive features of oocyst: wide M and horizontal striations in the oocyst wall.

Description of sporocyst and sporozoites: Sporocyst shape: ellipsoidal; L × W: 15.5 × 6–7 (original description); SB: present, large; SSB, PSB: both absent; SR: present; SR characteristics: granules of varying sizes scattered among or over SZs; SZ: banana-shaped, lying head-to-tail in sporocyst, with a clear RB at one end, ~4 wide, and a central N. Distinctive features of sporocyst: large SB and SR consisting of granules of varying size.

Prevalence: Unknown in type host; in 12/45 (26%) *S. floridanus* in Iowa (Carvalho, 1943); 13/139 (9%) *S. floridanus* in Pennsylvania (Wiggins et al., 1980); 10/32 (31%) *S. floridanus* in Illinois (Ecke and Yeatter, 1956); and 10/161 (6%) *S. floridanus* in Wisconsin (Dorney, 1962).

Sporulation: Exogenous. 55–80 hr, average 65 hr (Carvalho, 1943); in ~3 days (Carini, 1940); 96 hr at 20°C, 72 hr at 25°C, 60 hr at 30°C in 2.5% (w/v) aqueous $K_2Cr_2O_7$ solution (Hsu, 1970); or 2–3 days (Mandal, 1976).

Prepatent and patent periods: Prepatency is 6–8 days (Hsu, 1970), and patent period is 6–18 days (Carini, 1940), 6–10 days (Hsu, 1970) or 7–8 days and 14–18 days (Carvalho, 1943).

Site of infection: Small intestine (Carvalho, 1943), in the epithelial cells in both the distal and proximal parts of the villi and in the lamina propria of the jejunum and ileum (Hsu, 1970).

Endogenous stages: Hsu (1970) divided intestines from rabbits he infected into six equal parts (1–6) from the pylorus to the ileo-cecal valve and found two asexual generations. The first-generation meronts, most numerous in sections 2–4 of the intestine, were seen first in cells at 24 hr PI, they matured 4–5 days PI, measured 3–4 × 4–7, and produced 2–14 curved, banana-shaped merozoites that were 9 × 2 (7–14 × 1.5–3), pointed at both ends, and were usually arranged parallel to each other in their meronts, but sometimes appeared randomly arranged; in both instances, there was no RB in the mature first-generation meront. Most first-generation meronts were present 96–120 hr PI, while some were still present at 144 hr PI. When released from this meront, the merozoites penetrated other epithelial cells to give rise to second-generation meronts, most numerous in sections 4–6 of infected intestines. Mature second-generation meronts varied greatly in size (in tissue sections) and were 20 × 14 (12–35 × 6–23) and produced 3–86 merozoites that were 12 × 3 (10–15 × 2–4), and were most prevalent 7–8 d PI. Mature gamonts, like the second-generation meronts, were most numerous in sections 4–6 of infected intestines. Microgametocytes were spheroidal, 10–26 × 9–25, and produced many bi-flagellated microgametes, ~3 × 0.4. Mature macrogametocytes were found 6 days PI, and measured 15–26 × 10–12. Multiple parasitism of three to five gamonts was commonly seen in a single host cell, but no endogenous stages ever were found in the cecum, colon, or other organs examined (Hsu, 1970)

Cross-transmission: Hsu (1970) was unable to transmit this species from *S. floridanus* to the domestic rabbit, *O. cuniculus*.

Pathology: Carvalho (1943) saw no signs of pathology in *S. floridanus* infected with this species. Hsu (1970) said that its pathogenicity was "nil to slight," but noted slight diarrhea and some traces of mucus in one cottontail infected with 225,000 sporulated oocysts. In the experimentally infected rabbits he used to study endogenous development, he said, "bloody fluid containing red blood cells and other cells were found in the small intestinal lumen 6 and 7 days after inoculation with 350,000 oocysts," and "hyperemia of the villi and laminal (sic) propria or subepithelial layer, destruction of much (sic) superficial epithelial cells by sexual stages of the parasite, and dearrangement (sic) and necrosis of intestinal epithelium were observed."

Material deposited: None.

Remarks: The original description of *E. sylvilagi* by Carini (1940) was marginal at best, although he did provide a line drawing. Subsequent studies (e.g., Carvalho, 1943) provided significantly more detail. Carvalho (1943) described the oocysts as 29 × 17.5 (22.5–31 × 15–20), L/W 1.6, lacking an OR and a PG, with ellipsoidal sporocysts 14.5 × 7, with a SB and a SR. Hsu (1970) said sporulated oocysts were ovoidal to ellipsoidal with two wall layers, ~1.7 thick, and measured 30 × 20 (26–35 × 18–21), L/W ratio 1.7, with a M, ~6 wide; OR present, but PG absent; sporocysts ellipsoidal, 18 × 9 (16–20 × 9–10), with SR as a round mass of large granules; SB present; SZ banana-shaped, pointed at one end, 15 × 2.4 (13–17 × 2–3) each with a posterior RB, central N, and a smaller anterior RB. Wiggins et al. (1980) examined *S. floridanus* from Pennsylvania and said they found oocysts of *E. sylvilagi* that were 29.6 × 18.3 (24–37 × 16–26.5) with a M, ~5.4 wide, and lacking an OR or a PG; their sporocysts were 15.1 × 7.0 (12–19.5 × 5.5–7.5) with a SR, ~4.3 wide, and their SZ had a posterior RB, ~5 wide, but they made no mention about the presence/absence of a SB, SSB, or PSB. Hsu (1970) reported that an OR is present soon after sporulation, but that is gets smaller and later disappears.

Gill and Ray (1960) said they found *E. sylvilagi* in the Indian hare, *Lepus nigricollis*, in India, but this seems unlikely to us, and their report must be verified. Likewise, Mandal (1976) said he recovered oocysts of *E. sylvilagi* in *L. nigricollis* from Kashipur, India, with ovoidal oocysts that were 21–39 × 16.2, without an OR, and had ovoidal to ellipsoidal sporocysts,

16 × 7, with a SR. Its true identity and validity in *Lepus* species needs to be verified either by cross-transmission studies or by gene sequencing or both; thus, at present, we do not believe *E. sylvilagi* to be a valid parasite of any *Lepus* species.

SPECIES INQUIRENDAE (3)

Type IV *Eimeria* Herman and Jankiewicz, 1943

Original host: *Sylvilagus audubonii* (Baird, 1858), Desert cottontail.

Remarks: Herman and Jankiewicz (1943) examined 44 *S. a. vallicola* on the San Joaquin Experimental Range in California for parasites. They reported 43/44 (98%) to be infected with one or more intestinal species of *Eimeria*. They said this form, "differs from species reported in rabbit-like mammals (Lagomorpha) and only distantly resembles *E. exigua* Yakimoff." The oocysts were reported to be delicate pink, ovoidal, 17.5—20.0 × 13.9—15.6, with an inconspicuous, convex M, lacked an OR, and sporulated completely in 30 hr. It is not clear to us whether the oocysts actually sporulated completely since they only said that, "each sporocyst contains many residual granules," which indicates to us they were not fully developed.

Type V *Eimeria* Herman and Jankiewicz, 1943

Original host: *Sylvilagus audubonii* (Baird, 1858), Desert cottontail.

Remarks: As noted, Herman and Jankiewicz (1943) examined 44 *S. a. vallicola* on the San Joaquin Experimental Range in California for parasites. Their Type V oocysts differed from their Type IV by lacking a M and being narrower. They found no OR in "the jelly of the sporulated oocysts," which they reported to measure

15.3—21.0 × 11.6—13.9. No other information was provided.

Type VI *Eimeria* Herman and Jankiewicz, 1943

Original host: *Sylvilagus audubonii* (Baird, 1858), Desert cottontail.

Remarks: Herman and Jankiewicz (1943) said these oocysts were recovered from the caeca of 3/44 (7%) *S. a. vallicola* on the San Joaquin Experimental Range in California. They described these oocysts as light amber with the end opposite the M to be narrow or pointed. These oocysts measured 33 × 21 (29—37 × 19—22.5), lacked an OR, sporocysts had a distinct, rounded SR and the oocysts sporulated in 48 hr. They said this type VI, "differs from *E. irresidua* of the domestic rabbit in being smaller, slightly narrower, more rapid in its sporulation, and in possessing rounded intra-residual bodies." No other information was provided.

DISCUSSION AND SUMMARY

Coccidia and Coccidiosis in Cottontail Rabbits

Of the 17 eimerians known to be able to infect cottontail rabbits, we know nothing about the ability, or not, of 10 of these to cause pathology because we don't know where they live in the gut (*E. audubonii*, *E. azul*, *E. honessi*, *E. irresidua*, *E. minima*, *E. neoirresidua*, *E. paulistana*, *E. perforans*, *E. pintoensis*, *E. poudrei*). Four species, which have *Sylvilagus* species as their type hosts, can either be pathogenic, even to the point of causing death (*E. environ* and *E. neoleporis*), or non-pathogenic (*E. maior*, *E. sylvilagi*), as far as we know now. Finally, of the three species that have *O. cuniculus* as their type host, but also can infect *Sylvilagus*, one seems to be completely non-pathogenic (*E. media*), while the other two can cause serious pathology and even death (*E. magna*, *E. stiedai*).

Site and Host Specificity in *Sylvilagus*

Site Specificity

Of the 12 eimerians for which *Sylvilagus* species are the type host, we know precious little about site specificity. We know only that *E. environ* and *E. minima* both live "in the small intestine" (Carvalho, 1943), but nothing else. We also know that *E. sylvilagi* localizes in the small intestine (Carvalho, 1943) and, according to Hsu (1970), within the epithelial cells of both the distal and proximal parts of the villi and in the lamina propria of the jejunum and ileum.

We also know that *E. neoleporis* localizes in the ileo-cecal valve and apical process of the cecum in experimentally infected domestic rabbits (Carvalho, 1944). We know nothing about where the other eight species live within the gut of cottontails.

Host Specificity

Metcalf (1929) was among the first to suggest that parasites, especially coccidia like *Eimeria* and *Plasmodium* species, could be a potential tool in helping to assess evolutionary and phylogenetic affinities of their hosts because of their (presumed) relatively high degree of host specificity. For example, Duszynski et al. (1977) used structural characters of oocysts from muskoxen, *Ovibos moschatus*, and suggested eimerian oocysts from muskoxen were more similar to those described from members of the subfamily Caprinae (antelope, goats, ibex, sheep, serow, others) than to those described from other subfamilies of the Bovidae; this put them in agreement with Simpson's (1945) placement of muskoxen within the Caprinae, a view that is still maintained today (Wilson and Reeder, 2005).

Samoil and Samuel (1981) were the first to suggest the use of coccidia as indicators of phylogenetic relationships specifically for members of the order Lagomorpha. Using their review of the literature through 1980, they explored the possibility that the *Eimeria* species found in the lagomorph genera *Lepus*, *Sylvilagus*, and *Oryctolagus* might be useful indicators of the phylogenetic relationships between these host genera. This was long before the development of molecular tools, so at the time there only were three useful characteristics that could be used to distinguish between *Eimeria* species: oocyst morphology, host specificity, and the location, number, structure, and timing of endogenous stages in the definitive host. Since the location of endogenous stages and their timing were known for only a couple of rabbit coccidia at that time, they concentrated their comparative efforts on oocyst morphology and host specificity.

Samoil and Samuel (1981) compared both qualitative (presence/absence) and quantitative (mensural) oocyst and sporocyst characters from 12 *Eimeria* species, each from *Sylvilagus* and *Oryctolagus*, and 19 *Eimeria* species from *Lepus*, and found "no clear-cut characters of oocysts of *Eimeria* that connect one host genus more closely with another ... within the leporids." However, their review of the literature on cross-transmission work between host species in different lagomorph genera, although mixed (and incomplete), was potentially more informative.

Becker (1933) was able to infect *S. floridanus* with oocysts of *E. magna* from the domestic rabbit, *O. cuniculus*; Carvalho (1943) confirmed Becker's work and extended it to other species from the domestic rabbit (*E. irresidua*, *E. media*, *E. perforans*) that he successfully transmitted to *Sylvilagus*. Jankiewicz (1941) experimentally transmitted *E. stiedai* from the domestic rabbit to four young *S. audubonii* and all acquired the parasite, but Carvalho (1943) was unable to confirm this. Herman and Jankiewicz (1943) also said they transferred *E. stiedai* to cottontails. Thus, to date, five eimerians, *E. irresidua*,

E. magna, E. media, E. perforans, and *E. stiedai,* have been transmitted to *Sylvilagus* species from the tame rabbit (*O. cuniculus*), but only one, *E. neoleporis,* has been transmitted from *Sylvilagus* to *O. cuniculus* (Table 7.1). On the other hand, at least seven *Eimeria* species (*E. environ, E. honessi, E. maior, E. minima, E. paulistana, E. pintoensis, E. sylvilagi*) collected from *Sylvilagus* were unable to infect the domestic rabbit, *O. cuniculus.* To our knowledge, there are no reports yet of *Eimeria* species from *Sylvilagus* (for which they are the type host) being successfully transmitted to any *Lepus* species, although a number of attempts have been made (see Samoil and Samuel, 1981, for partial review).

From earlier studies available to them, Samoil and Samuel (1981), and others (Carvalho, 1943, 1944; Pellérdy, 1956), suggested that the genus *Sylvilagus* is more closely related phylogenetically to *Oryctolagus* than either is to *Lepus.* Interestingly, recent molecular work using retroposon insertions have shown that, indeed, *Sylvilagus* and *Oryctolagus* are more closely related to each other than either is to *Lepus,* which appears to be its own monophyletic group (Kriegs et al., 2010). Given the ability of at least one eimerian from *Sylvilagus* to readily infect *Oryctolagus* and at least five species from *Oryctolagus* being able to infect *Sylvilagus,* one wonders why these (and/or other) eimerians are not regularly/naturally found in species of both genera. Carvalho (1943) suggested it may simply be due to lack of contact between the species of both genera.

Prevalence of Eimerians in Cottontails

The coccidia are an extremely ancient lineage of protists, estimated to have diverged about 824 million years ago, long before the divergence of reptiles, birds, and mammals (Escalante and Ayala, 1995). When the progenitors of vertebrates finally arrived on land, they likely brought their coccidian parasites with them; thus, these parasites and their hosts have had

a long time to evolve together. Some vertebrates, like rodents, underwent rapid speciation, and their parasites either kept pace or they disappeared. Other vertebrates, like the lagomorphs, remained morphologically and chromosomally conservative, which may have given time for speciation of their parasites to outpace that of their hosts. Based on numerous surveys conducted with and by his students and colleagues, Duszynski (1986) was the first to notice that coccidia infections in mammals seemed to differ from host group to host group, which allowed him to place hosts into three categories according to the prevalence and intensity of their coccidia infections: (1) hosts with a high prevalence of infection, with most individuals harboring two or more coccidia whenever they were examined; host groups in this category include rabbits and moles; (2) hosts with a low prevalence of infection, with most hosts in a population being uninfected, while only a few of them are found to harbor only one coccidium when examined; host groups in this category include bats, shrews, rodents and carnivores; (3) hosts with intermediate values for both infection rate and single or multiple species infections when individuals in any population were examined; the only mammals placed in this category, to date, are the shrew-mole genera (Duszynski, 1986).

For the purpose of this discussion, we need to concern ourselves only with the lagomorphs that were placed in "Category 1" by Duszynski (1986). Surveys suggest that wild lagomorphs are always heavily infected with multiple *Eimeria* species. Nieschulz (1923), Morgan and Waller (1940), Bull (1953), Ecke and Yeatter (1956), Lechleitner (1959), and many others (see text for all species surveyed) reported 90–100% prevalence with several coccidia species in various lagomorph surveys. Herman and Jankiewicz (1943) reported six eimerian species in 43/44 (98%) cottontails on the San Joaquin experimental range in California, and Duszynski and Marquardt (1969) reported all 100 *S.*

audubonii near Ft. Collins, Colorado, to be infected with one to six eimerians. The reasons for these high infection rates, and usually with multiple species, can only be speculated upon, but some insight may be gained by looking at the evolutionary and ecological history of the host group being examined.

Numerous factors, from genes to host nutritional state, play a role in determining host susceptibility/resistance and all are affected by environmental considerations. "Category 1" hosts (Duszynski, 1986) all are ancient lineages of mammals, tend to be morphologically and chromosomally conservative, show lower heterozygosity than many other groups of mammals (as far as is known), and have not undergone the extensive radiation seen in many other mammalian orders (e.g., bats). They also are either fossorial, semi-fossorial, or, at least, associated with burrows or crevices (Anderson and Jones, 1967) where they defecate. Such habitats provide discharged oocysts with stable, moist environments where they can concentrate, and in time, become viable and readily available to the hosts that occupy them. This, in part, may contribute to the high infection rates seen in hosts such as rabbits and moles. If we consider the potential for increased survival of oocysts discharged mainly in burrows, holes, and crevices as their "external environment," we might predict that the coccidia species of such hosts each need to produce fewer oocysts to reach new hosts; lowered endogenous reproduction could, in turn, put less stress on the host immune system by not stimulating a strong anamnestic response upon subsequent infections. Those coccidia species discharged by hosts occupying more open habitats, where desiccation and exposure to UV radiation are important factors in their survival (or not), would be driven to have endogenous development to produce more oocysts, which, in turn, should stimulate a stronger host immune response to subsequent infections with such species. Because of the primeval nature of the coccidia, the genetically conservative nature of lagomorphs, and the likely association they have shared for millennia, we should expect a high degree of phylogenetic relatedness among these parasite species. Thus, hosts that always have many coccidians may have species that are more closely related to each other than are the coccidia found in hosts that usually have only one coccidian at any point in time. How such relationships contribute to host specificity is unknown.

TABLE 7.1   All Reported Coccidia Species[1] (Apicomplexa: Eimeriidae) from Cottontail Rabbits in the Genus *Sylvilagus* (Family Leporidae)

Sylvilagus spp.	*Eimeria* spp.	Natural (N) or experimental (E)	References
audubonii	*audubonii*	N	Duszynski & Marquardt, 1969
	environ	N	Duszynski & Marquardt, 1969
	honessi	N	Duszynski & Marquardt, 1969
	maior	N	Duszynski & Marquardt, 1969; Herman & Jankiewicz, 1943
	neoirresidua	N	Duszynski & Marquardt, 1969
	neoleporis	N	Herman & Jankiewicz, 1943
	perforans	N	Herman & Jankiewicz, 1943
	poudrei	N	Duszynski & Marquardt, 1969

(Continued)

TABLE 7.1 All Reported Coccidia Species[1] (Apicomplexa: Eimeriidae) from Cottontail Rabbits in the Genus *Sylvilagus* (Family Leporidae) (*cont'd*)

Sylvilagus spp.	*Eimeria* spp.	Natural (N) or experimental (E)	References
	stiedai	E	Jankiewicz, 1941; Herman & Jankiewicz, 1943
brasiliensis	*paulistana*	N	de Fonseca, 1933
	perforans	N	Carini, 1940
	pintoensis	N	de Fonseca, 1932
	sylvilagi	N	Carini, 1940
floridanus	*audubonii*	N	Wiggins et al., 1980
	azul	N	Wiggins & Rothenbacher, 1979; Wiggins et al., 1980
	environ	N/E	Carvahlo, 1943; Dorney, 1962; Ecke & Yeatter, 1956; Wiggins et al., 1980
	exigua	N (?)	Morgan & Waller, 1940
	honessi	N	Carvahlo, 1943; Dorney, 1962; Ecke & Yeatter, 1956; Wiggins et al., 1980
	irresidua	N/E	Carvahlo, 1943; Morgan & Waller, 1940
	leporis	N (?)	Morgan & Waller, 1940; Waller & Morgan, 1941
	magna	N/E	Becker, 1933; Carvahlo, 1943; Morgan & Waller, 1940
	maior	N	Carvahlo, 1943; Dorney, 1962; Wiggins et al., 1980
	media	N/E	Carvahlo, 1943; Dorney, 1962; Ecke & Yeatter, 1956
	minima	N	Carvahlo, 1943; Wiggins et al., 1980
	neoirresidua	N	Wiggins et al., 1980
	neoleporis	N/E	Carvahlo, 1942, 1943, 1944; Dorney, 1962; Ecke & Yeatter, 1956; Wiggins et al., 1980
	perforans	N/E	Carvahlo, 1943; Ecke & Yeatter, 1956; Morgan & Waller, 1940
	pintoensis[2]	N	Ecke & Yeatter, 1956
	stiedai	N/ E	Hsu, 1970; Harkema, 1936
	sylvilagi	N	Carvalho, 1943; Dorney, 1962; Hsu, 1970; Wiggins et al., 1980
nuttallii	*environ*	N	Honess, 1939; Honess & Winter, 1956
	honessi	N	Honess, 1939; Honess & Winter, 1956
	maior	N	Honess, 1939; Honess & Winter, 1956
	stiedai	N	Honess, 1939; Honess & Winter, 1956
Four host species	19 *Eimeria* (17 valid)		

[1] 19 Eimeria *species reported and/or described from the lagomorph genus* Sylvilagus. *We believe that all of these eimerians are valid parasites of this host genus* except *E. exigua and E. leporis (see text).*
[2] E. pintoensis *likely is a valid species of* Sylvilagus, *but Ecke and Yeatter (1956) based their identification of this species in S. floridanus only on unsporulated oocysts!*
(?) not considered a valid parasite of *Sylvilagus*.

Sarcocystidae Poche, 1913, the Predator-Prey Coccidia in Rabbits: *Besnoitia, Sarcocystis, Toxoplasma*

The Biology and Identification of the Coccidia (Apicomplexa) of Rabbits of the World
http://dx.doi.org/10.1016/B978-0-12-397899-8.00008-1

INTRODUCTION

Within the true coccidia (Order Eucoccidiorida Leger and Duboseq, 1910) is a second major family, Sarcocystidae Poche, 1913, that is closely related to the Eimeriidae. All members of this family, however, are **heteroxenous.** That is, two or more hosts must be involved (except for *Toxoplasma*; see below) for transmission to be completed: an **intermediate** (prey) host in which asexual development takes place in the tissues to eventually produce cysts, which then must be ingested by the **definitive** (predator) host where sexual reproduction and oocyst formation occur, usually in the intestinal epithelium. The oocysts are all "*Isospora*-like" in that they contain two sporocysts, each with four sporozoites (Levine, 1988; Dubey et al., 1989). This family is divided into two subfamilies, Sarcocystinae Poche, 1913 and Toxoplasmatinae Biocca, 1957. Members placed in the Sarcocystinae are distinguished by being obligatorily heteroxenous with asexual multiplication within the prey host culminating

with its last-generation meronts (called sarcocysts) that first form **metrocytes**, that give rise to **bradyzoites**, which are infectious for the predator host when the host tissue and sarcocysts are consumed (Levine, 1988). Only sexual stages are found in the intestinal epithelial cells of the definitive host; the oocysts formed after fertilization sporulate within the gut epithelial cells, but have very thin walls, which often rupture during transit through the gut, resulting in sporocysts predominating in the carnivore's fecal material. Interestingly, these "naked" sporocysts closely resemble oocysts of *Cryptosporidium* species (see Chapter 9). A genus of sarcocystinae found reasonably often in rabbits is *Sarcocystis*.

Parasitic protists placed within the Toxoplasmatinae also have obligate heteroxenous life cycles, but the asexual stages usually are transmissible from one intermediate host to another, metrocytes are not formed, there can be both asexual stages (with multiple types of meronts) and sexual stages in the definitive host's gut cells (usually a feline), and oocysts must leave the definitive host before their sporulation occurs (Levine, 1988). Two Toxoplasmatinae genera found with some frequency in rabbits are *Besnoitia* and *Toxoplasma*.

Unlike the format used in prior chapters, in which we listed the information known about each coccidian species under each host and genus of lagomorph, in this chapter we will include what little is known about the predator-prey coccidia by parasite subfamily and genus and then the coccidians known from each rabbit host genus/species from which that parasite has been reported.

SARCOCYSTINAE POCHE, 1913

SARCOCYSTIS LANKESTER, 1882, IN RABBITS

The history of the discovery of *Sarcocystis* was reviewed by Dubey et al. (1989, Chapter 1, pp. 1–2) and later by Duszynski and Upton (2009, Chapter 9, pp. 223–224) and need not be repeated here in detail. Suffice it to say that *Sarcocystis* species are found commonly as white, rice-like cysts (sarcocysts) in the skeletal and cardiac muscles of many vertebrate species (e.g., cattle, sheep, wild mammals like rabbits and mice, etc.) that act as intermediate hosts in which asexual reproduction (merogony) occurs. Between its initial discovery in the house mouse by Miescher (1843) and 1972, numerous "species" were named based only on the structure of the sarcocyst in host muscle tissue. In 1972, however, Fayer published a seminal paper demonstrating the sexual stages of a *Sarcocystis* species in tissue culture to establish the coccidian nature of the parasite that suggested a two-host, predator-prey, life cycle. Several years later, Fayer and Kradel (1977) were among the first to demonstrate that sarcocysts in the muscles of cottontail rabbits would produce sporocyst production in cats, when these were still new and exciting discoveries. To reiterate, after

infected muscle is ingested by a suitable carnivore, merozoites from the tissue cysts are liberated in the gut of the carnivore and enter intestinal epithelial cells to undergo gamogony (gamete formation), which leads to fertilization, oocyst formation, and sporogony within these cells. Some believe that *Sarcocystis* species are more host specific for their intermediate hosts than for their definitive hosts, but other investigators, including some who work with rabbit sarcocystids, point out great similarities in the ultrastructure of sarcocysts found in widely disparate host species, suggesting—at least to them—that the alternative may be true. When the fully sporulated, isospora-like oocysts leave the confines of the intestinal epithelial cells, their very thin walls rupture, releasing fully formed sporocysts into the feces of the carnivore host.

Manz (1867) was the first to note a *Sarcocystis* species in rabbits in southwest Germany. Almost 50 years later, Brumpt (1913) placed all of the sarcosporidia known at the time into the genus *Sarcocystis* Lankester, 1882, and listed 23 species, including 16 in mammals, six in birds, and one in a reptile. In the mammalian sarcocystid "species," he placed his name after the specific epithet for two of them: *Sarcocystis darling* Brumpt, 1913, from an opossum (*Didelphis* sp.) from Panama, and for which he gave minimal measurements (cysts, 1.5–2 mm; merozoites, 8–12 × 2–4 μm) and *Sarcocystis cuniculi* Brumpt, 1913, which he said occurred in the domestic rabbit (*Lepus domesticus* [?]; probably *L. cuniculus* L., 1758, later changed to *Oryctolagus cuniculus* [L., 1758]) and the wild rabbit (*Lepus timidis* L., 1758), but gave no description(s) and, thus, created a *nomem nudum*. Crawley (1914) then described a second species, *S. leporum*, from a "very old male rabbit" shot in Bowie, Maryland; according to Erickson (1946), a letter from E.W. Price of the U.S. Bureau of Animal Industry (Philadelphia) stated that the Maryland material was from *Lepus sylvaticus* (Bachman, 1837), which is now known to be *Sylvilagus floridanus* (J.A. Allen, 1890). In spite of the dubious nature and validity of *S. cuniculi*, both names, *S. cuniculi* for domestic and *S. leporum* for wild rabbit sarcocystids, have been accepted (more-or-less) and used by all subsequent authors except Odening et al. (1996). Fayer and Kradel (1977), who determined the first sarcocystan life cycle in rabbits (*S. leporum*), also made the first argument that the two species are independent; prior to that, earlier studies with rabbits dealt either with the discovery or the morphology of the sarcocyst stage in rabbit muscles (e.g., Vande Vusse, 1965). Odening et al. (1994c, 1996), however, who differentiated the wild European rabbit (*O. cuniculus* agriot.) from the domestic European rabbit (*O. cuniculus* hemerot.), used both light (LM) and transmission electron microscopes (TEM) to suggest there were six *Sarcocystis* species identifiable by the morphology of their sarcocysts in the European hare (*L. europaeus*). They said, "because of the distinct ultrastructural differences between *S. cuniculi* and *S. leporum*, it is clear that they are separate species" (Odening et al., 1994), and in 1996 they proposed a new name, *S. cuniculorum*, to replace *S. cuniculi*, because of its long-standing status as a *nomen nudum*. However, although Odening et al. (1996) may be correct in their interpretation of the International Code of Zoological Nomenclature (ICZN) that a *nomen nudum* is not an available name since it has no description tied to it, we find it unnecessary to clutter (and perhaps confuse) the literature with a second name for one that has been used by all other authors since 1913, who have just accepted *S. cuniculi* as the name for the species in *O. cuniculus*. In addition, Tadros and Laarman (1977a, b) finally attached a description to this name almost 20 years before Odening et al. (1996) wanted to change it.

In 1980 there were 93 named species of *Sarcocystis* and in 2011 there were about 130 named species. Revision of the taxonomy of this genus is ongoing, and it is possible that the currently recognized species may be a much smaller number that can infect multiple intermediate

hosts. For the most updated list, which is out of date, see Levine and Tadros (1980).

FAMILY OCHOTONIDAE THOMAS, 1897

HOST GENUS OCHOTONA LINK, 1795

Sarcocystis dogeli (Matschoulsky, 1947b) Tadros, 1980 and Levine

Definitive type host: Unknown.

Type locality: ASIA: Russia: Siberia, near northern shore of Lake Baikal.

Other definitive hosts: Unknown.

Intermediate type host: *Ochotona dauurica* (Pallas, 1776), Daurian pika.

Other intermediate hosts: Unknown.

Geographic distribution: ASIA: Russia.

Description of sporulated oocyst: Unknown.

Description of sporocyst and sporozoites: Unknown.

Prevalence: Not given.

Sporulation: Unknown, but likely endogenous with infective sporocysts shed in the feces of the definitive host.

Prepatent and patent periods: Unknown.

Site of infection, definitive host: Unknown, presumably the epithelium of the small intestine.

Site of infection, intermediate host: Skeletal muscles.

Endogenous stages, definitive host: Unknown.

Endogenous stages, intermediate host: Sarcocysts are bag-shaped, 0.8–1.2 × 0.4–0.5 mm.

Cross-transmission: None to date.

Pathology: Unknown.

Material deposited: None.

Remarks: Levine and Tadros listed this as one of 93 named species of *Sarcocystis* which they compiled in a paper in 1980. Matschoulsky's (1947b) paper was not available to us, so the information on this species was taken from Odening et al. (1994c).

Sarcocystis galuzoi Levit, Orlov and Dymkova, 1984

Definitive type host: Unknown.
Type locality: ASIA: Kazakhstan.
Other definitive hosts: Unknown.
Intermediate type host: *Ochotona alpina* (Pallas, 1773), Alpine pika.
Other intermediate hosts: Unknown.
Geographic distribution: ASIA: Kazakhstan.
Description of sporulated oocyst: Unknown.
Description of sporocyst and sporozoites: Unknown.
Prevalence: Not given.
Sporulation: Unknown, but likely endogenous with infective sporocysts shed in the feces of the definitive host.
Prepatent and patent periods: Unknown.
Site of infection, definitive host: Unknown, presumably the epithelium of the small intestine.
Site of infection, intermediate host: Cells of muscle connective tissue.
Endogenous stages, definitive host: Unknown.
Endogenous stages, intermediate host: Sarcocysts are ovoidal, 0.5–2.0 mm long with a smooth wall that is ~3–5 μm thick.
Cross-transmission: None to date.
Pathology: Unknown.
Material deposited: None.
Remarks: Odening et al. (1994c) noted that the very modest description offered by Levit et al. (1984) didn't correspond to any other sarcocysts described previously from lagomorphs. The validity of this species needs to be clarified.

FAMILY LEPORIDAE G. FISCHER, 1817

HOST GENUS *LEPUS* LINNAEUS, 1758

**Sarcocystis germaniaensis sp. n.

Definitive type host: Unknown.
Type locality: EUROPE: Germany.

Other definitive hosts: Unknown.
Intermediate type host: *Lepus* europaeus Pallas, 1778, European hare.
Other intermediate hosts: Unknown.
Geographic distribution: EUROPE: Germany, Poland.
Description of sporulated oocyst: Unknown.
Description of sporocyst and sporozoites: Unknown.
Prevalence: Found in 26/205 (13%) of the type host in Germany (Witzmann, 1982; Witzmann et al., 1983), but the prevalence varied in four hunting areas from 2% to 22%; 11/219 (5%) in hares in Poland (Odening et al., 1996).
Sporulation: Unknown, but likely endogenous with infective sporocysts shed in the feces of the definitive host.
Prepatent and patent periods: Unknown.
Site of infection, definitive host: Unknown, presumably the epithelium of the small intestine.
Site of infection, intermediate host: Muscle fibers of the masseter (tongue) and thigh.
Endogenous stages, definitive host: Unknown.
Endogenous stages, intermediate host: Sarcocysts are ovoidal to cigar-shaped, with inconspicuous compartmentalization, up to 835 × 138 μm in tissue sections. Cyst wall is thin and smooth, ~1 × 0.5 μm, without villar protrusions, but provided with small invaginations in the region of the primary cyst wall (type 1 of Dubey et al., 1989). Cystozoites are squat, fusiform, small, 5.6 × 1.9 (4–7 × 2.5–2).
Cross-transmission: None to date.
Pathology: Unknown.
Material deposited: Paraffin blocks of fixed sarcocysts in muscle tissue were stored in the Institute for Zoo Biology and Wildlife Research, Berlin, Germany (see Witzmann et al., 1983; Odening et al., 1996).
Remarks: Witzmann et al. (1983) reported a *Sarcocystis* species in *L. europaeus* from Thuringia and Brandenburg, Germany, a paper resulting from thesis work published a year earlier by Witzmann (1982), but they did not name the

species. Paraffin blocks of fixed sarcocysts in muscle tissue were stored in the Institute for Zoo Biology and Wildlife Research, Berlin, Germany, and Odening et al. (1996) re-examined these and described sarcocysts with both LM and TEM, but they also did not name this species. They emphasized how similar the sarcocysts of this species are to those of *S. sebeki* that they (1994a) found in the European badger, as well as to those described in the raccoon in Germany by Stolte et al. (1996). The sarcocysts of *S. sebeki* were said to occur in *Apodemus sylvaticus* and experimentally in *Mus musculus* (Tadros and Laarman, 1976) according to Odening et al. (1996), and this led them to speculate that *S. sebeki* can occur in intermediate hosts other than in the Muridae.

Along with describing the cysts of this species from *L. europaeus*, they also saw what they thought were some cysts of *S. cuniculi*. Odening et al. (1996) attributed the overall low prevalence and intensity of *S. cuniculi* in hares to the argument that the definitive host is the cat and that wild felids are not very prevalent in Poland. With type specimens archived, we believe there is sufficient information available to name this species as new.

Etymology: The specific epithet is taken from the Latinized origin of Germany, Germania and −*ensis* (L., originating from).

HOST GENUS *ORYCTOLAGUS* LILLJEBORG, 1874

Sarcocystis cuniculi (Brumpt, 1913) Tadros and Laarman, 1977a

Synonym: *Sarcocystis cuniculorum* Odening, Wesemeier and Bockhardt, 1996.

Definitive type host: *Felis silvestris* Schreber, 1775 (syn. *Felis catus* L., 1758; *Felis domesticus* Erxleben, 1777), Domestic cat.

Type locality: EUROPE: Germany.

Other definitive hosts: Unknown.

Intermediate type host: *Oryctolagus cuniculus* (L., 1758) (syn. *Lepus cuniculus*), European (domestic) rabbit.

Other intermediate hosts: *Lepus europaeus* Pallas, 1778, European hare; *Lepus timidis* L., 1758, Mountain hare (?).

Geographic distribution: EUROPE: Czech Republic, Germany, Netherlands, Poland; AUSTRALIA: Tasmania; NEW ZEALAND.

Description of sporulated oocyst: L × W: 18.5−19.6 × 12.9−14 (Černá et al., 1981).

Description of sporocyst and sporozoites: 10.5 × 7.9 (12−14.5 × 9−10), 13 × 10 (11.6−14.5 × 8.7−10), or 13−14.4 × 9.2−10.4; L/W ratio 1.4; SR: present; SB, SSB, PSB: all absent.

Prevalence: Munday et al. (1978) found what likely was *S. cuniculi* in 23/109 (21%) *O. cuniculus* in Australia. Collins and Charleston (1979) reported it in16% of rabbits they examined in New Zealand and sarcocysts were found in 6/8 (75%) rabbits in northern Tasmania (Munday et al., 1980). Černá et al. (1981), using serological samples examined by IFAT from rabbits raised on a farm, found that 28/117 (29%) were positive, mostly in serum diluted 1:10. Sarcocysts were reported in 12/65 (18.5%) *O. cunuculis* collected from the area around Bonn, Germany (Elwasila et al., 1984); interestingly, of the German rabbits, 11/29 (38%) were infected when collected from July to September, 1982, but only 1/36 (3%) collected January to March, 1982 was infected. Černá et al. (1981) reported a prevalence of 36%. Odening et al. (1994c) found it in 3/4 (75%) free-ranging wild rabbits that originated from the Research Station of the Polish Hunting Association in Czempiń (near Poznań, Poland) and were collected near the area of the zoo, Tierpark Berlin-Friedrichsfelde.

Sporulation: Endogenous; infective sporocysts shed in the feces of the definitive host.

Prepatent and patent periods: Prepatency was 9−12 days in the cat and patency continued for up to 105 days PI (Levine, pers. com.; Munday et al., 1980; Tadros and Laarman, 1982). Černá et al. (1981) found that kittens fed infected meat discharged sporulated oocysts and sporocysts in their feces 14 days after infection.

Site of infection, definitive host: Gametocytes were found in goblet cells and the subepithelial lamina propria of the small intestine of young kittens, 9–12 hr PI. Černá et al. (1981) reported both non-sporulated and sporulated oocysts in the duodenal and jejunal scrapings at 18 DPI.

Site of infection, intermediate host: Sarcocysts were found mostly in the striated muscles of infected rabbits. Tadros and Laarman (1978) used sarcocysts from the abdominal wall muscles of a wild, naturally infected European rabbit shot in the Netherlands to study its ultrastructure. Černá et al. (1981) found sarcocyst formation mostly in the muscles of the esophagus and diaphragm of rabbits. Munday et al. (1978) said the sarcocysts in rabbits in Australia were in the diaphragm and tongue, but never the hearts of rabbits, and none was macroscopically visible. Odening et al. (1994c) reported cysts in the muscle fibers around the esophagus, tongue, ribs, thorax, diaphragm, thigh, and loin.

Endogenous stages, definitive host: Sporulated sporocysts were found in the lamina propria of kitten intestines, especially the jejunum (Munday et al., 1980).

Endogenous stages, intermediate host: Sarcocysts were elongate, up to several mm long and 0.5–0.9 mm wide, compartmented (Munday et al., 1980), contained many fine projections up to 11 high and tightly packed to form a "luxuriant pile" (Taros and Laarman, 1977b, 1978); the cysts had striated walls, ~8–11 thick. Merozoites in sarcocysts were rounded, 11–16 × 4–6. Munday et al. (1980) said that the sarcocysts of this species were indistinguishable from those of *S. leporum*. They also found that it took 93 or more days PI for the sarcocysts in rabbits to become infective for cats. Under TEM, the sarcocyst wall was seen to consist of long, closely packed slender protrusions. Each protrusion had a central core of about 100 tightly packed **microfilaments** which continued into the ground substance of the sarcocyst wall proper.

Merozoites were contained in compartments whose **trabeculae** could be seen only with TEM. Dubey et al. (1989) separated sarcocysts from various *Sarcocystis* species into 24 distinct structural "types," based on the ultrastructural (TEM) reports published at that time. Sarcocysts of *S. cuniculi* were placed into their type 10 category, consisting of those with tightly packed villar protrusions that are conical or tongue-shaped, with a core containing **electron-dense** granules and microtubules that extend from the villar tips into the granular layer. Elwasila et al. (1984) found muscle sarcocysts that were 1.0–1.5 mm long × 125–200 μm wide. Each sarcocyst was subdivided into typical compartments by **septa** and contained metrocytes and banana-shaped merozoites. The striated wall was up to 10 μm thick. The cysts were bounded by an electron-dense primary cyst wall which bordered the numerous finger-like projections that appeared as striations by LM. The projections contained numerous fibrillar elements, which originated in granular ground substance and extended longitudinally to the tips of the projections. Sarcocysts were divided into compartments in which metrocytes and **merozoites** were located. Metrocytes were few compared to merozoites and were seen at the periphery of the cyst. Merozoites were 10–12 × 2–3 and each was surrounded by a pellicle consisting of three membranes. Odening et al. (1994c), using TEM, showed that cyst walls had tightly packed long, slim finger-like protrusions; their tapering distal ends were rounded on top and were a little narrowed at their base. The villar protrusions mostly had a polygonal outline in cross section, although a few were round in cross section.

Cross-transmission: This species, from the skeletal muscles of a European rabbit shot in the Netherlands, completed its gametogonic development in cats, but not in dogs, foxes, weasels, or kestrels (Tadros and Laarman, 1976, 1977a, b, 1982; Munday et al., 1980). Munday et al. (1980) were unable to transmit *S.*

cuniculi from one rabbit to another by feeding sarcocyst-infected muscles.

Pathology: Černá et al. (1981) infected kittens by feeding them rabbit tissues infected with sarcocysts. One week after infection, all kittens showed symptoms of indigestion.

Material deposited: None.

Remarks: It was found by Tadros and Laarman (1977a) and Collins (1979), and confirmed by Munday et al. (1980), that the cat is the definitive host of this species and the domestic rabbit the intermediate host. The latter authors showed that after dosing rabbits with sporocysts from cats, they developed sarcocysts, and these became infective for cats at not less than 93 DPI. Infected muscle from these experimental rabbits was not infective for other rabbits. They also said that, microscopically, sarcocysts in the European rabbits were morphologically indistinguishable from those in cottontail rabbits, *S. floridanus*, except that sarcocysts in the latter were larger, up to 300 μm wide. Elwasila et al. (1984) compared the structure of *S. cuniculi* from the European rabbit (*O. cuniculus*) and *S. leporum* from the cottontail rabbit (*S. floridanus*) by both LM and TEM and noted "only minor differences in cyst size, dimensions and shape of the cyst wall projections and the amount of ground substance." Clearly, cross-transmission experiments are needed to determine the relationship between all of the *Sarcocystis* species known from rabbits.

Černá and Kvašňovska (1986), in the Czech Republic, reported that a heterologous system using *S. dispersa* from mice was positive in titers 1:10—1:40 in rabbits experimentally infected with sporocysts of *S. cuniculi* and in the titers of 1:10—1:160 in spontaneously infected rabbits. Lukešová et al. (1984) and Donát (1989) also found and described this species from rabbits in the Czech Republic, while Unger (1977) reported sarcocysts that likely represented this species in *O. cuniculus* agriot from Land Brandenburg, Germany. Dubey et al. (1989) placed this species, along with *S. leporum* (below), into

their type 10 ultrastructural wall type. Odening et al. (1994c) noted that *S. cuniculi*, therefore, "unites features type 9 (microtubules reaching into the ground substance) and type 10 (tightly packed villar protrusions)."

HOST GENUS *SYLVILAGUS* GRAY, 1867

Sarcocystis leporum Crawley, 1914

Definitive type host: *Felis silvestris* Schreber, 1775 (syn. *Felis catus* L., 1758; *Felis domesticus* Erxleben, 1777), Domestic cat.

Type locality: NORTH AMERICA: USA: Maryland.

Other definitive host: *Procyon lotor* L., 1758, Raccoon.

Intermediate type host: *Sylvilagus floridanus* (J.A. Allen, 1890), Eastern cottontail.

Other intermediate hosts: *Sylvilagus nuttallii* (Bachmann, 1837), Mountain cottontail; *Sylvilagus palustris* (Bachmann, 1837), Marsh rabbit.

Geographic distribution: NORTH AMERICA: USA: Alabama, Arkansas, Georgia, Illinois, Iowa, Maryland, Michigan, Minnesota, Mississippi, Missouri, New York, North Carolina, Pennsylvania, South Carolina, South Dakota; Virginia; EUROPE: Germany.

Description of sporulated oocyst: Not described, but most likely a typical sarcocystan oocyst with a very thin, membranous wall.

Description of sporocyst and sporozoites: Ellipsoidal, 14 × 9 (13—17 × 9—11), L/W ratio 1.6; SB, SSB, PSB: all absent; SR: present (Fayer and Kradel, 1977); or 13 × 10, L/W ratio 1.3; SB, SSB, PSB: all absent; SR: present (Crum and Prestwood, 1977), both from cats. In the raccoon: 13 × 9, (11—14 × 9—11), L/W ratio 1.4; SB, SSB, PSB: all absent; SR: present (Crum and Prestwood, 1977).

Prevalence: Le Dune (1936) found sarcocysts of this species in 36/204 (18%) cottontails in New York. New York State Annual Reports to the

Legislature by the New York Conservation Department (Section on Pathological Examination of Game) for 1935—36 and 1938—40 reported that sarcocysts were found in 121 of 1,438 (8%) of the cottontails examined. Morgan and Waller (1940) reported this species (which they called *S. cuniculi*) in the femoral and lumbar muscles of 20/210 (9.5%) *S. floridanus* examined from 12 counties in Iowa; Bell and Chalgren (1943) found sarcocysts in 4/45 (9%) cottontails from New York, 1/67 (1%) from Virginia, and 9/65 (14%) from Missouri; Erickson (1946) found sarcocysts in 30/78 (38%) cottontails in South Dakota; Vande Vusse (1965) found this species in the muscles of 42/173 (24%) *S. floridanus* from Iowa; it was found in the muscles of 15/18 (83%) *S. floridanus* near University Park, Pennsylvania (Fayer and Kradel, 1977); in the muscles of 29/260 (11%) *S. floridanus* from eight southeastern states in the USA (Andrews et al., 1980); in the muscles of 100/185 (54%) *S. floridanus* in Pennsylvania (Cosgrove et al., 1982); and Elwasila et al. (1984) found sarcocysts in 8/12 (66%) cottontails collected near Berrien Springs, Michigan.

Sporulation: Endogenous. Sporulated sporocysts were found in the feces of the cat and raccoon.

Prepatent and patent periods: In the cat the prepatent period is 10—25 days (Crum and Prestwood, 1977; Fayer and Krandel, 1977), while in the raccoon it is reported to be 13 days. The patent period in the cat is 3 to > 46 days (Fayer and Krandel, 1977) or up to 69 days (Crum and Prestwood, 1977).

Site of infection, definitive host: Unknown, but presumably in the intestinal epithelium. However, Crum and Prestwood (1977), who did experimental infections of cats, did not see sexual stages of *Sarcocystis* upon histological examination of cat intestine.

Site of infection, intermediate host: In the skeletal muscle of *S. floridanus*. Erickson (1946) said that cysts were most often seen in the muscles of the hind legs, flanks, and loins and Hugghins (1961) wrote that the muscles of *S.*

floridanus in South Dakota "were found to have a profusion of white streaks visible on the surface." Cosgrove et al. (1982) found two of 100 infected cottontails with cysts in esophageal skeletal muscle, but none in cardiac muscles; all 100 rabbits had cysts in skeletal muscles of the fore and hind legs.

Endogenous stages, definitive host: Unknown, but presumably in the intestinal epithelium.

Endogenous stages, intermediate host: According to Crawley (1914), the cysts are short, delicate threads or rods lying in the muscles, ~2 × 0.2—0.25 mm, compartmented, with a striated wall 5—6 thick, that contained a great number of papilliform processes (**cytophaneres**) that project into the cyst from the basement membrane; the inner layer produced septa dividing the lumen of the cyst into compartments within which many banana-shaped merozoites resided, each about 13—15 × 5 µm. The merozoites each had two "clear-cut oval areas," a N in the posterior half and a clear region at the opposite tip. Erickson (1946) said that mature muscle cysts placed between a cover glass and slide were 4—5 × 0.4 mm, in paraffin sections they were 0.1—0.4 mm wide, and when the cyst was crushed the merozoites were banana-shaped, 12.2 × 4 µm. Hugghins (1961) found two heavily infected *S. floridanus* from South Dakota and said that sarcocysts were pointed at each end and measured 4—8 × 0.3—0.4 mm, with crescent-shaped "spores" (= merozoites), about 11 × 4 µm. Crum and Prestwood (1977) said that sarcocysts from *S. floridanus* were 243 µm wide in paraffin tissue sections. Cosgrove et al. (1982) said that most cysts of *S. leporum* were compartmentalized, although immature cysts might not be septate. Elwasila et al. (1984) found spindle-shaped sarcocysts in *S. floridanus* from Bonn, Germany, that measured 1—5 mm × 100—400 µm, metrocytes were ovoidal and bounded by a three-membrane pellicle, and the merozoites were nearly identical to those of *S. cuniculi*, banana-shaped, 8—13 × 2—4, with a broader posterior part containing the N. Sarcocysts of *S. leporum* were placed into the type 10 category by Dubey et al. (1989); that is, those with tightly packed villar protrusions that were conical or tongue shaped, with a core containing electron-dense granules and microtubules that extended from the villar tips into the granular layer.

Cross-transmission: Fayer and Kradel (1977) fed muscle cysts from *Sarcocystis*-infected cottontails, *S. floridanus* from Pennsylvania, to four dogs, *Canis familiaris*, and to seven cats, *F. silvestris*. None of the four dogs ever shed sporocysts in their feces during the 24 days of observation, but 6/7 (86%) cats discharged sporulated oocysts, and the prepatent period ranged from 10 to 25 days after feeding. Five 3–6-mo-old domestic rabbits, *O. cuniculus*, were infected with 200, 1,000, 25,000, 50,000 or 75,000 sporocysts from cats; rabbits with the two smallest doses were killed 34 DPI and those with the three largest doses were killed at 86, 50, and 30 DPI, respectively. None of the five rabbits showed any signs of illness and no lesions were observed in histological sections of their tissues (Fayer and Kradel, 1977). Crum and Prestwood (1977) fed heavily infected rabbit muscle from one *S. floridanus* caught in Virginia to two cats, and sporocysts were detected in their feces from 14 to 69 DPI. They observed that the number of sporocysts shed in cat feces was small compared to how heavily infected the rabbit muscle appeared to be and they speculated that this low production suggested that the domestic cat "may be rather poor hosts for *S. leporum*." At the end of their (1977) paper they appended an addendum stating that since their paper had been accepted for publication they had had the opportunity to conduct an additional experiment. Tissues from another heavily infected cottontail (*S. floridanus*) were fed to a captive raccoon (*P. lotor*), which shed large numbers of sporocysts and had a prepatent period of 14 days.

Pathology: Cosgrove et al. (1982) did not observe tissue reaction around intact sarcocysts, but did report an intense inflammatory response near ruptured cysts in two rabbits. This inflammatory reaction, consisting mostly of **heterophils**, appeared to be a response to a degenerating cyst, possibly as a result of host defense mechanisms against it.

Material deposited: None.

Remarks: Although *S. leporum* was described in 1914, Babudieri (1932) synonymized *S. cuniculi* with it even though *S. cuniculi* was described a year earlier (1913), thus violating the principle of priority of the ICZN. American authors never adopted this synonymy, so the name has remained (e.g., Erickson, 1946; Vande Vusse, 1965; Crum and Prestwood, 1977; Fayer and Kradel, 1977). Crum and Prestwood (1977) suggested that the raccoon was the "normal" host, since they obtained many more sporocysts in it than from the cat, but the possibility must be considered that they were dealing with another species of *Sarcocystis*. In their study, Cosgrove et al. (1982) compared infection rates of juvenile (21%) vs. adult (69%) *S. floridanus* and concluded that adults have a much higher prevalence of *Sarcocystis*.

SARCOCYSTIS SPECIES INQUIRENDAE (13)

FAMILY OCHOTONIDAE THOMAS, 1897

HOST GENUS OCHOTONA LINK, 1795

Sarcocystis sp. of Rausch, 1961

Original host: *Ochotona collaris* (Nelson, 1893) (syn. *Lagomys collaris*), Collared pika.

Remarks: Rausch (1961) collected 73 collared pikas in Alaska from 1950 to 1961. In July, 1954, while working in the Talkeetna Mountains, he watched an ermine carrying a pika. He collected the ermine, an important predator on pikas in south-central Alaska, but did not examine it for intestinal stages of *Sarcocystis*. He did report, however, that "the striated muscles of collared pikas, particularly in the Talkeetna Mountains, often contain *Sarcocystis* sp."

Sarcocystis sp. of Barrett and Worley, 1970

Original host: *Ochotona princeps* (Richardson, 1828) (syn. *Lepus* [*Lagomys*] *princeps*), American pika.

Remarks: Barrett and Worley (1970) found what they said were cysts of a *Sarcocystis* species "in association with the musculature and connective tissue in the peritoneal cavity" of two American pikas, 1/9 (11%) from Park County and 1/45 (2%) from Gallatin County in Montana. The musculature and mesenteries from these pikas were examined grossly and cysts and zoites were measured "microscopically." Mature cysts were reported to be 0.8–0.9 mm long by 0.5–0.7 mm wide with zoites (which they called spores) typically banana-shaped, 12.6 × 3.2 (11–14 × 2.5–5) μm. They made no attempt to transmit the cysts or to further identify this form, but said, "it was similar morphologically to *S. leporum* Crawley, 1914 (= *S. cuniculi* Brumpt, 1913) … reported from the cottontail rabbit in Minnesota by Erickson (1946)."

Sarcocystis sp. of Grundmann and Lombardi, 1976

Original host: *Ochotona princeps* (Richardson, 1828) (syn. *Lepus* [*Lagomys*] *princeps*), American pika.

Remarks: Grundmann and Lombardi (1976) reported "a species of *Sarcocystis*" in 4/12 (33%) American pikas in Utah.

Sarcocystis sp. of Levit, Orlov and Dymkova, 1984

Original host: *Ochotona alpina* (Pallas, 1773), Alpine pika.

Remarks: Levit et al. (1984) found sarcocysts in pikas from Kazakhstan. Sarcocysts were found in muscle cells (presumably striated muscles?) and were reported to be thread-like, 0.9–3.3 mm long with a smooth wall that was 0.5–1.5 μm thick.

Sarcocystis sp. 1 of Fedoseenko, 1986

Original host: *Ochotona alpina* (Pallas, 1773), Alpine pika.

Remarks: Fedoseenko (1986) found cysts in muscle cells (presumably striated muscles?) of pikas from Kazakhstan. He called them "macrocysts" and said they were round, up to 3 mm wide. Odening et al. (1994) noted that the very modest description offered by Fedoseenko (1986) didn't correspond to any other sarcocysts described previously from lagomorphs. However, much more information is needed on this form before a valid name can be applied to it.

Sarcocystis sp. 2 of Fedoseenko, 1986

Original host: *Ochotona alpina* (Pallas, 1773), Alpine pika.

Remarks: Fedoseenko (1986) found cysts in muscle cells (presumably striated muscles?) of pikas from Kazakhstan. He called them "microcysts" and said they were threadlike.

FAMILY LEPORIDAE G. FISCHER, 1817

HOST GENUS *LEPUS* LINNAEUS, 1758

Sarcocystis cf. *bertrami* Doflein, 1901 of Odening, Wesemeier, Pinkowski, Walter, Sedlaczek and Bockhardt, 1994c

Original host: *Lepus europaeus* Pallas, 1778, European hare.

Remarks: Odening et al. (1994c) found sarcocysts in the muscle fibers of the loin of

L. europaeus from the hunting district of Czempiń, near Poznań, Poland. The cyst wall of this form had elongated leaflet-like villar protrusions that arose at irregular and spacious distances from each other from the surface of the cyst wall and were "often clinging to it like scales." The undulating margins of these protrusions were "coarser" than the primary cyst wall because of many invaginations between the protrusions. In the middle of these protrusions was a dense core of microtubules that penetrated deeply into the ground substance in continuation of its longitudinal axis. Coarse **osmiophilic** granules surrounded the bundle of microtubules. The authors felt this form could not be distinguished from *S. bertrami* on the basis of its ultrastructure, stating "It is noteworthy that this species has been found hitherto in equids only (the dog is the definitive host)." They went on to say that the ultrastructure of the form they found in *L. europaeus* corresponded well to those found in horses (Europe, USA), zebras (Africa), and donkeys (Egypt), all of which are synonyms of *S. bertrami* (also see Odening et al., 1995a). These sarcocysts belonged to the type 11 ultrastructure of Dubey et al. (1989), but we agree with Odening et al. (1994c) that additional work must be done to prove that the form found in *L. europaeus* in Poland is the same as the species from the horse.

Sarcocystis cf. *cuniculi* Brumpt, 1913 of Odening, Wesemeier, Pinkowski, Walter, Sedlaczek and Bockhardt, 1994c

Original host: *Lepus europaeus* Pallas, 1778, European hare.

Remarks: Odening et al. (1994c) found sarcocysts in the muscle fibers of the diaphragm, thigh, and loin of *L. europaeus* from the hunting district of Czempiń, near Poznań, Poland. They described the cyst wall with tightly packed, long, slim-looking finger-like villar protrusions. They showed that these protrusions had a slightly tapering distal portion with a rounded top and were a little narrowed at their base. These protrusions were always round to oval in ultrathin cross section. Microtubules in their core ran from the top into the ground substance of the cyst wall, where they changed their direction, running a short distance parallel to the cyst wall. They (1994c) decided to call the form from the hare *S.* cf. *cuniculi* because the cross sections of the villar protrusions were ultrastructurally different (primarily round) than those of *S. cuniculi* (primarily polygonal), although they admitted that those from *O. cuniculus* "sometimes had a transitional appearance." Thus, they thought it "most likely that the form from the hare is identical with *S. cuniculi*," with the slight differences, perhaps, being a form of host modification. However, the identity of the forms in *Oryctolagus* and *Lepus* should be corroborated by further study.

Sarcocystis sp. 1 of Odening, Wesemeier, Pinkowski, Walter, Sedlaczek and Bockhardt, 1994c

Original host: *Lepus europaeus* Pallas, 1778, European hare.

Remarks: Odening et al. (1994c) found sarcocysts in the muscle fibers of the heart and thighs of *L. europaeus* from the hunting district of Czempiń, near Poznań, Poland. The cyst wall of this form was of irregular thickness with "unstable hair-like villar protrusions occurring singularly or sometimes in more or less closely packed groups" that tapered continuously to their top. Inside these protrusions were numerous fine and a few large granules. In some cases, they noticed that these protrusions arose from dome-shaped bases on the primary cyst wall. In their argument to call this a separate species of *Sarcocystis*, but not name it, they said they couldn't proceed "on the assumption that all the *Sarcocystis* species found in hares are specific parasites of lagomorphs or even merely of *L. europaeus*." To support their statement, they argued that of the four different *Sarcocystis* species found in badgers (*Meles*), only one was believed to be specific for the badger, while two others are believed to be species from roe deer (*Capreolus*), citing their own work (Odening et al., 1994a, b, 1995 a, b) to substantiate their claim. Thus, they suggested this species from *L. europaeus* resembled *S. capreolicanis* (roe deer), *S. cruzi* (cattle, other bovids), *S. arieticanis* (sheep), *S. hircicanis* (goat), and *S.* sp. Yamada et al. (1993, from the horse in Japan). Finally, they noted that the cyst wall of this form belongs to the type 7 ultrastructure of Dubey et al. (1989), all of which are characterized by hair-like villar protrusions of the cyst wall.

Sarcocystis sp. 2 of Odening, Wesemeier, Pinkowski, Walter, Sedlaczek and Bockhardt, 1994c

Original host: *Lepus europaeus* Pallas, 1778, European hare.

Remarks: Odening et al. (1994c) found sarcocysts in the muscle fibers around the esophagus and trachea of *L. europaeus* from the hunting district of Czempiń, near Poznań, Poland. The cyst wall of this form had irregularly-shaped, highly ramified, cauliflower-like villar protrusions with undulated margins; these arose from the surface of the cyst wall and had a relatively thin and short stalk. Microtubules in the core of the villar protrusions appeared to be "interrupted" in places and had obvious rows of large granules. They argued that the ultrastructure of the cyst wall of this form had not yet been described in rabbit sarcocysts, although it had a "remote resemblance" to the type 21 ultrastructure of Dubey et al. (1989). Odening et al. (1996) said that the ultrastructure of the sarcocysts is similar to that of *S. hirsata* from cattle and other Bovinae (cf. Odening et al., 1995a). Given the striking difference in ultrastructure of the sarcocysts, it is not clear why they chose not to formally name this species.

Sarcocystis sp. 3 of Odening, Wesemeier, and Bockhardt, 1996

Original host: *Lepus europaeus* Pallas, 1778, European hare.

Remarks: Odening et al. (1996) found only one sarcocyst in the muscle fibers of the tongue. In tissue sections it was 385 × 135 μm with a cyst wall that had "irregularly stub-shaped villar protrusions," ~2.4 × 3.6 μm. The cyst contained fusiform cystozoites, 10.3 × 2.8 (9–11.5 × 2.5–3). They said that this cyst was similar to a *Sarcocystis* species from the Mongolian gazelle (*Procapra gutturosa*) in Mongolia, that they had reported in an abstract at the Eighth International Congress of Parasitology (1994, p. 252) held 10–14 October in Izmir-Turkey.

HOST GENUS *SYLVILAGUS* GRAY, 1867

Sarcocystis sp. of Stiles, 1894

Original host: *Sylvilagus floridanus* (J.A. Allen, 1890) (syn. *Lepus sylvaticus*), Eastern cottontail).

Remarks: Stiles (1894) reported a *Sarcocystis* species in a cottontail from Maryland, but did not describe or name it.

Sarcocystis sp. of Bell and Chalgren, 1943

Original host: *Sylvilagus* sp. (presumably, *Sylvilagus floridanus* (J.A. Allen, 1890), Eastern cottontail).

Remarks: Bell and Chalgren (1943) found sarcocysts in 8/45 (18%) cottontails from upstate New York State, in 1/67 (1.5%) cottontails from Virginia, and in 9/65 (14%) rabbits from Missouri. They noted that "the muscles of a few rabbits were heavily infected by *Sarcocystis*, but in no case could we be sure that the infection had been inimical to the health of the animals."

TOXOPLASMATINAE BIOCCA, 1957

BESNOITIA HENRY, 1913, IN RABBITS

Besnoitia is one of the more obscure and little-studied parasitic protist genera, and because of this, we know very little about it. Although it has been known and studied for 100 years, our knowledge of the life history of *Besnoitia* species in wild animals is still incomplete, and there are no definitive morphological criteria to distinguish between *Besnoitia* species. We do know that members of this genus are cyst-forming coccidia with a complex two-host (predator-prey) life cycle, similar to those of *Sarcocystis* and *Toxoplasma*, but with their own unique nuances. Intermediate hosts ingest sporulated oocysts when they forage and these oocysts release sporozoites in their gut, which penetrate various host cells and undergo merogony. First-generation meronts develop in the endothelial cells of blood vessels, while later generations develop primarily in **fibroblasts** of various organs and tissues, producing large (up to 1 mm) intracellular cysts (= pseudocysts), without subdivisions, that are surrounded by a thick, laminated, nucleated cyst wall enclosing numerous small bradyzoites. Thus, the most distinguishing morphological characteristic of this genus is the large bradyzoite-filled cyst that occurs in the connective tissues of the herbivorous intermediate hosts. When cyst-containing tissue is eaten by the definitive carnivore host, the bradyzoites are released, penetrate enterocytes of their intestinal tract and undergo (presumably) only one merogonous generation, which leads to gamete formation that gives rise to oocysts after fertilization occurs. Interestingly, the genus *Besnoitia* seems to be unique in at least two respects: (1) the ability of oocysts to initiate gametogenesis in the definitive host has been lost, the completion of the sexual cycle being dependent on the ingestion of tissue cysts from suitable intermediate hosts, and (2) these species may be successfully propagated asexually by mechanical transmission by blood-sucking arthropods.

Other than these features, little else is known about the life history, epidemiology or ecology of the species in this genus (Leighton and Gajadhar, 2001). We do know that the disease besnoitiosis, caused by *Besnoitia* species, has been recognized only in intermediate hosts and the prevalence of infection is high in some host populations where clinical disease can and does occur; for example, besnoitiosis in cattle is of considerable economic importance in some parts of the world (e.g., Africa), but the relationship of these parasites to the population dynamics and health of other herbivorous hosts (both wild and domesticated) is not known (Leighton and Gajadhar, 2001). Probably the least distinguishing morphological characteristic of *Besnoitia* species is the structure of their sporulated oocysts, which have been documented on only a handful of occasions. They are similar to those of *Toxoplasma* by being small (< 20 µm), having two sporocysts each with four sporozoites, and being shed in the feces in an unsporulated condition. Levine (1988) listed what he considered to be seven valid species of *Besnoitia*: *B. bennetti*, *B. besnoiti*, *B. darlingi*, *B. jellisoni*, *B. paraguayensis*, *B. tarandi*, and *B. wallacei*, while Leighton and Gajadhjar (2001) listed seven named species, but added *B. caprae*, while omitting *B. paraguayensis*. Unfortunately, the primary basis for distinguishing between species of *Besnoitia* is the structure of the tissue stage(s) and the intermediate host(s) in which they are found. No molecular work yet has been done to help distinguish between these species; thus, all taxonomy within the genus should be considered provisional until some gene sequences can be used to more rigorously clarify their specific status. *Besnoitia* species from cattle have been transmitted to the European rabbit, *O. cuniculus*, on a number of occasions, but have only been isolated from naturally infected rabbits in rabbitries a few times.

FAMILY OCHOTONIDAE THOMAS, 1897

HOST GENUS OCHOTONA LINK, 1795

As far as we know, there are no *Besnoitia* species reported to occur naturally or from experimental infections in any *Ochotona* species to date.

FAMILY LEPORIDAE G. FISCHER, 1817

HOST GENUS *LEPUS* LINNAEUS, 1758

As far as we know, there are no *Besnoitia* species reported to occur naturally or from experimental infections in any *Lepus* species to date.

HOST GENUS ORYCTOLAGUS LILLJEBORG, 1874

To our knowledge, only the domestic rabbit has been implicated occasionally as a host that is naturally infected with *Besnoitia* species; in most instances, *O. cuniculus* has been an experimental host to which various species from cattle have been transmitted. First, we believe it is useful here to review some of the history of the discovery of *Besnoitia*, then list those "species" that have been experimentally transmitted to domestic rabbits, and finally cite the few cases in which a *Besnoitia* species has been isolated from naturally infected rabbits.

History of Discovery

Besnoit and Robin (1912) first recognized what they thought were a large number of *Sarcocystis*-like cysts in the connective tissue of the skin of cows from the Pyrenees Mountains, France, which they called *Sarcocystis* sp. Marotel (1912) recognized that nothing similar had been found previously in cattle and thought that the senior author should be recognized for his discovery, so he proposed the combination *Sarcocystis besnoiti* for the cause of these cysts, and Henry (1913) thought this organism was better placed in a new genus and, thus, *B. besnoiti* became the type species. It is known to parasitize the skin and cornea of cattle in Europe and Kazakhstan (Peteshev et al., 1974, citing the work of other authors) and cattle, impala, wildebeest, and sheep in Africa (Pols, 1960; Neuman, 1962; Tadros and Laarman, 1976; others).

In a seemingly unrelated study at the time, Wallace and Frenkel (1975) discovered some small oocysts in the feces of domestic cats in Hawaii during a survey for *Toxoplasma*. The following year, Tadros and Laarman (1976) named this organism *Isospora wallacei* in honor of Dr. Gordon Wallace; the name was later corrected to *Besnoitia wallacei* by Dubey (1977a, b) and by Frenkel (1977). During their original discovery and experimental work, Wallace and Frenkel (1975) learned that sporulated oocysts were not infectious to cats, because they saw no oocyst shedding over a period of 42–56 days PI, and Frenkel (1977) also could not produce patent infections in four cats fed sporulated oocysts. Similarly, Mason (1980) inoculated two cats with 500,000 sporulated oocysts and monitored their feces for 64 days PI. Neither cat developed a patent infection and sporulated oocysts administered *per os* to four non-immune cats failed to produce oocyst shedding in 45–56 days of observation PI, reinforcing the supposition that cats do not become infected by ingesting oocysts, only by ingesting tissue cysts from an intermediate host (Tadros and Laarman, 1976).

To date, both of these *Besnoitia* species have been transmitted experimentally to rabbits.

Experimental Transmission of *Besnoitia* to rabbits

Besnoitia besnoiti (Marotel, 1912) Henry, 1913

Synonyms: *Sarcocystis besnoiti* Marotel, 1912; *Gastrocystis robini* Brumpt, 1913; *Gastrocystis besnoiti*; (Marotel, 1912) Brumpt, 1913; *Globidium besnoiti* (Marotel, 1912) Wenyon, 1926; *Isospora besnoiti* (Marotel, 1912) Tadros and Larman, 1976.

Intermediate hosts: *Oryctolagus cuniculus* (L., 1758) (syn. *Lepus cuniculus*), European (domestic) rabbit (experimental), and many others.

Remarks: Pols (1954a) passed infected cattle blood to mice, rats, guinea pigs, and rabbits and noted that the former three species remained perfectly healthy 2 mo PI, although no attempt was made to establish whether or not an "inapparent (sic) form of the disease" was present in them. However, he was able to establish the "high susceptibility" of rabbits to infection by injecting two rabbits with "proliferative forms" of *B. besnoiti* from the blood of infected oxen. Both developed clinical symptoms after 13 and 16 days, and subsequent examination of blood smears from the rabbits revealed small numbers of trophozoites similar to those seen in the cattle; he (1954a) then serially passed the organism to six more generations of rabbits. Pols (1954b) continued his sub-inoculations by intravenous (IV), subcutaneous (SC), and intraperitoneal (IP) routes, infecting 145 rabbits during 19 serial passages, which he noted did not modify the incubation period, "but a more protracted course generally characterized the later passages." In blood, lung, and testis smears, *B. besnoiti* zoites were 5–9 × 2–5, generally ovoidal and slightly pointed at one end. These zoites were seen extracellularly as well as in **monocytes** and sometimes in **neutrophils**. In histological sections,

the initial stages of cyst formation were observed as early as 16–18 days PI and by 9 wk PI the cysts were 80–120 × 50–200, about one-quarter to one-third of those seen in cattle (Pols, 1954b). Summarizing his work with *B. besnoiti* from cattle blood that was serially passed in rabbits, Pols (1960) employed 125 serial passages in 440 rabbits. Of these, 11 died from causes other than besnoitiosis, 352 developed typical reactions accompanied by microscopically demonstrable parasites, 68 only showed hyperthermia and/or subcutaneous swellings, and nine failed to reveal any symptoms. Neuman (1962) transmitted this species to mice following passage in rabbits. Bigalke (1967) transmitted *B. besnoiti* from chronically infected cattle to susceptible cattle and rabbits by inoculation with cystozoites from tissue cysts and also established that its tissue cysts retained their viability up to 9 yrs, implicating chronically infected cattle as important natural reservoirs for arthropod transmission to clean cattle and, perhaps, other susceptible intermediate hosts (rabbits?). Basson et al. (1970) artificially infected rabbits with both bovine and antelope strains of *B. besnoiti*. Peteshev et al. (1974) fed *B. besnoiti* cysts from the skin of naturally infected cattle to domestic dogs, wolves, foxes, ground squirrels, goats, lambs, rabbits, hedgehogs, white mice, rooks, domestic cats, and a wild spotted cat, *Felis lybica*. They reported that the dogs, wolves, foxes, hedgehogs, and rooks did not become infected; the ground squirrel died 10–12 days PI of severe generalized besnoitiosis, and the goats, lambs, and rabbits became febrile 9–13 days PI for a period of 2–9 days. They (1974) detected endozoites of *Besnoitia* in blood smears of the lambs and goats, but not in the rabbits, and found *Besnoitia* cysts at 2 mo PI in one goat and one lamb, and brain cysts 19–35 wide in the mice. Daily examination of the feces of all animals showed that only the wild cat and 2/2 domestic cats shed oocysts in their feces. Neuman et al. (1979) used rabbits, golden hamsters, guinea pigs, and gerbils, in experimental infections with *B. besnoiti* from cattle, and were able to find the parasite in the blood of the rabbits after an incubation period of 12 days. McKenna and Charleston (1980) fed infected muscle, heart, and brain from rats dosed with sporulated oocysts 78 days earlier to two 10-wk-old kittens; both passed unsporulated oocysts 11 days PI. They (1980) then fed ~200,000 sporulated oocysts to two rabbits, and spheroidal *Besnoitia* tissue cysts were found in both rabbits when they were killed 57 and 91 days PI.

Metelkin (1935) was among the first to suggest that wild flies (e.g., *Musca domestica*, *Calliphora erythrocephala*, *Lucilia caesar*, *Cynomyia mortuorum*, *Stomoxys calcitrans*, *Phormia groenlancica*) that were fed on fecal suspensions containing coccidial oocysts might play a role in spreading coccidia in nature from one host to another. In a similar line of reasoning, Bigalke (1960) showed that *B. besnoiti* could be transmitted by the sucking fly, *Glossina brevipalpis*, many tabanids, and *Stomoxys calcitrans*, which he regarded as a viable mode of mechanical transmission. Huan (1968; reference unavailable, from Tadros and Laarman, 1976) showed that cattle in Kazakhstan became infected during winter when flying diptera were completely absent and suggested that ixodid ticks may play a role in its transmission to cattle. Interestingly, none of the other related tissue-cyst coccidians (i.e., *Sarcocystis*, *Toxoplasma*) is known to be capable of having arthropods transmit tissue stages from intermediate host to intermediate host.

In rabbits that received blood from naturally infected cattle, a well-defined **febrile** reaction developed in nearly all experimentally infected rabbits (Pols, 1960), often accompanied by severe swelling of the head and body, and in males scrotal swelling was often the first symptom seen (Pols, 1954b); their fever of 103° to 107°F lasted from 3 to 7 days. During their fever, rabbits became listless, refused to feed, and lost weight rapidly; during the middle of their fever, hot, painful, subcutaneous edematous swellings were evident on the ears, head, limbs, and scrotum of most animals, and some rabbits died within 2–5 days after the initial rise in temperature (Pols, 1954a, b). In some males the swelling progressed to a complete **necrosis** of the scrotum and testes (Pols, 1960). Basson et al. (1970), using rabbits infected with bovine and antelope strains of *B. besnoiti*, demonstrated necrotic and degenerative vascular lesions, mainly of the medium and smaller veins and some arteries during the acute stage of the disease. These lesions coincided with the parasitism of cells in the walls of the vessels where *B. besnoiti* proliferated prior to the development of the cyst stage, and the lesions led to edema, degeneration, and infarction. Kaggwa et al. (1979) infected 3–6-mo-old rabbits with a strain of *B. besnoiti* isolated from cattle in Onderstepoort, Pretoria, South Africa, and allowed infections to progress for 1 yr with the intention of examining the ability of two serological tests, indirect **immunofluorescent antibody technique (IFAT)**, and **enzyme-linked immunosorbent assay (ELISA)**, to diagnose *Besnoitia* infections in chronically infected animals. They found that, in rabbits, *B. besnoiti* produced more acute symptoms than it did in cattle, goats, blue wildebeest, impala, and kudus, which are considered to be the "natural" intermediate hosts in Africa. In their (1979) study 24/26 (92%) infected rabbits showed fever, subcutaneous edema, nasal and **lachrymal** discharge, and 12/24 (50%) symptomatic rabbits died in 15–24 days PI. Without going into detail of how titers fluctuated during the year, the IFAT gave negative results throughout the year of infection when confronted with a **heterologous** antigen (*B. jellisoni*), indicating that it may be a suitable test to use in determining possible latent infections. The ELISA, however, showed a certain degree of cross-reaction using the same heterologous antigen, which allowed them to conclude it was unsuitable in the specific diagnosis of latent infections. Tissues of four of the rabbits that had reacted strongly in their serological tests were used to try to re-isolate *B. besnoiti* by harvesting their organs and sub-inoculating into four non-infected rabbits. None of the four showed any clinical signs of infection, and they concluded that the donor rabbits no longer harbored *B. besnoiti*. Neuman and Nobel (1981) infected 227 rabbits via IP, SC, conjunctival, IV, intracerebral, and/or intratesticular routes using emulsions of *B. besnoiti* tissue cysts collected from naturally infected bulls in an endemic area (Israel?) after slaughter. Clinical signs from these infections included increased temperature, edema of the ears, head, limbs, testes and penis, keratitis, abortion in pregnant females, **epistaxis**, and **paraplegia**, with 80% mortality. Histologically, they (1981) observed exudative dermatitis, an ulcerative **desquamation** of the epithelium, **pyknotic** nuclei, cytoplasmic vesicle formation, and the infiltration of large mononuclear cells. The rabbit may be one of several "natural" intermediate hosts for *B. besnoiti*, harboring tissue cysts that could be transmitted to feral and wild cats when eaten as prey items. This species is generally asymptomatic in the cat definitive host. For additional information and summaries, see Dubey (1976), Tadros and Laarman (1976), Frenkel (1977), Levine (1988), Bigalke and Prozesky (1994), and Leighton and Gajadhar (2001), but, as currently defined, *B. besnoiti* may consist of more than one species (Peteshev et al., 1974; Diesing et al., 1988; Leighton and Gajadhar, 2001).

Besnoitia jellisoni Frenkel 1955a

Synonyms: None.

Intermediate hosts: *Oryctolagus cuniculus* (L., 1758) (syn. *Lepus cuniculus*), European (domestic) rabbit (experimental), and many others.

Remarks: This species was found in the white-footed deer mouse (*Peromyscus maniculatus artemisiae*) near the upper Salmon River in Lemhi County, Idaho, USA, and was able to be experimentally transmitted to many small mammals (Frenkel, 1955a). It also was transmitted to rabbits by Frenkel (1955b, 1965) and others (Bigalke, 1967, 1968; Basson et al., 1970; Ito et al., 1978).

Besnoitia wallacei (Tadros and Laarman, 1976) Dubey, 1977

Synonyms: *Besnoitia* sp. of Wallace and Frenkel, 1975; *Isospora wallacei* of Tadros and Laarman, 1976; non *Isospora wallacei* of Dubey, 1976; *Besnoitia* sp of McKenna and Charleston, 1980; *Isospora bigemina* Large Type of Ito et al., 1978 and of Ito and Shimura, 1986.

Intermediate host: *Oryctolagus cuniculus* (L., 1758) (syn. *Lepus cuniculus*), European (domestic) rabbit (experimental), and many others.

Remarks: Ito et al. (1978) and Ito and Shimura (1986) reported the isolation of the "larger type" oocysts of *Isospora bigemina* from feral cats collected in Ibaraki Prefecture, Honshu, Japan; after looking at biological characteristics that included structure and size of the oocysts and various developmental stages of the life cycle, as well as comparing it to the strain of *B. wallacei* from Hawaii studied by Frenkel (1977), they concluded, "Therefore, the differences between *I. bigemina* large type of cats and *B. wallacei* are no longer recognized." Ito and Shimura (1986) also compared the Hawaii and NIAH strains serologically by agar gel diffusion and direct fluorescent antibody (FA) tests, and immunologically, by cross-immunity infections, and in every instance they were found not to be different from each other.

Much is known about the biology and various stages of this species including: description of the sporulated oocysts (Wallace and Frenkel, 1975; Mason, 1980; McKenna and Charleston, 1980; Ng'ang'a et al., 1994); oocyst discharge by stray/feral cats (Wallace and Frenkel, 1975; McKenna and Charleston, 1980; Ng'ang'a et al., 1994); prepatent and patent periods in cats (Wallace and Frenkel, 1975; Frenkel, 1977; Mason, 1980; McKenna and Charleston, 1980; Ng'ang'a et al.,1994); site of infection in both the definitive (Wallace and

Frenkel, 1975; Tadros and Laarman, 1976; Frenkel, 1977) and some intermediate hosts (Wallace and Frenkel, 1975; McKenna and Charleston, 1980); the endogenous stages in both the definitive (Frenkel, 1977) and some intermediate hosts (Wallace and Frenkel, 1975; Frenkel, 1977; Mason, 1980; Ito and Shimura, 1986); and the pathology produced in some of the intermediate hosts (Mason, 1980; Dubey, 1977b).

Ng'ang'a et al. (1994) orally inoculated 32 mice, 12 rats, and two rabbits with 1×10^6 sporulated oocysts; all animals were killed 78 days PI and tissues were collected from their skin, intestines, liver, kidneys, skeletal muscles, lungs, heart, brain, and spinal cord. All inoculated animals had numerous white, sand-like cysts on the serosal surface of the cecum and ileum, with fewer cysts on the anterior colon, jejunum, duodenum, mesentery, **omentum**, peritoneum, and the heart, and a few cysts were found in the scleral conjunctiva, subcutis, and fascia in the two rabbits. Microscopically, the cysts were spheroidal to ovoidal, found in tissue sections of the heart, kidneys, mesentery, intestines, and lungs of all three groups of mammals. These cysts ($N = 25$) were 208.5 × 164.1 (122−225 × 81−210); the PAS-positive cyst wall was 2.5−25 thick. The host cell N were hypertrophic and hyperplastic and, together with the cytoplasm, were pushed peripherally to form part of the cyst wall, but Ng'ang'a et al. (1994) said the infection in the intermediate hosts provoked little or no inflammatory reaction.

NATURAL INFECTIONS OF *BESNOITIA* IN RABBITS

To date, only two natural infections have been reported in rabbits. Mbuthia et al. (1993) discovered cysts of a *Besnoitia* in the interalveolar lung tissue secretions of a rabbit that had died suddenly in a rabbitry in Kenya. Venturi et al. (2002) found *Besnoitia* tissue cysts in the thorax, abdomen, and subcutaneous tissues of 5/5 (100%) rabbits from a small farm in Argentina. However, without substantive life-history information from either author, these both must be classified as *species inquirendae*.

HOST GENUS *SYLVILAGUS* GRAY, 1867

As far as we know, there are no *Besnoitia* species reported to occur naturally or from experimental infections in any *Sylvilagus* species to date.

BESNOITIA SPECIES INQUIRENDAE (2)

Besnoitia sp. of Mbuthia et al., 1993

Original intermediate type host: *Oryctolagus cuniculus* (L., 1758) (syn. *Lepus cuniculus*), European (domestic) rabbit.

Remarks: Mbuthia et al. (1993) reported what may be the first documented natural infection of a rabbit with *Besnoitia*. An adult female rabbit (presumably *O. cuniculus*) was submitted for necropsy to their veterinary clinic in Nairobi, Kenya, Africa, after it had suddenly died. Upon examination, they found *Besnoitia* cysts in the pulmonary interalveolar tissue that measured 127 × 185 wide and provoked a mild mononuclear inflammatory reaction. The rabbit came from a rabbitry, which at the time was not experiencing any clinical signs of disease in any of its rabbits. The dead rabbit was pale and emaciated with a rough hair coat, the hairs around the eyes, nostrils, and mouth were wet, and there was a generalized hyperemia in all organs, mild subcutaneous edema, and excessive fluid in the serosal cavities. The rabbit apparently died as a result of concurrent acute bronchopneumonia and

pyelonephritis, but probably not from its *Besnoitia* infection.

Besnoitia sp. of Venturini et al., 2002

Original intermediate type host: *Oryctolagus cuniculus* (L., 1758) (syn. *Lepus cuniculus*), European (domestic) rabbit.

Remarks: Venturini et al. (2002) found *Besnoitia* tissue cysts in several tissues of 5/5 (100%) rabbits from a breeder in La Plata, Argentina, South America. The pinhead-sized cysts were found in the fascia and serosal layers of the thorax, abdomen, and subcutaneous tissues, normally on the surface of these tissues. Bradyzoites released from macroscopic tissue cysts from the rabbits were inoculated onto bovine tissue culture monocytes and into interferon gamma gene knockout (KO) mice. *Besnoitia* tachyzoites were seen in the peritoneal exudate of KO mice on day 10 PI and these tachyzoites were infective to other KO mice. Tachyzoites grown in cell culture were infective to four gerbils, *Meriones unguiculatus*, with tachyzoites found in several tissues examined 12–20 days PI and tissue cysts seen in the lungs, heart, intestinal mucosa, kidneys, and skeletal muscle of three of the gerbils examined 33–47 days PI. Theirs was the first report of a *Besnoitia* infection in any host in Argentina.

TOXOPLASMA NICOLLE AND MANCEAUX, 1909, IN RABBITS

Toxoplasma gondii (Nicolle and Manceaux, 1908) Nicolle and Manceaux, 1909, is unique among the parasites of vertebrates because it is ubiquitous; that is, it seems capable of infecting virtually all species of all vertebrate classes, except snakes (see Duszynski and Upton, 2009). Within this myriad of susceptible host species worldwide, it can be found in almost any cell type of any host organ system. Yet,

despite its almost non-existent host specificity, it seldom manifests clinical signs (disease symptoms) in an infected host, unless there are extenuating circumstances (stress, crowding, shipping, etc.) mostly due to domestication, pet trade, and shrinking habitats.

The history, general biology, and life cycle of *Toxoplasma* has been detailed several times (Dubey and Beattie, 1988; Dubey, 2010) and reviewed briefly by Duszynski and Upton (2009). Suffice it to reiterate that only felids can act as definitive hosts in which the sexual stages (gamogony, the formation of male and female gametes) and fertilization occur in their intestinal epithelium, resulting in the production of an unsporulated oocyst that leaves the felid when it defecates. Once outside the host, if appropriate moisture and oxygen are present, sporogony (the formation of sporocysts with infective sporozoites) occurs within the oocyst, at which time it is infective to virtually any vertebrate that may ingest it, including cats (which distinguishes it from sporulated *Besnoitia* oocysts, which are not infective to their feline definitive host). All vertebrates, other than cats, serve as intermediate hosts in which only asexual processes occur (endodyogeny, merogony), which result in tachyzoites (in pseudocysts or clones) or bradyzoites (in tissue cysts). In cats, the oocysts, if ingested, will excyst in the intestine of the cat and go through the sexual process of gamogony and fertilization typical of the definitive host. The infective stages (tachyzoites and bradyzoites) produced by asexual means are infective both to cats and to other vertebrates via carnivorism and cannibalism, but the parasite also can be transmitted transplacentally to fetuses *in utero* if a naive mother becomes infected during pregnancy. This can happen by ingesting sporulated oocysts or infective tachyzoites in tissue cysts with food or water or by blood transfusion or exposure of mucous membranes to blood or other bodily fluids that may be contaminated with various zoites.

Toxoplasma gondii frequently has been found in the domestic rabbit (*O. cuniculus*), cottontail rabbits (*Sylvilagus* spp.), and in hares (*Lepus* spp.), which are only a few of more than 200 known vertebrate intermediate hosts (Dubey and Beatttie, 1998; Dubey, 2010). With the demand for rabbit meat for human consumption increasing globally each year, toxoplasmosis in rabbits could be of epidemiological significance. We provide just a few examples, but for a more thorough analysis see Dubey (2010).

FAMILY OCHOTONIDAE THOMAS, 1897

HOST GENUS OCHOTONA LINK, 1795

We know of no reports of *T. gondii* in any pika species in this genus to date.

FAMILY LEPORIDAE G. FISCHER, 1817

HOST GENUS *LEPUS* LINNAEUS, 1758

Surveying hares (*Lepus* spp.) for *T. gondii* in Europe, Frölich et al. (2003) examined tissues and sera from 321 hunter-killed European brown hares (*L. europaeus*) from Germany. They recorded antibodies by the Sabin Feldman Dye Test (SFDT; cutoff, 1:16) and *T. gondii* antigen by immunohistochemistry (IHC) staining. Antibodies were found in 146/318 (46%) hares and *T. gondii* antigen was demonstrated in 115/201 (57%) hares. Interestingly, they reported no cases of clinical toxoplasmosis. These results (Frölich et al., 2003) are in contrast to the report of Gustafsson and Uggla (1994), who found no antibodies to *T. gondii* in 176 brown hares (*L. europaeus*) from Sweden; this led them to suggest that *L. europaeus* may be highly susceptible to *T. gondii* and, perhaps, the disease is fatal to most of them that become infected. Conversely, it also could suggest that this hare species may exhibit some natural immunity. Using the Ouchterlony agar diffusion test in Bulgaria, Arnaudov et al. (2003) found *T. gondii* antibodies in 9/58 (15.5%) wild hares from that country. Edelhoffer et al. (1989) in Austria, using the IHA test (cutoff, 1:32), reported only 62/3,124 (2%) hares to show antibodies to *T. gondii*. Hejlíček et al. (1997) found SFDT antibodies (cutoff, 1:4) in 7/164 (4%) hares from the Czech Republic.

HOST GENUS *ORYCTOLAGUS* LILLJEBORG, 1874

Almeria et al. (2004) surveyed 456 wild rabbits (*O cuniculus*) from five regions of Spain from 1992–2003 and found 65/456 (14%) had antibodies in their blood by the Modified Agglutination Test (MAT; cutoff, 1:25). Seroprevalence was the highest (14/26, 54%) for rabbits from Catalonia (forests of northeast Spain) and lowest (9/148, 6%) for rabbits from Cadiz and other, drier areas; they also reported that the prevalence in young rabbits (< 7 mo) was similar (11/67, 16%) to that in older (> 7 mo) rabbits (40/363, 11%), indicating to them that congenital infection or early postnatal infection had occurred. Stutzin et al. (1989) found *T. gondii* antibodies using IHA (cutoff, 1:16) in 4/50 (8%) wild rabbits from Chile. Figuerosa-Castillo et al. (2006) detected *T. gondii* antibodies in 77/286 (27%) rabbits from three rabbit farms in Mexico; they reported that seroprevalence was lower (19%) on a well-managed farm, but much higher on two poorly managed farms (40% and 33%). Hejlíček et al. (1981), in the Czech Republic using the SFDT, found antibodies (cutoff, 1:4) in only 8% of the rabbits they surveyed. Pokorný et al. (1981), also in the Czech Republic, examined 423 domestic and free-living animals representing 28 species and found two *O. cuniculus* to be naturally infected via the SFDT (cutoff, 1:4). Hejlíček and Literák (1994), again using the SFDT (cutoff, 1:4), examined 366 rabbits for antibodies from 48 properties in Strakonice, Czech Republic. Antibodies were found in 54/366 (18%) sera samples. Homogenates from portions of diaphragm, brain, and liver (pooled for each rabbit) from 304 rabbits were inoculated IP into two mice each and viable *T. gondii* was isolated from the tissues of 54 (18%) of these rabbits. Interestingly, of the 54 rabbits with demonstrable *T. gondii*, 18 were seronegative by the SFDT. Sroka et al. (2003), using the MAT (cutoff, 1:8,000), found *T. gondii* antibodies in 2/9 (22%) rabbits from a farm in Poland, where cases of human toxoplasmosis were detected. They isolated viable *T. gondii* from the brain of one of two (50%) seropositive rabbits and *T. gondii* tachyzoites were found in tissues of the mice inoculated with brain tissues. This strain was mildly pathogenic to mice. Hughes et al. (2008) detected *T. gondii* DNA in the brains of 39/57 (68%) wild rabbits from the UK. Most recently, Dubey et al. (2011) isolated viable *T. gondii* from domestic *O. cuniculus* in Brazil. Examining the serum and brains, they found antibodies in 2/21 (10%) rabbits by the modified agglutination test (titer 1:25 or higher); also, viable *T. gondii* was isolated from both seropositive rabbits by bioassay in mice from one rabbit and by bioassay in a cat from the other. Their rabbit strain (TgRabbitBr1) was highly virulent for out-bred mice; those fed one infective oocyst died of acute toxoplasmosis. Their rabbit isolate of *T. gondii* was grown in CV1 cell culture and tachyzoite-derived DNA was genotyped using 10 PCR-restriction fragment length polymorphism markers (SAG1, SAG2, SAG3, BTUB, GRA6, c22–8, c29–2, L358, PK1 and Apico). They noted that their rabbit isolates were genetically most similar to isolates from cats from Brazil.

HOST GENUS *SYLVILAGUS* GRAY, 1867

Using the SFDT, Smith and Frenkel (1995) found *T. gondii* antibodies in 3/12 (25%) cottontails (*S. floridanus*) from east central Kansas, USA.

CLINICAL TOXOPLASMOSIS

Dubey and Beattie (1988) cited several reports of epizootics in rabbits and hares (Christiansen and Siim, 1951; Møller, 1958; Anderson and Garry-Anderson, 1963; Harcourt, 1967), mainly from Scandinavia, and these epizootics have been recognized since the early 1950s (Christiansen and Siim, 1951; Gustafsson et al., 1989). In one study, 264/2,812 (9%) hares examined from 1935–1950 died of acute toxoplasmosis (Christiansen and Siim, 1951). Epizootics were highest during winter months (January, February), lowest during summer (June, July), and hares died of hepatitis, pneumonia, and mesenteric lymph node adenitis (Christiansen and Siim, 1951; Møller, 1958; Anderson and Garry-Anderson, 1963; Harcourt, 1967). There also is a report of an outbreak of toxoplasmosis among hares (*Lepus timidus ainu*) in Sapporo, Japan (Shimizu, 1958), but Dubey (2010) said that these epizootics have not been seen in other countries. There are two reports of acute toxoplasmosis involving *O. cunculus* in Georgia, Massachusetts, and Texas, from the USA (Dubey, 2010, Table 18.2, p. 232); the predominant lesions were in the spleen.

Experimentally, both domestic rabbits and hares can be made susceptible to clinical toxoplasmosis, particularly by the oral inoculation of oocysts (Gustafsson et al., 1997; Dubey, 2010). Seven hares and nine rabbits fed 50 sporulated oocysts of *T. gondii* (Swedish isolate TgSwef) were killed 7 or 8 days PI, before they became severely ill. All hares had histological evidence of acute visceral toxoplasmosis characterized by necrosis, whereas lesions in rabbits were mainly inflammatory (Gustafsson et al., 1997). Sedlák et al. (2000) conducted a similar experiment in the Czech Republic; 12 rabbits and 12 hares were fed 10, 1000, or 100,000 *T. gondii* oocysts (four animals for each dose) and observed for 33 days PI. All hares died of acute visceral toxoplasmosis between 8 and 19 days PI; hares fed 10 oocysts died 12–19 days PI. Three rabbits (one in each dose) died of concurrent bacterial infection, but the remainder survived and were found to have tissue cysts. Both hares and rabbits seroconverted between 7 and 12 days PI. Rabbits have been used in the laboratory to study the pathogenesis of toxoplasmic chorioretinitis, because they are highly susceptible to *T. gondii* infection and have eyes big enough for manipulation (Garwig et al., 1998).

TABLE 8.1 Apicomplexa: Sarcocystidae (*Besnoitia, Sarcocystis, Toxoplasma*) Known from all Lagomorpha

Besnoitia, Sarcocystis, Toxoplasma spp.	Definitive host(s)	Intermidiate host(s)	References
B. besnoiti	*Felis silvestris*	*Oryctolagus cuniculus*	Besnoit & Robin, 1912; Marotel, 1912; Henry, 1913; Pols, 1960; Neuman, 1962; Peteshev et al., 1974; Wallace & Frenkel, 1975; Dubey, 1976; Tadros & Laarman, 1976; Frenkel, 1977; others, see text
B. jellisoni	Unknown	*Peromyscus maniculatus* *Ory. cuniculus*	Frenkel, 1955b, 1965; Bigalke, 1967, 1968; Basson et al., 1970; Ito et al., 1978
B. wallacei	*F. silvestris*	*Ory. cuniculus*	Wallace & Frenkel, 1975; Frenkel, 1977; Ito et al., 1978; Mason, 1980; McKenna & Charleston, 1980; Ito & Shimura, 1986; Ng'ang'a et al., 1994
B. sp.	Unknown	*Ory. cuniculus*	Mbuthia et al., 1993
B. sp.	Unknown	*Ory. cuniculus*	Venturini et al., 2002
S. dogeli	Unknown	*Ochotona dauurica*	Matschoulsky, 1947b; Levine and Tadros, 1980; Odening et al., 1994c
S. galuzoi	Unknown	*Och. alpine*	Levit et al., 1984; Odening et al., 1994c
S. germaniaensis	Unknown	*Lepus europaeus*	Witzmann, 1982; Witzmann et al., 1983; Odening et al., 1996
S. cuniculi	*F. silvestris*	*Ory. cuniculus* *L. europaeus* *L. timidis* (?)	Brumpt, 1913; Tadros and Laarman, 1977; Collins, 1979; Munday et al., 1980
S. leporum	*F. silvestris* *Procyon lotor*	*Sylvilagus floridanus* *S. nuttallii* *S. palustris*	Crawley, 1914; Le Dune, 1936; Morgan and Waller, 1940; Bell and Chalgren, 1943; Vande Vusse, 1965; Crum and Prestwood, 1977; Fayer and Krandel, 1977; Cosgrove et al., 1982; Elwasilia et al., 1984

S. sp. cf. *bertrami*	Unknown	*L. europaeus*	Doflein, 1901; Odening et al., 1994c
S. sp. cf. *cuniculi*	Unknown	*L. europaeus*	Brumpt, 1913; Odening et al., 1994c
S. sp.	Unknown	*Och. collaris*	Rausch, 1961
S. sp.	Unknown	*Och. princeps*	Barrett and Worley, 1970
S. sp.	Unknown	*Och. princeps*	Grundmann and Lombardi, 1976
S. sp.	Unknown	*Och. alpine*	Levit et al., 1984
S. sp. #1	Unknown	*Och. alpine*	Fedoseenko, 1986; Odening et al., 1994c
S. sp. #2	Unknown	*Och. alpine*	Fedoseenko, 1986
S. sp. #1	Unknown	*L. europaeus*	Odening et al., 1994c
S. sp. #2	Unknown	*L. europaeus*	Odening et al., 1994c
S. sp. #2	Unknown	*L. europaeus*	Odening et al., 1996
S. sp.	Unknown	*S. floridanus*	Stiles, 1894
S. sp.	Unknown	*S. floridanus* (?)	Bell and Chalgren, 1943
T. gondii	*F. silvestrris*	*Lepus* spp. *Ory. cuniculus* *Sylvilagus* spp.	Numerous authors, see text

Cryptosporidium and Cryptosporidiosis in Rabbits

INTRODUCTION

All known members of the genus *Cryptosporidium* possess features of both the coccidia and the gregarines, and as more evidence accumulates it seems clear that they represent a distantly related lineage of the Apicomplexa (Barta and Thompson, 2006; but also see Bull et al., 1998; Butaeva et al., 2006; Valigurová et al., 2007, 2008). Earlier studies based only upon small-subunit (SSU) rRNA sequences suggested that *Cryptosporidium* was a lineage that emerged very early among the Apicomplexa (Zhu et al., 2000). Later re-evaluations for SSU rRNA, fused SSU/large-subunit (LSU) rRNA, and six protein sequences using phylogenetic reconstruction

methods of distance-based neighbor-joining, maximum-parsimony, and maximum-likelihood methods showed a trend for the very early emergence of *Cryptosporidium* at the base of the Apicomplexa (Zhu et al., 2000); also, Slow-Fast analysis, which focuses upon the slowly evolving positions within sequences, and is useful if a long-branch attraction (LBA) is suspected, showed no role in this placement (Zhu et al., 2000). We know that the eimeriid and sarcocystid coccidia are intracellular and intracytoplasmic, whereas the cryptosporidids are intracellular, but extracytoplasmic; thus, there are distinct biological differences. Dr. G. Zhu (Texas A&M University, pers. comm.) asks the reasonable question: "Are we ready to remove

The Biology and Identification of the Coccidia
(Apicomplexa) of Rabbits of the World
http://dx.doi.org/10.1016/B978-0-12-397899-8.00009-3

the *Cryptosporidium* group out of the coccidia-proper to form a new group (Class?), perhaps on an equal status with the Haematozoa and the Gregarina?" We are convinced that someone needs to take the time and have the courage to make this change; the sooner, the better.

The genus *Cryptosporidium* was first defined by Tyzzer (1907, 1910) when he described the type species, *C. muris*, in the stomach of a mouse; 2 years later (1912) he described *C. parvum* in the mouse small intestine. We now know that members of this protozoan genus infect all classes of vertebrates and at least 260 vertebrate species (Shi et al., 2010). In 2000, Fayer et al. said there were 10 valid species in this genus, but currently 20 valid *Cryptosporidium* species are reported (Robinson and Chalmers, 2009), of which 14 infect mammals, three infect birds, two infect reptiles, and one infects amphibians; valid species names for the cryptosporidia in fish are lacking (Fayer, 2007, 2010). In addition, there are more than 60 *Cryptosporidium* genotypes of uncertain species status (Shi et al., 2010). All species are obligate, intracellular (but extracytoplasmic) protists that undergo endogenous development culminating in the production of an encysted stage discharged in the feces. Thus, transmission is fecal–oral, either through direct contact with infected hosts or through multiple other vehicles that include recreational (lakes, streams, etc.) and drinking water, food, or **fomites** (Casemore, 1990). *Cryptosporidium* species are widespread among vertebrates, causing mainly gastrointestinal disease in mammals and reptiles and enteric, renal, and respiratory disease in birds. Respiratory disease is also very occasionally seen in mammals, including humans. Although infection has been reported in fishes, amphibians, and snakes, disease is not well described in these hosts (Fayer, 2007, 2010; Duszynski and Upton, 2009).

In human cryptosporidiosis, symptoms can last for up to 3 wks, but are usually self-limiting; however, disease can be prolonged and life-threatening for immune-compromised

patients (Hunter and Nichols, 2002). Understanding the hosts, sources of infection, and transmission routes is vital for control of these parasites, for which specific treatment options are limited and vaccines are lacking. Zoonotic transmission has long been recognized; the first human cases were reported in 1976 (Meisel et al., 1976; Nime et al., 1976) and, although the source could not be confirmed, both cases lived on cattle farms. Another 11 cases were reported over the next 6 yr, and the parasite now has been reported from > 90 countries on six continents (Ugar, 1990). By the early 1980s, outbreaks among veterinary students and workers demonstrated the risk to animal handlers (Current et al., 1983; Jokipii et al., 1983; Chalmers and Giles, 2010). Historically, clinical laboratory testing by animal handlers (e.g., veterinarians) is only for the presence/absence of the parasite, and most infections in companion animals and humans were ascribed to *C. parvum*. A legacy of this lack of species differentiation has been an assumption that all animal hosts pose an equal risk to humans. Today, genetic analysis can be undertaken to identify the infecting species and their subtypes, to ascertain parasite–host adaptation, track sources of infection, and evaluate the risk of zoonotic transmission. Fayer (2007, 2010) warned that species validity requires sufficient morphological, host range, and genetic data; genetically distinct isolates lacking sufficient data are referred to as "genotypes" and these are often named for the animal from which they were originally found. Some degree of host adaptation is evident: *Cryptosporidium* species infecting reptiles and amphibians, do not appear to infect mammals or birds.

CRYPTOSPORIDIUM IN RABBITS

It appears that rabbits are susceptible to infection with *Cryptosporidium* "rabbit genotype" (now a junior synonym of *C. cuniculus*; see below), *C. parvum* and *C. meleagridis*. All of these species are human pathogens, and the role of rabbits as a potential source of zoonotic *Cryptosporidium* must be considered as a possible risk to public health. Our current understanding and knowledge of *Cryptosporidium* in rabbits is limited. Since the first mention of *Cryptosporidium* in rabbits by Tyzzer (1912), the parasite has been reported from farm, pet, laboratory, and wild rabbits (Iseki et al., 1989; Zhang et al., 1992; Pavlásek et al., 1996; Mosier et al., 1997; Sturdee et al., 1999; Xiao et al., 2002; Ryan et al., 2003; Shiibashi et al., 2006; Robinson and Chalmers, 2009). Recently, new zoonotic risks seem to have emerged that document at least one outbreak in humans, caused by a *Cryptosporidium* sp. of a rabbit genotype (Robinson and Chalmers, 2009; Chalmers and Giles, 2010). Both rabbits and humans are the only known major hosts of the rabbit genotype, and no minor hosts have yet been identified (see Robinson and Chalmers, 2009, for summary). These studies reinforce the need to characterize infecting and contaminating isolates to ensure appropriate interventions. In particular, it is important to describe and characterize the risks of zoonotic cryptosporidiosis by detailing the hosts acting as a potential reservoir, the risks of transmission to humans, outbreaks in animal-associated settings, and guidance for control. However, the biological mechanisms for host adaptation, pathogenicity, and virulence are not well understood. The prevalence of cryptosporidiosis in rabbits may be much higher than indicated by the number of reports in the literature, and the true prevalence of *Cryptosporidium* in rabbits is affected by several factors, including population density and structure, the age of the host, and exposure to infective oocysts. Many infections may be missed because of a number of factors, including the miniscule size of the parasite, the presence of small numbers of organisms, and possible confusion with *Eimeria* as has been done previously (Tyzzer, 1912; see Ryan et al., 1986).

FAMILY LEPORIDAE G. FISCHER, 1817

HOST GENUS ORYCTOLAGUS LILLJEBORG, 1873

Cryptosporidium cuniculus Inman and Takeuchi, 1979

Synonym: *Cryptosporidium* sp. Tyzzer, 1929; *Cryptosporidium* rabbit genotype.

Type host: *Oryctolagus cuniculus* (L., 1758) (syn. *Lepus cuniculus*), European (domestic) rabbit.

Type locality: NORTH AMERICA: USA: Washington DC.

Other hosts: *Homo sapiens* L., 1758, Humans; *Sylvilagus floridanus* (J.A. Allen, 1890), Eastern cottontail (?).

Geographic distribution: ASIA: China; AUSTRALIA: New South Wales, Victoria; EUROPE: Belgium, England, Czech Republic, Hungary, Netherlands; NEW ZEALAND; NORTH AMERICA: USA: California, Illinois (?), Massachusetts, Texas, Washington DC.

Description of sporulated oocyst: Oocyst shape: subspheroidal to ellipsoidal; number of walls: 1; wall characteristics: clear, thick, and smooth with a suture at one pole; L × W: 6.0 × 5.4 (5.6–6.4 × 5.0–5.9); L/W ratio: 1.1; OR: present; OR characteristics: composed of fine granules and a large spherical globule, ~1.4, in each oocyst; M, PB: both absent. Distinctive features of oocyst: very small, typical cryptosporidial oocyst.

Description of sporozoites: four naked SZ are present in each oocyst.

Prevalence: Inman and Takeuchi (1979) said that they saw several cases at Walter Reed Army Institute of Research, Washington DC, and *C. cuniculus* was found in 2/6 (33%) healthy New Zealand female rabbits from California. Chalmers et al. (2009a, b) isolated "rabbit genotype" profiles (see Robinson et al., 2010, who synonymized this genotype with *C. cuniculus*) from 23/34 (68%) human stool samples that were laboratory-identified cases of cryptosporidiosis in Northamptonshire, England. All case-patients lived in the area affected by a contaminated water supply incident and had onset dates consistent with exposure by drinking water or by person-to-person spread (Chalmers et al., 2009a, b). In a systematic review of the literature using electronic searches of seven online databases (e.g., Pub-Med, Web of Knowledge, etc.) and "gray literature" (i.e., not published in peer-review journals), Robinson and Chalmers (2009) said that *Cryptosporidium* in European rabbits was reported from 13 countries and in 8/11 studies of wild rabbits (see their Table 1); unfortunately, most of the studies were of incidental findings rather than population-based prevalence studies, so they could not be sure of the species or genotype causing most natural infections in rabbits as a result of a lack of molecular characters. Nolan et al. (2010) used molecular tools to detect *C. cuniculus* in rabbits in four locations in Victoria, Australia, by PCR-coupled sequencing and phylogenetic analysis of sequence data for loci within the SSU rRNA of nuclear rRNA and the 60-kDa glycoprotein gene (GP60); they detected *C. cuniculus* in 12/176 (7%) fecal samples. For the SSU rRNA gene, all 12 sequences were identical to each other and to that of *C. cuniculus* (of Robinson et al., 2010) and for the GP60 gene, all sequences matched the known genotype Vb and were classified as subgenotype VbA23R3 ($N =$ 11) and VbA26R4 ($N = 1$). They noted that present evidence indicated that genotype Vb is limited to rabbits. Also see various surveys and prevalence studies under *Cryptosporidium* "rabbit genotype" below (e.g., Shi et al., 2010).

Sporulation: Endogenous.

Prepatent and patent periods: Prepatency is 4–7 days; patency lasts 7 days (Robinson et al., 2010).

Site of infection: Brush border of the villar epithelial cells in the small intestine, mostly the ileum (Inman and Takeuchi, 1979; Robinson et al., 2010), but also in the jejunum (Rehg et al., 1979). Developmental stages were never seen in areas below the base of adjacent microvilli.

Endogenous stages: Meronts contained eight merozoites and a residuum.

Cross-transmission: Experimental infections were established in weanling rabbits, immune-suppressed Mongolian gerbils (*Meriones unguiculatus* Milne-Edwards, 1867), and immune-suppressed Porton-strain laboratory (house) mice (*Mus musculus* L., 1758), but not in neonatal mice.

Pathology: Very little. The infected rabbit from which this species first was described (Inman and Takeuchi, 1979) did not have diarrhea and seemed healthy. However, histologically, infected areas of the ileum had short, blunt villi with a moderate decrease in the villus-crypt ratio, slight edema in the lamina propria, and dilated lacteals. Although many lymphocytes were seen in the epithelium, there was no increase in the number of either acute or chronic inflammatory cells (Inman and Takeuchi, 1979). When examined by both transmission (TEM) and scanning electron microscopy (SEM), the endogenous developmental stages were found closely attached to the host cell surface displacing or replacing the microvilli, which were absent wherever the organisms were located. Some microvilli near the parasites were slender, elongated, and closely apposed to them, while others had a bleb at their tip. Rehg et al. (1979) noted that microvilli were disrupted at the site of the organisms, but adjacent microvilli were either longer or shorter than those of non-infected cells; occasionally, a conical elevation or a depression of the infected host cell surface also was observed at the parasite's attachment site. Robinson et al. (2010) saw no clinical signs in rabbits, immune-suppressed Mongolian gerbils, or Porton-strain mice, although they did say that occasionally they saw infiltrations of eosinophils in the base of the lamina propria and the tips of the villi. A number of other reports commented on the lack of clinical signs and the low numbers of parasites present, particularly in weaned and older rabbits (Inman and Takeuchi, 1979; Rehg et al., 1979; Peeters, 1988; Pavlasek et al., 1996; Cox et al., 2005; Shii-bashi et al., 2006). Experimental infection of 3-day-old rabbits, with a rabbit-derived isolate, caused liquid diarrhea and high mortality, whereas weaned rabbits suffered no mortality and little diarrhea (Peeters, 1988). Similarly, Pavlasek et al. (1996) found that in weaned rabbits > 50 days old, the occurrence of *Cryptosporidium* decreased significantly and that cryptosporidiosis was particularly pathogenic to rabbits between 30 and 40 days old. An interesting suggestion from these latter two sets of authors was that the source of infection in these young rabbits was from their mothers, who seemed to excrete oocysts sporadically, shortly before parturition and several days thereafter (Robinson and Chalmers, 2009).

Material deposited: A phototype, purified oocysts, genomic DNA, paraffin-embedded and fixed tissue on slides are archived in the UK *Cryptosporidium* Reference Unit, Swansea, Wales, UK. GenBank accession numbers for the type strain rabbit isolate (W17211) at SSU rRNA, HSP70, GP60, BIB13, COWP and actin genes, respectively, are: FJ262725, FJ262728, FJ262731, GU327781-GU327783.

Remarks: In various publications, Tyzzer (1907, 1912, 1929) remarked, in passing, that he had found a *Cryptosporidium* species similar to *C. parvum* (of the laboratory mouse) in the rabbit's intestine, but did not discuss it further. It was again found in laboratory rabbits by Inman and Takeuchi (1979), who finally named it, but did so based on morphology (**LM, SEM, TEM**), **epidemiology**, and their belief at the time of strict host-specificity for *Cryptosporidium* species, a concept we now know is not true. Rehg et al. (1979) found *C. cuniculus* in two apparently healthy rabbits from a rabbitry in California. Ryan et al. (1986) found *Cryptosporidium* in a wild cottontail rabbit, *S. floridanus*, found dead in a live trap in Monticello, Illinois, USA, but it is unknown whether this was *C. cuniculus*,

C. parvum, or another (?) "rabbit genotype." Finally, Robinson et al. (2010), based on very small molecular differences combined with more distinct biological differences, settled the issue (for now) by presenting a reasonable case for synonymizing the "rabbit genotype" from previous studies in China, the Czech Republic, and England, with *C. cuniculus* first named by Inman and Takeuchi (1979). The measurements for the oocyst description used above are from Robinson et al. (2010), because Inman and Takeuchi (1979) didn't provide measurements and said only that the oocysts they saw were, "similar to that of *C. wrairi* from the guinea pig," but did not otherwise describe them (also see Jervis et al., 1966; Vetterling et al., 1971). The measurements given by Robinson et al. (2010) are slightly larger than those of *C. parvum* and *C. hominis*, although the shape of the oocysts of all three species is similar. Measurements given by Shi et al. (2010) for oocysts they measured in China of *Cryptosporidium* "rabbit genotype" (now known to be *C. cuniculus*) were 5.1×4.8 ($4.9–5.4 \times 4.5–5.1$) with L/W ratio 1.07 (1.0–1.1), while *C. hominis* oocysts, as reported by Morgan-Ryan et al. (2002), were 5.2×4.9 ($4.4–5.9 \times 4.4–5.4$).

The prevalence of cryptosporidiosis in rabbits may be much higher than indicated by the number of reports in the literature. For example, Robinson and Chalmers (2009) suggested that in rabbits *Cryptosporidium* infection may be as high as 5%, but with the caveat that more large-scale, population-based epidemiological studies are needed. A second caveat (Robinson and Chalmers, 2009) was that not all *Cryptosporidium* detected in rabbits are necessarily *C. cuniculus*, because *C. parvum* and *C. meleagridis* also have been documented to infect rabbits. In fact, Robinson et al. (2010) inferred a close phylogenetic relationship between *C. cuniculus* and *C. parvum* from molecular analyses at the SSU rRNA, 70-kDa heat shock protein (HSP70), actin, *Cryptosporidium* oocyst wall protein (COWP), and 60-kDa glycoprotein (GP60) genes, and at a region encoding a product of unknown function (LIB13). Many infections may be missed because of a number of factors, including the miniscule size of the parasite's oocyst stage, the presence of small numbers of organisms, and possible confusion with *Eimeria*, as had been done by Tyzzer (1912; see Ryan et al., 1986). Both pet and wild rabbits are a potential source of human cryptosporidiosis and, as such, good hygiene is recommended during and after handling rabbits or exposure to their feces.

Cryptosporidium meleagridis Slavin, 1955

Type host: *Meleagris gallopavo* L., 1758, Common turkey.

Type locality: EUROPE: Scotland: Lasswade.

Other hosts: *Coturnix coturnix* L., 1758, Common quail (Chalmers and Giles, 2010, no reference given); *Gallus gallus* L., 1758, Domestic chicken; *Peromiscus* sp., deer mouse (Chalmers and Giles, 2010, no reference given); *Canis familiaris*, Domestic dog (Chalmers and Giles, 2010, no reference given); *Alectoris rufa* L., 1758, Red-legged partridge (Chalmers and Giles, 2010, no reference given); *Psittacula kramerii* Soopoli, 1789, Rose-ringed parakeet (Chalmers and Giles, 2010, no reference given), *Oryctolagus cuniculus* (L., 1758) (syn. *Lepus cuniculus*), European (domestic) rabbit.

Geographic distribution: Cosmopolitan.

Description of sporulated oocyst: Oocysts: ovoidal; L × W: 4.5 × 4.0 and appear indistinguishable from those of *C. parvum*.

Description of sporozoites: Presumably, four SZ occur within each oocyst. Although Slavin (1955) saw enormous numbers of oocysts in smears, SZ could not be identified within them.

Prevalence: Low.

Sporulation: Endogenous (?); however, Slavin (1955) never saw oocysts with four SZ inside them. He explained this as follows: "as the zygote develops to the oocyst stage it quickly loses its hold on the epithelial surface and is swept along with the intestinal contents; hence, not many sporulated forms would be seen in tissue preparations. Sporulation of the oocyst within the host is not prominent in *C. meleagridis* and although enormous numbers of oocysts have been seen in rectal smears no sporulated forms have been found."

Prepatent and patent periods: Unknown.

Site of infection: Developmental stages of a parasite that conformed to those found by Tyzzer (1907) from the small intestine of mice were found on the villus epithelium in the terminal one-third of the small intestine of turkeys in Scotland, and the parasite did not appear to invade host tissues (Slavin, 1955).

Endogenous stages: Based on dried smears stained by the MacNeal-modified Romanowsky method, the cytology and measurements of merozoites, trophozoites, meronts, gametes, and oocysts were done and the organism was given this name as a new species (Slavin, 1955). Merozoites were slender, 5 × 1.1, and tapered toward the ends, with one end blunter than the other; they had an ovoidal N that was sub-terminal. Young meronts (which Slavin, 1955, called trophozoites) were ovoidal, ~4—5 wide, and had an eccentric ovoidal N. They attached to the villi of the intestinal epithelium in enormous numbers, almost to form a "covering sheath." Slavin (1955) also said that these trophozoites "have been seen in the goblet cells and between cells as far down as the basement membrane." Meronts were ovoidal, 4 × 5, with eight evenly dispersed small chromatin bodies that later formed merozoites. Macrogametocytes were roughly ovoidal structures, 4.5—5.0 × 3.5—4.0, with an eccentric N. Microgametocytes were rounded or ovoidal and ~4.0 wide at their greatest diameter, with 16 N around the periphery. Macrogametocytes were seen on the surface of epithelial cells, deep in the mucosa, within the N of epithelial cells, and seemed to emerge from the N. Microgametes were small, solid chromatin rods, 1.0 × 0.3, with rounded ends, but no flagella were seen (Slavin, 1955). Slavin (1955) wrote about the juxtaposition of male and female elements, called syzygy, and said it "is characteristic of the *Cryptosporidium* described here." He went on to report that it was common for him to find developing microgametocytes and macrogametocytes "close together in twos or in fours … and to find a microgametocyte beside a fertilized macrogamete. The latter have not been seen with more than one microgamete attached." This allowed him to conclude that microgametes of *C. meleagridis* are most likely nonmotile and that "their unvarying shape confirms this."

Cross-transmission: No attempts were made by Slavin (1955) to transmit this parasite to other hosts, although subsequent studies have demonstrated that turkeys and chickens were susceptible to infection after oral inoculation with *C. meleagridis* oocysts (Lindsay and Blagburn, 1990). This species also was shown to infect neonate rabbits (Mosier et al., 1997; Darabus and Olariu, 2003).

Pathology: Illness with diarrhea and a low death rate in 10–14-day-old turkey poults was associated with the parasite (Slavin, 1955). There also is evidence for association of this species with diarrhea in children in a shanty town in Peru (Cama et al., 2008).

Materials deposited: None.

Remarks: Lindsay et al. (1989) gave measurements of 5.2 × 4.6 (4.5–6.0 × 4.2–5.3) for viable oocysts from turkey feces. Interestingly, this intestinal species is the only bird species known to infect many mammalian hosts, including humans and rabbits. Molecular analysis of a turkey isolate in North Carolina, USA, and a parrot isolate in Australia at the SSU rRNA, HSP70, COWP, and actin loci demonstrated the genetic uniqueness of *C. meleagridis* (Xiao et al., 1999, 2000; Morgan et al., 2000; Sulaiman et al., 2000, 2002). Viable oocysts measured 5.1 by 4.5 (Morgan et al., 2000). When the morphology, host specificity, and organ location of *C. meleagridis* from a turkey in Hungary were compared with those of a *C. parvum* isolate, phenotypic differences were small, but statistically significant (Sreter et al., 2000). Oocysts of *C. meleagridis* were successfully transmitted from turkeys to immune-suppressed mice and from mice to chickens. Sequence data for the SSU rRNA gene of *C. meleagridis* isolated from turkeys in Hungary were found to be identical to the sequence of a *C. meleagridis* isolate from North Carolina. Even though it has been suggested *C. meleagridis* may be *C. parvum* (Champliaud et al., 1998; Sreter and Varga, 2000), results of molecular and biological studies have confirmed that *C. meleagridis* is a distinct species (Xiao et al., 1999, 2000, 2002; Morgan et al., 2000, 2001; Sreter et al., 2000; Sulaiman et al., 2000, 2002). *Cryptosporidium meleagridis* is known to infect other avian hosts (for example, parrots), not just turkeys (Morgan et al., 2000, 2001). It also is the third most common *Cryptosporidium* parasite in humans (Pedraza-Diaz et al., 2000; Xiao et al., 2001). Several subtypes of *C. meleagridis* have been described based on multilocus analysis (Glaberman et al., 2001).

Cryptosporidium parvum Tyzzer, 1912

Type host: *Mus musculus* (L., 1758), House mouse.

Type locality: NORTH AMERICA: USA: Massachusetts, Cambridge.

Other hosts: > 150 known mammalian species (Robinson and Chalmers, 2009) including *Homo sapiens* L., 1758, Humans; ruminants including cattle, sheep, goats; laboratory rodents, etc.

Geographic distribution: Cosmopolitan.

Description of sporulated oocyst: Oocyst shape: spheroidal; number of walls: 1; wall characteristics: smooth, with a suture along one side; L × W: 5.2 × 4.6 (5–6 × 4–5); L/W ratio: 1.1–1.2; M: absent; OR: present; OR characteristics: spheroidal; PG: absent. Distinctive features of oocyst: very small size and lack of sporocysts.

Description of sporozoites: four SZ occur within the oocysts.

Prevalence: Highly variable and dependent upon the host and the circumstances of infection. For example, cryptosporidiosis may occur sporadically or as outbreaks following zoonotic transmission from farm animals, person-to-person spread, or the contamination of water supplies (Casemore, 1991). Clinical symptoms following the consumption of contaminated water may affect hundreds or even thousands of consumers, as outbreaks in Devon, UK (Anonymous, 1995, 2008) and Milwaukee, USA (MacKenzie et al., 1994) have shown.

Sporulation: Endogenous.

Prepatent and patent periods: Prepatent period is generally 4 days and patency generally lasts 7—10 days in immune-competent hosts, but can become extensively prolonged in immune-suppressed animals.

Site of infection: Preferred site of infection is the ileum, although other sites also can be colonized. Each generation can develop and mature in as little as 14—16 hr. Due to the rapidity of the life cycle and to the auto-infective cycles, huge numbers of organisms can colonize the intestinal tract in several days. The ileum soon becomes crowded, and secondary sites such as the duodenum and large intestine are often infected. In immune-suppressed individuals, parasites sometimes are found in the stomach, biliary and pancreatic ducts, and respiratory tract.

Endogenous stages: Merogony resulted in the formation of eight merozoites within the meront; these are termed Type I meronts. The merozoites penetrate new cells and undergo another merogony. Type I merozoites are thought to be capable of recycling indefinitely; thus, the potential exists for new Type I meronts to arise continuously (Duszynski and Upton, 2001). Some Type I merozoites were triggered into forming a Type II meront, which contained only four merozoites. When liberated, Type II merozoites appeared to form the sexual stages. Some Type II merozoites entered cells, enlarged into the macrogametocyte, and then formed macrogametes. Others entered cells to become microgametocytes that undergo multiple fission to form 16 non-flagellated microgametes. Microgametes ruptured from the microgametocyte and penetrated macrogametes, forming a zygote. Sporogony occurred, resulting in the production of four SZ; thus, sporulated oocysts are passed in the feces into the environment (Duszynski and Upton, 2001). About 20% of the oocysts produced in the gut failed to form an oocyst wall, and only a series of membranes surround the developing SZs. These "oocysts," devoid of a true wall, are sometimes termed "thin-walled oocysts." It is believed that the SZs produced from these thin-walled oocysts can excyst while still within the gut and infect new cells. Thus, *C. parvum* appears to have two auto-infective cycles: the continuous recycling of Type I merozoites and the SZs rupturing from thin-walled oocysts.

Cross-transmission: Numerous examples exist in the literature (see Duszynski and Upton, 2001 for some examples).

Pathology: Significant illness due to *C. parvum* rarely has been reported from wild-caught animals, but it is common once the animals become captive, when crowding and additional stress become factors. Acute infection causes moderate to severe diarrhea, weight loss, abdominal cramping, lethargy, lack of appetite, and occasionally fever. Severe dehydration may occur that can result in electrolyte imbalance and death. Some animals (e.g., chinchillas, foals, hamsters, primates, many ruminants) are more susceptible than others, and in these significant mortality may occur, especially if exposed to additional stress. Infections are associated with villus atrophy and villus fusion, infiltration of the lamina propria by inflammatory cells, and sloughing and degeneration of individual enterocytes. Metaplasia of the surface epithelium to cuboidal or low columnar cells also has been observed, and crypts may become dilated and filled with necrotic debris. These changes result in reduced absorption of vitamins and sugars, and it is likely that enterocyte damage results in impaired glucose-stimulated Na^+ and water absorption. Since disaccharidase activity is thought to decrease due to the loss of mature enterocytes, it has been postulated that the un-degraded sugars also may allow for bacterial over-growth and a change in osmotic pressure due to the formation of volatile free fatty acids (Duszynski and Upton, 2001). In humans, healthy individuals may clear the infection in less than a month, but those who are immunosuppressed, particularly AIDS patients, suffer prolonged and potentially fatal episodes of diarrhea (Casemore et al., 1997).

Material deposited: None.

Remarks: Because so many mammalian species can be infected with *C. parvum*, a large potential zoonotic reservoir exists in wild, domestic, and companion animals such as pet rabbits. Numerous studies have demonstrated or correlated transmission of this parasite from various other animals to humans, especially to children, and such cases usually involve direct exposure to infected animals or their feces or exposure to contaminated raw milk, food, or water (see Duszynski and Upton, 2001, for review). Sturdee et al. (1999) found 2/28 (7%) wild *O. cuniculus* in central mainland Britain to be passing oocysts of *C. parvum*. Their finding, along with finding *C. parvum* in other wild mammals (primarily rodents), highlights the potential for transmission between host species and warns of the possibility of direct exposure for anybody using the countryside for professional or recreational purposes. In Belgium, Peeters et al. (1986) found *Cryptosporidium* species in 7/29 (24%) rabbitries examined during an epidemiological survey. They did not attempt to identify the species of *Cryptosporidium*, but from the pathology observed, it seems more likely to be *C. parvum* then *C. cuniculus*; this is only a "best guess" on our part. In unweaned rabbits, experimental infection caused liquid diarrhea and high mortality, while only discrete lesions and no mortality were found in weaned rabbits, but their weight gain was reduced 32% in the second week PI. Light and TEM of histological sections showed severe atrophy of intestinal villi (Peeters et al., 1986). Infection with *C. parvum* is of concern for human and other animal health because the ingestion of even low numbers of oocysts can cause severe diarrhea for reasons explained above (and see Angus, 1987). The oocysts survive well outside the host and are resistant to chlorination levels used for disinfection of tap water. Many wild mammals have been identified as reservoirs of *C. parvum*, and the infection in most of these mammals appears to be asymptomatic, but oocysts shed in the feces may pose a threat to human health in urban and rural environments, particularly for individuals who may have frequent contact with animal feces such as field and farm workers or veterinarians. Wild rabbits share habitats with farm animals or travel through grazing land used by them, enabling ample opportunity for transmission of *C. parvum* from animal to animal and providing a link between rural and urban foci of the disease. This could contribute to many of the sporadic human cases of cryptosporidiosis in towns and cities. Rabbits leave their many small droppings wherever they forage, thus contaminating human and animal feed stores and accommodations. There are still no reliably effective treatments to combat cryptosporidiosis (Fayer et al., 1990, 1997).

SPECIES INQUIRENDAE (2)

Cryptosporidium sp. of Tyzzer, 1929

Original host: *Oryctolagus cuniculus* (L., 1758) (syn. *Lepus cuniculus*), European (domestic) rabbit.

Remarks: Tyzzer (1929) commented in passing that he had found a *Cryptosporidium* similar to *C. parvum* from the laboratory mouse in the intestine of a rabbit, but he did not discuss it further. It is now accepted that the form first seen by Tyzzer (1929) likely is a synonym of *C. cuniculus*.

Cryptosporidium "rabbit genotype"

Type host: *Oryctolagus cuniculus* (L., 1758) (syn. *Lepus cuniculus*), European (domestic) rabbit.

Remarks: Prior to one drinking-waterborne outbreak caused by a rabbit genotype in Northamptonshire, England, in July, 2008 (Health Protection Report, 2008), outbreaks of human cryptosporidiosis were thought to be caused only by *C. parvum* or *C. hominis*. However, enhanced surveillance established by the health protection team in the same affected area documented the rabbit genotype as a new human pathogen (Chalmers et al., 2009a, b). They detected *Cryptosporidium* oocysts in the large bowel contents of a rabbit carcass found in a tank at a water treatment plant in Northamptonshire. During late July and August, 2008, human stool samples from 34 local laboratory-identified cases of cryptosporidiosis in the affected area were typed. *Cryptosporidium* species were identified by bidirectional sequencing of PCR products generated by nested PCR for the SSU rRNA gene. Samples from 23/34 (68%) cases were homologous with isolates from rabbits in the People's Republic of China (Xiao et al., 2002) and the Czech Republic (Ryan et al., 2003); of the other 11 samples, six were not confirmed, two were *C. hominis*, one was *C. parvum*, and two could not be typed. All case-patients lived in the area affected by the water supply incident and had onset dates consistent with exposure by drinking water consumption or by person-to-person spread. However, it is not known whether this rabbit genotype is identical or not with *C. cuniculus*.

More recently, Shi et al. (2010) examined 1,081 fecal samples from rabbits on eight farms in five different areas in Hunan Province, China. Although the infection rate was relatively low, 37/1,081 (3.4%), they found a significant association between prevalence and the age of animals, with the prevalence of infection highest (10.9%) in 1–3-mo-old rabbits. No apparent clinical signs of cryptosporidiosis were seen in the rabbits at the time of specimen collection. The *Cryptosporidium* species in positive fecal samples were genotyped by sequence analyses of the 18S rRNA, 70-kDa heat shock protein (HSP70), oocyst wall protein (COWP), and actin genes and were subtyped by sequence analysis of the 60-kDa glycoprotein (GP69) gene. Only *Cryptosporidium* rabbit genotype was identified, with 100% sequence identity to published sequences of the 18S rRNA, HSP70, COWP, and actin genes, and the strains belonged to three GP60 subtypes (VbA36, VbA35, VbA29). The authors emphasized that, "In view of the recent finding of the *Cryptosporidium* rabbit genotype in human outbreak and sporadic cases, the role of rabbits in the transmission of human cryptosporidiosis should be reassessed" (Shi et al., 2010). Finally, as noted above, Robinson et al. (2010) presented the case for synonymizing the "rabbit genotype" from previous studies in China, the Czech Republic, and England, with *C. cuniculus* first named by Inman and Takeuchi (1979).

ZOONOTIC POTENTIAL OF RABBIT COCCIDIA

Close contact of humans with companion, wild, or domestic animals provides a good opportunity for direct transmission, as the oocysts are already mature and infective when shed in the feces of these animals. Since there are no effective drugs against cryptosporidiosis (Cacciò and Pozio, 2006), the identification and genetic characterization of *Cryptosporidium* from rabbits (and other wild and domesticated animals) are essential to assessing sources of infection, modes of transmission, and the potential risk of infection to those who work with them. Because of the lack of host specificity, it is not possible to reliably identify or differentiate *Cryptosporidium* species or genotypes based on host origin or by conventional coprological techniques, including staining and immunestaining (Jex et al., 2008; Jex and Gasser, 2010). Thus, tools based on the PCR reaction are now widely employed to define species and/or genotypes and to assess the risk that *Cryptosporidium*-infected animals pose as reservoirs for human infection (Saiki et al., 1988; Cacciò and Pozio, 2006; Smith et al., 2006; Nolan et al., 2010). In particular, PCR-sequencing of genetic loci, including SSU nuclear rRNA and the 60-kDA glycoprotein (GP60) gene, has been shown to be a practical approach for the genetic identification and characterization of *Cryptosporidium* (Jex et al., 2008; Nolan et al., 2010). These and other molecular-epidemiological studies have linked outbreaks of cryptosporidiosis in humans to livestock and a wide range of wild animals, including rabbits (*O. cuniculus*), in Europe, North America, China, and elsewhere in the world (Hunter, 1999; Ong et al., 1999; Chalmers et al., 2009a, b; Nolan et al., 2010). The recent discovery that the "rabbit genotype" of *Cryptosporidium* (now *C. cuniculus*; see above) has been linked to diarrheal disease in humans has raised considerable awareness about the importance of investigating rabbits as a source of *Cryptosporidium* transmissible to humans (Robinson and Chalmers, 2009a, b; Smith et al., 2010). Investigating the presence and distribution of genotypes in wild rabbits globally is imperative, not only to determine the significance of this host group as reservoirs of *Cryptosporidium* for transmission to humans, but also to establish the mode(s) of transmission and the real host range of this suite of parasites.

TABLE 9.1 All Known *Cryptosporidium* Species (Apicomplexa: Cryptosporididae) from Lagomorpha

Cryptosporidium spp.	Type host spp. infected	Rabbit/other host	References
C. cuniculus	*Oryctolagus cuniculus*	*Homo sapiens* *Sylvilagus floridanus*	Jervis et al., 1966; Vetterling et al., 1971; Inman and Takeuchi, 1979; Rehg et al., 1979; Ryan et al., 1986; Peeters, 1988; Pavlasek et al., 1996; Cox et al., 2005; Morgan-Ryan et al., 2002; Shiibashi et al., 2006; Chalmers et al., 2009a, b; Robinson and Chalmers, 2009; Nolan et al., 2010; Robinson et al., 2010; Shi et al., 2010
C. meleagridis	*Meleagris gallopavo*	*Coturnix coturnix* *Gallus gallus* *Peromiscus sp.* *Canis familiaris* *Alectoris rufa* *Psittacula kramerii* *Oryctolagus cuniculus*	Slavin, 1955; Mosier et al., 1997; Lindsay et al., 1989; Lindsay and Blagburn, 1990; Champliaud et al., 1998; Xiao et al., 1999, 2000, 2001, 2002; Morgan et al., 2000, 2001; Pedraza-Diaz et al., 2000; Sreter and Varga, 2000; Sreter al., 2000; Sulaiman et al., 2000, 2002; Darabus and Olariu, 2003; Cama et al., 2008; Chalmers and Giles, 2010

(Continued)

TABLE 9.1   All Known *Cryptosporidium* Species (Apicomplexa: Cryptosporididae) from Lagomorpha (*cont'd*)

Cryptosporidium spp.	Type host spp. infected	Rabbit/other host	References
C. parvum	*Mus musculus*	*O. cuniculus* >150 other mammal species	Tyzzer, 1912; Peeters et al. 1986; Angus, 1987; Casemore, 1991; MacKenzie et al., 1994; Anonymous, 1995, 2008; Casemore et al., 1997; Sturdee et al.,1999; Robinson and Chalmers, 2009
C. "rabbit genotype"	*O. cuniculus*	*H. sapiens*	Xiao et al., 2002; Ryan et al., 2003; Health Protection Report, 2008; Chalmers et al., 2009a, b; Shi et al., 2010

Strategies for Management, Control, and Chemotherapy

INTRODUCTION

As we've seen in Chapters 1–9, at least six genera of apicomplexan "coccidia" infect wild and domesticated rabbits. Species of *Eimeria*, *Isospora*, and *Cryptosporidium* have direct life cycles with sporulated oocysts from one rabbit directly infective to a second, susceptible rabbit. To become infected with species of *Besnoitia*, *Sarcocystis*, and *Toxoplasma*, oocysts also are involved, but not from another rabbit because species in these genera have an obligate two-host, indirect life-cycle that involves a carnivore definitive host which discharges oocysts into the environment, while rabbits which ingest these oocysts become intermediate hosts harboring only asexual tissue stages, and eventually cysts, of these parasites.

MANAGEMENT

All wild animals have multiple species of coccidia that can, and do, infect them from time to time. Some are highly pathogenic, others moderately so, and still others are practically nonpathogenic. Infections with only a single species are rare in nature, so any observed effects in wild populations is the result of the combined actions of the particular mixture of coccidia and other parasites present, together with the modifying effects of the nutritional condition of the host, and environmental factors such as weather and management practices (Levine, 1973a, b).

Coccidiosis, caused by intestinal and hepatic coccidia (*Cryptosporidium*, *Eimeria*, *Isospora*), in rabbits is self-limiting, both in the individual and in the local population. Because of the ability of sporulated oocysts to resist environmental extremes, they also are persistent in wild populations. Although coccidia are the most common gut parasites found in rabbits, they rarely cause clinical signs of disease in adult rabbits; thus, finding oocysts in the stool of rabbits is very common and should not necessarily prompt drug treatment (Quesenberry and Carpenter, 2010). However, juveniles, < 6-mo-old, are more susceptible and may become ill. In a typical outbreak of coccidiosis, signs of disease appear in only a few animals at first, the number of affected animals can increase rapidly to a peak in a few days to about a week, and then the disease subsides spontaneously. In general, rabbit intestinal coccidia are immunogenic and natural resistance occurs that is highly species specific. Symptoms in young, immune-compromised, or highly susceptible adults include diarrhea that may contain mucus and/or blood, weight loss, dehydration, and death in severe infections. Animals that survive the disease state develop species-specific and, in some instances, strain-specific immunity that may last for life (Bhat et al., 1996; Quesenberry and Carpenter, 2010). Both humoral and cell-mediated immunity are involved, with the latter playing a predominant role.

Those coccidia that cause disease by encysting within the tissues of rabbits (*Besnoitia*, *Sarcocystis*, *Toxoplasma*) are not self-limiting, and the infections generally stay with the animal for its lifetime. As noted in Chapter 8, some members of these genera can cause pathology, even death, in wild rabbits if they become heavily infected. In natural populations of rabbits, nothing can be done to manage either of these situations once rabbits have ingested oocysts to which they are susceptible.

In rabbitries, most raisers do little during the early stages of intestinal infections because they initially think the condition is unimportant and will soon be over. Once more animals become affected and losses occur, it takes some time to establish a diagnosis (fecal oocysts, liver lesions), so treatment often is not started until the outbreak has reached its peak. Under such circumstances, it matters little what treatment is used because the disease will subside and be over before the chemotherapy can have any effect. This course of events is what is usually encountered by the small-animal veterinarian. The patient (pet rabbit) with coccidiosis is not taken to the vet until it is already sick. By that time, it is too late for any anti-coccidial drug to be of value, although supportive treatment and control of secondary infections may be helpful.

It also must be kept in mind that the mere presence of these intestinal coccidia does not mean that they caused the observed enteritis. Many rabbits carry a few intestinal coccidia without suffering any noticeable effects. For example, Lund (1951, p. 14) considered intestinal coccidia to be the cause of enteritis in only 80/1,541 (0.5%) infected rabbits. The etiological agent of an enteritis also can be difficult to diagnose because the ascospores of the rabbit-specific yeast, *Cyniclomyces guttulatus* (syn. *Saccharomycopsis guttilata*), may be difficult to differentiate from oocysts by an inexperienced technician (van Praag, 2011).

Likewise, infection of pet rabbits, and those managed in rabbitries, with tissue-stage coccidia is virtually undetectable at the time of infection, as symptoms don't begin to occur until well after the zoites have migrated to the tissues, the cysts begin to develop and mature, and the host manifests an immune response to the parasite(s), or not.

GENERIC METHODS OF CONTROL

Currently, prevention of contact with infective oocysts from all six apicomplexan genera that can infect rabbits is the only practical method of control. Duszynski and Upton

(2009, pp. 304–306), in their book on snake coccidia, detailed important steps in prevention and early detection of coccidial infections and outlined management and sanitation techniques that need not be repeated here in detail. Thus, our comments will be directed toward specific control measures that can help to limit exposure of rabbits to infective stages of these parasites.

Feeders and water dispensers or water crocks should be designed so that they do not become contaminated with rabbit droppings, and they should be kept clean with disinfectant (e.g., 10% aqueous (v/v) ammonia solution). Hutch floors should be self-cleaning or should be cleaned frequently and then kept dry, and manure should be removed frequently. During disinfection treatment of the environment, rabbits should be kept in another part of the home to avoid the danger of contact with the cleaning product(s) and possible intoxication therefrom.

In commercial operations, rabbits should be kept in uncrowded, stress-free environments, handled as little as possible, and care should be taken not to contaminate the animals or their food, utensils, or equipment. In addition, the rabbitry should be kept as free as possible of insects, rodents, and carnivores, especially felids. For example, Metelkin (1935) was among the first to demonstrate that laboratory-bred and wild flies (*Musca domestica, Calliphora erythrocephala, Lucilia caesar, Cynomyia martuorum, Stomoxys calcitrans*, and *Phormia groenlandica*) would feed on rabbit fecal suspensions containing oocysts and both ingest them and get oocysts on their body hairs. All fly species ingested oocysts, which remained unaltered and viable in their intestinal tracts for up to 24 hr, and in their feces until they dried. He suggested that these flies played an important epidemiological role in the mechanical transmission of coccidial oocysts to rabbits. Similar experiments confirmed the early work of Metelkin (1935). Wallace et al. (1972) thought that flies could assist the transmission of *Toxoplasma*

oocysts, because he passed them through the digestive tract of laboratory-bred flies and they were then infective for mice; Smith and Frenkel (1978) showed that, in the laboratory, cockroaches acted as experimental vectors of murine *Sarcocystis* sporocysts; and Marcus (1980) showed that sporocysts from two bovine *Sarcocystis* species were not affected by passage through the gut of experimentally reared flies as judged by the ability of their sporocysts to excyst.

Rabbits should be given dry rather than moist pellets, washed fresh vegetables, and plenty of fresh water. Branches and leaves rich in tannin (willow, hazelnut, oak, ash, fruit trees, etc.) are excellent in helping to prevent coccidiosis. Before a rabbit is given a twig to chew, it is important to check that it's picked from a tree that is not toxic to rabbits and the tree must not have been exposed to chemicals or pollution from busy roads (van Praag, 2011). When rabbits are housed together, it is recommended to avoid putting food on the ground. Carnivores, especially felids and canids, should be excluded from rabbitries and from access to, and contact with, their feed, water, and bedding; uncooked meat or offal should never be fed to pet cats or dogs where rabbits are kept. Freezing can drastically reduce or eliminate infectious sarcocysts in meat, and heating meat to 55°C or higher for 20 min kills sarcocysts. Dead animals that may have sarcocysts in their tissues should be buried or incinerated, rather than left in the field where carnivores have access to them.

IMMUNOLOGICAL CONTROL

There are no vaccines, yet, to protect rabbits against coccidia, but experimental work in many other domesticated animal systems (cattle, poultry) shows they can be immunized by low dosages of live and/or attenuated oocysts and/or sporocysts. In fact, immunity normally plays a major role in the control of coccidiosis in domesticated meat animals, and a few oocysts may induce strong immunity. For example, single oocysts of *E. maxima* given to chickens on successive days for a period of 1 week produced stronger immunity than obtained by many times this number of oocysts given in a single inoculation (Reid, 1975).

Current research on development of an immune-prophylactic control for *Eimeria* species in rabbits is broadly representative of that being done on other coccidia species, especially in poultry. Vaccines using virulent strains of *Eimeria* species already are in use against poultry coccidiosis (e.g., Coccivac, Immucox). Such types are live, multivalent vaccines; although they provide effective protection, they run the risk of provoking symptoms or outbreaks of the disease. Another more recent vaccine for chickens, Paracox, uses a precocious strain; a similar approach for the development of live-attenuated vaccine using precocious strains is being followed in *Eimeria* species parasitizing commercial rabbits (Licois et al., 1990; Braun et al., 1992; Akpo et al., 2012). These strains are strongly immunogenic and cause very little pathological changes because they go through one or two fewer merogonous cycles of replication than do the wild-type strains. Using these does not lead to sterile immunity, but considerably reduces pathological changes and morbidity (Williams, 1994). For example, Akpo et al. (2012) used 35-day-old rabbits that were vaccinated with a mixture of precocious lines of *E. magna* and *E. media* and, 3 wks after challenge inoculation, no case of diarrhea was recorded in the two groups of vaccinated rabbits vs. the non-vaccinated rabbits that did develop diarrhea. In addition, during patency, oocyst output of vaccinated rabbits was significantly lower than that of control animals, confirming a good immunogenic stimulation by the precocious lines. Precocious attenuated strains of *Eimeria* may be useful as vaccines for some years to come, but, in the long run, subunit vaccines made by recombinant DNA technology appear most attractive as attempts to develop

recombinant vaccines against poultry coccidia are continuing. The vaccines are likely to consist of either purified coccidial antigens obtained from the large-scale fermentation culture of an appropriate recombinant micro-organism and intended for injection, or alternatively as live micro-organisms that, during the course of replication, will synthesize and release coccidial antigens within the host (Ellis and Tomley, 1991; Bhat et al., 1996). Development of a genetically engineered subunit vaccine using recombinant protein for immunoprophylaxis of rabbit coccidiosis is at a preliminary stage (Braun et al., 1992).

CHEMOTHERAPY/ CHEMOPROPHYLAXIS

History of Drug Development for Coccidiosis

Levine (1973b) reviewed much of the early history of anti-coccidial drug development. He pointed out that the first compound found effective against coccidia was sulfur, which was introduced by Herrick and Holmes (1936). Later, Hardcastle and Foster (1944) introduced borax, but neither of these proved to be satisfactory anti-coccidial drugs. Sulfur interferes with calcium metabolism, causing a condition known as sulfur rickets in chickens, while borax is only partially effective and is toxic in therapeutic doses.

P.P. Levine (1939) introduced the first practical anti-coccidial drug, a sulfonamide called fulfanilamide. Since then many different drugs have been used, particularly against the most pathogenic chicken coccidians, especially *E. tenella*. These included both sulfonamides and derivatives of **phenylarsonic acid, diphenylmethane, diphenyldisulfide, nitrofuran, triazine, carbamilide (urea), imidazole**, and **benzamide** (modified from Levine, 1973b). Because of the importance of coccidiosis in the commercial poultry industry worldwide, many thousands of papers have been published on coccidiostatic drugs, and their use in poultry production is so pervasive today that it is difficult to obtain a commercial feed which does not contain one or more coccidiostats. However, they are used to a considerably smaller degree for other classes of meat animals (e.g., cattle, rabbits).

No coccidiostatic drugs will cure a case of coccidiosis once signs of the disease have appeared because they are all prophylactic. That is, they must be administered either before, or at the time of exposure, or soon thereafter, to be effective against the endogenous developmental stages. Most coccidiostats act to disrupt the metabolic pathways of the meronts and their merozoites, and occasionally act against the sporozoites, and thus, prevent the life cycle from being completed. They are not, however, very effective against the gamonts. Hence, since exposure in nature and in commercial ventures is continuous, any drugs that are administered must be given continuously by mixing them with feed and/or water.

Successful anti-coccidials block metabolic pathways of the endogenous stages to prevent their further development. In some cases, the biochemical pathways and enzyme systems involved are known and differences between host cell and parasite requirements are recognized. In other cases, little is known of the biochemical activity of anti-coccidials that are discovered by blind screening techniques (Levine, 1973b). Reid (1975) offered a classification scheme of 13 different modes of action based on knowledge of the time or the stage in the eimerian life cycle when the anti-coccidial was most active against poultry coccidia (see his Table 1).

Drugs for Enteric and Hepatic (*Eimeria* spp.) Coccidiosis

Early in the development of our knowledge about drug therapy, some of the sulfonamides were found to be helpful in preventing coccidiosis if given continuously in feed or drinking water. These included **sulfadiazine**, **sulfamerazine**, **sulfamethazine**, **sulfaquinoxaline**, and **succinylsulfathiazole** (see Levine, 1973b). Long-term, continuous feeding of drugs is not particularly desirable, nor is it usually necessary. If the hosts are exposed to coccidia during the drug-feeding period, as they usually are, an aborted infection results and this usually is sufficient to induce immunity. The drug then can be safely stopped.

Peeters et al. (1987) said that control of rabbit coccidiosis was entirely dependent on prophylactic chemotherapy and listed some coccidiostats used then to control this disease. They emphasized, and we now know this to be dogma, that no coccidiostatic drugs, then or now, will cure a case of coccidiosis once signs of disease have appeared; they all are prophylactic and must be administered at the time of exposure to the parasite or soon thereafter to be effective. Since exposure in nature is continuous, these drugs must be used regularly in the non-immune hosts, and this is usually done by mixing them with feed or water. Due to the continuous exposure of the host to these drugs, the appearance of the drug-resistant strains has been rapid, and this has limited the useful life of any chemical coccidiostats (Peeters et al., 1987).

Robenidine hydrochloride, used more-or-less systematically in many countries, is an efficient anti-coccidial against rabbit coccidiosis (Coudert and Provôt, 1988). Some **ionophore antibiotics** are now widely used as anti-coccidials in the poultry industry, but most of them have proved to be toxic for rabbits. Coudert and Provôt (1988) were among the first to study **Lasalocid** (an ionophore) in rabbits to determine its level of toxicity and its efficacy in the prevention of *Eimeria* species infections in the rabbit. They used daily weight gain as their toxic parameter and reported that Lasalocid had a growth-depressive effect that disappeared after 2 weeks at the lowest dose administered (25 ppm), but persisted at the highest dose (125 ppm; Table 10.1). The anti-coccidial efficacy of Lasalocid was very different for the two *Eimeria* species they studied. It was extremely effective against *E. flavescens* as soon as the 25 ppm was administered; on the other hand, the efficacy against *E. intestinalis* was only moderate at 75 ppm and "above this level the interaction with the depressive effect due to the product does not allow a clear interpretation of the results." At 90 ppm, robenidine hydrochloride completely controlled the "disease," as assessed by weight gain and mortality, while output of *E. intestinalis* oocysts in rabbit feces remained unchanged.

Niedźwiadek et al. (1990) used two groups of 400 rabbits (200 does, 200 bucks) that all (100%) were infected with coccidia. The experimental group was fed pelleted feed with the additive of 20 ppm **salinomycin**. They stated that this drug "proved a very effective preparation and was helpful in almost completely doing away with coccidiosis." In addition to virtually eliminating coccidiosis, the average body weight of rabbits in the drug-treatment group was 225 g heavier than non-treated controls. Coudert (1990) studied the efficacy of **diclazuril** in weanling SPF rabbits infected with high doses of *E. flavescens* and *E. intestinalis*. A dose of 0.25 ppm in pelleted feed was efficacious in controlling *E. intestinalis* oocyst output, while it took 1.0 ppm to achieve similar results with *E. flavescens*. They pointed out that diclazuril was "well-accepted by rabbits and a growth-depressive effect could be only observed with more than 30 times the recommended dose."

Polozowski (1993) gave rabbits each of the following for 5 weeks: lasalocid (Avatec), **maduramycin** (Cygro), robenidine (Cycostat), salinomycin (Sacox), **monensin** (Elancoban), **clopidol** + **methylbenzoate** (Lerbek) and **narasin** (Monteban). He found that the best results in controlling coccidiosis were obtained with salinomycin at dosages of 25, 35, and 50 ppm and maduramycin at 2 and 3 ppm. Equally effective were clopidol + methylbenzoate at 216.7 ppm, lasalocid at 90 and 125 ppm, and monensin at 20 ppm. Robendine was very effective against intestinal coccidia at 66 ppm, and it is well tolerated by rabbits, but its regular preventive use over the last 20 yrs has raised resistance in some eimerians (e.g., *E. media*, *E. magna*) toward this compound (van Praag, 2011). Narasin and maduramycin (4.5 ppm) were toxic to rabbits. In field trials with does and their progeny, salinomycin (25 ppm), maduramycin (1.5 ppm), and monensin (20 ppm) were reported to be "highly effective" by Polozowski (1993).

For hepatic coccidiosis, drug treatment is difficult and coccidiostats are best administered to all rabbits suspected of having *E. stiedai* for a minimum of 5–6 days and then repeated after 5 days (van Praag, 2011). Peeters and Geeroms (1986) reported that toltrazuril (Bay Vi 9142) administered by continuous exposure in drinking water (Table 10.1) was highly effective in reducing oocyst output of *E. stiedai* (and four intestinal eimerians), and in preventing clinical signs and macroscopic lesions, while sporulation of excreted oocysts was not affected. They (1986) also said that toltrazuril administered during late merogony or gamogony allowed development of immunity against reinfection with each homologous species. As noted above, robendine is well tolerated by rabbits and is very effective against several intestinal coccidia (although drug resistance continues to develop in some eimerians), but was only weakly effective on *E. stiedai* infections (Polozowski, 1993).

Unfortunately, recurring drug-resistance of coccidians requires continuous development of new anti-coccidial drugs, involving high costs. Together with increasing concern for the possible presence of drug residues in products for human consumption, this has led to an interest in alternative means of control. There is an urgent need for development of an imunoprophylactic control to replace the chemotherapy currently in use (Bhat et al., 1996).

Drugs for Enteric Cryptosporidiosis

To date, there is no effective treatment for cryptosporidiosis in humans, bovines, or rabbits (Quesenberry and Carpenter, 2010), with most of the work being done in the former two host-parasite systems for obvious reasons. Patenburg et al. (2009) focused on what we now know about anti-parasitic therapy of cryptosporidiosis, but of necessity focused on immune-competent and immune-compromised humans with the disease. Of course, the role of anti-parasitic drugs still remains controversial and no anti-parasitic drug has proven particularly effective in severely immune-compromised individuals and, certainly, nothing has been accomplished in this arena with rabbit *Cryptosporidium* species. Their (2009) attention focused on recent literature on four major drugs: **nitazoxanide**, **paromomycin**, **azithromycin**, **rifaximin**, and a few combinations thereof. Briefly, they reported the following: nitazoxanide is a nitrothiazolyl-salicylamide derivative, originally developed as a veterinary anthelmintic and was USFDA approved for treatment of cryptosporidiosis and giardiasis in humans who were not HIV-positive. The drug has a broad antibiotic spectrum that includes helminths, parasitic protists, and viruses. Rossignol et al. (2001) studied immune-competent adults and children and found that treatment with nitazoxanide led to a cure (no diarrhea, no oocysts in stool) in 67% of those infected and a second trial of infected adults and adolescents had a cure rate of 93%; dosages were 100 mg for children 1–3 yrs old, 200 mg for children 4–11 yrs old, and 500 mg for those > 12 yrs old, with all doses administered twice daily for 3 days. Paromomycin is an oral nonabsorable aminoglycoside approved in the 1960s to treat amebiasis. Hashmey et al. (1997) and Stockdale et al. (2008) used it to treat cryptosporidiosis in adults and children with various levels of success; generally, oocysts were seldom eliminated from stools and relapse of diarrhea occurred in > 50% of the patients. However, an earlier study (White et al., 1994), using 25–35 mg/kg/day for 2 wks, found that paromomycin significantly decreased oocyst shedding and improved diarrheal symptoms, even though only one of 10 patients completely cleared their infections. Another drug that remains controversial in the treatment of cryptosporidiosis is azithromycin, an antibiotic used to treat respiratory infections and bacterial gastrointestinal infections. In a pilot study, Allam and Shehab (2002) treated Egyptian school children with 500 mg/day for 3 wks and said they saw a 91% cure rate and a 99% reduction in oocyst discharge in stool samples. In contrast, however, Blanshard et al. (1997) said that the frequency of resolution of diarrhea in patients treated with azithromycin was unremarkable, a view later supported by others (Blagburn and Soave, 1997; Stockdale et al., 2008). After their thorough review of the most recent literature on treatment of cryptosporidiosis, Pantenburg et al. (2009) concluded that the only anti-parasitic drug proven to be effective against *Cryptosporidium* in children is nitazoxanide (100 mg for children 1–3 yrs old, 200 mg for children 4–11 yrs old, 500 mg for those > 12 yrs old) administered 2 times/day with food. More recently, De Waele et al. (2010) used **halofuginone lactate** to try to control cryptosporidiosis in 32 naturally infected neonatal calves on a dairy farm in Ireland. Halofuginone lactate was shown to be effective in reducing clinical signs of cryptosporidiosis and environmental contamination, but it did not delay the onset of diarrhea and it did not reduce the risk of infection among calves reared together in a highly contaminated environment. On the other hand, when halofuginone lactate was used in calves in combination with good hygiene, it was effective in reducing the risk of cryptosporidiosis among 7–13-day-old calves. They concluded that control of the parasite could be achieved by the combination of using effective preventive drugs and good animal husbandry procedures. Nothing along the lines of human and bovine drug trials has been done with rabbits infected with a *Cryptosporidium* species; perhaps someone will attempt such experimental work with these drugs in the future.

Drugs for Sarcocystid Tissue Cyst Coccidia

All of the attention to chemotherapy in mammals infected with tissue cysts of *Sarcocystis* species has focused on domestic meat animals, especially cattle, sheep, and goats. Rommel et al. (1981) developed a *Sarcocystis muris* model in mice for research on the chemotherapy of acute sarcocystosis of domestic animals. In their mouse model, they defined four phases of infection: sporozoite migration (2–10 DPI), merogony (11–17 DPI), merozoite migration (18–27 DPI), and sarcocyst formation (28–50 DPI), the timing of which could vary in different host-parasite systems. They found that anti-coccidials were most effective against merogony (Rommel et al., 1981). Of 12 anticoccidials tested in their mouse/*S. muris* model, many "standard" drugs (e.g., amprolium, monensin, sulfaquinoxaline, others) had no anti-sarcocystid activity. Others (e.g., lasalocid, halofuginone) had only marginal anti-*S. muris* activity, whereas Zoalene (dinitolmide)®, sulfaquinoxaline, and a few others had excellent anti-*S. muris* activity. Zoalene is a broad-spectrum anti-coccidial in chickens, leaves no residue in tissues, and can be used to prevent coccidiosis in domestic rabbits. Of particular interest in the study by Rommel et al. (1981) was the activity of sulfaquinoxaline plus pyrimethamine against immature and mature sarcocysts, because no other drugs previously were known to be effective against tissue sarcocysts (see Dubey et al., 1989). The most important results with the murine *S. muris* model were that outcomes with one *Sarcocystis* species may not be directly applicable to another species of *Sarcocystis*. However, perhaps the prophylactic

use of standard anti-coccidials in other hosts (including chickens) may be a practical method of controlling sarcocystosis in rabbits.

For example, several anti-coccidial drugs used to treat poultry coccidia are effective against sarcocystosis in cattle, sheep, and goats when administered continuously for a month, starting at or before inoculation with *Sarcocystis* (Dubey et al., 1989). Amprolium (100 mg/kg body weight), a thiamine analog, administered from 0 to 30 DPI, prevented acute disease or death in cattle infected with *S. cruzi* and in sheep infected with *S. tenella*, and calves treated with the same dose 21–35 DPI with 100,000 sporocysts had only mild signs of acute sarcocystosis, indicating to them that the second-generation meronts and all subsequent asexual stages preceding sarcocyst development were affected. Halofuginone (Stenorol) also reduced or prevented acute sarcocystosis in goats infected with *S. capracanis*, and in sheep infected with *S. tenella*, when it was administered in feed from day 5 before inoculation and continued up to 36 DPI. However, many standard drugs (e.g., Lasalocid, Monensin, others), each incorporated into feed at 33 mg/kg, had minimal or no effect on the course of acute sarcocystosis in cattle inoculated with 250,000 *S. cruzi* sporocysts when the medicated food was given from day 7 before inoculation through 80 DPI (Dubey et al., 1989).

Dubey and Beattie (1988) summarized the literature on treatment regimens used for therapy of toxoplasmosis. They recommended two drugs that were the most widely used at the time, sulfonamides (e.g., sulfadiazine, sulfamethazine, sulfamerazine) and pyrimethamine (Daraprim®) and said that together they acted synergistically by blocking the metabolic pathway involving *p*-aminobenzoic acid and the folic-folinic acid cycle, respectively. These drugs were well-tolerated and were effective against toxoplasmosis, at least in humans; they were most beneficial in the acute stage of the disease, when *Toxoplasma* was undergoing extensive asexual multiplication in the tissues and, although they restrained the growth of tissue cysts, they did not eradicate the infection. There are no specific treatments outlined for rabbits that are serologically positive for, or having symptoms of, *Toxoplasma*, but even more than two decades later, sulfamethoxazole and pyrimethamine are recommended for treating rabbits exhibiting toxoplasmosis (Quesenberry and Carpenter, 2010).

TABLE 10.1 Some Recommended Therapies/Coccidiostats Found to be Efficacious for the Treatment of Coccidiosis in Rabbits

Drug trade name (generic name or chemical group)	Treatment regime/dosage in feed or water/time	References
Amprolium	9.6% in drinking water (0.5 ml/500 ml)	Quesenberry & Carpenter, 2010
	100 mg/kg body wt from 0 to 30 DPI	Dubey et al., 1989
Baycox (toltrazuril)	2.5–5 mg/kg, 2 times @ 5 d intervals	van Praag, 2011
Baycox (toltrazuril [Bay Vi 9142])	10–15 ppm, drinking water, continuously	Peeters & Geeroms, 1986
Bio-Cox (salinomycine)	??	van Praag, 2011
Clinicox (diclazuril)	??	van Praag, 2011
Coban (monensin, ionophore)	50 ppm	Coudert & Nouzilly, 1981
Coban (narasin, ionophore)	20, 25, 40, 60 ppm	Coudert & Nouzilly, 1981; Polozowski, 1993
Coyden (clopidon, meti-chlorpindol, 25%)	200 ppm	Coudert & Nouzilly, 1981; Peeters et al.,1982
Clopidol + methylobenzoate	216.7 ppm	Polozowski, 1993
Decox (decoqinate)	100–160 ppm	Coudert, 1972, 1979b, 1981; Coudert & Nouzilly, 1981
Diclazuril	0.25 ppm in feed	Coudert, 1990
Emtryl (dimetridazole)	250 ppm in water	Coudert & Nouzilly, 1981
Formosulfathiazole	750 ppm	Coudert, 1979b, 1981
Lasalocid (avatec, ionophore)	25–125 ppm	Coudert & Provôt, 1988; Polozowski, 1993
Lerbek (clopidol + methyl-benzoate	216.7 ppm	Polozowski, 1993; Joyner et al., 1983
Lerbek (metichlorpindol, 25%; methylbenzoate, 1.67%)	200 ppm	Coudert, 1979b; Coudert & Nouzilly, 1981
Maduramycin	2, 3 ppm	Polozowski, 1993
Monensin	20 ppm	Polozowski, 1993
Pancoxin (amprolium, 25%; ethopabat, 1.6%)	200 ppm	Coudert, 1979b; Coudert & Nouzilly, 1981

(Continued)

TABLE 10.1 Some Recommended Therapies/Coccidiostats Found to be Efficacious for the Treatment of Coccidiosis in Rabbits (*cont'd*)

Drug trade name (generic name or chemical group)	Treatment regime/dosage in feed or water/time	References
Pancoxin (amprolium 20%; ethopabat, 1%; sulfa-quinoxaline, 12%; pyri-methamine, 1%)	200 ppm	Coudert & Nouzilly, 1981
Ponazuril	20 mg/kg 1×/d	Quesenberry & Carpenter, 2010
Robenidine	16.4–33–50 ppm	Coudert, 1979a; Licois & Coudert, 1980,
	66 ppm	Polozowski, 1993 (nine intestinal eimerians)
	100 ppm	Coudert, 1979b, 1981
Ridzol (ronidazol)	30 ppm	Coudert & Nouzilly, 1981
Sacox (salinomycin, ionophore)	25–35–50 ppm	Coudert & Nouzilly, 1981; Polozowski, 1993
	1–2 mg/kg body wt	Dubey et al., 1989
Sodium sulfaquinoxaline	3.5 g/3.785 l, 3 wk	Levine, 1973b (p. 236)
	1 g/l water	van Praag, 2011
Statyl (methylbenzoate, 20%)	20 ppm	Coudert & Nouzilly, 1981
Stenorol (halofuginone)	1–3 ppm	Coudert & Nouzilly, 1981
	0.22 mg/kg body wt	Dubey et al., 1989
Sulfamethazine	128 mg/kg for 1–3 d	Gološin & Tešić, 1963; Gološin et al., 1963
Sulfadimethoxine	0.5–0.7 g/l water	van Praag, 2011
	75 mg/kg/d for 7d	Quesenberry & Carpenter, 2010
	15 mg/kg per os every 12 hr for 10d	
Sulfaquinoxaline	1 g/l drinking water	van Praag, 2011
Sulfadimethoxine-diaveridine (3:1)	3:1/100 g feed, 3 d 100 ml/l water, 8 d	Dürr & Lammler, 1970
Sulfadimerazine	2 g/l water	van Praag, 2011
	0.02% in water, 7 d	Quesenberry & Carpenter, 2010
Toltrazuril (Bay Vi 9142)	7–10 mg/kg 1×/d for 7 d	Quesenberry & Carpenter, 2010
	10–15 ppm in water for 5 wk	Peeters & Geeroms, 1986
	2.5–5.0 mg/kg, 2×, repeat after 5 d	van Praag, 2011
Trimethoprim-sulfamethoxazole	30 mg/kg each 12 hr for 10 d	Quesenberry & Carpenter, 2010
Whitsyn 10 (Sulfa-quinoxaline, 8.33%; pyrimethamine, 0.85%)	80–100 ppm	Coudert & Nouzilly, 1981
Zoalene® (coccidine A, zoamix)	125 or 250 ppm	Rommel et al., 1981

11

Summary and Conclusions

INTRODUCTION

Our knowledge of lagomorph coccidia is perhaps the best of any mammalian order, but it is nevertheless **very** poor; a great deal remains to be learned. At this point in time, we know there are at least 87 Apicomplexa species that we believe to be valid identifications, from the five extant lagomorph genera (of 13) (38%) that have been examined for them: three *Besnoitia* species (in *Oryctolagus*, all experimental), three *Cryptosporidium* species (from *Oryctolagus*), 73

Eimeria species (*Brachylagus* [1], *Lepus* [30], *Ochotona* [15], *Oryctolagus* [15], *Sylvilagus* [12]), two *Isospora* species (from *Ochotona*), five *Sarcocystis* species (*Ochotona* [2], *Lepus* [1], *Oryctolagus* [1], *Sylvilagus* [1]), and *Toxoplasma gondii* (in *Lepus*, *Oryctolagus*, *Sylvilagus*) (Table 11.1). Each of these host-parasite associations is unique in many ways. Surveys have shown that virtually all vertebrate species, and a high percentage of lagomorph species, are infected with one or more coccidia when examined; the prevalence of infection has been reported as low as 4%

The Biology and Identification of the Coccidia (Apicomplexa) of Rabbits of the World
http://dx.doi.org/10.1016/B978-0-12-397899-8.00011-1

263

(Soveri and Valtonen, 1983) but, more often, near or at 100% (Duszynski and Marquardt, 1969). This generalization refers *only* to intestinal eimeriid coccidians (*Eimeria, Isospora*), not the other four genera documented here. Infected animals almost always are infected with as few as two to as many as nine intestinal species. A number of species, *E. intestinalis, E. piriformis, E. coecicola* (intestinal forms), *E. stiedae* (liver, bile duct), and a few others, are known to be highly pathogenic (Cheissin, 1947a, b, 1948; Pellérdy and Dürr, 1970; Coudert et al., 1993), but these are found primarily in Old World and domesticated rabbits (*O. cuniculus*). The three lagomorph genera from which > 80% of all known coccidia species are described are *Lepus, Oryctolagus*, and *Sylvilagus*. However, Andrews et al. (1980) noted that the lack of salient coccidiosis lesions, in conjunction with an absence of any cases of clinical coccidiosis in numerous surveys of wild rabbits, suggests that the pathogenicity of coccidian parasites in wild rabbit populations may be "overestimated." Although some of the *Eimeria* that infect these genera have been shown by cross-infection studies to be species specific (e.g., *E. sculpta*, see Carvalho, 1943), other species (e.g., *E. neoleporis*; see Carvalho, 1942, 1944; *E. stiedai*; see Table 11.2) can be transmitted between genera.

A lot of interesting work has been done with rabbit coccidia, which has biological implications beyond the individual organisms used. Licois et al. (1990) were able to develop a precocious line of *E. intestinalis* by selection for early developmental oocysts after six consecutive passages in domestic rabbits, *O. cuniculus*; Pakandl et al. (1996b) also developed a precocious strain of *E. media*. Such studies may lead to the development of immunity via attenuated pathogenicity in highly pathogenic strains (see Chapter 9, and Jeffers, 1975; McDonald et al., 1986). Aly (1993) was able to infect dexamethaxone-treated mice with *E. stiedai* and achieved a patent infection in 20/25 (80%) mice; the prepatent period was from 30 to 35 days and patency lasted at least 60

DPI. And perhaps host transfer occurs more often than we know under natural conditions, where many random cross-transmission events must occur daily among **syntopic** hosts, especially those that have similar nutritional requirements and can provide coccidia with intestinal milieus similar to one another.

Finally, molecular tools have started to be used to diagnose coccidia species in veterinary parasitology (see Combes et al., 1996 for review). Among the first to demonstrate this use with mammalian coccidia were Cere et al. (1995), who studied inter- and intraspecific variation of *Eimeria* spp. in rabbits using random amplified polymorphic DNA (RAPD) assays. Although profiles differed significantly between nine eimerian species, species-specific fingerprints were obtained that might prove useful in species diagnosis.

OOCYST SIZE, STRUCTURES AND THEIR STABILITY

Cheissin (1947a, b) said that oocysts he recovered early during experimental infections often were smaller and paler than those discharged later during patency. This is now known to be a rather common phenomenon, as many others have documented that oocyst size sometimes changes considerably within a single species. This has been noted for species of rabbit coccidia, as well as those known from chickens, sheep, and rodents (Boughton, 1930; Fish, 1931a, b; Jones, 1932; Cheissin, 1940, 1947b, 1948, 1957a; Becker et al., 1955, 1956; Kogan, 1962, 1965; Duszynski, 1971, Joyner, 1982; Parker and Duszynski, 1986; Gardner and Duszynski, 1990; Upton et al., 1992; Seville and Stanton, 1993; others). Differences in the sizes of oocysts discharged during patency have even been observed from infections of the host by a single oocyst, to help alleviate the suspicion that several species of parasites might have been present in the original inoculum. Cheissin (1967) suggested that this variability in oocyst size depended on a series of factors, including the development of macrogametes. He and others observed that the average length and width of oocysts of *E. magna* and *E. intestinalis* from rabbits decreased when the inoculation dosage was increased (Jones, 1932; Cheissin, 1947b, 1957a). The suggestion is that during large infecting doses, several macrogametes are forced to develop in each host cell, with the ultimate size each can attain being severely limited by the size of the host cell. As a result, each gamete accumulates a smaller amount of nutritive material before reaching its full size and becomes fertilized, after which they yield undersized oocysts. This phenomenon has not been observed in other species. Another explanation for smaller oocysts appearing at the end of patency, vs. those seen when patency begins, is that there may be a certain retardation of macrogamete growth when the rabbit is developing a protective reaction to the early endogenous stages of that species (Cheissin, 1967). Kogan (1962, 1965) said that the host's diet may influence the sizes of the oocysts of *E. necatris* in chickens.

Both the oocyst (OR) and sporocyst (SR) residua are qualitative (presence/absence) structures in the oocysts of many species of coccidia and vary in shape, size, and composition (a few scattered granules to large homogeneous globules) from species to species. Thus, like oocyst and sporocyst size, they are important determinants of species identification. Pérard (1924a, b) first suggested that the SR was nutritive material used for maintenance of SZ viability, because he saw it decrease gradually in size in the external environment. Cheissin (1958a, b, 1959, 1967) studied both the fate and chemical composition of the OR and SR over time when he examined the sporulated oocysts of *E. intestinalis*, *E. irresidua*, *E. magna*, and *E. stiedae* from rabbits (*Oryctolagus* sp.), those of *Isospora laidlawi* from ferrets (*Mustela* sp.), *I. rastegaievi* from the hedgehog (*Erinaceus* sp.), and *I. felis* and *I. vulpis* from foxes (*Vulpes* sp.).

Sporulated oocysts were kept in 2% (w/v) aqueous potassium dichromate ($K_2Cr_2O_7$) solution at $18-25°C$. He (1959, 1967) said the SR of *E. intestinalis* and *E. magna* sporocysts changed "considerably" after 20 mo, being reduced in size by about 75%; after $3-4$ mo the SR "lost the ability to take an iodine stain, indicating the disappearance of glycogen," while the fat droplets were larger and became more scattered. He also demonstrated that $16-18$-mo-old oocysts had SRs that "practically disappeared, and lost their infectivity," as determined by experimental inoculation. In *E. stiedai*, the SR decreased by $>$ 50%, from 9 to 4 μm wide after 19 mo, while in *E. irresidua* the SR decreased from 12 to $5-7$ μm within 20 mo, and it could no longer be stained with iodine, demonstrating that the glycogen part of the SR is used before the stored lipid(s). These changes in the SR, under both aerobic and anaerobic conditions, suggested to Cheissin (1959, 1967) that the SR contains food substances used by the SZs while their oocysts are in the external environment.

Duszynski and Marquardt (1969) used the presence or absence of both the OR and SR as a main criterion to distinguish between four morphologically similar oocysts (*E. environ*, *E. honessi*, *E. neoirresidua*, *E. poudrei*), which they termed their "environ group," from cottontails (*S. audubonii*) collected in Colorado, USA. It was important to their analysis to verify (or not) that both the OR and SR remained unchanged over a reasonable period of time. Their initial observation was that 9-mo-old sporulated oocysts showed no structural changes in oocyst size or in the presence/absence or size of their OR and SR when compared to 1-mo-old sporulated oocysts. To reinforce their observation, fresh oocysts of these four eimerians that sporulated after 5 days at 20°C were then placed into incubators for 12 days at 37°C to "age" them. Five hundred sporulated oocysts and their OR and SR from each species were measured immediately after sporulation and after 12 days at 37°C; they found no significant changes could be detected between the oocysts, sporocysts, or their respective residua. On the one hand, their observations support those of Cheissin (1967), who said the OR in the oocysts of *E. magna* and *E. intestinalis* persisted for 20–21 mo in oocysts found in the external environment without changing significantly. Thus, it may not be used as an energy source for maintaining the viability of the living SZ within the sporocysts. On the other hand, Cheissin's reports (1959, 1967) of the disappearance of the SR over time in the external environment are in contrast to those of Duszynski and Marquardt (1969), who saw no significant changes in the SR of four cottontail eimerians that had been "aged" for 12 days at 37°C.

Hsu (1970) studied the process of sporulation of *E. sylvilagi* oocysts, which he subjected to different temperatures, and noted that as he increased the temperature from 8°C to 32°C, oocysts sporulated more rapidly; at 8°C it took several months before he could find a few sporulated oocysts, at 20°C they sporulated in 72 hr, at 32°C oocysts sporulated within 24 hr, at 35–37°C sporulation was "abnormal and somewhat suppressed," while at 39° C no oocysts sporulated. Hsu (1970) looked carefully at the sporulation process of oocysts he maintained at 25°C for 60 hr and concluded that with the formation of SZs, sporulation was completed. At this stage, he noted a small round OR, which he said "disintegrated into a few granules or disappeared ... after several days at 25°C."

Kvičerová et al. (2007) worked with *E. cahirinensis* in spiny mice (*Acomys* spp.), and said its oocysts formed two distinct OR during sporulation, a globule consisting of many small granules and one with only several smooth vacuoles; they reported that the first OR type was typical of oocysts ≤ 15 days old, whereas only vacuoles occurred in older oocysts. Kvičerova (pers. com., unpublished) also said that sporulated oocysts of *E. citelli*, from *Spermophilus citellus*, and a new eimerian from *Habromys lophurus*, both possessed a distinct OR when young, but that these disappeared entirely when the oocysts aged. Finally, more recently, Williams et al. (2010) examined long-term storage of five chicken eimerians which had SRs. Oocysts were stored 23 yrs in 2.5% (w/v) aqueous potassium dichromate solution in the dark at 4 ± 2°C; their detailed microscopic examinations showed that the internal structures of the oocysts and sporocysts remained intact. Photomicrographs of 24-yr-old oocysts of these five species still had some visible SR granules (Williams et al., 2010).

One of the many questions still unresolved in coccidian biology is why the oocysts of some species have ORs and/or SRs, while others do not. It is interesting to note that all eimerian species of chickens, cattle, and sheep lack an OR.

LOCALIZATION IN THE RABBIT HOST: TISSUE, CELLS, AND PARASITE DEVELOPMENT

Among the intestinal coccidia of rabbits, several species have development which occurs in one part of the intestine in one stage and in a different area of the intestine during the next stage (e.g., *E. coecicola*, *E. intestinalis*). Usually coccidia are located in a single organ, and the greatest concentration of the endogenous stages is found in a single part of that organ. For example, *E. media*, *E. irresidua*, and *E. perforans* each develop only in certain parts of the small intestine, but *E. piriformis* can be found in various parts of the entire large intestine. In general, however, each species has a characteristic and limited part of a given organ where the greatest number of endogenous stages develop. In rabbit coccidia, as is seen in other species in domestic animals (e.g., chickens), one sees that the more intense the infection, the wider the area in which the endogenous stages of the species is likely to settle (Cheissin, 1967).

Cheissin (1967) found that the localization of endogenous stages of at least *E. magna*, *E. media*, and *E. irresidua* do not change when the pH and other factors of the intestinal lumen are altered by various diets. Coccidia are usually found in epithelial cells, and many species complete their entire life cycles in the epithelium. However, in some cases when asexual generations and gamonts develop in the epithelium, the oocysts submerge in the underlying connective tissue where development is completed (e.g., *E. coecicola* in the appendix of rabbits; see Cheissin, 1967). And in other species (e.g., *E. media*, *E. irresidua*) only merogony occurs in the epithelium, while gamogony, fertilization, and oocyst formation take place in the subepithelium.

In some cases where two species are in the same place in the host gut, Cheissin (1967) reported that an antagonism in their development occurs. For example, he found that when a rabbit was simultaneously inoculated with oocysts of *E. coecicola* and *E. intestinalis*, which have overlapping locations in the small intestine, the development of both species was suppressed. The prepatent period was retarded for 1 or 2 days, and the productivity of the parasite was considerably lower than that occurring in separate development; however, it is possible that this was associated with the more intense asexual reproduction of *E. intestinalis*. The meronts of this species hinder, to some degree, the development of the meronts of *E. coecicola* (also see Duszynski, 1972, on concurrent infections with two *Eimeria* species it *Rattus*).

Clearly, all intracellular parasites have an intimate relationship with their host cells and derive all of their nutrients therefrom; the coccidia, in particular, are no exception to this rule, and it is during endogenous development that they must accumulate whatever reserve materials they may need to survive once the oocyst leaves the protective confines of the host epithelium to venture to the external environment. Unfortunately, this complex aspect of the host-parasite relationship has not been studied very much. Beyer and Ovchinnikova (1964) looked at the quantitative measurements of the cytoplasmic RNA during macrogametogenesis of *E. intestinalis* and *E. magna* in rabbits and found a steady rise of RNA in the macrogamete until the formation of the zygote. Once fertilization occurred, the character of parasite metabolism changed from anaerobic to aerobic (Beyer, 1962, 1963). Later, Beyer and Ovchinnikova (1966) measured cytoplasmic RNA in zygotes (oocysts) of *E. intestinalis* and determined that the quantity of cytoplasmic RNA in them was less than in mature macrogametes. They also said that cytoplasmic volume of both mature macrogametes and new zygotes (still inside the host cell) was the same, that the cytoplasmic volume of recently discharged oocysts was significantly greater than that of the younger, intracellular stage, but that oocysts both inside and outside host cells have similar amounts of RNA. They concluded that a certain portion of the RNA is "spent" during wall formation, after which time the oocyst breached its active relationship with the host cell.

The phenomenon of two types of meronts in the asexual development of rabbit coccidia in intestinal epithelial cells has been noted in nearly all domestic rabbit (*O. cuniculus*) coccidia, including *E. stiedai* (Péllerdy and Dürr, 1970), *E. perforans* (Streun et al., 1979; Licois et al., 1992b), *E. intestinalis* (Pellérdy, 1953; Licois et al., 1992c), *E. coecicola* (Pakandl et al., 1996a), *E. magna* (Ryley and Robinson, 1976; Pakandl et al., 1996c), *E. media* (Rutherford, 1943; Pellérdy and Babos, 1953; Licois et al., 1992a; Pakandl et al., 1996b), *E. flavescens* (Norton et al., 1979; Pakandl et al., 2003), and *E. piriforis* (Pakandl and Jelínková, 2006). See Chapter 6 for specific details.

PREPATENT PERIOD

Small fluctuations in the timing of the prepatent period depend on several causes. Cheissin (1947a, 1967) noted that the prepatent period of *E. magna* and *E. intestinalis* in rabbits was decreased by several hours by the use of a heavy infecting dose of oocysts as compared with a small dose. When a rabbit was inoculated with one oocyst of *E. magna*, the prepatent period was 9 days, but a large dose reduced it to 7–8 days; he observed the same shortening of prepatency with *E. irresidua*. In *E. intestinalis*, this difference did not involve more than 24–48 hr, and in *E. media* and *E. perforans* it was no more than 8 hr. In other eimerians from domestic rabbits, the prepatent period seems to be quite precise for each species. It is 16–18 days for *E. stiedai*, 5–6 days for *E. perforans* and *E. media*, 7 days for *E. magna*, and 9–11 days for all remaining intestinal species.

The length of the intestine, and the place where endogenous development occurs, both influence, to some degree, the time when oocysts appear in the feces. Oocysts usually appeared in the intestinal lumen a few hours earlier than they are found in the feces, because they have to pass the remaining length of the intestine, and some are retained for a period of time in the cecum (Cheissin, 1967). For example, the first oocysts of *E. media* appeared in the duodenum 115–118 hr PI, but they did not appear in the feces until 120–124 hr PI. The oocysts of *E. irresidua*, from the middle region of the small intestine, appeared in the feces after a lag of 2–8 hr, while the oocysts of *E. magna*, which develop in the lower half of the small intestine at 175–180 hr PI, did not appear in the feces until 182–185 hr PI. Because they form in the depths of the crypts of the large intestine and cannot be immediately eliminated from the crypts, oocysts of *E. piriformis* also lag by 3–4 hr before they are found in the feces (Cheissin, 1967).

The length of the prepatent period may change within small limits depending on the age of the host. In year-old rabbits, the prepatent period of *E. irresidua* and *E. media* was 24–48 hr longer than in younger animals. Cheissin (1967) explained the increased prepatent period in adult rabbits by "the development of immunity

which results in a suppression of certain functions of the parasite." In some cases, the longer prepatent period is associated with a greater number of asexual generations preceding gamogony. For example, the prepatent period of *E. magna* is longer than that of *E. irresidua*. In the former, the third-generation merozoites begin gametogenesis in 142 hr, but in the latter species it is the second-generation which yields gamonts after 96–120 hr. The first oocysts of *E. magna* appear at about 180 hr PI, while those of *E. irresidua* appear at 150 hr PI (Cheissin, 1967).

The prepatent period also depends on the length of both asexual and sexual development, with the former usually taking much more time, while the latter is usually completed in 24–48 hr. In rabbits, for example, merogony takes 96 hr for *E. media*, 136–140 hr for *E. magna*, and 7–8 days for *E. intestinalis*, while gamete development is 48 hr for *E. coecicola*, 30 hr for *E. irresidua*, 32–38 hr for *E. intestinalis* and *E. magna*, and 20–22 hr for *E. media* (Cheissin, 1967).

PATENCY

The length of the patent period also fluctuates from several to 30 days, or more, in rabbits. The shortest patent period is in *E. perforans*, 6–7 days, while it is 5–10 days in *E. intestinalis*, 6–10 days in *E. media*, 7–9 days in *E. coecicola*, 11–12 days in *E. piriformis*, up to 16 days in *E. irresidua*, and 15–19 days in *E. magna*. Thus, in the intestinal eimerians, the patent phase lasts from 5–35 days, whereas in *E. stiedai* it is 21–30 days and mortality generally occurs during this period.

SPORULATION

Once outside the host, the oocyst must sporulate before it is infective to another suitable host, and it is important to remember that no entrance of nutritional materials from the environment can occur through the virtually impervious oocyst wall. Thus, the process of sporulation can only occur by using energy from the reserve materials within it—materials (glycogen, lipids, proteins) sequestered from the host cell during the parasite's endogenous development. The presence of oxygen, moisture, shade (direct exposure to ultraviolet radiation—sunlight—will kill oocysts quickly) and, generally, a temperature less than the body temperature of the host, all are necessary for oocyst survival. In controlled conditions, an incubation temperature of 28–30°C has been found to be optimum for sporulation. Sporulation of oocysts will not, and cannot, occur in an environment lacking oxygen, and any factor that removes oxygen from the immediate vicinity of the oocysts will impede sporogony. Thus, it is important to eliminate bacteria, and the putrefaction they produce, from the oocyst's environment because of their negative effect on sporulation. Through trial and error over time it has been learned that the best medium for culturing and sporulating oocysts is a 2–3% (w/v) aqueous solution of potassium dichromate ($K_2Cr_2O_7$). In this solution, oocysts have complete access to oxygen and bacteria do not develop. Other solutions that suppress putrefaction have been used for culturing coccidial oocysts (e.g., 2% [v/v] sulfuric acid, H_2SO_4), but they do not seem to work as well. If these conditions are met, the sporulation process will continue. Once completed, the fully formed oocyst and its sporocysts are resistant to environmental extremes and the sporozoites therein are infective to the next suitable hosts that may ingest them. However, Cheissin (1967) cautioned that after the sporozoites are formed, another few days are required for the oocysts to become infective. For example, oocysts of *E. intestinalis* and *E. magna*, which completed sporulation in 72 hr, did not cause an infection in rabbits, but after 120 hr the same oocysts were infective (Cheissin, 1947a, 1948). Edgar (1954,

1955) had similar results with *E. maxima* (in chickens); he said that oocysts with formed sporozoites were not infective immediately, but that they required another few hours to become infective.

Among oocysts of the same species, the small ones seem to sporulate before the large ones. For example, large *E. magna* oocysts, 35 × 22, completed sporulation at 22°C in 65–72 hr, but smaller oocysts, 25–28 × 16–18, sporulated in 48–52 hr (Cheissin, 1967). This led Cheissin (1967) to conclude that oocysts of various sizes in one portion of feces do not all sporulate at the same time, that sporulation begins sooner in small oocysts, and that large ones of the same species may lag as much as a day behind. In rabbit feces collected at one time, Cheissin (1967) noted the sporulation of small oocysts of *E. media* and *E. intestinalis* within 18 hr, the oocysts of *E. magna* and the small oocysts of *E. irresidua* within 24 hr, and the large oocysts of *E. irresidua* and *E. intestinalis* at the end of 48 hr.

DETECTION AND IDENTIFICATION: MORPHOLOGICAL, IMMUNOLOGICAL, MOLECULAR

Morphological Diagnosis

The end product of endogenous intracellular development within an infected host is a resistant propagule, the oocyst, which leaves the host, usually in the feces. This is the structure most readily available to the practitioner (veterinarian, wildlife biologist, pet owner) who wants to identify the species, in most cases without killing the host. Thus, the vast majority (> 98%) of coccidian species are defined only from this life-cycle stage, the sporulated oocyst, with little or no knowledge of the temporal (e.g., pre- and patent periods), or spatial (location in the host body) biology, or the number and structure of the endogenous (asexual, sexual) stages.

In most cases, the unique combination of host species, and of the size and specialized structures of the sporulated oocyst, is sufficient to separate it from all others that are known from that host group. However, sporulated oocysts have only a limited number of structural characters (< 20) and the fewer characters available, the more bothersome the (morphological) species problem becomes. In some cases, oocysts from unrelated host species look very nearly identical and cannot be reliably differentiated by morphology and size alone (Joyner, 1982). Other times, individual coccidia species may produce oocysts that vary substantively in size and/or appearance during patency (Parker and Duszynski, 1986; Gardner and Duszynski, 1990); thus, we are sometimes haunted by the question, "What is this species with which I am dealing?"

Before oocysts can be identified to genus or species, they first must be found in contaminated feces. Larger oocysts (most *Eimeria*, *Isospora* species) are easier to detect in contaminated feces than are very small oocysts or sporocysts (e.g., *Besnoitia*, *Cryptosporidium*, *Toxoplasma*, *Sarcocystis*). In previous contributions, Duszynski and Wilber (1997) and Duszynski and Upton (2009, Chapter 2) outlined techniques necessary to collect and preserve feces, store samples with coccidian oocysts, and identify and measure the key structural features to be able to make reasonably accurate identifications, so these need not be repeated here. However, very small oocysts (e.g., *Cryptosporidium*, *Toxoplasma* species) and isolated sporocysts (*Sarcocystis* species) need more labor-intensive, microscopical methods that may include both concentration and staining of fecal smears before they can be seen. Differential staining methods include safranin-methylene blue stain (Baxby et al., 1984), Kinyoun (Ma and Soave, 1983), Ziehl-Neelsen (Henricksen and Pohlenz, 1981), and DMSO-carbol fuchsin (Pohjola et al., 1984), which all stain the oocysts red and counterstain the

background. These methods, however, vary in sensitivity and specificity and can be time-consuming (Fayer et al., 2000). Fluorochrome stains (Campbell et al., 1992; Fayer et al., 2000) are more sensitive, but non-oocyst-like structures in the fecal debris often also take up the stain.

Immunological Diagnosis

Various immunological-based (IB) techniques are available for detection of very small oocysts, but most of these have been developed to detect oocysts of *Cryptosporidium*, most often in human stools. Some of the IB techniques include polyclonal fluorescent antibody tests, direct immunofluorescent antibody tests (IFAT), latex agglutination reactions, immnofluorescence (IF) with monoclonal antibodies, enzyme-linked immunosorbent assays (ELISA), reverse passive hemagglutination (RPH) immunoserology using IF detection and ELISA, and solid-phase qualitative immunochromatographic assays (see Fayer et al., 2000, p. 1313, for summary and literature citations).

Molecular Diagnosis using DNA

Finally, the development and continued refinement of molecular techniques, especially polymerase chain reaction (PCR), now offer reasonably quick, highly sensitive, accurate, and inexpensive alternatives to conventional diagnosis of the coccidia of humans and domesticated and wild animals. This molecular tool provides information on genetic variability of various isolates of presumed "species" and can demonstrate that what was once thought to be a single species is not genetically uniform, but consists of several distinct genotypes or cryptic species. Most genetic studies use parasite-specific PCR primers to overcome the problem of oocysts recovered from environmental/fecal specimens that contain many contaminants. Applications via PCR include isolating DNA fragments from genomic DNA by selective amplification of a specific region of the target DNA to produce a single or a few copies of a piece of DNA and generate multiple copies of a particular DNA sequence. Sequence analysis examines all bases at a particular locus and has become the "gold standard" of genotyping studies (Fayer et al., 2000). Comparisons between various genotype sequences from oocysts of rabbit origins include sequences of the small subunit ribosomal (SSU rRNA), 70-kDa heat shock protein (hsp70), actin, *Cryptosporidium* oocyst wall protein (COWP), genomic sequence of unknown coding function (Lib13), and 60-kDa glycoprotein (gp60) loci that group into two subtype families Va and Vb.

PCR amplifies regions of DNA that it targets and can be used to analyze extremely small amounts of a sample, including a few or even a single oocyst. Once DNA pieces have been cloned and sequenced, DNA phylogenies can be constructed if similar gene regions are known from other related species. Since the base sequences of only a few loci have been examined and detailed for only a small number of mammalian (and other) coccidia, further research on the molecular characterization and speciation of these organisms is desperately needed (Fayer et al., 2000).

The random amplified polymorphic DNA (RAPDs) technique was developed in the 1990s (Welsh and McClelland, 1990; Williams et al., 1990), and such specific fingerprints have been used to differentiate species or strains of parasitic protists, including *Plasmodium* and *Leishmania* (Tibayrenc et al., 1993), trypanosomes (Steindel et al., 1993; Waitumbi and Murphy, 1993), and rodent eimerians (Hnida and Duszynski, 1999a, b). This technique enables the examination of a large number of independent loci randomly distributed throughout the genome.

Cere et al. (1995) did a comparative analysis of RAPDs and generated fingerprints using 11 arbitrary primers on nine rabbit *Eimeria* species: *E. coecicola, E. exigua, E. flavescens, E. intestinalis, E. magna, E. media, E. perforans, E. piriformis,*

E. vejdovskyi, two strains of *E. intestinalis,* and four strains of *E. media* from different geographic regions. The profiles obtained differed considerably according to the species; these species-specific fingerprints showed that RAPD assays might be useful for diagnostic purposes. These same species also can be distinguished on the basis of other criteria, including the location in the intestine and their macroscopic lesions, the endogenous stages of their life cycles, their prepatent periods, their pathogenicity, and on the morphology of their oocysts, with the latter being more routinely used. However, none of these criteria alone is perfect for identification of species and the considerable variation of oocyst dimensions in some species makes only the observation of oocysts of poor value for specific diagnosis of some species. Moreover, although some *Eimeria* species exhibit host specificity, several different eimerians also may infect the same host and, sometimes, the same intestinal segment. In their study, Cere et al. (1995) developed a new approach for the construction of a coccidia diagnostic tool.

This tool is based on the Digoxigenin (DIG, Boehringer)-radiolabeled isolation (DIG-labeling) of a species-specific probe after a RAPD and determination of two primers in this probe for a PCR, which is more specific than RAPD (Cere et al., 1996). A specific fragment of 800 bp was isolated after RAPD and then cloned and DIG, Boehringer-radiolabeled. Sequencing the 3' and 5' ends of this probe enabled the determination of the two primers that could be used in a PCR reaction. The amplified product of 750 bp was specific for *E. media* and seems to be conserved in *E. media*, because it was amplified in multiple strains tested from different countries (France, Balearic Isles, Poland, Guadalupe in the French West Indies). Their technique allowed the detection of as few as 10 oocysts, and the efficiency of the amplification was not changed when two species were mixed. The threshold of detection of oocysts in fecal matter was as few as 30 oocysts.

Thus, the RAPD very quickly provides a great number of species markers, as they showed (Cere et al., 1995, 1996). Their work showed a high degree of interspecific variability in *Eimeria* from the rabbit (which is why RAPDs work in these instances) and a low extent of intraspecific variability that was characterized by the presence of several common bands in all the strains and lines of *E. media*. In addition, because of its rapidity, requirement of only minute quantities of DNA, and no requirement of sequence information on the genome to be fingerprinted, the RAPD technique can prove helpful in searching for genetic markers involved in various biological characteristics of the coccidia such as drug resistance, virulence, or the trait of precociousness (for precocious lines).

Rabbit (and most other) coccidians usually are identified based on observation of the morphology of sporulated oocysts, but several eimerians from domestic rabbits, *O. cuniculus*, cannot be easily identified by non-specialists based only on this morphology. For example, it is possible to distinguish neither *E. perforans* from *E. media* nor *E. flavescens* from *E. irresidua* only by oocyst form and measurements. The differentiation of these species can be difficult and time-consuming because the life cycle (site of development in the intestine, prepatent period, etc.) or other criteria (clinical signs, etc.) must be studied. Thus, newer methods, like RAPDs, are needed for direct, rapid, and efficient identification of these (and other) intestinal coccidians as demonstrated on *Eimeria* spp. of the domestic fowl, rat, mice, cattle, sheep, and rabbit (Mac Pherson and Gajadhar, 1993; Procunier et al., 1993; Cere et al., 1995; Hnida and Duszynski, 1999a, b). However, this method cannot be used when the different species are mixed or in very small quantities in feces. In this case, amplification of specific DNA sequences by PCR and hybridization of these fragments with oligonucleotide probes provide a highly

sensitive and specific tool for the detection of coccidia directly from feces.

Finally, we must be cautious in our reliance on DNA. Recently, there seems an attitude among some of our molecular colleagues (anon., pers. comm.) that the organism itself is no longer important and that all we need is some of its DNA to figure out what it is. We hope that this viewpoint is not universal among reductionists. Here we offer the caution expressed by Chargaff (1997) that the phenomenon of life, or any living entity, is much more than just molecules obeying the laws of physics and chemistry.

OOCYST SURVIVAL

The oocyst wall, composed of two or more layers in most coccidian species, is the parasite's first line of defense in surviving the external environment. Monné and Hönig (1954), working with rabbit (*E. stiedai*), chicken, and cat coccidian oocysts, investigated the optical and chemical properties of the wall. They concluded that the most exterior layer of the wall was quinone-tanned protein, while the inner layer(s) consisted of a lipid coat firmly associated with a protein lamella. The outer part of the inner layer had a minor amount of lipid, while the innermost part of the inner layer had a major amount of lipid in its protein lamella. These chemical constituents of the oocyst wall certainly play an important role in its highly resistant nature of protecting the developing sporoplasm within.

Our understanding of the survival of oocysts in the external environment, and the mechanisms by which they reach an appropriate definitive host, is minimal and requires additional study. We know that the availability of oxygen is critical to sporulation and that moisture, temperature, and direct exposure to sunlight all influence the ability of oocysts to sporulate in the external environment (or not). The interactions of these and other factors (e.g. mechanical vectors such as invertebrates) are not well understood. Wagenbach and Burns (1969) measured respiration of *E. stiedai* oocysts polarographically. An early increase in respiratory rate was followed by a depression in rate that correlated with the appearance of the early spindle stage. The rate again increased and then decreased toward a base rate during and after completion of sporulation.

In general, oocysts that leave the host unsporulated (e.g., *Eimeria*, *Isospora*, *Toxoplasma*, *Besnoitia* species) sporulate more rapidly at higher temperatures and slower at lower temperatures; exposure to temperatures less than 10°C or greater than 50°C is usually lethal to unsporulated oocysts. Oocysts of *E. magna* and *E. perforans* did not sporulate below 10°C (Becker and Crouch, 1931). Sibalić (1949) looked at the effect of temperature and moisture on sporulation of *E. stiedai* and *E. perforans* oocysts and found that sporulation was achieved between 10 and 35°C, but that the optimal temperature for sporulation was 20–27°C. For *E. stiedae*, complete sporulation occurred in oocysts at 40°C for 3 days, at 50°C for 6 hrs, and even at 60°C for 30 min. Sibalić (1949) also reported that when he held oocysts of both *E. stiedai* and *E. perforans* at 0°C for 135 days and then raised the temperature to 27°C, the oocysts would sporulate normally. On the other hand, Cheissin (1967) said that oocysts of *E. magna* began sporulation when kept at 10–12°C, but did not completely develop in 226–325 hr, and that 90–100% of these oocysts later degenerated. At 20–22°C about 95% of *E. magna* oocysts sporulated in 65–73 hr, while those of *E. perforans* sporulated within 48 hr at the same temperature (Cheissin, 1967). At 25°C, ~50% of the oocysts of *E. magna* sporulated completely in 48 hr, 40% formed sporoblasts, and 10% degenerated (Becker and Crouch, 1931). At that temperature, *E. perforans* required 48 hr for complete sporulation. At 33°C, ~80% of the oocysts of *E. magna* completed sporulation in

72 hr, but *E. perforans* required only 48 hr. At 35°C, ~41% of *E. magna* oocysts developed within 96 hr, but at 36°C sporulation began, but was not completed, while 10–14% of *E. perforans* oocysts sporulated completely at 35°C, but none developed at any higher temperature. Oocysts of both species were killed within 10 min at 51°C (Becker and Crouch, 1931).

Many authors have looked at the survival of oocysts of avian coccidia in the soil, but only a few bear mention here to illustrate the extremes of their conflicting results. Koutz (1950) worked with a culture of four *Eimeria* species and found differences in survival between species and trials. While monocultures of *E. tenella* oocysts, in 14 ft^2 wire-enclosed outdoor pens, remained infective for 272 days from September through June, its oocysts did not survive in mixed cultures with three other species when exposed to "a severe winter and partial summer up to 322 days even though heavy weeds had grown during the spring to afford some protection," while oocysts of the other three species did survive and remained infective. In another experiment with the mixed culture of four species, two species survived a year's exposure while two did not. In one pen left in direct sunlight and not protected by grass or weeds, no oocysts of any of the four species survived. Doran and Vetterling (1969) froze sporulated oocysts of two species of poultry eimerians (*E. tenella*, chickens; *E. meleagrimitis*, turkeys) in media containing 7% dimethylsulfoxide, frozen to −80°C at 1 degree/min and stored them above liquid nitrogen for up to 4 mo, at which time oocysts were uninfective to chicks and turkey poults, but after 3 mo sporocysts of both species produced infections. More recently, Williams et al. (2010) used a chilled 2.5% (w/v) potassium dichromate solution as a long-term preservative of sporulated oocysts for six *Eimeria* species of chickens and stored these oocysts in the dark at 4 ± 2°C; after 23 yrs, oocysts were not viable, as measured by their inability to infect chickens, but after 24 yrs it was still possible to recover their DNA, and after almost 25 yrs the oocyst walls and internal structures remained well-preserved in 83–98% of the oocysts of all six species examined.

Oocysts of *Cryptosporidium* species can remain viable in the environment for many months, and we know they are resistant to chlorine disinfection, which further facilitates their transmission (Fayer, 2007). Additional factors, such as ammonia, influence survival of all oocysts in stored animal wastes. These factors and the combined effects of soil matrix, vegetation, and precipitation on the transfer of oocysts were reviewed by King and Monis (2007). The temperatures and times given are mostly for oocysts of *C. parvum*, but likely would be similar for *C. cuniculus* (from Fayer et al., 2000). Oocysts held for 6 mo at 20°C were still infectious for suckling mice, but higher temperatures resulted in a rapid loss of viability, as some oocysts held at 25 and 30°C were infections for only 3 mo. Warming oocysts from 9 to 55°C over 20 min resulted in loss of infectivity for suckling mice, while oocysts held at 59.7°C for 5 min had very low infectivity, and others held at 71.7°C for only 5 sec were killed (Harp et al., 1996). Freezing kills *Cryptosporidium* oocysts. Snap freezing and programmed freezing to −70°C resulted in immediate killing of *C. parvum* oocysts even in the presence of a variety of cryoprotectants (Fayer et al., 2000). At higher temperatures oocysts survived longer; some held at −20°C were viable for up to 8 hr, but not at 24 hr, and some held at −10°C were infectious for mice up to 1 wk after storage; those held at −5°C remained viable for up to 2 mo. Fayer et al. (2000) suggested that these results indicate that fluids within *Cryptosporidium* oocysts offer minimal cryoprotection to the sporozoites. Desiccation is lethal to *Cryptosporidium* oocysts. Only 3% of oocysts were found viable after 2 hr of desiccation and 100% killing was seen at 4 hr (Fayer et al., 2000).

Between the extremes noted above, the sporulation of oocysts in a field-collected fecal sample is dependent upon at least the following factors: the coccidian species, the time and temperature between collection and arrival of the sample at the laboratory, the medium in which the sample was stored, the amount of molecular oxygen available to the stored oocysts, and the concentration of oocysts in the sample. Under optimal laboratory conditions, sporulation of oocysts from mammals occurs best between 20 and 25°C, but this will vary among vertebrate classes (Duszynski and Wilber, 1997). Delapland and Stuart (1935) observed live oocysts of *E. tenella* in the soil of shaded meadow plots 1.5 yr after removal from the chicken. Cheissin (1959, 1967) found the oocysts of rabbit coccidia were viable after 2 yr in $K_2Cr_2O_7$ solution.

Interestingly, Černá (1974) reported an anomaly in the sporulation of *I. lacazei* from sparrows (*Passer domesticus*) in Prague. She found that some *I. lacazei* oocysts resembled oocysts of *Caryospora* (one sporocyst with eight sporozoites) even after "normal" sporulation at 20–25°C, while a few others were intermediate between the genera. Matsui et al. (1989) confirmed this sporulation abnormality (?) when they saw that a few oocysts (< 1%) of some *Eimeria* species (normally, four sporocysts each with two sporozoites) can be induced to change into *Isospora*-like oocysts (two sporocysts each with four sporozoites) when fresh, unsporulated oocysts are first heated to 50°C for 30–60 seconds before incubation at 25°C for 1 wk.

Fish (1931a, b) noted that ultraviolet (UV) irradiation killed oocysts of *E. tenella*. Farr and Wehr (1949) looked at the survival on soil of three *Eimeria* species from chickens, under various field conditions including direct sunlight, partial shade, and deep shade, but their results were equivocal both between species and between test plots; however, they said that infective oocysts of *E. acervulina* were recovered from the plots for up to 86 wk after they were seeded. Cheissin (1967, p. 172) reported a series of studies by Litver (1935,

1938; papers not available to us) that examined the effects of UV light on rabbit coccidia in more detail. When irradiated for 3–60 min with a mercury-quartz lamp, some unsporulated oocysts of *E. perforans* failed to develop and died. After a 3-min exposure, 4% failed to develop and a 60-min exposure killed 93% of the oocysts. As the UV dosage was increased, there was a rise in the percentage of undeveloped oocysts. When *E. perforans* oocysts were irradiated in feces within 12 hr after removal from rabbits, different results occurred. At low levels (< 12 min exposure), a comparatively low percentage of oocysts, 4–15%, failed to develop. Increasing the dosage did not raise the death rate. However, with *E. stiedai* oocysts, 60-min exposure to UV resulted in only 33% of the oocysts failing to develop, demonstrating that the *E. stiedai* oocysts are more resistant than those of *E. perforans* under similar conditions of exposure to UV irradiation. That the oocysts of *E. perforans* were less sensitive to UV irradiation 12 hr after emerging than were newly emerged oocysts seems to indicate that during sporulation there are developmental stages that have a greater or lesser sensitivity to harmful types of light rays. The sensitivity of *E. stiedai* oocysts did not change when they were kept at temperatures of from 1 to 20°C for 3 hr prior to irradiation; that is, there was no effect on sensitivity to UV rays. However, the temperature regime after irradiation was of special significance to the repair of oocysts after they have been damaged by UV light. Following 4 min of irradiation at 14 to 26°C, 86–90% of the oocysts died and in some oocysts, sporulation ended within 62–88 hr (Cheissin, 1967).

PREVENTION OF CONTAMINATION

Decontamination of animal housing/bedding is difficult, at best, owing to the resistant nature of oocysts and the ineffectiveness of bleach-based disinfectants, especially after oocysts are sporulated. Hot washing of surfaces and utensils with detergent and rinsing, followed by complete drying, is recommended. Washing may be followed by ammonia-based disinfectants, steam-cleaning, or hydrogen peroxide with either peroxyacetic acid (PAA) or silver nitrate (Chalmers and Giles, 2010). However, some of these procedures may be impractical for most routine implementations.

TRANSMISSION: ENTRY OF OOCYSTS INTO THE HOST

Regardless of which organ the endogenous stages end up locating in, the oocysts always enter the host passively *per os*. Feces deposited on the ground are subjected to wind, water, and invertebrates transporting oocysts across or through the soil. In some cases, humans and other animals contribute to the movement of oocysts. To initiate infection, sporulated oocysts must be ingested with food, water, or by close personal contact with infected people, animals, or contaminated surfaces. Fayer et al. (2000) discussed the possible routes of transmission of *Cryptosporidium* oocysts, but these principles apply to all coccidian oocysts. Routes of transmission can be host-to-host through direct contract, water-borne through drinking water or immersion activity (e.g., swimming, wading), food-borne, and possibly airborne. The mechanisms of such transmission can be more complex than originally thought.

Fecal contamination of soil and natural bodies of water ultimately can lead to contamination of fresh foods, drinking water, and rivers, ponds and lakes, but the movement of oocysts from feces on land surfaces and ground water has been little studied. We know that oocysts can be picked up passively by birds, pass unharmed through their digestive tracts, and deposited at great distances from their source of origin. Oocysts also may be disseminated by earthworms, cockroaches, house flies (Metelkin, 1935), and dung beetles on land, and by rotifers in lakes, ponds, puddles, moss, and damp soils (see Fayer et al., 2000, for review and references; see Conn et al., 2007, for mechanical dissemination of *Cryptosporidium* oocysts by houseflies). Oocysts, especially those of *Cryptosporidium* species, have been found on the surface of raw vegetables from the market place (Fayer et al., 2000). Thus, rabbits may ingest oocysts along with contaminated hay, oats, and other food from the external environment where the oocysts arrive with the feces, other excretory products, and/or via contamination from various invertebrates. The less often that rabbit hutches are cleaned, the greater the possibility that rabbits will become infected. The dose of ingested oocysts will vary in every instance; sometimes only isolated oocysts will be ingested, but sometimes a large number of oocysts will be ingested at one time. Some domestic animals, especially rabbits, are infected with multiple species of coccidia most of the time. Because these animals tend to be congregated in large groups when raised on farms, almost all animals are likely to be infected and, in fact, it is very difficult to raise a coccidia-free rabbit.

EXCYSTATION OF OOCYSTS IN THE DIGESTIVE TRACT

The excystation of sporozoites from their sporocysts and from the oocyst that contains them takes place in different parts of the digestive tract. This is largely determined by the structure of the digestive tract and the functions of individual sections. There is a great deal of evidence on where excystation of eimerians occurs in the gut of chickens, but very little information has been established for rabbits. It is thought that the process does not occur in the stomach of rabbits (Smetana, 1933a). Within 15–20 min after introduction *per os* of a large dose of sporulated *E. intestinalis* oocysts, intact

oocysts were observed in the duodenum. Within 30 min, free sporozoites were found in the upper part of the small intestine and were later found along the entire length of the intestine (Smetana, 1933a). Infection of rabbits by various species of coccidia has been successful in all cases in which intact, sporulated oocysts were introduced directly into the duodenum and the stomach was bypassed. This suggests the action(s) of pepsin is/are not necessary for excystation (Cheissin, 1967). Smetana (1933a) observed excystation of small numbers of *E. stiedai* oocysts in pancreatic juice at a temperature of 37°C and explained that the activating agent for excystation was trypsin, but pancreatic enzymes may also play a role (Cheissin, 1967).

ENDOGENOUS DEVELOPMENT AND SOME PECULIARITIES

Once sporozoites have been released into the intestinal lumen from their confinement within the sporocyst, they actively seek out and invade epithelial cells. Within the cell(s) of their "choice" they take up residence in a parasitophorous vacuole (PV), often in the host cell cytoplasm, although a few species may reside within the host's nucleoplasm. Within the PV, they begin to grow and undergo merogony; nuclear division and cytokinesis occur rather rapidly, and numerous first-generation merozoites are produced. When mature, these merozoites kill and exit from the host cell to invade other gut cells. This process continues for a genetically programmed number of times, with the final generation of merozoites entering new cells to produce gamonts. There are at least two peculiarities that have been noted in the development of coccidia within rabbits, endodyogeny and the formation of multinucleate merozoites; both deserve further exploration.

Endodyogeny is a process typical of heteroxenous coccidia (e.g., *Besnoitia*, *Toxoplasma*, *Sarcocystis*), but it also is known to occur in some monoxenous eimerians of rabbits. Scholtyseck (1973) first recorded the formation of "cytomeres" from concentrically arranged vesicles of the endoplasmic reticulum within the meronts of *E. stiedai*, and he observed a process corresponding to endodyogeny within them. Later, Scholtyseck and Ratanavichien (1976) reported endodyogeny in the merozoites of *E. stiedai*, and Pakandl (1989) observed a process corresponding to endodyogeny in the first-generation meronts of *E. coecicola*. Endodyogeny thus seems to be a common phenomenon in the asexual endogenous development of rabbit coccidia.

In general, multinucleate merozoites are an exceptional phenomenon in the life cycle of coccidia, but their occurrence in rabbit coccidia is quite common. Multinucleate merozoites have been recorded in *E. perforans* (Streun et al., 1979), *E. magna* (Cheissin, 1940; Sénaud and Černá, 1969; Speer et al., 1973a, b; Danforth and Hammond, 1972; Ryley and Robinson, 1976), *E. intestinalis* (Pellérdy and Dürr, 1970; Heller, 1971), *E. flavescens* (Norton et al., 1979), *E. stiedai* (Černá and Sénaud, 1971), and in *E. media* and *E. vejdovskyi* (Pakandl, 1988). Multinucleate merozoites seem to be a common component of the life cycle of rabbit coccidia. In no other host group are they found to occur in so many *Eimeria* species (Pakandl, 1988). It appears the multinucleate merozoites possess some properties of a merozoite and some of a meront. From this viewpoint, the strict differentiation of asexual stages into meronts and merozoites need not be always correct (Pakandl, 1989).

Although interesting, there is still no satisfactory explanation for the role of multinucleate merozoites in coccidian life cycles in which they are found. Whether they further divide inside the same host cell or they leave the host cell and infect another one is unknown. Also, the possible sexual determination of two meront types in each generation still remains unsolved (Pakandl, 1988). The quantitative ratio suggests that meronts with multinucleate merozoites may give rise to microgamonts, whereas the mononucleate merozoites give rise to meronts, also with mononucleate merozoites or macrogamonts (Pakandl, 1988).

Interestingly, there are no other intestinal coccidia in the genera *Caryospora*, *Cyclospora*, *Tyzzeria*, and *Wenyonella* documented from rabbits, and only two *Isospora* species have been reported in *Ochotona* species, but these are rare reports and nothing is known about their endogenous development. Eberhard et al. (2000) inoculated rabbits (*O. cuniculus*), eight other mammal species (e.g., mice, rats, etc.), and three primate species with sporulated oocysts of *Cyclospora cayetanensis*, but were unable to achieve infection in any of them.

HOST SPECIFICITY AND CROSS-TRANSMISSION

Some of the coccidia reviewed here are host specific and some are not. For example, of the 17 eimerians known to be capable of infecting *Sylvilagus* species (Table 7.1), 12 were found in naturally infected populations, while five (*E. irresidua*, *E. magna*, *E. media*, *E. perforans*, *E. stiedai*) were successfully transmitted experimentally from *O. cuniculus*, their natural host. All of the cross-transmission work done to date is summarized in Table 11.2. Host specificity, as it pertains to each host genus, is discussed within the relevant chapter. This is certainly a rich area for future exploration.

PATHOLOGY

Coccidiosis is one of the most frequent and prevalent parasitic diseases of domesticated food animals, including rabbits, and is accompanied by weight loss, mild intermittent to severe diarrhea, feces containing mucus or blood, dehydration, and decreased (rabbit) breeding (Peeters et al., 1984; Bhat and Jithendran, 1995; Jithendran and Bhat, 1996).

Coudert et al. (1995) classified *E. exigua*, *E. perforans*, and *E. vejdovskyi* as slightly pathogenic. The pathogenicity may be connected, at least partially, to the localization of these coccidia, in most instances, in the upper parts of the villi (Streun et al., 1979; Pakandl and Coudert, 1999). In contrast, the most pathogenic rabbit coccidia, *E. intestinalis* and *E. flavescens*, parasitize the crypts of the lower part of the small intestine or cecum, respectively (Norton et al., 1979; Licois et al., 1992c; Pakandl et al., 2003). The intestinal epithelium is apparently more heavily damaged if the parasite destroys its stem cells located in the crypts (Jelínková et al., 2008). Endogenous development also causes desquamation of intestinal mucosa, capillary rupture, and bleeding into the intestinal lumen with catarrhal or hemorrhagic enteritis (Peeters et al., 1984; Bhat and Jithendran, 1995). Besides damaging intestinal mucosa, coccidia may cause a general reaction of the host with consequent changes in blood, urine, and feces (Licois et al., 1978a, b; Jithendran and Bhat, 1996; Kulišić et al., 1998; Tambur et al., 1998a, b, c, 1999). Blood taken from rabbits experimentally infected with coccidia revealed significant changes in the activity of GOT and alkaline phosphatase, and in the amount of bilirubin (Sherkov et al., 1986).

Kulišić et al. (2006) infected 52-day-old chinchilla rabbits (*O. cuniculus*) with a mixture of either 2×10^5 or 4×10^5 sporulated oocysts of *E. flavescens* (7%), *E. matsubayashii* (9%), *E. magna* (12%), *E. neoleporis* (19%), *E. perforans* (21%), and *E. media* (32%); rabbits were bled immediately before inoculation (day 0) and on days 4, 7, and 10 PI. During their infection, rabbits developed what they termed "mild coccidiosis," and presented symptoms that included **polydipsia**, bristling hair, decreased appetite, and moderate body weight loss. In their infected rabbits, the number of white blood cells (WBCs) never increased significantly over uninfected controls. Neutrophils in both infected groups increased significantly over control values on days 4, 7, and 10 PI, apparently because of the inflammation.

Percentages of basophils and eosinophils remained at mostly similar levels throughout the experiment. Monocytes rose significantly only on day 10 PI. Lymphocyte numbers decreased significantly in infected rabbits. They (2006) suggested that the inflammatory process locally recruited leukocytes (lymphocytes?) that were lost through the damaged intestinal mucosa, and this was the probable reason for their count decrease. The count increase in the later phase of infection was attributed to hemoconcentration as a result of fluid loss (Kulišić et al., 2006).

Finally, it is difficult to find any brood of rabbits without also finding coccidian oocysts in their feces. These infections attack young rabbits more severely, especially those from 2 to 6 mo old (Gres et al., 2003). Older animals having recovered from any disease-related symptoms acquire immunity, but they are still important as carriers.

PHYLOGENETIC RELATIONSHIPS OF RABBIT EIMERIANS

Our knowledge about the evolution of morphological and biological (e.g., host specificity, endogenous development, pathogenicity, etc.) traits within the *Eimeria* is trivial, as is our understanding of how these traits may relate to the parasite's evolutionary history with its host. One of the first studies to address this issue was by Reduker et al. (1987) working with rodent eimerians. Using cladistics and phenetic analysis of isozyme banding patterns, the morphology of sporulated oocysts, and several life-cycle traits, they recognized two different *Eimeria* lineages. Later, Hnida and Duszynski (1999a, b) reported similar results; that is, two different eimerian lineages, using phylogenetic analyses of molecular sequence and bioprinting data. Zhao and Duszynski (2001a, b) were the first to split rodent *Eimeria* species into two distinct lineages, based partially on oocyst size and shape, but primarily on the presence or absence of one unique structure, the oocyst residuum (OR). Using plastid ORF 470 and 23S rDNA and nuclear 18S rDNA sequences, they showed distinct phylogenetic relationships among rodent eimerians in three families based on the presence/absence of the OR. No similar studies were done with rabbits until Kvičerová et al. (2008) examined the phylogenetic relationships among *Eimeria* species infecting domesticated rabbits in the Czech Republic. They used 11/16 (69%) of the valid eimerians known from *O. cuniculus*, namely *E. coecicola*, *E. exigua*, *E. flavescens*, *E. intestinalis*, *E. irresidua*, *E. magna*, *E. media*, *E. perforans*, *E. piriformis*, *E. stiedai*, and *E. vejdovskyi*. Sporulated oocysts of these species are relatively heterogeneous relative to their oocyst morphology (e.g., presence/absence of M, OR, SR, etc.) and size (e.g., oocyst, sporocyst dimensions), as is the location of their endogenous stages in the host (e.g., various locations in the gut vs. liver), high (*E. coecicola*, *E. intestinalis*, *E. irresidua*, *E. magna*), moderate (*E. flavescens*, *E. media*, *E. stiedai*) or low pathogenicity (*E. exigua*, *E. matsubayashii*, *E. neoleporis*, *E. perforans*, *E. piriformis*), and so on. Using partial sequences of 18S rDNA from their 11 rabbit eimerians they did a phylogenetic analysis using all *Eimeria* sequences available in GenBank and found that the rabbit eimerians were a well-formed, monophyletic cluster possessing a relatively long common branch. Prior to their work (2008), only three molecular studies were known on rabbit coccidia and they focused on diagnostics and inter- and intraspecific variation using RAPD techniques (Ceré et al., 1995, 1996, 1997). Recently, Oliveira et al. (2011), using nucleotide sequences of ITS1 ribosomal DNA, developed species-specific molecular assays for the identification of *Eimeria* species infecting the domestic rabbit. Their work showed good reproducibility and presented a consistent sensitivity with three different brands of amplification enzymes. Unfortunately, however, they did not have sufficient molecular data to provide a clear phylogenetic signal.

The study by Kvičerová et al. (2008), however, did provide a clear phylogenetic signal, indicating that the rabbit-specific *Eimeria* form a monophyletic species group/cluster. And, of all the morphological and biological characters used in their analysis, only the presence or absence of an OR strictly followed the phylogenetic division into two monophyletic sister lineages, which made it impossible to decide which state is **plesiomorphic** for their rabbit eimerians. The only conclusion they could make was that the presence/absence of an OR seemed to be an evolutionarily conserved feature. They wisely acknowledged, however, that in such a species-rich group as the *Eimeria*, any analysis of its evolutionary history is compromised by the samples that are available to each investigator (Kvičerová et al., 2008); that is, the 60 or so *Eimeria* sequences deposited in GenBank to date represent only a small fraction of known species in mammals and a miniscule fraction of all species yet to be discovered from all vertebrates. For example, there are about 5,416 mammal species (Wilson and Reeder, 2005) organized into 1,229 genera, 53 families and 29 orders. Members of only 5/29 (17%) orders have been examined for coccidia, from which <15% of their species have been examined for coccidia. Using these and other available data on the extant number of vertebrates and their known coccidians on Earth, Duszynski (2011) estimated there may be, minimally, 124,300 coccidia species that parasitize all vertebrates, most of which are *Eimeria* species. The 1,800 coccidia species currently known is only 1.4% of the number of species that likely exist in Earth's vertebrates! Clearly, a much more complete dataset of hosts, eimerians, and sequenced genes must be available before the evolution of various traits within *Eimeria* species can be seriously addressed (Kvičerová et al., 2008). This was clearly pointed out by Kvičerová et al. (2008) in the lack of congruence they found between any of their phylogenies and most of the morphological and biological traits of the 11 *Eimeria* species they examined from *O. cuniculus*.

TABLE 11.1   Alphabetical list of All Coccidian Parasites Covered in this Book and the Lagomorph Hosts from Which They Have Been Reported. All parasites are reported from naturally infected rabbits unless indicated as experimental (E).

Besnoitia besnoiti **(Marotel, 1912) Henry, 1913** (E)

  Oryctolagus cuniculus (L., 1758), European (domestic) rabbit

Besnoitia jellisoni **Frenkel 1955a** (E)

  Oryctolagus cuniculus (L., 1758), European (domestic) rabbit

Besnoitia wallacei **(Tadros and Laarman, 1976) Dubey, 1977** (E)

  Oryctolagus cuniculus (L., 1758), European (domestic) rabbit

Cryptosporidium cuniculus **Inman and Takeuchi, 1979**

  Oryctolagus cuniculus (L., 1758), European (domestic) rabbit

  Sylvilagus floridanus (J.A. Allen, 1890), Eastern cottontail (?)[1]

Cryptosporidium meleagridis **Slavin, 1955**

  Oryctolagus cuniculus (L., 1758), European (domestic) rabbit

Cryptosporidium parvum **Tyzzer, 1912**

  Oryctolagus cuniculus (L., 1758), European (domestic) rabbit

Eimeria americana **Carvalho, 1943**

  Lepus townsendii (Bachman, 1839), White-tailed jackrabbit

Eimeria athabascensis **Samoil & Samuel, 1977a**

  Lepus americanus (Erxleben, 1777), Snowshoe hare

Eimeria audubonii **Duszynski & Marquardt, 1969**

  Sylvilagus audubonii (Baird, 1858), Desert cottontail

  Sylvilagus floridanus (J.A. Allen, 1890), Eastern cottontail

Eimeria azul **Wiggins and Rothenbacher 1979**

  Sylvilagus floridanus (J.A. Allen, 1890), Eastern cottontail

Eimeria bainae **Aoutil, Bertani, Bordes, Snounou, Chabaud & Landau, 2005**

  Lepus granatensis (Rosenhauer, 1956), Granada hare

Eimeria balchanica **Glebezdin, 1978**

  Ochotona rufescens (Gray, 1842), Afghan pika

Eimeria banffensis **Lepp, Todd & Samuel, 1973**

  Ochotona collaris (Nelson, 1893), Collared pika

  Ochotona curzoniae (Hudgson, 1858), Plateau pika

  Ochotona hyperborea (Pallas, 1811), Northern pika

  Ochotona princeps (Richardson, 1828), American pika

TABLE 11.1 Alphabetical list of All Coccidian Parasites Covered in this Book and the Lagomorph Hosts from Which They Have Been Reported. All parasites are reported from naturally infected rabbits unless indicated as experimental (E). *(cont'd)*

Eimeria barretti **Lepp, Todd & Samuel, 1972**

  Ochotona collaris (Nelson, 1893), Collared pika

  Ochotona princeps (Richardson, 1828), American pika

Eimeria brachylagia **Duszynski, Harrestien, Couch & Garner, 2005**

  Brachylagus idahoensis (Merriam, 1891), Pygmy rabbit

Eimeria cabareti **Aoutil, Bertani, Bordes, Snounou, Chabaud & Landau, 2005**

  Lepus europaeus (Pallas, 1778), European hare

  Lepus granatensis (Rosenhauer, 1956), Granada hare

Eimeria calentinei **Duszynski & Brunson, 1973**

  Ochotona collaris (Nelson, 1893), Collared pika

  Ochotona curzoniae (Hudgson, 1858), Plateau pika

  Ochotona hyperborea (Pallas, 1811), Northern pika

  Ochotona princeps (Richardson, 1828), American pika

Eimeria campania **(Carvalho, 1943) Levine & Ivens, 1972**

  Lepus europaeus (Pallas, 1778), European hare

  Lepus nigricollis (F. Cuvier, 1823), Indian hare

  Lepus townsendii (Bachman, 1839), White-tailed jackrabbit

Eimeria circumborealis **Hobbs & Samuel, 1974**

  Ochotona collaris (Nelson, 1893), Collared pika

  Ochotona hyperborea (Pallas, 1811), Northern pika

  Ochotona princeps (Richardson, 1828), American pika

Eimeria coecicola **Cheissin (Kheysin), 1947**

  Oryctolagus cuniculus (L., 1758), European (domestic) rabbit

Eimeria coquelinae **Aoutil, Bertani, Bordes, Snounou, Chabaud & Landau, 2005**

  Lepus europaeus (Pallas, 1778), European hare

  Lepus granatensis (Rosenhauer, 1956), Granada hare

Eimeria cryptobarretti **Duszynski & Brunson, 1973**

  Ochotona collaris (Nelson, 1893), Collared pika

  Ochotona curzoniae (Hudgson, 1858), Plateau pika

  Ochotona hyperborea (Pallas, 1811), Northern pika

  Ochotona princeps (Richardson, 1828), American pika

(Continued)

TABLE 11.1   Alphabetical list of All Coccidian Parasites Covered in this Book and the Lagomorph Hosts from Which They Have Been Reported. All parasites are reported from naturally infected rabbits unless indicated as experimental (E). (*cont'd*)

Eimeria daurica Matschoulsky, 1947a

　　Ochotona dauurica (Pallas, 1776), Daurian pika

Eimeria environ Honess, 1939

　　Sylvilagus audubonii (Baird, 1858), Desert cottontail

　　Sylvilagus floridanus (J.A. Allen, 1890), Eastern cottontail

　　Sylvilagus nuttallii (Bachman, 1837), Mountain cottontail

Eimeria erschovi Matschoulsky, 1949

　　Ochotona dauurica (Pallas, 1776), Daurian pika

　　Ochotona pallasi (Gray, 1867), Mongolian pika

Eimeria europaea Pellérdy, 1956

　　Lepus capensis (L., 1758), Cape hare

　　Lepus timidus (L., 1758), Mountain hare

　　Lepus europaeus (Pallas, 1778), European hare

Eimeria exigua Yakimoff, 1934

　　Lepus arcticus (Ross, 1819), Arctic hare (?)

　　Lepus capensis (L., 1758), Cape hare (?)

　　Lepus europaeus (Pallas, 1778), European hare (?)

　　Lepus timidus (L., 1758), Mountain hare (?);

　　Oryctolagus cuniculus (L., 1758), European (domestic) rabbit

　　Sylvilagus floridanus (J.A. Allen, 1890), Eastern cottontail (?)

Eimeria flavescens Marotel & Guilhon, 1941

　　Oryctolagus cuniculus (L., 1758), European (domestic) rabbit

Eimeria gantieri Aoutil, Bertani, Bordes, Snounou, Chabaud & Landau, 2005

　　Lepus granatensis (Rosenhauer, 1956), Granada hare

Eimeria gobiensis Gardner, Saggerman, Batsaikan, Ganzorig, Tinnin & Duszynski, 2009

　　Lepus tolai (Pallas, 1778), Tolai hare

Eimeria groenlandica Madsen, 1938 emend. Levine & Ivens, 1972

　　Lepus arcticus (Ross, 1819), Arctic hare

Eimeria haibeiensis Yi-Fan, Run-Roung, Jian-Hua, Jiang-Hui & Duszynski, 2009

　　Ochotona curzoniae (Hudgson, 1858), Plateau pika

Eimeria holmesi Samoil & Samuel, 1977a

　　Lepus americanus (Erxleben, 1777), Snowshoe hare

TABLE 11.1 Alphabetical list of All Coccidian Parasites Covered in this Book and the Lagomorph Hosts from Which They Have Been Reported. All parasites are reported from naturally infected rabbits unless indicated as experimental (E). (*cont'd*)

Eimeria honessi **(Carvalho, 1943) emend. Levine and Ivens, 1972 and Pellérdy, 1974**

Sylvilagus audubonii (Baird, 1858), Desert cottontail

Sylvilagus floridanus (J.A. Allen, 1890), Eastern cottontail

Sylvilagus nuttallii (Bachman, 1837), Mountain cottontail

Eimeria hungarica **Pellérdy, 1956**

Lepus capensis (L., 1758), Cape hare

Lepus europaeus (Pallas, 1778), European hare

Lepus granatensis (Rosenhauer, 1956), Granada hare

Lepus nigricollis (F. Cuvier, 1823), Indian hare

Lepus timidus (L., 1758), Mountain hare (?)

Eimeria intestinalis **Cheissin, 1948**

Lepus sp. (?)

Oryctolagus cuniculus (L., 1758), European (domestic, tame) rabbit

Eimeria irresidua **Kessel & Jankiewicz, 1931**

Lepus europaeus (Pallas, 1778), European hare

Lepus nigricollis (F. Cuvier, 1823), Indian hare (?)

Oryctolagus cuniculus (L., 1758), European (domestic, tame) rabbit

Sylvilagus floridanus (J.A. Allen, 1890), Eastern cottontail

Eimeria keithi **Samoil & Samuel, 1977a**

Lepus americanus (Erxleben, 1777), Snowshoe hare

Eimeria klondikensis **Hobbs & Samuel, 1974**

Ochotona collaris (Nelson, 1893), Collared pika

Ochotona hyperborea (Pallas, 1811), Northern pika

Ochotona princeps (Richardson, 1828), American pika

Eimeria lapierrei **Aoutil, Bertani, Bordes, Snounou, Chabaud & Landau, 2005**

Lepus europaeus (Pallas, 1778), European hare

Lepus granatensis (Rosenhauer, 1956), Granada hare

Eimeria leporis **Nieschulz, 1923**

Lepus americanus (Erxleben, 1777), Snowshoe hare

Lepus arcticus (Ross, 1819), Arctic hare

Lepus capensis (L., 1758), Cape hare

(*Continued*)

TABLE 11.1   Alphabetical list of All Coccidian Parasites Covered in this Book and the Lagomorph Hosts from Which They Have Been Reported. All parasites are reported from naturally infected rabbits unless indicated as experimental (E). (cont'd)

Lepus europaeus (Pallas, 1778), European hare

Lepus granatensis (Rosenhauer, 1956), Granada hare

Lepus nigricollis (F. Cuvier, 1823), Indian hare

Lepus timidus (L., 1758), Mountain hare (?)

Lepus tolai (Pallas, 1778), Tolai hare

Oryctolagus cuniculus (L., 1758), European (domestic) rabbit (?)

Eimeria macrosculpta Sugár, 1979

Lepus europaeus (Pallas, 1778), European hare

Lepus granatensis (Rosenhauer, 1956), Granada hare

Eimeria magna Pérard, 1925b

Lepus californicus (Gray, 1837), the Black-tailed jackrabbit (?)

Lepus capensis (L., 1758), Cape hare (?)

Lepus europaeus (Pallas, 1778), European hare (?)

Lepus nigricollis (F. Cuvier, 1823), Indian hare (?)

Lepus timidus (L., 1758), the Mountain hare (?)

Oryctolagus cuniculus (L., 1758), European (domestic) rabbit

Sylvilagus floridanus (J.A. Allen, 1890), Eastern cottontail (E)

Eimeria maior Honess, 1939

Sylvilagus audubonii (Baird, 1858), Desert cottontail

Sylvilagus floridanus (J.A. Allen, 1890), Eastern cottontail

Sylvilagus nuttallii (Bachman, 1837), Mountain cottontail

Eimeria matsubayashii Tsunoda, 1952

Lepus sp. (?)

Oryctolagus cuniculus (L., 1758), European (domestic) rabbit

Eimeria media Kessel, 1929

Lepus californicus (Gray, 1837), the Black-tailed jackrabbit (?)

Lepus capensis (L., 1758), Cape hare (?)

Lepus nigricollis (F. Cuvier, 1823), Indian hare (?)

Lepus timidus (L., 1758), Mountain hare (?)

Oryctolagus cuniculus (L., 1758), European (domestic) rabbit

Sylvilagus floridanus (J.A. Allen, 1890), Eastern cottontail (E)

TABLE 11.1   Alphabetical list of All Coccidian Parasites Covered in this Book and the Lagomorph Hosts from Which They Have Been Reported. All parasites are reported from naturally infected rabbits unless indicated as experimental (E). (*cont'd*)

Eimeria metelkini **Matschoulsky, 1949**

Ochotona dauurica (Pallas, 1776), Daurian pika

Eimeria minima **Carvalho, 1943**

Sylvilagus floridanus (J.A. Allen, 1890), Eastern cottontail

Eimeria nagpurensis **Gill & Ray, 1960**

Lepus sp. (?)

Oryctolagus cuniculus (L., 1758), European (domestic) rabbit

Eimeria neoirresidua **Duszynski and Marquardt, 1969**

Sylvilagus audubonii (Baird, 1858), Desert cottontail

Sylvilagus floridanus (J.A. Allen, 1890), Eastern cottontail

Eimeria neoleporis **Carvalho, 1942**

Oryctolagus cuniculus (L., 1758), European (domestic) rabbit (E)

Sylvilagus audubonii (Baird, 1858), Desert cottontail

Sylvilagus floridanus (J.A. Allen, 1890), Eastern cottontail

Eimeria nicolegerae **Aoutil, Bertani, Bordes, Snounou, Chabaud & Landau, 2005**

Lepus granatensis (Rosenhauer, 1956), Granada hare

Eimeria ochotona Matschoulsky, 1949

Ochotona dauurica (Pallas, 1776), Daurian pika

Eimeria oryctolagi **Ray & Banik, 1965b**

Oryctolagus cuniculus (L., 1758), European (domestic) rabbit

Eimeria paulistana **da Fonseca, 1933**

Sylvilagus brasiliensis (L., 1758), Tapeti or Brazilian cottontail

Eimeria perforans **(Leuckart, 1879) Sluiter & Schwellengrebed, 1912**

Lepus americanus (Erxleben, 1777), Snowshoe hare (?)

Lepus californicus (Gray, 1837), Blacktailed jack rabbit (?)

Lepus capensis (L., 1758), Cape hare (?)

Lepus europaeus (Pallas, 1778), European hare (?)

Lepus nigricollis (syn. *Lepus ruficaudatus*) (F. Cuvier, 1823), Indian hare (?)

Lepus timidus (L., 1758), Mountain hare (?)

Oryctolagus cuniculus (L., 1758), European (domestic) rabbit

(Continued)

TABLE 11.1   Alphabetical list of All Coccidian Parasites Covered in this Book and the Lagomorph Hosts from Which They Have Been Reported. All parasites are reported from naturally infected rabbits unless indicated as experimental (E). (*cont'd*)

Sylvilagus brasiliensis (L., 1758), Tapeti

Sylvilagus floridanus (J.A. Allen, 1890), Eastern cottontail (E)

Eimeria pierrecouderti Aoutil, Bertani, Bordes, Snounou, Chabaud & Landau, 2005

Lepus granatensis (Rosenhauer, 1956), Granada hare

Eimeria pintoensis da Fonseca, 1932

Sylvilagus brasiliensis (L., 1758), Tapeti or Brazilian cottontail

Eimeria piriformis Kotlán & Pospesch, 1934

Lepus capensis (L., 1758), Cape hare (?)

Lepus europaeus (Pallas, 1778), European hare

Oryctolagus cuniculus (L., 1758), European (domestic) rabbit

Eimeria poudrei Duszynski and Marquardt, 1969

Sylvilagus audubonii (Baird 1858), Desert cottontail

Eimeria princepsis Duszynski & Brunson, 1973

Ochotona collaris (Nelson, 1893), Collared pika

Ochotona hyperborea (Pallas, 1811), Northern pika

Ochotona princeps (Richardson, 1828), American pika

Eimeria punjabensis Gill & Ray, 1960

Lepus nigricollis (F. Cuvier, 1823), Indian hare

Eimeria qinghaiensis Yi-Fan, Run-Roung, Jian-Hua, Jiang-Hui & Duszynski, 2009

Ochotona curzoniae (Hudgson, 1858), Plateau pika

Eimeria reniai Aoutil, Bertani, Bordes, Snounou, Chabaud & Landau, 2005

Lepus europaeus (Pallas, 1778), European hare

Lepus granatensis (Rosenhauer, 1956), Granada hare

Eimeria robertsoni (Madsen, 1938) Carvalho, 1943

Lepus americanus (Erxleben, 1777), Snowshoe hare

Lepus arcticus (Ross, 1819), Arctic hare

Lepus capensis (L., 1758), Cape hare

Lepus europaeus (Pallas, 1778), European hare

Lepus nigricollis (syn. *Lepus ruficaudatus*) (F. Cuvier, 1823), Indian hare

Lepus timidus (L., 1758), Mountain hare

Lepus townsendi (Bachman, 1839), White-tailed jackrabbit

TABLE 11.1 Alphabetical list of All Coccidian Parasites Covered in this Book and the Lagomorph Hosts from Which They Have Been Reported. All parasites are reported from naturally infected rabbits unless indicated as experimental (E). (*cont'd*)

Eimeria rochesterensis **Samoil & Samuel, 1977a**

Lepus americanus (Erxleben, 1777), Snowshoe hare

Eimeria roobroucki **Grés, Marchandeau & Landau, 2002**

Oryctolagus cuniculus (L., 1758), European (domestic) rabbit

Eimeria rowani **Samoil & Samuel, 1977a**

Lepus americanus (Erxleben, 1777), Snowshoe hare

Eimeria ruficaudati **Gill & Ray, 1960**

Lepus americanus (Erxleben, 1777), Snowshoe hare

Lepus granatensis (Rosenhauer, 1956), Granada hare

Lepus nigricollis (F. Cuvier, 1823), Indian hare

Eimeria sculpta **Madsen, 1938**

Lepus arcticus (Ross, 1819), Arctic hare

Lepus europaeus (Pallas, 1778), European hare

Lepus townsendii (Bachman, 1839), White-tailed jackrabbit

Eimeria semisculpta **(Madsen, 1938) Pellérdy, 1956**

Lepus arcticus (Ross, 1819), Arctic hare

Lepus europaeus (Pallas, 1778), European hare

Lepus timidus (L., 1758), Mountain hare (?)

Lepus townsendii (Bachman, 1839), White-tailed jackrabbit

Eimeria septentrionalis **Yakimoff, Matschoulsky & Spartansky, 1936**

Lepus arcticus (Ross, 1819), Arctic hare

Lepus europaeus (Pallas, 1778), European hare

Lepus timidus (L., 1758), Mountain hare

Lepus townsendii (Bachman, 1839), White-tailed jackrabbit

Eimeria stefanskii **Pastuszko, 1961a**

Lepus europaeus (Pallas, 178), European hare

Eimeria stiedai **(Lindemann, 1865) Kisskalt and Hartmann, 1907**

Lepus americanus (Erxleben, 1777), Snowshoe hare

Lepus californicus (Gray, 1837), Black-tailed jackrabbit

Lepus capensis (L., 1758), Cape hare

(*Continued*)

TABLE 11.1 Alphabetical list of All Coccidian Parasites Covered in this Book and the Lagomorph Hosts from Which They Have Been Reported. All parasites are reported from naturally infected rabbits unless indicated as experimental (E). (*cont'd*)

Lepus europaeus (Pallas, 1778), European hare (E)

Lepus timidus (L., 1758), Mountain hare

Oryctolagus cuniculus (L., 1758), European (domestic) rabbit

Sylvilagus audubonii (Baird, 1858), Desert cottontail (E)

Sylvilagus floridanus (J.A. Allen, 1890), Eastern cottontail (E)

Sylvilagus nuttallii (Bachman, 1837), Mountain cottontail

Eimeria sylvilagi Carini, 1940

Lepus nigricollis (F. Cuvier, 1823), Indian hare (?)

Sylvilagus brasiliensis (L., 1758), Tapeti or Brazilian cottontail

Sylvilagus floridanus (J.A. Allen, 1890), Eastern cottontail

Eimeria tailliezi Aoutil, Bertani, Bordes, Snounou, Chabaud & Landau, 2005

Lepus europaeus (Pallas, 1778), European hare

Eimeria townsendi (Carvalho, 1943) Pellérdy, 1956

Lepus americanus (Erxleben, 1777), Snowshoe hare

Lepus europaeus (Pallas, 1778), European hare

Lepus timidus (L., 1758), Mountain hare

Lepus townsendii (Bachman 1839), White-tailed jackrabbit

Eimeria vejdovskyi (Pakandl, 1988) Pakandl & Coudert, 1999

Oryctolagus cuniculus (L., 1758), European (domestic) rabbit

Eimeria worleyi Lepp, Todd & Samuel, 1972

Ochotona hyperborea (Pallas, 1811), Northern pika

Ochotona princeps (Richardson, 1828), American pika

Isospora marquardti Duszynski & Brunson, 1972

Ochotona collaris (Nelson, 1893), Collared pika

Ochotona hyperborea (Pallas, 1811), Northern pika

Ochotona princeps (Richardson, 1828), American pika

Isospora yukonensis Hobbs & Samuel, 1974

Ochotona collaris (Nelson, 1893), Collared pika

Sarcocystis cuniculi (Brumpt, 1913) Tadros and Laarman, 1977a

Lepus europaeus (Pallas, 1778), European hare

Lepus timidis (L., 1758), Mountain hare (?)

Oryctolagus cuniculus (L., 1758), European (domestic) rabbit

TABLE 11.1 Alphabetical list of All Coccidian Parasites Covered in this Book and the Lagomorph Hosts from Which They Have Been Reported. All parasites are reported from naturally infected rabbits unless indicated as experimental (E). (*cont'd*)

Sarcocystis dogeli **(Matschoulsky, 1947b) Levine and Tadros, 1980**

Ochotona dauurica (Pallas, 1776), Daurian pika

Sarcocystis galuzoi **Levit, Orlov and Dymkova, 1984**

Ochotona alpina (Pallas, 1773), Alpine pika

Sarcocystis germaniaensis **Duszynski and Couch, 2013**

Lepus europaeus (Pallas, 1778), European hare

Sarcocystis leporum **Crawley, 1914**

Sylvilagus floridanus (J.A. Allen, 1890), Eastern cottontail

Sylvilagus nuttallii (Bachmann, 1837), Mountain cottontail

Sylvilagus palustris (Bachmann, 1837), Marsh rabbit

Toxoplasma gondii **(Nicolle and Manceaux, 1908) Nicolle and Manceaux, 1909**

Lepus europaeus (Pallas, 1778), European hare

Oryctolagus cuniculus (L., 1758), European (domestic) rabbit

Sylvilagus floridanus (J.A. Allen, 1890), Eastern cottontail

SPECIES INQUIRENDAE (33)

Besnoitia sp. **of Mbuthia et al., 1993**

Oryctolagus cuniculus (L., 1758), European (domestic) rabbit

Besnoitia sp. **of Venturini et al., 2002**

Oryctolagus cuniculus (L., 1758), European (domestic) rabbit

Cryptosporidium **sp. of Tyzzer, 1929**

Oryctolagus cuniculus (L., 1758), European (domestic) rabbit

Cryptosporidium **"rabbit genotype"**

Oryctolagus cuniculus (L., 1758), European (domestic) rabbit

Eimeria babatica **Sugár, 1978**

Lepus europaeus (Pallas, 1778), European hare

Eimeria belorussica **Litvenkova, 1969**

Lepus europaeus (Pallas, 1778), European hare

Eimeria gresae **Aoutil, Bertani, Bordes, Snounou, Chabaud & Landau, 2005**

Lepus europaeus (Pallas, 1778), European hare

Lepus granatensis (Rosenhauer, 1956), Granada hare

(Continued)

TABLE 11.1   Alphabetical list of All Coccidian Parasites Covered in this Book and the Lagomorph Hosts from Which They Have Been Reported. All parasites are reported from naturally infected rabbits unless indicated as experimental (E). (*cont'd*)

Eimeria mazierae **Aoutil, Bertani, Bordes, Snounou, Chabaud & Landau, 2005**

    Lepus granatensis (Rosenhauer, 1956), Granada hare

Eimeria pallasi **(Svanbaev, 1958) Lepp, Todd & Samuel, 1972**

    Ochotona pallasi (Gray, 1867), Mongolian pika

Eimeria pellerdi **Coudert, 1977a, b**

    Oryctolagus cuniculus (L., 1758), European (domestic) rabbit

Eimeria shubini **(Svanbaev, 1958) Lepp, Todd & Samuel 1972**

    Ochotona pallasi (Gray, 1867), Mongolian pika

Eimeria **sp. 1 Gvozdev, 1948**

    Lepus tolai (Pallas, 1778), Tolai hare

Eimeria **sp. 2 Gvozdev, 1948**

    Lepus tolai (Pallas, 1778), Tolai hare

Eimeria **sp. of Golemanski, 1975**

    Lepus europaeus (Pallas, 1778), European hare

Eimeria **sp. of Svanbaev, 1958**

    Ochotona pallasi (Gray, 1867), Mongolian pika

Eimeria **spp. of Barrett & Worley, 1970**

    Ochotona princeps (Richardson, 1828), American pika

Eimeria **type IV of Herman and Jankiewicz, 1943**

    Sylvilagus audubonii (Baird, 1858), Desert cottontail

Eimeria **type V of Herman and Jankiewicz, 1943**

    Sylvilagus audubonii (Baird, 1858), Desert cottontail

Eimeria **type VI of Herman and Jankiewicz, 1943**

    Sylvilagus audubonii (Baird, 1858), Desert cottontail

Isospora **sp. of Barrett & Worley, 1970**

    Ochotona princeps (Richardson, 1828), American pika

Sarcocystis **cf.** *bertrami* **Doflein, 1901 of Odening, Wesemeier, Pinkowski, Walter, Sedlaczek and Bockhardt, 1994c**

    Lepus europaeus (Pallas, 1778), European hare

TABLE 11.1   Alphabetical list of All Coccidian Parasites Covered in this Book and the Lagomorph Hosts from Which They Have Been Reported. All parasites are reported from naturally infected rabbits unless indicated as experimental (E). (*cont'd*)

Sarcocystis cf. *cuniculi* **Brumpt, 1913 of Odening, Wesemeier, Pinkowski, Walter, Sedlaczek and Bockhardt, 1994c**

Lepus europaeus (Pallas, 1778), European hare

Sarcocystis **sp. 1 of Fedoseenko, 1986**

Ochotona alpina (Pallas, 1773), Alpine pika

Sarcocystis **sp. 1 of Odening, Wesemeier, Pinkowski, Walter, Sedlaczek and Bockhardt, 1994c**

Lepus europaeus (Pallas, 1778), European hare

Sarcocystis **sp. 2 of Fedoseenko, 1986**

Ochotona alpina (Pallas, 1773), Alpine pika

Sarcocystis **sp. 2 of Odening, Wesemeier, Pinkowski, Walter, Sedlaczek and Bockhardt, 1994c**

Lepus europaeus (Pallas, 1778), European hare

Sarcocystis **sp. 3 of Odening, Wesemeier, and Bockhardt, 1996**

Lepus europaeus (Pallas, 1778), European hare

Sarcocystis **sp. of Stiles, 1894**

Sylvilagus floridanus (J.A. Allen, 1890), Eastern cottontail

Sarcocystis **sp. of Barrett and Worley, 1970**

Ochotona princeps (Richardson, 1828), American pika

Sarcocystis **sp. of Bell and Chalgren, 1943**

Sylvilagus floridanus (J.A. Allen, 1890), Eastern cottontail

Sarcocystis **sp. of Grundmann and Lombardi, 1976**

Ochotona princeps (Richardson, 1828), American pika

Sarcocystis **sp. of Levit, Orlov and Dymkova, 1984**

Ochotona alpina (Pallas, 1773), Alpine pika

Sarcocystis **sp. of Rausch, 1961**

Ochotona collaris (Nelson, 1893), Collared pika

---

[1] *(?) These reports are questionable and, likely, are not valid identifications from that host.*

TABLE 11.2 Cross-Transmission Studies Done with *Lepus*, *Oryctolagus*, and *Sylvilagus* Species, Through 2012, with Various Rabbit Coccidia

Coccidian/original host	Inoculating dose	Recipient host	Results (+/−)	References
Besnoitia besnoiti				
Cattle	Proliferative forms in blood	*O. cuniculus*	2 adults +	Pols, 1954a
O. cuniculus	Subinoculations IV, SC, IP, 19 serial passes	*O. cuniculus*	145 rabbits +	Pols, 1954b
O. cuniculus	Subinoculatons, 125 serial passages	*O. cuniculus*	431 rabbits + 9 rabbits −	Pols, 1960
Cattle	Emulsified cyst, from slaughtered cow	*O. cuniculus*	75 passages + in rabbits	Neuman, 1962
Cattle	Emulsified cyst, from slaughtered bull	*O. cuniculus*	227+ rabbits +	Bigalke, 1967; Neuman & Nobel, 1981
Antelope, cattle	Blood forms	*O. cuniculus*	++	Basson et al., 1970
Cattle	Skin cysts	*O. cuniculus*	+	Peteshev et al., 1974
Cat	200,000 oocysts	*O. cuniculus*	2 adults +	McKenna & Charleston, 1980
Cattle strain, serial passed in rabbits	Unknown, presumably tissue cysts from rabbits	*O. cuniculus*	24/26 3- to 6-mo-olds +	Kaggawa et al., 1979
B. jellisoni				
Peromyscus maniculatus	Tissue cysts	*O. cuniculus*	+	Frenkel, 1955b, 1965; Biglake, 1967, 1968; others
B. wallacei				
Felis domestica	1×10^6 sporulated oocysts	*O. cuniculus*	2 adults +	Ng'ang'a et al., 1994
Eimeria americana				
Lepus townsendii	Sporulated oocysts	*O. cuniculus*	−	Carvalho, 1943
		S. floridanus	−	

E. campania				
L. townsendii	Sporulated oocysts	*O. cuniculus*	–	Carvalho, 1943
		S. floridanus	–	
E. environ				
S. floridanus	80,000	*S. floridanus*	5 juv. +	Carvalho, 1943
	100,000	*O. cuniculus*	5 juv., –	
			2 adults –	
E. europaea				
L. europaeus	Sporulated oocysts	*O. cuniculus*	–	Pellérdy, 1956
E. exigua				
O. cuniculus	Sporulated oocysts	*L. europaeus*	2 adults –	Pellérdy, 1956
E. honessi				
S. floridanus	"Light dose"	*O. cuniculus*	1 rabbit +	Carvalho, 1943
E. hungarica				
L. europaeus	Sporulated oocysts	*O. cuniculus*	31 juv. –	Pellérdy, 1954a, b, 1956
L. timidus	Sporulated oocysts	*O. cuniculus*	–	Burgaz, 1973
E. irresidua				
O. cuniculus	Sporulated oocysts	*S. floridanus*	+	Carvalho, 1943
		L. europaeus	–	Pellérdy, 1954a
E. leporis				
L. europaeus	Sporulated oocysts	*O. cuniculus*	–	Nieschulz, 1923; Pellérdy, 1956; Lucas et al., 1959
E. magna				
L. timidus	Sporulated oocysts	*O. cuniculus*	–Burgza, 1973	
O. cuniculus	Sporulated oocysts	*S. floridanus*	+	Becker, 1933; Carvalho, 1943
	Sporulated oocysts	*L. europaeus*	–	Pellérdy, 1954a
	Sporulated oocysts	*L. timidus*	+ (?)	Burgaz, 1973
E. maior				
S. floridanus	50,000	*O. cuniculus*	3 juv. –	Carvalho, 1943
			2 adult –	

(Continued)

TABLE 11.2 Cross-Transmission Studies Done with *Lepus*, *Oryctolagus*, and *Sylvilagus* Species, Through 2012, with Various Rabbit Coccidia (*cont'd*)

Coccidian/original host	Inoculating dose	Recipient host	Results (+/−)	References
E. media				
O. cuniculus	Sporulated oocysts	*S. floridanus*	1 rabbit +	Carvalho, 1943
L. timidus	Sporulated oocysts	*O. cuniculus*	+(?)	Burgaz, 1973
E. minima				
S. floridanus	Sporulated oocysts	*O. cuniculus*	2 juv. −	Carvalho, 1943
E. neoleporis				
S. floridanus	≤600,000	*O. cuniculus*	65+ rabbits +	Carvalho, 1942, 1943
S. floridanus	150,000*	*S. floridanus*	3 juv. +	
E. paulistana				
S. brasiliensis	Sporulated oocysts	*O. cuniculus*	1 rabbit −	da Fonseca, 1933
E. perforans				
O. cuniculus	Sporulated oocysts	*S. floridanus*	+	Carvalho, 1943
		L. europaeus	−	Pellérdy, 1954a
		L. timidus	−	Burgaz, 1973
E. pintoensis				
S. brasiliensis	Sporulated oocysts	*O. cuniculus*	1 rabbit −	da Fonseca, 1932
E. piriformis				
O. cuniculus	Sporulated oocysts	*L. europaeus*	−	Pellérdy, 1954a, 1956
		L. timidus	−	Burgaz, 1973
E. robertsoni				
L. townsendii	Sporulated oocysts	*O. cuniculus*	−	Carvalho, 1943
		S. floridanus	−	
L. europaeus	Sporulated oocysts	*O. cuniculus*	−	Pellérdy, 1956
L. timidus	Sporulated oocysts	*O. cuniculus*	−	Burgaz, 1973
L. americanus	100, 30,600, 50,000 oocysts	*O. cuniculus*	7 adults −	Samoil & Samuel, 1977b
E. sculpta				
L. townsendii	Sporulated oocysts	*O. cuniculus*	−	Carvalho, 1943
		S. floridanus	−	

Species	Host source	Stage	Host exposed	Infectivity	References
E. semisculpta					
L. europaeus		Sporulated oocysts	O. cuniculus	−	Carvalho, 1943; Pellérdy, 1956
L. timidus		Sporulated oocysts	O. cuniculus	−	Burgaz, 1973
E. septentrionalis					
L. townsendii		Sporulated oocysts	O. cuniculus	−	Carvalho, 1943
			S. floridanus	−	
E. stiedai					
O. cuniculus		Sporulated oocysts	L. europaeus	9 adults + / 6 juv. +	Varga, 1976; Scholtyseck et al., 1979; Entzeroth & Scholtyseck, 1977; Burgaz, 1973
			L. timidus	+	Jankiewicz, 1941; Herman & Jankiewicz, 1943
			S. audubonii	+	
			S. floridanus	+	Hsu, 1970
E. sylvilagi					
S. floridanus		Sporulated oocysts	O. cuniculus	−	Hsu, 1970
E. townsendii					
L. townsendii		Sporulated oocysts	O. cuniculus	−	Carvalho, 1943
L. europaeus			S. floridanus	−	Pellérdy, 1954
			O. cuniculus	−	
Sarcocystis cuniculi					
O. cuniculus		Tissue cysts	F. domestica	+	Tadros & Laarman, 1976, 1977a, b, 1982; Munday et al., 1980
S. leporum					
S. floridanus		Tissue cysts	Canis familiaris	4 adults −	Fayer & Kradel, 1977
			F. domestica	6/8 adults +	
Felis domestica		200–75,000 sporocysts	O. cuniculus	2 juv. + / 5 adults −	Crum & Prestwood, 1977; Fayer & Kradel, 1977
Toxoplasma gondii					
F. domestica		Sporulated oocysts	Lepus sp.	7 adults +	Gustafsson et al., 1997; Dubey, 2010; Sedlák et al., 2000
			O. cuniculus	9 adults +	
		10, 1000 or 100,000 sporulated oocysts	Lepus sp.	12 adults +	
			O. cuniculus	12 adults +	

* Same oocyst culture that was infective to the domestic rabbits.

Glossary and List of Abbreviations

Acid phosphatase (AcP) One of several acid hydrolases located in lysosomes and concentrated in the Golgi apparatus of the cell.

Alkaline phosphatase (AlkP) A broad-specificity enzyme that hydrolyzes many phosphoric ester compounds with an optimum activity in the basic pH range.

α-lipoprotein A lipoprotein that transports cholesterol in the blood; composed of a high proportion of protein and relatively little cholesterol.

Arborescent Treelike in shape or growth; branching.

Altricial Implies "requiring nourishment;" this refers to a growth and developmental pattern in organisms incapable of moving around on their own soon after birth. The term derives from the Latin root *alere-* ("to nurse or to nourish"), and refers to the need for young to be fed and taken care of for a long duration. Specifically, when referring to rabbits, it means the young are born blind and furless, in a fur-lined nest (warren), and are totally dependent upon their mother.

Anlagen An embryonic area capable of forming a structure: the primordium, germ, or bud.

Anorexia Loss of appetite and inability to eat.

Apical complex Dense ring and cone-like structure at the anterior end of an apicomplexan parasite (e.g., sporozoite) stage.

Apicomplexa A phylum containing organisms that possess a certain combination of structures, called an apical complex, that is distinguishable only by electron microscopy.

Archaea Any of a group of prokaryotic microorganisms that resemble bacteria, but are different from them in certain aspects of their chemical structure, such as the composition of their cell walls.

Atrophy A wasting away of the body or of an organ or part, as from defective nutrition or nerve damage.

Autapomorphic In cladistics, a character state uniquely defining a taxon.

Automorphic Patterned after one's self.

Azithromycin An antibiotic of the subclass macrolides used to treat bacterial infections.

Basal bodies A cytoplasmic organelle of animals and some protists from which cilia or flagella arise.

β-globulin A class of vertebrate plasma proteins.

β-lipoprotein A lipoprotein that transports cholesterol in the blood; composed of a moderate amount of protein and a large amount of cholesterol.

Benzamide An off-white solid organic chemical that is a derivative of **benzoic acid**, and is used in a wide variety of compounds, including analgesics, antiemetics, antipsychotics, and others.

Benzoic acid A simple aromatic carboxylic acid; its salts are used as a food preservative and it is an important precursor for the synthesis of many organic substances, including phenol and plasticizers. It also inhibits the growth of mold, yeast, and some bacteria.

Beringia The Bering land bridge, app. 1,600 km north to south, that joined present day Alaska and eastern Siberia during the Pleistocene ice ages during the Last Glacial Maximum (~16,500 years ago).

Bilirubinemia The presence of bilirubin in the blood.

Bradyzoites Small, slowly replicating stage in various tissue-cyst coccidians (e.g., *Toxoplasma*, *Sarcocystis* spp.) that develop inside the host tissue zoitocyst of that parasite.

Bromsulfophthalein metabolites Bromsulfophthalein is a dye used in liver function tests. Determining the rate of removal of the dye from the blood stream gives a measure of liver function.

Cachectic Characterized by physical wasting with loss of weight and muscle mass due to disease.

Calvarium Upper part of the cranial cavity containing the brain; the skull cap.

Carbamilide (urea) The main nitrogen-containing substance in the urine of mammals. It is widely used in fertilizers and is an important raw material in the chemical industry as a component of animal feed to provide a rather cheap source of nitrogen.

Catarrh Inflammation of a mucous membrane with free discharge.

cf. Compares to (confers with).

Clade A taxonomic group of organisms classified together on the basis of homologous features traced to a common ancestor.

Chromosome painting A technique for resolving complex abnormalities in the structure of chromosomes which are impossible or very difficult to detect by standard histological preparations using bright-field microscopy.

Chromosomal synapomorphies Traits on chromosomes shared by two or more taxa and their most recent common ancestor, whose own ancestor in turn does not possess the trait.

Conoid Truncated cone of spiral fibrils located within the polar rings of some Apicomplexa.

Clostridial overgrowth An excessive growth of *Clostridium* species, which are soil bacteria that can enter the body via contaminated food or through a puncture wound; the result can be a life-threatening infection.

Coccidiosis Any of a series of specific infectious diseases caused by epithelial protozoan parasites from the Coccidia, such as *Eimeria* and *Isospora* species.

Commensal Of an animal, plant, fungus, etc., living with, on, or in another, without injury to either.

Cytomeres Cytoplasmic "regions" each with numerous nuclei on their periphery; characteristic in the development of some microgamonts.

Cytophaneres A radial spine seen in certain muscle cysts of *Sarcocystis* species.

Cytostome A part of a cell specialized for phagocytosis, usually in the form of a microtubule-supported funnel or groove.

Crepuscular Animals that are primarily active during twilight times, dawn and/or dusk.

Definitive host Host in which a parasite achieves sexual maturity.

Desquamation The shedding of epithelial elements, particularly of the skin, in scales or sheets.

Diclazuril A coccidiostat—an antiprotozoal agent that acts upon coccidia parasites, it is a benzeneacetonitrile derivative, the mode of action of which is not precisely known.

Diphenylmethane An organic compound of methane where two hydrogen atoms are replaced by two phenyl groups.

Diphenyl disulfide A colorless crystalline compound that is one of the most popular organic disulfides used in organic systhesis.

Doxapram hydrochloride A respiratory stimulant, usually administered intravenously.

Ectomerogony Multiple fission (asexual reproduction) of a meront to produce daughter cells (merozoites), at the surface of the meront or by infolding into the meront.

Edematous Characterized by or pertaining to edema (swelling).

Electron-dense In electron microscopy, having a density that prevents electrons from entering.

Endogenous stages Refers to the asexual and sexual stages of the coccidian life cycle that take place within epithelial or endothelial cells of the gastrointestinal tract or associated organs (e.g., liver, bile ducts).

Endopolyogeny Formation of daughter cells, each surrounded by its own membrane, while still in the mother cell.

Enzyme-linked immunoabsorbent assay (ELISA) Immunodiagnostic test designed to detect the presence of fixed antibody through linkage with an enzymatic reaction.

Eosinophilia The formation and accumulation of an abnormally large number of eosinophils in the blood.

Epidemiology The study of disease; the branch of medicine dealing with the incidence and prevalence of disease in large populations and with detection of the source and cause of epidemics of infectious disease.

Epistaxis A nosebleed or hemorrhage from the nose due to the rupture of small blood vessels.

Eukaryotic organisms Living organisms composed of eukaryotic cells—cells with a distinct nucleus and membrane-bound organelles.

Euryoecious Organisms that can have a broad variety of ecological living conditions.

Eutherian mammals Those mammals that have a placenta (e.g., humans), as compared to marsupial mammals (e.g., kangaroos) that do not.

Exogenous stages Refers to the oocysts discharged in the feces from an infected animal.

Extravasation Leaking of blood capillaries resulting in accumulation of blood in the tissues.

Febrile Pertaining to fever; feverish.

Fibroblasts An immature fiber-producing cell of connective tissue.

Fomites Inanimate objects capable of carrying infective propagules (spores, oocysts, etc.) from an infected person to another person (e.g., clothes, bedding).

Forms A term for the nest of a rabbit, generally referring to members of the genus *Sylvilagus*.

γ-globulin A group of plasma proteins which have sites of antibody activity.

Genotype Generally refers to the specific genetic makeup of a cell, organism or individual. However, it is used by *Cryptosporidium* researchers to distinguish genetically distinct populations of oocysts for which there is insufficient information to assign to species status. This term is used almost exclusively for *Cryptosporidium* species that have very small oocysts with virtually no structural differences between species.

Glucosophosphate isomerase (GPI) An enzyme that catalyzes the conversion of glucose-6-phosphate into fructose 6-phosphate in the second step of glycolysis.

Glutamic oxaloacetic transaminase (GOT) An enzyme that is normally present in liver and heart cells. It catalyzes the reversible transfer of an α-amino group between aspartate and glutamate and, as such, is an important enzyme in amino acid metabolism.

Glutamic pyruvic transaminase (GPT) An enzyme that catalyzes the transfer of an amino group from alanine to α-ketoglutarate, the products of this reversible transamination reaction being pyruvate and glutamate.

Glutamic transaminase Another name for the enzyme glutamic pyruvic transaminase (GPT).

Gregarine A member of the Apicomplexa phylum that parasitizes invertebrates, primarily annelids and arthropods.

Halofuginone lactate A chemical used against parasitic diseases caused by protozoa, specifically *Cryptosporidium* spp.

Heterologous Made up of tissue not normal to the part.

Heterophils A granular leukocyte represented by neutrophils in humans, but characterized in other mammals by granules which have variable sizes and staining characteristics.

Heteroxenous A parasite life cycle where two or more hosts are involved; that is, there is a mandatory intermediate host within which biological development of the parasite must take place for the life cycle to be continued.

Histiocytosis A condition marked by an abnormal appearance of histiocytes (macrophages) in the blood.

Holarctic Throughout the northern continents of the world and includes the Palearctic and the Nearctic.

Homoplasy Correspondence in form or structure, owing to a similar environment.

Hypermotility Increased motility of the gut resulting in diarrhea.

Hyperperistalsis Excessive rapidity of the passage of food through the stomach and intestine.

Hyperemia (or hyperemic) Increased blood flow to a body part resulting in engorgement.

Hypoglycemia A deficiency of glucose concentration in the blood.

Hypomotility Diminished motility of the gut resulting in constipation.

Hypoproteinemia A deficiency of protein in the blood.

Icteric Jaundiced.

IFAT Immunofluorescent antibody technique.

Imidazole Colorless, organic solid that is soluble in water. Its derivatives are important biological building blocks of many drugs such as antifungals (e.g., nitroimidazole, which has an imidazole ring).

Intermediate host A host in which a parasite develops to some extent, but not to sexual maturity.

Intussusception A condition that occurs when a segment of the intestine is forced by hyperperistalsis to invaginate into itself in a manner similar to the way a section of a telescope slides into the next larger section ahead of it. This can often result in obstruction.

Ionophore antibiotic A drug that increases the permeability of cell membranes to a specific ion.

Karyosome Also called chromocenter, any of several masses of chromatin in the reticulum of a cell nucleus; a chromosome.

Koch's postulates The four conditions/criteria designed to establish a causal relationship between a parasitic microbe and a disease. These conditions are: (1) the microbe must be found in all hosts suffering from the disease; (2) the microbe must be isolated from the diseased organism in pure culture; (3) the microbe should cause the same disease/infection conditions when introduced into a healthy, non-infected host, generally of the same host species; and (4) the microbe must be reisolated from the inoculated, experimental host and be identical in form and structure to the original causative agent.

Lachrymal Pertaining to tears.

Lactate dehydrogenase (LDH) An enzyme that catalyzes the interconversion of pyruvate and lactate with concomitant interconversion of NADH and NAD^+.

Lamina propria The moist linings corresponding to the mucous membranes (mucosa) that line various tubes in the vertebrate body.

Lasalocid (Avatec) An antibacterial agent and a coccidiostat that is produced by strains of the bacterium *Streptomyces lasaliensi*.

Lerbek Clopidol + methylbenzoate.

Lipidemia/lipemia An elevated concentration of any or all of the lipids in plasma.

LM Light microscope.

Maduramycin (Cygro) An ionophore coccidiostat that works primarily on *Eimeria* spp.

Merogony The process of merozoite formation via asexual multiple fission.

Merozoite A daughter cell resulting from merogony.

Metrocytes Cells accumulating inside *Sarcocystis* species' tissue cyst wall and eventually giving rise to infective bradyzoites.

Microfilaments Any of the submicroscopic filaments composed chiefly of actin and found in the cytoplasmic matrix of almost all cells.

Micronemes Slender, convoluted bodies that join a duct system with the rhoptries, opening at the tip of a sporozoite or merozoite.

Micropore(s) Opening at the side of a sporozoite, functioning in food uptake.

Microspectrophotometer A microscope and spectrophotometer combined.

Microtubule(s) A hollow cylindrical structure in the cytoplasm of most cells, involved in intracellular shape and transport.

Molecular clock analysis A technique in molecular evolution that uses fossil constraints and rates of molecular change to deduce the time in geological history when two species or other taxa diverged.

Monensis (Elancoban) A polyether antibiotic isolated from *Streptomyces cinnamonensis*.

Monocytes A mononuclear, phagocytic leukocyte.

Moribund In a dying state; near death; on the verge of extinction or termination.

Myxosporans (= Myxozoans) Protozoan parasites of invertebrates and vertebrates (mostly fish), characterized by spores of multicellular origin.

Narasin (Monteban) A coccidiostat and antibacterial agent.

Nearctic All of North America, Greenland, and the highlands of Mexico.

Necrosis Morphological changes indicative of cell death caused by progressive enzymatic degradation.

Neutrophils A granular leukocyte having a nucleus with three to five lobes connected by threads of chromatin, and cytoplasm with very fine granules.

Nitraoxanide An antiprotozoal agent.

Nitrofuran(s) A class of drugs with antibiotic or antimicrobial activity.

Nitroimidazole An imidazole derivative with a nitro group that constitutes a class of antibiotics used to combat anaerobic bacterial and parasitic infections.

Nucleolus A small, dense region visible in the nucleus of nondividing eukaryotic cells; consists of rRNA molecules, ribosomal proteins, and loops of chromatin from which the rRNA molecules are transcribed.

Omentum A fold of peritoneum extending from the stomach to adjacent abdominal organs.

Osmophilic (osmiophilic) Having an affinity for solutions of high osmotic pressure.

Oxaloacetic transaminase Another name for the enzyme glutamic oxaloacetic transamine (GOT).

Palearctic The largest of Earth's eight ecozones, including terrestrial regions of Europe, Asia north of the Himalayan foothills, northern Africa, and the northern and central Arabian Peninsula.

Paraplegia Paralysis of the lower part of the body including the legs.

Parasitophorous vacuole (PV) Vacuole within a host cell that contains a parasite.

Paresis Slight or incomplete paralysis.

Paromomycin An aminoglycoside antibiotic, first isolated from *Streptomyces krestomuceticus*.

Perforatorium Another name for acrosome, the anterior prolongation of a spermatozoon that releases egg-penetrating enzymes.

Per os By mouth.

Phenylarsonic acid An organic derivative of arsenic acid that is a colorless solid. It is used as a precursor to other organoarsenic compounds, some of which are used in animal nutrition.

Pinocytosis Mechanism by which cells ingest extracellular fluid and its contents.

Plesiomorphic A primitive character trait that is shared with an ancestral clade.

Polydipsia A non-medical symptom in which an infected individual displays excessive thirst.

Precocial Pertaining to early development of mental or physical traits.

Prion A tiny proteinaceous particle, likened to viruses and viroids, but having no genetic component, thought to be an infectious agent in bovine spongiform encephalopathy, Creutzfeldt-Jakob disease, and similar encephalopathies.

Protist(s) The term currently accepted as the most convenient name for eukaryotic (nucleated) organisms that are relatively undifferentiated, and not true plants, animals or fungi.

Purulent Containing or forming pus.

Pyknotic Pertaining to the thickening or degeneration of a cell in which the nucleus has shrunk in size and the chromatin condenses to a solid, structureless mass.

Pyelonephritis Inflammation of the kidney and its pelvis due to bacterial infection.

Pyruvic transaminase Another name for the enzyme glutamic pyruvic transaminase (GPT).

Reticuloendothelial Pertaining to the reticuloendothelium or to the reticuloendothelial system (phagocytic cells such as macrophages).

Retroposon insertions DNA sequences with the capacity to move about within a genome, inserting site specificity into host DNA, but without requiring extensive DNA homology to do so.

Rhoptries Elongated, electron-dense bodies extending within the polar rings of an apicomplexan.

Rifaximin A semisynthetic antibiotic based on rifamycin. It is used primarily in the treatment of traveler's diarrhea and hepatic encephalopathy.

Robenidine hydrochloride (Cycostat) A drug used as an aid in the prevention of coccidiosis in chickens caused by various *Eimeria* spp.

Rough endoplasmic reticulum An extensive system of membranes coated with ribosomes that is present in most eukaryotic cells, dividing the cytoplasm into compartments and channels.

Salinomycine (Sacoxc) An antibacterial and coccidiostat ionophore therapeutic drug.

SEM Scanning electron microscope.

Septa Dividing walls or partitions.

Scut A short, erect tail in hares, rabbits, and deer.

Semiarboreal Often inhabiting and/or frequenting trees.

Speciose A term describing a group of organisms (order, class, family, genus) with many species.

SPF An acronym that refers to specific pathogen-free animals used in biomedical research.

Sporulation The process of sporocyst formation in oocysts.

Stenoecious Organisms limited in their ability to endure any variety in their ecological living conditions.

Succinylsulfathiazole A sulfonamide (sulfa) drug.

Sulfadiazine A sulfonamide (sulfa) drug that kills bacteria that cause infection by stopping the production of folic acid in the bacterial cell.

Sulfamerazine A sulfonamide (sulfa) drug; also an antibacterial that acts as a competitive inhibitor of dihydrofolate, an enzyme involved in folate synthesis.

Sulfamethazine A sulfonamide (sulfa) drug that inhibits bacterial synthesis of dihydrofolic acid, which ultimately is required for synthesis of purines in cell growth.

Sulfaquinoxaline A veterinary medicine that can be given to cattle and sheep to treat coccidiosis.

Synapomorphy A derived or specialized character (apomorphy) shared by two or more groups and which originated in their last common ancestor.

Syntopic Living together at the same locality.

TEM Transmission electron microscope.

Trabeculae A general term for the supporting or anchoring strands of connective tissue.

Triazine (not triazene) One of three organic chemicals that are isomeric with each other. Triazine compounds are often used as the basis of various herbicides.

Tunica propria A proper coat or layer of a part, as distinguished from an investing membrane.

Uremia (uraemia) The retention of excessive by-products of protein metabolism (urea, etc.) in the blood.

Vicariance (vicariant event) The geographical separation and isolation of a subpopulation, resulting in the original population's differentiation as a new variety or species.

Literature Cited (*)
and Related References

Abdel-Ghaffar, F., Marzouk, M., Ashour, M.B., Mosaad, M.N., 1990. Effects of *Eimeria labbeana* and *E. stiedai* infection on the activity of some enzymes in the serum and liver of their hosts. Parasitology Research 76, 440–443. *.

Agnarsson, I., Kuntner, M., 2007. Taxonomy in a changing world: seeking solutions for a science in crisis. Systematic Biology 56, 531–539. *.

Akpo, Y., Kpodékon, M.T., Djago, Y., Licois, D., Youssao, I.A.K., 2012. Vaccination of rabbits against coccidiosis using precocious lines of *Eimeria magna* and *Eimeria media* in Benin. Veterinary Parasitology 184, 73–76. *.

Al-Ghamdy, A.O., Shazly, M., AL-Rasheid, K.A.S., Mubarak, M., Bashtar, A.-R., 2005. Light and electron microscopy of *Eimeria magna* Pérard, 1925 infecting the house rabbit, *Oryctolagus cuniculus* from Saudi Arabia. II. Gamogony and oocyst wall formation. Saudi Journal of Biological Sciences 12, 114–125. *.

Allam, A.F., Shehab, A.Y., 2002. Efficacy of azithromycin, praziquantel and mirazid in treatment of cryptosporidiosis in school children. Journal of the Egyptian Society of Parasitology 32, 969–978. *.

Al-Mathal, E.M., 2008. Hepatic coccidiosis of the domestic rabbit *Oryctolagus cuniculus domesticus* L. in Saudi Arabia. World Journal of Zoology 3, 30–35. *.

Almería, S., Calvete, C., Pagés, A., Gauss, C., Dubey, J.P., 2004. Factors affecting the seroprevalence of *Toxoplasma gondii* infection in wild rabbits (*Oryctolagus cuniculus*) from Spain. Veterinary Parasitology 123, 265–270. *.

Aly, M.M., 1993. Development of *Eimeria stiedae* in a nonspecific host. Journal of the Egyptian Society of Parasitology 23, 95–100. *.

Anderson, P., Garry-Anderson, A.S., 1963. Eläinten toksoplasmoosista suomessa. Finsk Veterianärtdskrift. Suomen Eläinlääkärilehti 69, 159. *.

Anderson, S., Jones Jr., J.K. (Eds.), 1967. Recent Mammals of the World. Ronald Press Co., New York *.

Andrews, C.L., Davidson, W.R., Provost, E.E., 1980. Endoparasites of selected populations of cottontail rabbits (*Sylvilagus floridanus*) in the southeastern United States. Journal of Wildlife Diseases 16, 395–401. *.

Angermann, R., Flux, J.E.C., Chapman, J.A., Smith, A.T., 1990. Lagomorph classification. In: Chapman, J.A., Flux, J.E.C. (Eds.), Rabbits, Hares and Pikas: Status

Conservation Action Plan. International Union for Conservation of Nature and Natural Resources, Gland, Switzerland, pp. 7–13. *.

Angus, K.W., 1987. Cryptosporidiosis in domestic animals and humans. Practice 9, 47–49. *.

Anonymous, 1995. Second major Crypto outbreak hits Torbay area. ENDS Report 247, 6–7. *.

Anonymous, 2008. Outbreak of cryptosporidiosis associated with a water contamination incident in the East Midlands. Health Protection Report 2, 3–5 (Also available at. http://www.hpa.org.uk/hpr/archives/2008/hpr2908.pdf.). *.

Anonymous, 2012. List of mammals of Nigeria. http://en.wikipedia.org/wiki/ListofmammalsofNigeria#citenote-0 (cited 01/12/12). *.

Aoutil, N., Bertani, S., Bordes, F., Snounou, G., Chabaud, A., Landau, I., 2005. *Eimeria* (Coccidia: Eimeridea) of hares in France: Description of new taxa. Parasite 12, 131–144. *.

Arnastauskene, T.V., 1982. On eimeriasis in hares and its prophylaxis in the Lithuanian SSR. Aktualnye problemy parazitologii v Pribaltike 1–2, 7. * (in Russian).

Arnastauskene, T., Kazlauskas, J., 1970. On the fauna of the coccidia and helminths in the hare in the Lithuanian SSR. Acta Parasitologica Lituanica 10, 85–93. * (in Russian, English summary).

Arnaudov, D., Arnaudov, A., Kirin, D., 2003. Study on the toxoplasmosis among wild animals. Experimental Pathology and Parasitology 6/11, 51–54. *.

Asher, R.J., Meng, J., Wible, J.R., McKenna, M.C., Rougier, G.W., Dashzeveg, D., Novacek, M.J., 2005. Stem Lagomorpha and the antiquity of Gilires. Science 307, 1091–1094. *.

Audoin, F., 1986. Ossements animaux du Moyen Age au monastère de la Charité sur Loire. Publications de la Sorbonne (Université de Paris I, Histoire ancienne et medieval), Paris. *.

Aydin, Y., Ozkul, I.A., 1996. Infectivity of *Cryptosporidium muris* directly isolated from the murine stomach for various laboratory animals. Veterinary Parasitology 66, 257–262.

Ayeni, A.O., 1969. Zur Frage des Sauerstoffverbrauches der Oocysten von *Eimeria stiedai* (Lindemann, 1865) Kisskalt et Hartmann, 1907 (Sporozoa, Coccidia) während und nach der Sporulation (On the question of oxygen consumption of the oocysts of *Eimeria stiedai* [Lindemann, 1865] Kissalt and Hartmann, 1907 [Sporozoa, Coccidia] during and after sporulation). Faculty of Veterinary Medicine, Justus Liebig-Universität Geißen, Geißen, Gemany. * (in German, English summary).

Babudieri, R., 1932. I sarcosporidi e le sarcosporidiosi (Studio monografico). Archive für Protistenkunde 76, 421–580. *.

Bachman, G.W., 1930. Immunity in experimental coccidiosis in rabbits. American Journal of Hygiene 12, 641–649.

Balbiani, E.G., 1884. Lecons sur les Sporozoaires. Recuellies par Journal Pelletan, O. Dion (Ed.) Paris. *.

Baker, A.J., Eger, J.L., Peterson, R.L., Manning, T.H., 1983. Geographic variation and taxonomy of arctic hares. Acta Zoologica Fennica 174, 45–48.

Ball, S.J., Hutchison, W.M., Pittilo, R.M., 1988. Ultrastructure of microgametogenesis of *Eimeria stiedai* in rabbits (*Oryctolagus cuniculus*). Acta Veterinaria Hungarica 36, 229–232. *.

Ballarini, G., 1964. Elementi di coprologia clinica in *Lepus europaeus*. Veterinaria Italiana 17, 252–262.

Ballarini, G., 1966. Aspetti quantitative della dinamica di popolazioni di eimeriidae in cenosi con *Lepus europaeus* (Quantitative aspects in the dynamics of Eimeriidae populations in cenosis with *Lepus europaeus*). Veterinaria Italiana 17, 235–243 (in Italian).

Barrett, R.E., Worley, D.E., 1970. Parasites of the pika (*Ochotona princeps*) in two counties in south-central Montana, with new host records. Proceedings of the Helminthological Society of Washington 37, 179–181. *.

Barriga, O.O., Arnoni, J.V., 1979. *Eimeria stiedae*: Weight, oocyst output, and hepatic functions of rabbits with graded infection. Experimental Parasitology 48, 407–414. *.

Barriga, O.O., Arnoni, J.V., 1981. Pathophysiology of hepatic coccidiosis in rabbits. Veterinary Parasitology 8, 201–210. *.

Barta, J.R., Thompson, R.C.A., 2006. What is *Cryptosporidium*? Reappraising its biology and phylogenetic affinities. Trends in Parasitology 22, 463–468. *.

Barta, J.R., Martin, D.S., Liberator, P.A., Dashkevicz, M., Anderson, J.W., Feighner, S.D., Elbrecht, A., Perkins-Barrow, A., Jenkins, M.C., Danforth, H.D., Ruff, M.D., Profous-Juchelka, H., 1997. Phylogenetic relationships among eight *Eimeria* species infecting domestic fowl inferred using complete small subunit ribosomal DNA sequences. Journal of Parasitology 83, 262–271.

Bashtar, A.R., Al-Rasheid, K.A., Mobarak, M., Al-Ghamdy, A.A., 2003. Coccidiosis in rabbits from Saudia Arabia. 1. Endogenous stages of different *Eimeria* spp. Journal of the Egyptian German Society of Zoology 42 (D), 1–10.

Basson, P.A., McCully, R.M., Bigajki, R.D., 1970. Observation on the pathogenesis of bovine and antelope strains of *Besnoitia besnoiti* (Marotel, 1912) infection in cattle and rabbits. Onderstepoort Journal of Veterinary Research 37, 105–126. *.

Baxby, D., Blundell, N., Hart, C.A., 1984. The development and performance of a simple, sensitive method for the detection of *Cryptosporidium* oocysts in faeces. Journal of Hygiene (Cambridge) 92, 317–323. *.

Becker, E.R., 1933. Cross-infection experiments with coccidia of rodents and domesticated animals. Journal of Parasitology 19, 230–234. *.

Becker, E.R., 1934. Coccidia and Coccidiosis of Domesticated, Game and Laboratory Animals and of Man. Collegiate Press, Ames, Iowa.

Becker, E.R., Crouch, H.B., 1931. Some effects of temperature upon development of the oocysts of coccidia. Proceedings of the Society of Experimental Biology and Medicine 28, 529–530. *.

Becker, E.R., Patillo, W.H., Farmer, J.N., Van Doornick, M., 1955. Size of the oocyst of *Eimeria brunette* and *Eimeria necatrix*. Journal of Protozoology 41 (Suppl.), 18. *.

Becker, E.R., Zimmerman, W.J., Patillo, W.H., Farmer, J.N., 1956. Measurements of the unsporulated oocysts of *Eimeria acervulina, E. maxima, E. tenella* and *E. mitis*: coccidian parasites of the common fowl. Iowa State College Journal of Science 31, 79–84. *.

Bedrnik, P., Martinez, J., 1976. Effect of various anticoccidials on the intestinal coccidia species of the rabbit. Veterinaria SPOFA No 184, 165–174.

Bell, J.F., Chalgren, W.S., 1943. Some wildlife diseases in the eastern United States. Journal of Wildlife Management 7, 270–278. *.

Berg, C., 1981. Endoparasitism in wild Norwegian hares. Parasitologiska Institutionen Åbo Akademi Information 16, 88–89. *.

Besnoit, C., Robin, V., 1912. Sarcosporidiose cutanée chez une vache. Revue Vétérinaire 37, 649–663. *.

Beyer, T.V., 1960. Cytologic studies of the life cycle of coccidia of the rabbit. III. Studies of the phosphomonoesterases of *Eimeria intestinalis* and *E. magna*. In: Problems of Cytology and Protohistology, Moscow-Leningrad, pp. 277–280.

Beyer, T.V., 1961. Immunity in experimental coccidiosis of the rabbit caused by heavy infective doses of *Eimeria intestinalis*. International Congress on Protozoology 1, 448.

Beyer, T.V., 1962. O raspredelenii sukcindehydrazy v žiznennom cikle *Eimeria intestinalis* (On the distribution of succinic dehydrogenase in the life cycle of *Eimeria intestinalis*). Tsitologiya (Cytology) 4, 232–237. * (in Russian).

Beyer, T.V., 1963. Citohimičeskoe issledovanie tiolovyh soedinenij na raznyh stadijah razvitija *Eimeria intestinalis* (Cytochemical studies of thiols at different developmental stages of *Eimeria intestinalis*). Tsitologiya (Cytology) 5, 59–65. * (in Russian).

Beyer, T.V., Ovchinnikova, L.P., 1964. A cytophotometric investigation of the RNA content in the course of macrogametogenesis in two rabbit intestinal coccidia *Eimeria magna* and *E. intestinalis*. Acta Protozoologica 2, 329–337. *.

Beyer, T.V., Ovchinnikova, L.P., 1966. A cytophotometrical investigation of the cytoplasmic RNA content in the course of oocyst formation in the intestinal rabbit coccidia *Eimeria intestinalis* Cheissin, 1948. Acta Protozoologica 4, 75–81. *.

Bhat, T.K., Jithendran, K.P., 1995. *Eimeria magna*: The effect of varying inoculum size on the course of infection in Angora rabbits. World Rabbit Science 3, 163–166. *.

Bhat, T.K., Jithendran, K.P., Kurande, N.P., 1996. Rabbit coccidiosis and its control: A review. World Rabbit Science 4, 37–41. *.

Bigalke, R.D., 1960. Preliminary observations on the mechanical transmission of cyst organisms of *Besnoitia besnoiti* (Marotel, 1912) from a chronically infected bull to rabbits by *Glossina brevipalpis* Newstead, 1910. Journal of the South African Veterinary Medicine Association 31, 37–44. *.

Bigalke, R.D., 1967. The artificial transmission of *Besnoitia besnoiti* (Marotel, 1912) from chonically infected to susceptible cattle and rabbits. Onderstepoort Journal of Veterinary Research 34, 303–316. *.

Bigalke, R.D., 1968. New concepts on the epidemiological features of bovine besnoitiosis as determined by laboratory and field investigations. Onderstepoort Journal of Veterinary Research 35, 3–138. *.

Bigalke, R.D., Prozesky, L., 1994. Chapter 18: Besnoitiosis. In: Coetzer, J.A.W., Thomson, G.R., Tustin, R.C. (Eds.), Infectious disease of livestock with special reference to South Africa, vol. 1. Oxford University Press, Cape Town, pp. 245–252. *.

Blagburn, B.L., Soave, R., 1997. Prophylaxis and chemotherapy: human and animal. In: Fayer, R. (Ed.), *Cryptosporidium* and Cryptosporidiosis. CRC Press, Boca Raton, Florida, pp. 111–128. *.

Blanshard, C., Shanson, D.C., Gazzard, B.G., 1997. Pilot studies of azithromycin, letrazuril, and paromomycin in the treatment of cryptosporidiosis. International Journal of STD and AIDS 8, 124–129. *.

Blaustein, R., 2009. The Encyclopedia of Life: Describing Species. Unifying Biology 59, 551–556. *.

Bosc, M.F.-J., 1898. Formes microbiennes et forms de granulation de *Coccidium oviforme* en pullulation intracellulaire dans 'certaines tumeurs du foie du lapin (Microbial forms and granules of *Coccidium oviforme* outbreak in intracellular tumors of the liver of the rabbit). Comptes Rendus des Seances Socété de Biologie (Paris) 50, 1156–1158. * (in French).

Boughton, R.V., 1930. The value of measurements in the study of a protozoan parasite *Isospora lacazie*. American Journal of Hygiene 11, 212–226. *.

Boughton, R.V., 1932. The influence of helminth parasitism on the abundance of the snowshoe rabbit (*Lepus americanus*) in western Canada. Canadian Journal of Research 7, 524–527. *.

Bouvier, G., 1967. Les coccidies rencontrées en Suisse chez le lièvre gris (*Lepus europaeus*). Annales de Parasitologie Humumaine et Comparée 42, 551–559. *.

Braun, R., Eckert, J., Roditi, I., Smith, N., Wallach, M., 1992. Coccidiosis of poultry and farm animals. Parasitology Today 8, 220–221.

Brooks, D.R., 1979. Testing the context and extent of host-parasite coevolution. Systematic Zoology 28, 299–307. *.

Brooks, D.R., Hoberg, E.P., 2000. Triage for the biosphere: the need and rationale for taxonomic inventories and phylogenetic studies of parasites. Comparative Parasitology 67, 1–25. *.

Bruce, E.A., 1919. A preliminary note on a new coccidium of rabbits. Journal of the American Veterinary Medical Association 55, 620–621. *.

Brumpt, E., 1913. Précis de Parasitologie (Summary of Parasitology). In: Masson, et al. (Eds.), second ed. Entieèrement Remaniecè, Paris. *.

Bull, P.C., 1953. Parasites of the wild rabbit, *Oryctolagus cuniculus* L., in New Zealand. New Zealand Journal of Science and Technology. B. General Research Section 34, 341–372. *.

Bull, P.C., 1958. Incidence of coccidia (Sporozoa) in wild rabbits, *Oryctolagus cuniculus* (L.) in Hawke's Bay, New Zealand. New Zealand Journal of Science 1, 289–329. *.

Bull, P.C., 1960. Parasites of the European rabbit, *Oryctolagus cuniculus* (L.), on some Antarctic islands. New Zealand Journal of Science 3, 258–273. *.

Bull, S., Chalmers, R., Sturdee, A.P., Curry, A., Kennaugh, J., 1998. Cross-reaction of an anti- *Cryptosporidium* monoclonal antibody with sporocysts of *Monocystis* species. Veterinary Parasitology 77, 195–197. *.

Burgaz, I., 1970a. A report on the presence of endoparasites in *Lepus timidus* and *Lepus europaeus* in Sweden. Transactions of the IX International Congress of Game Biology, Moscow (1969), pp. 628–635. *.

Burgaz, I., 1970b. The endoparasites of the varying hare (*Lepus timidus*) and the European hare (*Lepus europaeus*). Zoological Review (Stockholm) 32, 61–66. *.

Burgaz, I., 1973. A cross infection experiment with coccidia concerning mountain hares (*Lepus timidus* L.) and tame rabbits. Norwegian Journal of Zoology 21, 323–330. *.

Butaeva, F., Paskerova, G., Enzeroth, R., 2006. *Ditrypanocystis* sp. (Apicomplexa, Gregarinia, Selenidiidae): The mode of survival in the gut of *Enchytraeus albidus* (Annelida, Oligochaeta, Enchytraeidae) is close to that of the coccidian genus *Cryptosporidium*. Tsitologiia 48, 695–704. *.

Cacciò, S.M., Pozio, E., 2006. Advances in the epidemiology, diagnosis and treatment of cryptosporidiosis. Expert Reviews in Anti Infection Therapy 4, 429–443. *.

Cama, V.A., Bern, C., Roberts, J., Cabrera, L., Sterling, C.R., Ortega, Y., Gilman, R.H., Xiao, L., 2008. *Cryptosporidium* species and subtypes and clinical manifestations in children, Peru. Emerging Infections Diseases 14, 1567–1574. *.

Campbell, A.T., Haggart, R., Robertson, L.J., Smith, H.V., 1992. Fluorescent imaging of *Cryptosopordium* using a cooled charge couple device (CCD). Journal of Microbiological Methods 16, 169–174. *.

Campbell III, T.M., Clark, T.W., Groves, C.R., 1982. First record of pygmy rabbits (*Brachylagus idahoensis*) in Wyoming. Great Basin Naturalist 42, 100. *.

Carini, A., 1940. Eimerias da lébre silvéstre do Brasil. Arquivos de Biologia (São Paulo) 24, 218–219. *.

Carvalho, J.C.M., 1942. *Eimeria neoleporis* n. sp. occurring naturally in the cottontail and transmissible to the tame rabbit. Iowa State College Journal of Science 16, 409–410. *.

Carvalho, J.C.M., 1943. The coccidia of wild rabbits of Iowa. I. Taxonomy and host-specificity. Iowa State College Journal of Science 18, 103–135. *.

Carvalho, J.C.M., 1944. The coccidia of wild rabbits of Iowa. II. Experimental studies with *Eimeria neoleporis* Carvalho, 1942. Iowa State College Journal of Science 18, 177–189. *.

Casemore, D.P., 1990. Epidemiological aspects of human cryptosporidiosis. Epidemiology and Infection 104, 1–28. *.

Casemore, D.P., 1991. The epidemiology of human cryptosporidiosis and the water route of infection. Water Science Technology 24, 157–164. *.

Casemore, D.P., Wright, S.E., Coop, R.L., 1997. Cryptosporidiosis—Human and animal epidemiology. In: Fayer, R. (Ed.), *Cryptosporidium* and Cryptosporidiosis. CRC Press, Boca Raton, Florida, pp. 65–92. *.

Catchpole, J., Norton, C.C., 1975. Coccidiosis in rabbits: Experimental infection with *Eimeria intestinalis*. Journal of Protozoology 22, 49A (abstract). *.

Catchpole, J., Norton, C.C., 1979. The species of *Eimeria* in rabbits for meat production in Britain. Parasitology 79, 249–257. *.

Ceré, N., Licois, D., Humbert, J.F., 1995. Study of the inter- and intraspecific variation of *Eimeria* spp. from the rabbit using random amplified polymorphic DNA. Parasitology Research 81, 324–328. *.

Ceré, N., Humbert, J.F., Licois, D., Corvione, M., Afanassieff, M., Chanteloup, N., 1996. A new approach for the identification and diagnosis of *Eimeria media* parasite of the rabbit. Experimental Parasitology 82, 132–138. *.

Ceré, N., Licois, D., Humbert, J.F., 1997. Comparison of the genomic fingerprints generated by the random amplification of polymorphic DNA between precocious lines and parental strains of *Eimeria* spp. from the rabbit. Parasitology Research 83, 300–302. *.

Černá, Ž, 1974. Une anomalie de la sporulation de la coccidia des Oiseaux: *Isospora lacazei* (Eimeriidae). Journal of Protozoology 21, 481–482. *.

Černá, Ž., Kvašňovska, Z., 1986. Life cycle involving bird-bird relation in *Sarcocystis* coccidia with the description of *Sarcocystis accipitris* sp. n. Folia Parasitologica (Praha) 33, 305–309.

Černá, Ž., Sénaud, J., 1971. Some peculiarities of the fine structure of merozoites of *Eimeria stiedai*. Folia Parasitologica (Praha) 18, 177–178. *.

Černá, Ž., Loučková, M., Nedvědová, H., Vávra, J., 1981. Spontaneous and experimental infection of domestic rabbits by *Sarcocystis cuniculi* Brumpt, 1913. Folia Parasitologica (Praha) 28, 313–318. *.

Çetindağ, M., Biyikoğlu, G., 1997. Ankara yöresi evcil tavsanlarinda *Eimeria türlerinin yayilisi*. Türkiye Parazitoloji Dergisi 21, 301–304.

Chalmers, R.M., Giles, M., 2010. Zoonotic cryptosporidiosis in the UK—challenges for control. Journal of Applied Microbiology 109, 1487–1497. *.

Chalmers, R.M., Sturdee, A.P., Bull, S.A., Miller, A.M., 1995. Rodent reservoirs of *Cryptosporidium*. In: Betts, W.B., Casemore, D., Fricker, C., Smith, C., Watkins, J. (Eds.), Protozoan Parasites and Water. The Royal Society of Chemistry Special Publication No. 168. Royal Society of Chemistry, Cambridge, pp. 63–66. *.

Chalmers, R.M., Elwin, K., Thomas, A.L., Guy, E.C., Mason, B., 2009a. Long-term *Cryptosporidium* typing reveals the aetiology and species-specific epidemiology of human cryptosporidiosis in England and Wales, 2000 to 2003. European Surveillance 14, 6–14. *.

Chalmers, R.M., Robinson, G., Elwin, K., Hadfield, S.J., Xiao, L., Ryan, U., Modha, D., Mallaghan, C., 2009b. *Cryptosporidium* sp. rabbit genotype, a newly identified human pathogen. Emerging Infectious Diseases 15, 829–830. *.

Champliaud, D., Gobet, P., Naciri, M., Vagner, O., Lopez, J., Buisson, J.C., Varga, I., Harly, G., Mancassola, R., Bonnin, A., 1998. Failure to differentiate *Cryptosporidium parvum* from *C. meleagridis* based on PCR amplification of eight DNA sequences. Applied Environmental Microbiology 64, 1454–1458.

Chang, K., 1935. A general survey of the protozoa parasitic in the digestive tract of Shantung mammals. Peking Natural History Bulletin 9, 151–159. *.

Chapman, J.A., Flux, J.E.C., 1990. Rabbits, Hares and Pikas: Status Survey and Conservation Action Plan. International Union for Conservation of Nature and Natural Resources, Gland, Switzerland. *.

Chargaffe, E., 1997. In dispraise of reductionism. Bioscience 47, 795–797. *.

Chaudhuri, S.K., 1989. Investigations on the Coccidian Parasites of Some Zoo and Wild Mammals of India. Ph.D. Dissertation. Berhampur University, Calcutta, India.

Cheissin, E.M., 1935a. Structure de l'oocyste et permeabilite de ses membranes ches les coccidies du lapin. Annales Parasitologie 13, 136–146. *. §

Cheissin, E.M., 1935b. Vom Einfluss anaerober Bedingungen auf verschiedene Sporulationstadien der Oocysten von *Eimeria magna* und *E. stiedae*. Archive für Parasitenkunde 85, 426–435. *.

Cheissin, E.M., 1937. Coccidia of rabbits. I. Some data on the effects of diets on infectability of rabbits by intestinal coccidia and on mortality from coccidiosis. Trudy Instituta imeni Pastera 3, 20–40 (in Russian).

Cheissin, E.M., 1939. Coccidia of rabbits. II. Duration of a coccidiosis invasion when rabbits are infected with the oocysts of *Eimeria magna*. Vestnik Mikrobiologii, Epidemiologii. Parazitologii 18, 201–207 (in Russian).

Cheissin, E.M., 1940. Koksidios krolikov. III. Tsikl razvitiya *Eimeria magna*. (Rabbit coccidiosis. III. The developmental cycle of *Eimeria magna*). Uchenye Zapiski Institute Imeni Gertsena 30, 65–91. * (in Russian, English summary).

Cheissin, E.M., 1946. The duration of the life cycle of rabbit coccidia. Doklady Akademii Nauk SSSR 52, 561–564 (in Russian).

Cheissin, E.M., 1947a. Kokcidii kišečnika krolika. (Coccidia of the intestine of the rabbit). Uchenye Zapiski Leningradskogo Gosudarstvennyi Pedagogicheskii Institut imeni Gertsena 5, 1–229. *.

Cheissin, E.M., 1947b. Mutability of the oocysts of *Eimeria magna*. Zoologische Zhurnnal 26, 17–29. * (in Russian).

Cheissin, E.M., 1947c. A new species of rabbit coccidia (*Eimeria coecicola* n. sp.). Comptes Rendus (Doklady) de l'Académie des Sciences de l'URSS 55, 177–179. * (in Russian).

Cheissin, E.M., 1948. Razvitie dvukh kišhečhnykh koktsidij krolika *Eimeria piriformis* Kotlán i Pospesch i *Eimeria intestinalis* nom. nov. (Development of two intestinal coccidia of the rabbit— *Eimeria piriformis* Kotlán and

Pospech and *Eimeria intestinalis* nom. nov). Uchenye Zapiski Karelo-Finskogo Universiteta 3, 179–187. * (in Russian).

Cheissin, E.M., 1957a. Variability of the oocysts of *Eimeria intestinalis*—the parasite of the domestic rabbit. Vestnik Leningradskogo Gosudarstvennogo Universiteta, seriya Biologiya 2, 43–52. * (in Russian).

Cheissin, E.M., 1957b. Topologičeskie različija soprjažennyh vidov kokcidij domašnego krolika (Topological differences of associated species of coccidia in domestic rabbits [*Oryctolagus cuniculi*]). Trudy Leningradskogo Obščhchestra Estestvoispytatelej 73, 150–158.

Cheissin, E.M., 1958a. Cytologische Untersuchungen verschiedener Stadien des Lebenszyklus der Kaninchencoccidien. I. *Eimeria intestinalis* E. Cheissin, 1948. (Cytological investigations of different stages of the life cycle of rabbit coccidia. I. *Eimeria intestinalis* E. Cheissin, 1948). Archiv für Protistenkunde 102, 265–290 (+ 73 Figs. on 5 plates)* (in German).

Cheissin, E.M., 1958b. The role of the residual body of spores and oocysts of coccidia of the genera *Eimeria* and *Isospora*. Journal of Protozoology 5 (Suppl.) 8. *.

Cheissin, E.M., 1959. Observations of the residual bodies of oocysts and spores of several species of *Eimeria* from the rabbit and *Isospora* from the fox, skunk and hedgehog. Zoologische Zhurnnal 38, 1776–1784. * (in Russian).

Cheissin, E.M., 1960. Cytologic studies of the life cycle of the coccidia of the rabbit. II. *Eimeria magna* Pérard, 1924. In: Problems of cytology and protistology (in Russian; English translation, 1961, Israel Program for Scientific Translations, Jerusalem), Moscow-Leningrad, pp. 258–276. *.

Cheissin, E.M., 1964. Electron microscope study of microgametes of *E. intestinalis* (Sporozoa, Coccidiida) (in Russian, English translation by the Translating Unit, Libraryu Branch, Division of Research Services, NIH, Bethesda, Maryland USA). Zoologische Zhurnnal 43, 647–650. *.

Cheissin, E.M., 1965. Electron microscopic study of microgametogenesis in two species of coccidia from rabbit (*Eimeria magna* and *E. intestinalis*). Acta Protozoologica 3, 215–224. *.

Cheissin, E.M., 1967. Žhiznennye Tsikly Koktsidii Domašhnikh Zhivotnykh (in Russian; English translation by Plous, Jr., F.K.; Todd Jr., K.S. [Eds.], 1972. Life Cycles of Coccidia of Domestic Animals. University Park Press, Baltimore, Maryland). Isdat Nauka, Leningrad. 1–192. *.

§ Evgeniî Mineevich Kheysin, perhaps more than any other Russian author, has spelled his name (and has had his name spelled by others) several different ways (Kheysin, Keissin, Cheyssin, Cheissin). Throughout this book we use his French-transliterated name, E.M. Cheissin, which seems to be most commonly used, simply for internal consistency. However, his Russian-alphabet last name is Хейсин, which transliterates to Kheysin.

Cheissin, E.M., 1968. On the distinctness of the species *Eimeria neoleporis* Carvalho, 1942 from the cottontail rabbit *Sylvilagus floridanus mearnsii* and *Eimeria coecicola* Cheissin, 1947 from the tame rabbit *Oryctolagus cuniculus*. Acta Protozoologica 6, 5—12. *.

Cheissin, E.M., Snigirevskaya, E.S., 1965. Some new data on the fine structure of the merozoites of *Eimeria intestinalis* (Sporozoa, Eimeriidae). Protistologica 1, 121—125 (+ 10 Figs. on 2 plates). *.

Christiansen, M., Siim, J.C., 1951. Toxoplasmosis in hares in Denmark. Serological identity of human and hare strains of *Toxoplasma*. Lancet 1, 1201—1203. *.

Chroust, K., 1979. Determination of the species composition of coccidia in hares. Veterinářství 29, 507—509. * (in Czech).

Chroust, K., 1984. Dynamics of coccidial infection in free living and cage-reared European hares. Acta Veterinaria (Brno) 53, 175—182. *.

Ci, H.X., Lin, G.H., Su, J.P., Cao, Y.F., 2008. Host sex and ectoparasite infections in the plateau pika (*Ochotona curzoniae*, Hodgson) on the Qinghai-Tibetan Plateau. Polish Journal of Ecology 56, 535—539.

Clapham, P.A., 1954. Disease in rabbits and hares. Veterinary Record 66, 100. *.

Claridge, M., 1995. Introducing systematics agenda 2000. Biodiversity and Conservation 4, 451—454.

Clubb, S.L., Frenkel, J.K., 1992. *Sarcocystis falcatula* of opossums: transmission by cockroaches with fatal pulmonary disease in psittacine birds. Journal of Parasitology 78, 116—124.

Cobb, N.A., 1914. Nematodes and their Relationships. In: U.S. Department of Agriculture Yearbook, 1914. Department of Agriculture, Washington, D.C., pp. 457—490. *.

Collins, G.H., 1979. Studies on *Sarcocystis* species. II. Infection in wild and feral animals— prevalence and transmission. New Zealand Veterinary Journal 27 135—135. *.

Collins, G.H., Charleston, W.A.G., 1979. Studies on *Sarcocystis* species. I. Feral cats as definitive hosts for Sporozoa. New Zealand Veterinary Journal 27, 80—84. *.

Colwell, R., 2008. Editorial: Organisms from molecules to the environment. Bioscience 58, 3.

Combes, C., 1996. Parasites, biodiversity and ecosystem stability. Biodiversity and Conservation 5, 953—962. *.

Conn, D.B., Weaver, J., Tamang, L., Graczyk, T.K., 2007. Synanthropic flies as vectors of *Cryptosporidium* and *Giardia* among livestock and wildlife in a multispecies agricultural complex. Vector Borne Zoonotic Diseases 7, 643—651. *.

Cooper, H., 1927. I. Coccidiosis with particular reference to bovine coccidiosis and its significance as an infection of cattle in India. The Agricultural Journal of India (reprinted in the Indian Veterinary Journal 4, 56—67). *.

Corbet, G.B., 1983. A review of classification in the family Leporidae. Acta Zoologica Fennica 174, 11—15. *.

Correa, A.D., 1931. Coccidiosis de conejo. Archives de la Sociedad de Biologia de Montevideo (Suppl.) 5, 1101—1113. *.

Cosgrove, M., Wiggins, J.P., Rothenbacher, H., 1982. *Sarcocystis* sp. in the eastern cottontail (*Sylvilagus floridanus*). Journal of Wildlife Diseases 18, 37—40. *.

Coudert, P., 1972. Essai de chimioprévention de la coccidiose hépatique du lapin par le Decoquinate (Chemoprevention trial of coccidiosis of the rabbit liver by Decoquinate). Les Cahiers de Médecine Vétérinaire 3, 1—12. * (in French, English summary).

Coudert, P., 1973a. Coccidioses du lapin. Symposium International sur les Coccidioses, 12—13, September, Tours-Nouzilly, France. 6 p. + 3 figs.

Coudert, P., 1973b. Effet de la temperature de sporulation sur le pouvoir pathogéne des *Eimeria*. Journées de Recherches avicoles et cunicoles December 43-47.

Coudert, P., 1976. Les coccidioses intestinales du lapin: comparaison du pouvoir pathogène d' *Eimeria intestinalis* avec trois autres *Eimeria* (Intestinal coccidiosis of the rabbit: comparison of the pathogenic power of *Eimeria intestinalis* with three other *Eimeria*). Comptes Rendus de l'Academie des Sciences (Paris). Série D (Sciences Naturelles) 282, 2219—2222. * (in French).

Coudert, P., 1977a. Isolation and description of a new *Eimeria* species in rabbit (*Oryctolagus cuniculus*): *E. pellerdyi* P. Coudert, 1977 (Sporozoa, Coccidia). Proceedings of the Fifth International Congress of Protozoology, June 26 — July 2, 1977, New York City, New York.*.

Coudert, P., 1977b. Isolement et description d'une nouvelle éspèce d' *Eimeria* chez le Lapin (*Oryctolagus cuniculus*): *Eimeria pellerdyi* P. Coudert, 1977 (Sporozoa, Coccidia) (Isolation and description of a new species of *Eimeria* in rabbits [*Orychtolagus cuniculus*]: *Eimeria pellerdyi* P. Coudert, 1977 [Sporozoa, Coccidia]). Comptes Rendus de l'Academie des Sciences (Paris). Série D (Sciences Naturelles) 285, 885—887. * (in French, English summary).

Coudert, P., 1978. Evaluation comparative de l'efficacité de 10 médicaments contre deux coccidioses graves du lapin. 2e Journées de la Recherche Cunicole, Toulouse, Communication No. 13.

Coudert, P., 1979a. Comparison of pathology of several rabbit coccidia species and their control with Robenidine. In: Proceedings of the International Symposium on Coccidia, November 28—30, 1979. Czech Republic, Prague, pp. 159—163. *.

Coudert, P., 1979b. Comparative estimation of the efficiency of ten drugs against two severe coccidioses in the rabbit. Annales de Zootechnic 28, 141. *.

Coudert, P., 1981a. Chemoprophylaxe von Darm- und Gallengangskokzidiosen beim Kaninchen. In: Illnesses of Fur Animals, Rabbits and Domestic Animals (H.-Ch. Löliger, Conference Leader). Fourth Conference, Speciality Group of Small Animal Disease, Institute for Small Animal Breeding and the German Group of WRSA, Deutsche Veterinarmedizinische Gesellschaft, Celle, 18–20 June, pp. 106–119. *.

Coudert, P., 1981b. Coccidioses et diarrhées du lapin a l'engraissement (Diarrhea and coccidiosis of rabbits for fasting). In: Conference prononcee au Seminaire FRGTV de Toulouse, France, November, pp. 109–122.

Coudert, P., 1989. Some peculiarities of rabbit coccidiosis. In: Yvoré, P. (Ed.), Coccidia and intestinal coccidiomorphs, Proceedings of the Vth International Coccidiosis Conference, 17–20 October, 1989. France, Tours, pp. 481–488. *.

Coudert, P., 1990. Efficacite du Diclazuril contre deux coccidioses graves du lapin et tolerance (Effectiveness of Diclazuril against two severe coccidiosis of the rabbit and tolerance). 5th Journees de la Recherche Cunicole, No. 26, 12–13 December, Paris. 8 p.* (in French, English summary).

Coudert, P., Norton, C.C., 1979. Eimeria flavescens Marotel G. and Guilhon J., 1941 is a true coccidium of the rabbit (Oryctolagus cuniculus). Proceedings of the International Symposium on Coccidia, November, 1979. Czech Republic, Prague, p. 164. *.

Coudert, P., Nouzilly, I.N.R.A., 1981. Chemoprophylaxe von Darm- und Gallengangskokzidiosen beim Kaninchen. Der Fachgruppe Kleintierkrankheiten der fal und der Deutschen Gruppe der WRSA (The Specialty Group of Small Animal Disease in Connection with the Institute for Small Animal Breeding and the German Group of WRSA). Thema: Krankheiten der pelztiere kaninchen und heimtiere (Subject: Illnesses of fur animals, rabbits and domestic animals) 18–21 June, 106–119. *.

Coudert, P., Provôt, F., 1973. Métabolisme d' Eimeria stiedai (Lindemann, 1865) Kisskalt et Hartmann 1907: Influence de la composition du milieu pendant la sporogonie (Metabolism of Eimeria stiedai [Lindemann, 1865] Kisskalt and Hartmann 1907: Influence of medium composition during sporogony). Annales de Recherches Véterinaires 4, 613–626 (in French, English summary).

Coudert, P., Provôt, F., 1974. Développement interne et schizogonie en cultures cellulaires d' Eimeria stiedai (Lindemann 1865) Kisskalt et Hartmann 1907 (Internal development of schizogony in cell culture of Eimeria stiedai [Lindemann 1865] Kisskalt and Hartmann 1907). Comptes Rendus de l'Academie des Sciences (Paris) 279, 911–913. * (in French).

Coudert, P., Provôt, F., 1988. Lasalocid: Tolerance for the rabbit and activity against E. flavescens and E. intestinalis.

In: Proceedings 4th World Rabbit Congress, October 10–14, 1988. Hungary, Budapest, pp. 418–427. *.

Coudert, P., Doll, G., Dürr, U., 1972. Zur ultraschallresistenz der oocysten von Eimeria stiedai (Sporozoa, Coccidia) (The resistance of Eimeria stiedai oocysts to ultrasonic waves[Sporozoa, Coccidia]). Annales de Recherches Véterinaires 3, 551–570. * (in German, English summary).

Coudert, P., Yvore, P., Provot, F., 1973. Sporogonie d' Eimeria stiedai (Lindemann, 1865) Kisskalt et Hartmann 1907—Influence de la temperature sur la respiration et sur la durée de la segmentation (Sporogony of Eimeria stiedai [Lindemann, 1865] Kisskalt and Hartmann 1907—the influence of temperature on respiration and the duration of segmentation). Annales de Recherches Véterinaires 4, 371–388. * (in French, English summary).

Coudert, P., Vaissaire, J., Licois, D., 1978. Étude de l'évolution de quelques paramètres sanguins chez des lapereaux atteints de coccidiose intestinale (Evolutiion of some blood parameters in young rabbits infested with intestinal coccidia). Recueil de Médecine Vétérinaire 154, 437–440. * (in French, English summary).

Coudert, P., Licois, D., Streun, A., 1979. Characterization of Eimeria species. I. Isolation and study of pathogenicity of a pure strain of Eimeria perforans (Leukart, 1879; Sluiter and Swellengrebel, 1912). Zeitschrift für Parasitnekunde 59, 227–234. *.

Coudert, P., Licois, D., Provôt, F., Drouet-Viard, F., 1993. Eimeria sp. from the rabbit (Oryctolagus cuniculus): pathogenicity and immunogenicity of Eimeria intestinalis. Parasitology Research 79, 186–190. *.

Coudert, P., Licois, D., Drouet-Viard, F., 1995. Eimeria species and strains of rabbit. In: Eckert, J., Braun, R., Shirley, M.W., Coudert, P. (Eds.), Guidelines on Techniques in Coccidiosis Research Luxembourg Office Official Publications, European Communities, Directorate-General XII. Science, Research and Development Environment Research Programme, pp. 52–73. *.

Coudert, P., Licois, D., Zonnekeyn, V., 2000. Epizootic rabbit enterocolitis and coccidiosis: A criminal conspiracy. In: Proceeding of the 7th World Rabbit Congress, 4–7 July, 2000, Valencia, Spain,. World Rabbit Science Supplement 1 (8), 215–218.

Cox, P., Griffith, M., Angles, M., Deere, D., Ferguson, C., 2005. Concentrations of pathogens and indicators in animal feces in the Sydney watershed. Applied Environmental Microbiology 71, 5929–5934.

Crawley, H., 1914. Two new sarcosporidia. Proceedings of the Academy of Natural Sciences, Philadelphia 66, 214–218.*.

Crum, J.M., Prestwood, A.K., 1977. Transmission of Sarcocystis leporum from a cottontail rabbit to domestic cats. Journal of Wildlife Diseases 13, 174–175. *.

Cuckler, A.C., Malanga, C.M., 1955. Studies on drug resistance in Coccidia. Journal of Parasitology 41, 302–311.

Current, W.L., Reese, N.C., Ernst, J.V., Bailey, W.S., Heyman, M.B., Weinstein, W.M., 1983. Human cryptosporidiosis in immunocompetent and immunodeficient persons. Studies of an outbreak and experimental transmission. New England Journal of Medicine 308, 1252–1257. *.

Danforth, H.D., Hammond, D.M., 1972. Stages of merogony in multinucleate merozoites of *Eimeria magna*. Journal of Protozoology 19, 454–457. *.

Darabus, G., Olariu, R., 2003. The homologous and interspecies transmission of *Cryptosporidium parvum* and *Cryptosporidium meleagridis*. Polish Journal of Veterinary Science 6, 225–228. *.

Darwish, A.I., Golemansky, V., 1991. Coccidian parasites (Coccidia: Eimeriidae) of domestic rabbits (*Oryctolagus cuniculus domesticus* L.) in Syria. Acta Protozoologica 31, 209–216. *.

Dawson, M.R., 1958. Later Tertiary Leporidae of North America. In: Vertebrate, 6. University of Kansas Paleontology Contribution, pp. 1–75.

Dawson, M.R., 1981. Evolution of modern leporids. In: Myers, K., Machines, C.D. (Eds.), Proceedings of the World Lagomorph Conference. University of Guelph Press, Guelph, Ontario, pp. 1–8.

Delaplane, J.R., Stuart, H.O., 1935. The survival of avian coccidia in soil. Poultry Science 14, 67–69. *.

Deng, M.Q., Cliver, D.O., Mariam, T.W., 1997. Immunomagnetic capture PCR to detect viable *Cryptosporidium parvum* oocysts using MACS MicroBeads and high gradient separation columns. Journal of Microbiological Methods 40, 11–17.

Devlin, T.M., 2005. Textbook of Biochemistry with Clinical Correlations, sixth ed. Wiley Liss, New York.

De Vos, A.J., 1970. Studies on the host range of *Eimeria chinchillae* (De Vos and Van der Westhuizen, 1968). Ondersteeport Journal of Veterinary Research 37, 29–36. *.

De Waele, V., Speybroeck, N., Berkvens, D., Mulcahy, G., Murphy, T.M., 2010. Control of cryptosporidiosis in neonatal calves: Use of halofuginone lactate in two different calf rearing systems. Preventive Veterinary Medicine 96, 143–151. *.

Diesing, L., Heydorn, A.O., Matuschka, F.R., Bauer, C., Pipano, E., DeWaal, D.R., Potgieter, F.T., 1988. *Besnoitia besnoiti*: Studies on the definitive host and experimental infections in cattle. Parasitology Research 75, 114–117. *.

Ding, X.T., He, X.Q., Cao, Y.Q., Dai, L., 1999. Reports on parasite infections on plateau pika. Sichuan Journal of Zoology 18, 34–43 (in Chinese).

Dobell, C., 1922. The discovery of the coccidia. Parasitology 14, 342–348. *.

Dobell, C., 1932. Antony van Leeuwenhoek and his "little animals". Harcourt, Brace & Co., New York. *.

Doležel, D., Koudela, B., Jirků, M., Hypša, V., Obornik, M., Votýpka, J., Modrý, D., Šlepta, J.R., Lukeš, J., 1999. Phylogenetic analysis of *Sarcocystis* spp. of mammals and reptiles supports the coevolution of *Sarcocystis* spp., with their final hosts. International Journal of Parasitology 29, 795–798. *.

Donát, K., 1989. The distribution of sarcosporidians in domestic rabbits. Veterinárství 39, 492–493. * (in Czech).

Donciu, I., Purcherea, A., Asmasan, B., Nesterov, V., Brinzoi, M., 1968. Contributii la studiul specificitatii coccidici " *Eimeria stiedae*." Lucrari Stiintifice. Serial C 11, 305–310.

Donnard, E., 1982. Recherches sur les Léporidés quaternaires (Pléistocène moyen et supérieur, Holocène). Thèse d'E'tat, Université de Bordeaux I.

Doran, D.J., Vetterling, J.M., 1969. Infectivity of two species of poultry coccidia after freezing and storage in liquid nitrogen vapor. Proceedings of the Helminthological Society of Washington 36, 30–33. *.

Dorney, R.S., 1962. Coccidiosis in Wisconsin cottontail rabbits in winter. Journal of Parasitology 48, 276–279. *.

Douzery, E.J.P., Delsuc, F., Stanhope, M.J., Huchon, D., 2003. Local molecular clocks in three nuclear genes: divergence times for rodents and other mammals and incompatibility among fossil calibrations. Journal of Molecular Evolution 57, S201–S213.

Drouet-Viard, F., Licois, D., Provôt, F., Coudert, P., 1994. The invasion of the rabbit intestinal tract by *Eimeria intestinalis* sporozoites. Parasitology Research 80, 706–707. *.

Dubey, J.P., 1976. A review of *Sarcocystis* of domestic animals and of other coccidia of cats and dogs. Journal of the American Veterinary Medical Association 169, 1061–1078. *.

Dubey, J.P., 1977a. Taxonomy of *Sarcocystis* and other coccidia of cats and dogs. Journal of the American Veterinary Medical Association 170, 778–782. *.

Dubey, J.P., 1977b. *Toxoplasma, Hammondia, Besnoitia, Sarcocystis*, and other tissue cyst-forming coccidia of man and animals. In: Krier, J.P. (Ed.), Parasitic Protozoa, vol. III. Academic Press, New York, pp. 101–237. *.

Dubey, J.P., 1993. Intestinal protozoa infections. Veterinary Clinics of North America Small Animal Practice 23, 37–55. *.

Dubey, J.P., 1998. *Toxoplasma gondii* oocyst survival under defined temperatures. Journal of Parasitology 84, 862–865. *.

Dubey, J.P., 2010. Chapter 18, Toxoplasmosis in rodents and small mammals. In: Dubey, J.P. (Ed.), Toxoplasmosis of Animals and Humans, second ed. CRC Press, Boca Raton, Florida, pp. 227–233. *.

Dubey, J.P., Beattie, C.P., 1988. Toxoplasmosis of Animals and Man. CRC Press, Boca Raton, Florida. *.

Dubey, J.P., Miller, N.L., Frenkel, J.K., 1970a. Characterization of the new fecal form of *Toxoplasma gondii*. Journal of Parasitology 56, 447–456.

Dubey, J.P., Miller, N.L., Frenkel, J.K., 1970b. The *Toxoplasma gondii* oocyst from cat feces. Journal of Experimental Medicine 132, 636–662.

Dubey, J.P., Speer, C.A., Fayer, R., 1989. *Sarcocystis* of Animals and Man. CRC Press, Boca Raton, Florida. *.

Dubey, J.P., Brown, C.A., Carpenter, J.L., Moore, J.J., 1992. Fatal toxoplasmosis in domestic rabbits in the USA. Veterinary Parasitology 44, 305–309.

Dubey, J.P., Passos, L.M.F., Rajendran, C., Ferreira, L.R., Gennari, S.M., Su, C., 2011. Isolation of viable *Toxoplasma gondii* from feral guinea fowl (*Numida meleagris*) and domestic rabbits (*Oryctolagus cuniculus*) from Brazil. Journal of Parasitology 97, 842–845. *.

Dubos, R., 1965. Man Adapting. Yale University Press, New Haven, Connecticut. *.

Dufour, L., 1828. Note sur la grégarine, nouveau genre de ver qui vit en troupeau dansles intestines de divers insects. Annales des Sciences Naturelles 13, 366–368.

Dunlap, J.S., Dickson, W.M., Johnson, L., 1959. Ionographic studies of rabbits infected with *Eimeria stiedae*. American Journal of Veterinary Research 20, 589–591. *.

Dürr, U., 1971. Zur Excystation und Bewegung von Sporozoiten der Kokzidienart *Eimeria stiedai*. Deutsche Tierärztliche Wochenschrift 78, 17–21. *.

Dürr, U., 1972. Life cycle of *Eimeria stiedai*. Acta Veterinaria Academiae Scientiarum Hungaricae 22, 101–103. *.

Dürr, U., Pellérdy, L.P., 1969. The susceptibility of suckling rabbits to infection with coccidia. Acta Veterinaria Academiae Scientiarum Hungaricae 19, 453–462. *.

Dürr, U., Lammler, G., 1970a. Zur Prüfung coccidiostatisch wirksamer Arzneimittel bein Kaninchen im Feldversuch. Berliner Muchener Tierärztliche Wochenschrift 83, 50–54.

Dürr, U., Lammler, G., 1970b. Prophylaxeversuche mit Sulfonamiden bei der Darmkokzidiose des Kaninchens. Zentralblatt für Veterinärmedizin. Reihe B 17, 554–563.

Dürr, U., Reiser, W., 1972. Zur resistenz von *Eimeria stiedai*-oocysten gegen röntgen- und gamma-bestrahlung (For the resistance of *Eimeria stiedai* oocysts against X-ray and gamma irradiation). Acta Veterinaria Academiae Scientiarum Hungaricae 22, 409–416. * (in German, no English summary).

Dürr, U., Heunert, H.H., Milthaler, B., 1971. Beobachtungen zur sporogonie *Eimeria stiedai* (Protozoa, Sporozoa) (Observations on the sporogony of *Eimeria stiedai* [Protozoa, Sporozoa]). Acta Veterinaria Academiae Scientiarum Hungaricae 21, 421–432. * (in German, no English summary).

Dürr, U., Heunert, H.H., Milthaler, B., Galle, H.-K., 1972. Beobachtungen zur excystation von sporozoiten der kokzidienart *Eimeria stiedai* (Observations on the excystation of sporozoites of the coccidian *Eimeria stiedai*). Acta Veterinaria Academiae Scientiarum Hungaricae 22, 169–185. * (in German, no English summary).

Duszynski, D.W., 1971. Increase in size of *Eimeria separata* oocystsd during patency. Journal of Parasitology 57, 948–952. *.

Duszynski, D.W., 1972. Host and parasite interactions during single and concurrent infections with *Eimeria nieschulzi* and *E. separata* in the rat. Journal of Protozoology 19, 82–88. *.

Duszynski, D.W., 1974. More information on the coccidian parasites (Protozoa: Eimeriidae) of the Colorado pika, *Ochotona princeps*, with a key to the species. Journal of Wildlife Diseases 10, 94–100. *.

Duszynski, D.W., 1986. Host specificity in the coccidia of small mammals: Fact or fiction? In: Bereczky, M. (Ed.), Advances in Protozoological Research. Symposia Biologica Hungarica, vol. 33. Akademiai Kiado, Budapest, pp. 325–337. *.

Duszynski, D.W., 1999. Revisiting the Code: Clarifying name-bearing types for photomicrographs of Protozoa. Critical Comment. Journal of Parasitology 85, 769–770.

Duszynski, D.W., 2002. Coccidia (Apicomplexa: Eimeriidae) of the Mammalian order Chiroptera. Special Publication of the Museum of Southwestern Biology, No.5, University of New Mexico, Albuquerque, NM. *.

Duszynski, D.W., 2011. *Eimeria*. In: Encyclopedia of life sciences (ELS). John Wiley & Sons, Ltd, Chichester, United Kingdom. www.els.net. *.

Duszynski, D.W., Brunson, J.T., 1972. The structure of the oocyst and the excystation process of *Isospora marquardti* sp. n. from the Colorado pika, *Ochotona princeps*. Journal of Protozoology 19, 257–259. *.

Duszynski, D.W., Brunson, J.T., 1973. Structure of the oocysts and excystation processes of four *Eimeria* spp. (Protozoa: Eimeriidae) from the Colorado pika, *Ochotona princeps*. Journal of Parasitology 59, 28–34. *.

Duszynski, D.W., Marquardt, W.C., 1969. *Eimeria* (Protozoa: Eimeriidae) of the cottontail rabbit *Sylvilagus audubonii* in northeastern Colorado, with descriptions of three new species. Journal of Protozoology 16, 128–137. *.

Duszynski, D.W., Upton, S.J., 2000. Coccidia (Apicomplexa: Eimeriidae) of the Mammalian Order Insectivora. Special publication of the Museum of Southwestern Biology, No. 4, University of New Mexico, Albuquerque, NM. *.

Duszynski, D.W., Upton, S.J., 2001. *Cyclospora, Eimeria, Isopora* and *Cryptosporidium* spp. In: Samuel, W.M., Pybus, M.J., Kocan, A.A. (Eds.), Parasitic Diseases of Wild Mammals, second ed. Iowa State University Press, Ames, Iowa, pp. 416–459. *.

Duszynski, D.W., Upton, S.J., 2009. The biology of the coccidia (Apicomplexa) of snakes of the world. A scholarly handbook for identification and treatment. https://www.CreateSpace.com/33388533 ISBN 1448617995 (revised 2010). *.

Duszynski, D.W., Wilber, P.G., 1997. A guideline for the preparation of species descriptions in the Eimeriidae. Journal of Parasitology 83, 333–336. *.

Duszynski, D.W., Samuel, W.M., Gray, D.R., 1977. Three new *Eimeria* spp. (Protozoa, Eimeriidae) from muskoxen, *Ovibos moschatus*, with redescriptions of *E. faurei, E. granulose,* and *E. ovina* from muskoxen and from a Rocky Mountain bighorn sheep, *Ovis canadensis*. Canadian Journal of Zoology 55, 990–999. *.

Duszynski, D.W., Patrick, M.J., Couch, L., Upton, S.J., 1992. Eimerians in harvest mice, *Reithrodontomys* spp., from Mexico, California and New Mexico, and phenotypic plasticity in oocysts of *Eimeria arizonensis*. Journal of Protozoology 39, 644–648. *.

Duszynski, D.W., Wilson, W.D., Upton, S.J., Levine, N.D., 1999. Coccidia (Apicomplexa: Eimeriidae) in the Primates and Scandentia. International Journal of Primatology 20, 761–797. *.

Duszynski, D.W., Harrenstien, L., Couch, L., 2005. A pathogenic new species of *Eimeria* from the pygmy rabbit, *Brachylagus idahoensis*, in Washington and Oregon, with description of the sporulated oocyst and intestinal endogenous stages. Journal of Parasitology 91, 618–623. *.

Duszynski, D.W., Upton, S.J., Bolek, M., 2007. Coccidia (Apicomplexa: Eimeriidae) of the Amphibians of the World. Zootaxa (Magnolia Press) 1667, 1–77. *.

Ebach, M.C., Holdrege, C., 2005a. DNA barcoding is no substitute for taxonomy. Nature 434, 697. *.

Ebach, M.C., Holdrege, C., 2005b. More taxonomy, not DNA barcoding. Bioscience 55, 822–823. *.

Eberhard, M.L., Ortega, Y.R., Hanes, D.E., Nace, E.K., Do, R.Q., Robl, M.G., Won, K.Y., Gavidia, C., Sass, N.L., Mansfield, K., Gozalo, A., Griffiths, J., Gilman, R., Sterling, C.R., Arrowood, M.J., 2000. Attempts to establish experimental *Cyclospora cayetanensis* infection in laboratory animals. Journal of Parasitology 86, 577–582. *.

Ecke, D.H., Yeatter, R.E., 1956. Notes on the parasites of cottontail rabbits in Illinois. Illinois Academy of Science Transactions 48, 208–214. *.

Eckert, J., Taylor, M., Licois, D., Coudert, P., Catchpole, J., Bucklar, H., 1995. Identification of *Eimeria* and *Isospora* species and strains. Morphological and biological characteristics. In: Eckert, J., Braun, R., Shirley, M.W., Coudert, P. (Eds.), Biotechnology. Guidelines on Techniques in Coccidiosis Research. Office for Official Publications of the European Communities, Luxembourg, pp. 103–119. *.

Edelhofer, R., Heppe-Winger, E.M., Hassl, A., Aspöck, H., 1989. *Toxoplasma*-Infektionen bei jagdbaren Wildtieren in Östösterreich. Mitteilungen Östösterreich Gesellshaft Tropenmedizin Parasitologie 11, 119–123 (in German).

Edgar, S.A., 1954. Effect of temperature on the sporulation of oocysts of the protozoan *Eimeria tenella*. Transactions of the American Microscopical Society 73, 237–242. *.

Edgar, S.A., 1955. Sporulation of oocysts at specific temperatures and notes on the prepatent period of several species of avian coccidia. Journal of Parasitology 41, 214–216. *.

Ellis, J., Tomley, F., 1991. Development of a genetically engineered vaccine against poultry coccidiosis. Parasitology Today 7, 344–346. *.

El-Shahawi, G.A., El-Fayomi, H.M., Abdel-Haleem, H.M., 2012. Coccidiosis of domestic rabbit (*Oryctolagus cuniculus*) in Egypt: light microscopy. Parasitology Research 110, 251–258.

Elwasila, M., 1984. Fine structure of the process of oocyst wall formation of *Eimeria maxima* (Apicomplexa: Eimeriina). Acta Veterinaria Hungarica 32, 159–163.

Elwasila, M., Entzeroth, R., Chobotar, B., Scholtyseck, E., 1984. Comparison of the structure of *Sarcocystis cuniculi* of the European rabbit (*Oryctolagus cuniculus*) and *Sarcocystis leporum* of the cottontail rabbit (*Sylvilagus floridanus*) by light and electron microscopy. Acta Veterinaria Hungarica 32, 71–78. *.

Entzeroth, R., Scholtyseck, E., 1977. The life cycle of *Eimeria stiedai* from rabbits in hares. Fifth International Congress of Protozoology, Abstracts, 18. *.

Erickson, A.B., 1946. Incidence and transmission of *Sarcocystis* in cottontails. Journal of Wildlife Diseases 10, 44–46. *.

Escalante, A.A., Ayala, F.J., 1995. Evolutionary origin of *Plasmodium* and other Apicomplexa based on rRNA genes. Proceedings of the National Academy of Sciences. USA 92, 5793–5797. *.

Farr, M.M., Wehr, E.E., 1949. Survival of *Eimeria acervulina, E. tenella,* and *E. maxima* oocysts on soil under various field conditions. Annals of the New York Academy of Sciences 52, 468–472. *.

Fayer, R., 1972. Gametogony of *Sarcocystis* sp. in cell culture. Science 175, 65–67. *.

Fayer, R., 2007. General Biology. In: Fayer, R., Xiao, L. (Eds.), *Cryptosporidium* and Cryptosporidiosis, second ed. CRC Press, Boca Raton, Florida, pp. 1–42. *.

Fayer, R., 2010. Taxonomy and species delimitation in *Cryptosporidium*. Experimental Parasitology 124, 90–97. *.

Fayer, R., Kradel, D., 1977. *Sarcocystis leporum* in cottontail rabbits and its transmission to carnivores. Journal of Wildlife Diseases 13, 170–173. *.

Fayer, R., Speer, C.A., Dubey, J.P., 1990. General biology of *Cryptosporidium*. In: Dubey, J.P., Speer, C.A.,

Fayer, R. (Eds.), *Cryptosporidium* of Man and Animals. CRC Press, Boca Raton, Florida, pp. 2–29.

Fayer, R., Speer, C.A., Dubey, J.P., 1997. General biology of *Cryptosporidium*. In: Fayer, R. (Ed.), *Cryptosporidium* of Man and Animals. CRC Press, Boca Raton, Florida, pp. 1–41. *.

Fayer, R., Morgan, U., Upton, S.J., 2000. Epidemiology of *Cryptosporidium*: transmission, detection and identification. International Journal for Parasitology 30, 1305–1322. *.

Fayer, R., Santin, M., Macarisin, D., 2010. *Cryptosporidium ubiquitum* n. sp. in animals and humans. Veterinary Parasitology 172, 23–32.

Fedoseenko, V.M., 1986. *Sarcocystis* sp. from the Altai pika (*Ochotona alpina*). Materialy X Konferentsii Ukrainskogo Obshch-va Parazitolcgov Part 2, 281. * (in Russian).

Feng, Y., Alderisio, K.A., Yang, W., Blancero, L.A., Kuhne, W.G., Nadareski, C.A., Reid, M., Ziao, L., 2007. *Cryptosporidium* genotypes in wildlife from a New York watershed. Applied Environmental Microbiology 73, 6475–6483.

Figuerosa-Castillo, J.A., Duarte-Rosas, V., Juárez-Accvedo, M., Luna-Pastén, H., Correa, D., 2006. Prevalence of *Toxoplasma gondii* antibodies in rabbits (*Oryctolagus cuniculus*). Journal of Parasitology 92, 394–395. *.

Findley, J.S., 1987. The Natural History of New Mexico Mammals. University of New Mexico Press, Albuquerque, New Mexico.

Fioramonti, J., Sorraing, J.M., Licois, D., Bueno, L., 1981. Intestinal motor and transit disturbances associated with experimental coccidiosis (*Eimeria magna*) in the rabbit. Annales de Recherches Vétérinaires 12, 413–420. *.

Fish, F.F., 1931a. The effect of physical and chemical agents on the oocysts of *Eimeria tenella*. Science 73, 292–293. *.

Fish, F.F., 1931b. Quantitative and statistical analysis of infection with *Eimeria tenella* in the chicken. American Journal of Hygiene 14, 560–576. *.

Fitzgerald, P.R., 1967. The effect of ionizing radiation on unsporulated oocysts of *Eimeria stiedae*. Journal of Protozoology 14 (Suppl.), 21.

Fitzgerald, P.R., 1970a. New findings on the life cycle of *Eimeria stiedae*. Proceedings of the 2nd International Congress of Parasitology, Washington, D.C. 100, 181 (abstract). *.

Fitzgerald, P.R., 1970b. New findings on the life cycle of *Eimeria stiedae*. Journal of Parasitology 56 (Suppl. Part I), 100–101 (abstract). *.

Fitzgerald, P.R., 1970c. Development of *Eimeria stiedai* in avian embryos. Journal of Parasitology 56, 1252–1253. *.

Fitzgerald, P.R., 1972. Transmission of *Eimeria stiedai* by blood transfusion. Journal of Parasitology 58 (Suppl.), 62 (abstract). *.

Fitzgerald, P.R., 1974. Results of blood transfusioons from donor rabbits infected with *Eimeria stiedai* to recipient coccidia-free rabbits. Journal of Protozoology 21, 336–338. *.

Flatt, R.E., Campbell, W.W., 1974. Cysticercosis in rabbits: incidence and lesions of the naturally occurring disease in young domestic rabbits. Laboratory Animal Science 24, 914–918 *.

Fleck, D.G., Chessum, B.S., Perkins, M., 1972. Coccidian-like nature of *Toxoplasma gondii*. British Medical Journal 3, 111–112.

Flux, J.E.C., 1983. Introduction to taxonomic problems in hares. Acta Zoologica Fennica 174, 7–10. *.

da Fonseca, F, 1932. *Eimeria pintoensis*, n. sp., parasita do coelho sylvestre (*Sylvilagus minensis*). Memórias do Instituto Butantan 7, 173–177. *.

da Fonseca, F, 1933. *Eimeria paulistana*, sp. n. encontrada na lebre, *Silvilagus* (sic) *minesis*, no estado de São Paulo. Boletim Biologico São Paulo 1, 60–61. *.

Francalancia, G., Manfredini, L., 1967. Diagnosis of species of the rabbit coccidiosis. Veterinaria Italiana 18, 304–309. *.

Francalancia, G., Manfredini, L., 1970. Ricerca degli agenti specifici della coccidiosi della lepre (*Lepus europaeus*). La Clinica Veterinaria 93, 130–135.

Freitas, F.L.C., Yamamoto, B.L., Freitas, W.L.C., Almeida, K.S., Alessi, A.C., Machado, R.Z., Machado, C.R., 2009. Aspectos anatomopathológicos e bioquímicos da coccidiose hepatica em coelhos. Revista de Patologia Tropical 38, 115–125. *.

Freitas, F.L.C., Yamamoto, B.L., Freitas, W.L.C., Almeida, K.S., Machado, R.Z., Machado, C.R., 2010. *Eimeria stiedai*: metabolism of lipids, proteins and glucose in experimentally infected rabbits, *Oryctolagus cuniculus*. Brazilian Journal of Veterinary Pathology 3, 37–40. *.

Frenkel, J.K., 1955a. Infections with organisms resembling *Toxoplasma*, together with the description of a new organism: *Besnoitia jellisoni*. Atti del VI Congresso Internazionale di Microbiologia, Roma 5. 426–434. *.

Frenkel, J.K., 1955b. Ocular lesions in hamsters with chronic *Toxoplasma* and *Besnoitia* infections. American Journal of Ophthalmology 39, 203–225. *.

Frenkel, J.K., 1965. The development of the cyst of *Besnoitia jellisoni*: usefulness of this infection as a biologic model. In: Proceedings of the Second International Conference on Protozoology, London, International Congress Series, No. 91. Excerpta Medica, Amsterdam, p. 125. *.

Frenkel, J.K., 1977. *Besnoitia wallacei* of cats and rodents: With a reclassification of other cyst-forming isosporoid coccidia. Journal of Parasitology 63, 611–628. *.

Frenkel, J.K., Dubey, J.P., 1972. Toxoplasmosis and its prevention in cats and man. Journal of Infectious Diseases 126, 664–673.

Frenkel, J.K., Dubey, J.P., 1973. Effects of freezing on the viability of *Toxoplasma* oocysts. Journal of Parasitology 59, 587–588. *.

Frenkel, J.K., Ruiz, A., Chinchilla, M., 1975. Soil survival of *Toxoplasma* oocysts in Kansas and Costa Rica. American Journal of Tropical Medicine and Hygiene 24, 439–443.

Frey, J.K., Yates, T.L., Duszynski, D.W., Gannon, W.L., Gardner, S.L., 1992. Designation and curatorial management of type host specimens (symbiotypes) for new parasite species. Journal of Parasitology 78, 930–932.

Frölich, K., Wisser, J., Schmüser, H., Fehlberg, U., Neubauer, H., Grunow, R., Nikolaou, K., Priemer, J., Thiede, S., Streich, W.J., Speck, S., 2003. Epizootiologic and ecological investigations of European brown hares (*Lepus europaeus*) in selected populations from Schleswig-Holstein, Germany. Journal of Wildlife Diseases 39, 751–761. *.

Gardner, S.L., Duszynski, D.W., 1990. Polymorphism of eimerian oocysts can be a problem in naturally infected hosts: an example from subterranean rodents in Bolivia. Journal of Parasitology 76, 805–811. *.

Gardner, S.L., Seggerman, N.A., Batsaikhan, N., Ganzorig, S., Tinnin, D.S., Duszynski, D.W., 2009. Coccidia (Apicomplexa: Eimeriidae) from the lagomorph *Lepus tolai* in Mongolia. Journal of Parasitology 95, 1451–1454. *.

Garwig, J.G., Kuenzli, H., Boehnke, M., 1998. Experimental ocular toxoplasmosis in naïve and primed rabbits. Ophthalmologica 212, 136–141. *.

Gill, B.S., Ray, H.N., 1960. The coccidia of domestic rabbit and the common field hare of India. Proceedings of the Zoological Society of Calcutta 13, 129–143. *.

Glaberman, S., Sulaiman, I.M., Bern, C., Limor, J., Peng, M.M., Morgan, U., Gilman, R., Lal, A.A., Xiao, L., 2001. A multilocus genotypic analysis of *Cryptosporidium meleagridis*. Journal of Eukaryotic Microbiology 2001 (Suppl.), 19S–22S. *.

Glebezdin, V.S., 1978. About the coccidia fauna of wild mammals of southwestern Turkmenistan. Academy of Sciences of Turkmenistan. SSR Biology Series 3, 71–78. * (in Russian).

Godfray, H.C.J., Knapp, S., 2004. Taxonomy for the twenty-first century—Introduction. Philosophical Transactions of the Royal Society of London B 359, 559–569. *.

Golemanski, V., 1975. On the coccidia (Sporozoa, Eimeriidae) of the European hare (*Lepus europaeus* L.) in Bulgaria. Acta Zoologica Bulgarica 3, 39–47. * (in Bulgarian).

Gološin, R., Tešić, D., 1963. (Coccidiostatic drugs for the prevention of coccidiosis in hares in transit. I.). Jugoslovenski Veterinarski Glasnik 17, 821–824. * (in Croatian).

Gološin, R., Tešić, D., Terzić, L., 1963. (Coccidiostatic drugs for the treatment of coccidiosis in hares in transit. II.). Jugoslovenski Veterinarski Glasnik 17, 851–856. * (in Croatian).

Gomez-Bautista, M., Rojo-Vazquez, F.A., Alunda, J.M., 1987. The effect of the host's age on the pathology of *Eimeria stiedai* infection in rabbits. Veterinary Parasitology 24, 47–57. *.

Gonzalez-Redondo, Finzi, P.A., Negretti, P., Micci, M., 2008. Incidence of coccidiosis in different rabbit keeping systems. Arquivo Brasileiro de Medicina Veterinária e Zootecnia 60, 1267–1270.

Goodwin, M.A., Waltman, W.D., 1996. Transmission of *Eimeria*, viruses, and bacteria to chicks: Darkling beetles (*Alphitobius diaperinus*) as vectors of pathogens. Journal of Applied Poultry Research 5, 51–55.

Gotelli, N.J., 2004. A taxonomic wish-list for community ecology. Philosophical Transactions of the Royal Society of London B 359, 585–597. *.

Gottschalk, C., 1973. Endoparasiten den Feldhasen in ihrer Rolle für die Niederwildjagd Ostthüringens. Angewandte Parasitologie 14, 44–54. *.

Gould, S.J., 1992. What is a species? Discover 13 (Fall), 40–45. *.

Gousseff, W.F., 1931. Zur Frage der Kaninchenkokzidien. Zeitschrift für Infektionskrankheiten. Parasitäre Krankheiten und Hygiene Der Haustiere 39, 265–271. *.

Gräfner, G., Graubmann, H.D., Benda, A., 1967. Die Verbreitung und Bedeutung der Hasenkokzidiose im Bezirk Schwerin. Monatshefte für Veterinärmedizin 22, 449. *.

Grant, B., 2007. Cataloging life. The Scientist (December), 36–42. *.

Grant, B., 2009. A fading field. The Scientist (June), 33–38. *.

Green, J.S., Flinders, J.T., 1980. *Brachylagus idahoensis*. Mammalian Species 125, 1–4. *.

Grès, V., Marchandeau, S., Landau, I., 2000. The biology and epidemiology of *Eimeria exigua*, a parasite of wild rabbits invading the host cell nucleus. Parasitologica 42, 219–225. *.

Grès, V., Marchandea, S., Landau, I., 2002. Description d'une nouvelle espéce d' *Eimeria* (Coccidia, Eimeridea) chez le lapin de garenne *Oryctolagus cuniculus* en France (Description of a new species of *Eimeria* [Coccidia, Eimeridea] in the wild rabbit *Oryctolagus cuniculus* in France). Zoosystema 24, 203–207. * (in French, English summary).

Grès, V., Voza, T., Chabaud, A., Landau, I., 2003. Coccidiosis of the wild rabbit (*Oryctolagus cuniculus*) in France. Parasite 10, 51–57. *.

Grundmann, A.W., Lombardi, P.S., 1976. Parasitism of the pika, *Ochotona princeps* Richardson (Mammalia: Lagomorpha), in Utah and Nevada with the description of *Eugenuris utahensis* sp. n. (Nematoda: Oxyuridae). Proceedings of the Helminthological Society of Washington 43, 39—46. *.

Gurpata, S., Khahra, S.S., 1997. Incidence of rabbit coccidian in Punjab state. Veterinary Parasitology 11, 7—10.

Gurwitsch, B.M., 1927. Materialien zum stadium der struktur des Coccids *Eimeria stiedae* Lindemann bei Kaninchen (Materials for the study of the structure of *Eimeria stiedae* coccids in rabbits). Archiv für Protistenkunde 59, 369—372 (in German).

Gustafsson, K., Uggla, A., 1994. Serologic survey for *Toxoplasma gondii* infection in the brown hare (*Lepus europaeus* P.) in Sweden. Journal of Wildlife Diseases 30, 201—204. *.

Gustafsson, K., Uggla, A., Svensson, T., Sjöland, L., 1989. Detection of *Toxoplasma gondii* in liver tissue sections from brown hares (*Lepus europaeus* P.) and mountain hares (*Lepus timidus* L.) using the peroxidase antiperoxidase (PAP) technique as a complement to conventional histopathology. Journal of Veterinary Medicine B 35, 402—407. *.

Gustafsson, K., Uggla, A., Järplid, B., 1997. *Toxoplasma gondii* infection in the mountain hare (*Lepus timidus*) and domestic rabbit (*Oryctolagus cuniculus*). 1. Pathology. Journal of Comparative Pathology 17, 351—360. *.

Gvéléssiani, I.D., Nadiradze, G.I., 1945. Materialy dlia izucheniia koktsidii krolikov *Eimeria piriformis* n. sp. (Data on the study of the rabbit coccidium, *Eimeria piriformis* n. sp.). Gruzinskogo Nauchno-Issledovatel'skogo Veterinarnogo Instituta 9, 31—40. * (in Russian).

Hake, T.G., 1839. A treatise on varicose capillaries, as constituting the structure of carcinoma of the hepatic ducts, and developing the law and treatment of morbid growths. With an account of a new form of pus globule. United Kingdom, London. *.

Halanych, K.M., Robinson, T.J., 1997. Phylogenetic relationships of cottontails (*Sylvilagus*, Lagomorpha): Congruence of 12S rDNA and cytogenetic data. Molecular Phylogenetics and Evolution 7, 294—302.

Halanych, K.M., Robinson, T.J., 1999. The utility of cytochrome b and 12S rDNA data for phylogeny reconstruction of leporid (Lagomorpha) genera. Journal of Molecular Evolution 48, 369—379. *.

Halanych, K.M., Demboski, J.R., van Vuuren, B.J., Klein, D.R., Cook, J.A., 1999. Cytochrome b phylogeny of North American hares and jackrabbits (*Lepus*, Lagomorpha) and the effects of saturation in outgroup taxa. Molecular Phylogenetics and Evolution 11, 213—221. *.

Hall, E.R., 1981. The Mammals of North America, vol. 2. John Wiley & Sons, New York. *.

Hammond, D.M., Scholtyseck, E., Miner, M.L., 1967. The fine structure of microgametocytes of *Eimeria perforans*, *E. stiedae*, *E. bovis*, and *E. auburnensis*. Journal of Parasitology 53, 235—247. *.

Hanada, S., Omata, Y., Umemoto, Y., Kobayashi, Y., Furuoka, H., Matsui, T., Maeda, R., Saito, A., 2003. Relationship between lier disorders and protection against *Eimeria stiedai* infection in rabbits immunized with soluble antigens from the bile of infected rabbits. Veterinary Parasitology 111, 261—266.

Harcourt, R.A., 1967. Toxoplasmosis in rabbits. Veterinary Record 81, 91—92. *.

Hardcastle, A.B., Foster, A.O., 1944. Notes on a protective action of borax and related compounds in cecal coccidiosis of poultry. Proceedings of the Helminthological Society of Washington 11, 60—64. *.

Harkema, R., 1936. The parasites of some North Carolina rodents. Ecological Monographs 6, 153—232. *.

Harp, J.A., Fayer, R., Pesch, B.A., Jackson, G.J., 1996. Effect of pasteurization on infectivity of *Cryptosporidium parvum* oocysts in water and milk. Applied Environmental Microbiology 62, 2866—2868. *.

Hashmey, R., Smith, N.H., Cron, S., Graviss, E.A., Chappell, C.L., White Jr., A.C., 1997. Cryptosporidiosis in Houston, Texas. A report of 95 cases. Medicine 76, 118—139. *.

Haydorn, A.-O., Rommel, M., 1972a. Contributions to the life-cycle of the Sarcosporidia. II. Developmental stages of *S. fusiformis* in the alimentary tract and in the faeces of dogs and cats. Berliner München Tierärztliche Wochenschrift 85, 121—123.

Haydorn, A.-O., Rommel, M., 1972b. Beiträg zum Lebenszyklus der Sarkosporidien. IV. Entwicklungsstadien von *Sarcocystis fusiformis* in der Dünndarmschleimhaut der Katze. Berliner München Tierärztliche Wochenschrift 85, 333—336.

Hays, D.W., 2001. Washington pygmy rabbit: Emergency action plan for species survival. Washington Department of Fish and Wildlife Program, Olympia, Washington. *.

Health Protection Report, 2008. Outbreak of cryptosporidiosis associated with a water contamination incident in the East Midlands 2, 29 (available from http://www.hpa.org.uk/hpr/archives/2008/hpr2908.pdf). *.

Hegner, R., Chu, H.J., 1930. A survey of protozoa parasites in plants and animals of the Philippine Island. Philippine Journal of Science 43, 451—482. *.

Hejlíček, K., Literák, I., 1994. Prevalence of toxoplasmosis in rabbits in south Bohemia. Acta Veterinaria (Brno) 63, 145—150. *.

Hejlíček, K., Literák, I., Nezval, J., 1997. Toxoplasmosis in wild animals from the Czech Republic. Journal of Wildlife Diseases 33, 480—485. *.

Hejlíček, K., Prošek, F., Tremi, F., 1981. Isolation of *Toxoplasma gondii* in free-living small mammals and birds. Acta Veterinaria (Brno) 50, 233—236. *.

Helgen, K.M., Cole, F.R., Helgen, L.E., Wilson, D.W., 2009. Generic revision in the holarctic ground squirrel genus *Spermophilus*. Journal of Mammalogy 90, 270–305. *.

Heller, G., 1971. Elektronmikroskopische Untersuchungen zur Schizogonie in den sog. Kleinen Schizonten von *Eimeria stiedai* (Sporozoa, Coccidia). Protistologica 7, 461–469. *.

Heller, G., Scholtyseck, E., 1971. Feinstruktur-untersuchungen zur Merosoitenbildung bei *Eimeria stiedai* (Sporozoa, Coccidia). Protistologica 37, 451–460. *.

Hennig, W., 1966. Phylogenetic systematics. University of Illinois Press, Urbana, Illinois.

Henricksen, S.A., Pohlenz, J.F.L., 1981. Staining of crypto-sporidia by a modified Ziehl-Neelsen technique. Acta Veternaria Scandanavia 22, 594–596. *.

Henry, A., 1913. Analysed'un travail de Besnoit et Robin, 1912. Recueil de Médecine Vétérinaire 90, 327–328. *.

Henry, D.P., 1932. Observations on the coccidia of small mammals in California, with descriptions of seven new species. University of California Publications in Zoology. 37, 279–290. *.

Herbert, P.D.N., Cywinska, A., Ball, S.L., de Waard, J.R., 2003. Biological identifications through DNA barcodes. Proceedings of the Royal Society of London B 270, 313–322. *.

Herbert, P., Stoekle, M., Zemlak, T., Francis, C.M., 2004. Identifications of birds through DNA barcodes. PLoS Biology 2, 1657–1668. *.

Herman, C.M., Jankiewicz, H.A., 1943. Parasites of cottontail rabbits on the San Joaquin experimental range, California. Journal of Wildlife Management 7, 395–400. *.

Herrick, C.A., Holmes, C.E., 1936. Effects of sulfur on coccidiosis in chickens. Veterinary Medicine 31, 390–391. *.

Hibbard, C., 1963. The origin of the P3 pattern of *Sylvilagus, Caprolagus, Oryctolagus,* and *Lepus*. Journal of Mammalogy 44, 1–15. *.

Hnida, J.A., Duszynski, D.W., 1999a. Taxonomy and systematics of some *Eimeria* species of murid rodents as determined by the ITS1 region of the ribosomal gene complex. Parasitology 199, 349–357. *.

Hnida, J.A., Duszynski, D.W., 1999b. Taxonomy and phylogeny of some *Eimeria* (Apicomplexa: Eimeriidae) species of rodents as determined by polymerase chain reaction/restriction-fragment-length polymorphism analysis of 18s rDNA. Parasitology Research 85, 887–894. *.

Hobbs, R.P., Samuel, W.M., 1974. Coccidia (Protozoa, Eimeriidae) of the pikas *Ochotona collaris, O. princeps,* and *O. hyperborea yesoensis*. Canadian Journal of Zoology 52, 1079–1085. *.

Hobbs, R.P., Twigg, L.E., 1998. Coccidia (*Eimeria* spp.) of wild rabbits in southwestern Australia. Australian Veterinary Journal 76, 209–210. *.

Hoenig, V., Girardot, J.M., Haegele, P., 1974. *Eimeria stiedai* infection in the rabbit: Effect on bile flow and brom-sulphthalein metabolism and elimination. Laboratory Animal Science 24, 66–71. *.

Hoffmann, R.S., Smith., A.T., 2005. In: Order Lagomorpha, Wilson, D.E., Reeder, D.M. (Eds.) Mammal Species of the World: A Taxonomic and Geographic Reference, third ed., vol. 1. Johns Hopkins University Press, Baltimore, Maryland, pp. 185–211. *.

Holmdahl, O.J., Morrison, D.A., Ellis, J.T., Huong, L.T., 1999. Evolution of ruminant *Sarcocystis* (Sporozoa) parasites based on small subunit rDNA sequences. Molecular Phylogenetics and Evolution 11, 27–37.

Holmes, J.C., 1996. Parasites as threats to biodiversity in shrinking ecosystems. Biodiversity and Conservation 5, 975–983. *.

Holtcamp, W., 2010. Silence of the pikas. Bioscience 60, 8–12. *.

Honess, R.F., 1939. The coccidia infesting the cottontail rabbit, *Sylvilagus nuttallii grangeri* (Allen), with descriptions of two new species. Parasitology 31, 281–284. *.

Honess, R.F., Winter, K.B., 1956. Diseases of wildlife in Wyoming for those interested in the disease and parasites of wild animals. Bulletin 9, Wyoming Game and Fish Commission, 79–82. *.

Horton, R.J., 1967. The route of migration of *Eimeria stiedai* (Lindemann, 1865) sporozoites between the duodenum and bile ducts of the rabbit. Parasitology 57, 9–17. *.

Hsu, C.-K., 1970. Life cycle and host specificity of *Eimeria sylvilagi* Carini, 1940 from the cottontail *Sylvilagus floridanus mearnsii*. Ph.D. Dissertation. University of Illinois, Urbana. *.

Hunter, P.R., 1999. Outbreak of cryptosporidiosis in north west England. European Surveillance 3, 1403. *.

Hunter, P.R., Nichols, G., 2002. Epidemiology and clinical features of *Cryptosporidium* infection in immunocompromised patients. Clinical Microbiology Reviews 15, 145–154. *.

Hugghins, E.J., 1961. *Sarcocystis* in cottontail rabbits. Proceedings of the South Dakota Academy of Science 40, 240–241. *.

Hughes, J.M., Thomasson, B., Craig, P.S., Georgin, S., Pickles, A., Hide, G., 2008. *Neospora caninum*: detection in wild rabbits and investigation of co-infection with *Toxoplasma gondii* by PCR analyses. Experimental Parasitology 120, 255–260. *.

Inman, L.R., Takeuchi, A., 1979. Spontaneous cryptosporidiosis in an adult female rabbit. Veterinary Pathology 16, 89–95. *.

Iseki, M., Maekawa, T., Mriya, K., Uni, S., Takada, S., 1989. Infectivity of *Cryptosporidium muris* (strain RN 66) in various laboratory animals. Parasitology Research 75, 218–222. *.

Issac, N.J.B., 2004. Taxonomic inflation: Its influence on macroecology and conservation. Trends in Ecology and Evolution 19, 464–469. *.

István, V., 1976. A mezeinyúl kisérleti fertőzése Eimeria stiedaivel (Experimental transmission of Eimeria stiedai to the hare). Magyar Allatorvosok Lapja. November, 726–730.

Ito, S., Shimura, K., 1986. The comparison of Isospora bigemina large type of the cat and Besnoitia wallacei. Japanese Journal of Veterinary Science 48, 433–435. *.

Ito, S., Tsunoda, K., Shimura, K., 1978. Life cycle of the large type of Isospora bigemina of the cat. National Institute of Animal Health Quarterly 18, 69–82. *.

Jankiewicz, H.A., 1941. Transmission of the liver coccidium, Eimeria stiedae from the domestic to the cottontail rabbit. Journal of Parasitology 27 (Suppl.), 28. *.

Jankiewicz, H.A., 1945. Dosage of Eimeria stiedae related to severity of liver coccidiosis. Journal of Parasitology 31 (Suppl.), 8.

Jeffers, T.K., 1975. Attenuation of Eimeria tenella through selection for precociousness. Journal of Parasitology 61, 1083–1090. *.

Jelínková, A., Licois, D., Pakandl, M., 2008. The endogenous development of the rabbit coccidium Eimeria exigua Yakimoff, 1934. Veterinary Parasitology 156, 166–172. *.

Jensen, J.B., Hammond, D.M., 1975. Ultrastructure of the invasion of Eimeria magna sporozoites into cultured cells. Journal of Protozoology 22, 411–415. *.

Jensen, J.B., Edgar, S.A., 1976. Possible secretory function of the rhoptries of Eimeria magna during penetration of cultured cells. Journal of Parasitology 62, 988–992. *.

Jervis, H.R., Merrill, T.G., Sprinz, H., 1966. Coccidiosis in the guinea pig small intestine due to a Cryptosporidium. American Journal of Veterinary Research 27, 408–414. *.

Jex, A.R., Gasser, R.B., 2010. Genetic richness and diversity in Cryptosporidium hominis and C. parvum reveals major knowledge gaps and a need for the application of "next generation" techniques. Biotechnology Advances 28, 17–26. *.

Jex, A.R., Smith, H.V., Monis, P.T., Campbell, B.E., Gasser, R.B., 2008. Cryptosporidium—biotechnological advances in the detection, diagnosis and analysis of genetic variation. Biotechnology Advances 26, 304–317. *.

Jirouš, J., 1979. On the coccidia of hares in central and northwestern Bohemia. Lesnictví 25, 1015–1027. * (in Czech).

Jithendran, K.P., Bhat, K.P., 1996. Subclinical coccidiosis in Angora rabbits—A field survey in Himachal Pradesh (India). World Rabbit Science 4, 29–32. *.

John, N.M., Rodriguez Zea, M.E., Kawano, T., Omata, Y., Saito, A., Toyoda, Y., Mikami, T., 1999. Identification of carbohydrates on Eimeria stiedai sporozoites and their role in the invasion of cultured cells in vitro. Veterinary Parasitology 81, 99–105. *.

Jokipii, L., Pohjola, S., Jokipii, A.M., 1983. Cryptosporidium: a frequent finding in patients with gastrointestinal symptoms. Lancet 322, 358–361. *.

Jones, E.E., 1932. Size as a species characteristic in coccidia: variation under diverse conditions of infection. Archiv für Protistenkunde 76, 130–170. *.

Joyner, L.P., 1982. Host and site specificity. In: Long, P.L. (Ed.), Coccidiosis of Man and Domestic Animals. University Park Press, Baltimore, Maryland, pp. 35–62. *.

Joyner, L.P., Catchpole, J., Berrett, S., 1983. Eimeria stiedai in rabbits: the demonstration of responses to chemotherapy. Research in Veterinary Science 34, 64–67.

Kaggwa, E., Weiland, G., Rommel, M., 1979. Besnoitia besnoiti and Besnoitia jellisoni: a comparison of the indirect immunofluorescent antibody test (IFAT) and the enzyme-linked immunosorbent assay (ELISA) in diagnosis of Besnoitia infections in rabbits and in mice. Bulletin of Animal Health Production of Africa 27, 127–137. *.

Kalyakin, V.N., Zasukhin, D.N., 1975. Distribution of Sarcocystis (Protozoa: Sporozoa) in vertebrates. Folia Parasitologia 22, 289–307.

Karaer, Z., 2001. Evcil Tavsanlarda (Oryctolagus cuniculus) Coccidiosis. In: Dincer, S (Ed.), Coccidiosis Türkiye Parasitoloji Derneği Yayinlari, No. 17. Meta Basim, Izmir, pp. 269–278. *.

Kasim, A.A., Al-Shawa, Y.R., 1987. Coccidia in rabbits (Oryctolagus cuniculus) in Saudi Arabia. International Journal for Parasitology 17, 941–944. *.

Kauffmann, W., 1847. Analecta ad tuberculorum et entozoorum cognitionem. Inaugural Dissertation Berol, Germany. *.

Kessel, J.F., 1929. The Eimeria of the domestic rabbits. Journal of Parasitology 16, 100 (abstract). *.

Kessel, J.F., Jankiewicz, H.A., 1931. Species differentiation of the coccidia of the domestic rabbit based on a study of the oöcysts. The American Journal of Hygiene 14, 304–324. *.

Kisskalt (=Kißkalt), K., Hartmann, M., 1907. Eimeria stiedae (Lindem). In: Praktikum der Bakteriologie und Protozoologie. Verlag von Gustav Fischer, Jena. *.

King, B.J., Monis, P.T., 2007. Critical processes affecting Cryptosporidium oocyst survival in the environment. Parasitology 134, 309–323. *.

Klesius, P.H., Kramer, T.T., Frandsen, J.C., 1976. Eimeria stiedai: delayed hypersensitivity response in rabbit coccidiosis. Experimental Parasitology 39, 59–68. *.

Kogan, Z.M., 1962. Variability of shape in oocysts of chicken coccidia and its biological significance. Zoologische Zhurnnal 41, 1317–1326. * (in Russian).

Kogan, Z.M., 1965. Variability of the oocysts of chicken coccidia Eimeria necatrix and factors which determine it. Zoologische Zhurnnal 44, 986–996. * (in Russian).

Korf, R.P., 2005. Reinventing taxonomy: a curmudgeon's view of 250 years of fungal taxonomy, the crisis in biodiversity, and the pitfalls of the phylogenetic age. Mycotaxon 93, 407–415. *.

Kotlán, S., Pellérdy, L.P., 1936. Kísérleti vizsgálatok a házinyúl májcoccidiosisáról. I. Közlemények az Összehasonlítóéletes Kórtan 26, 1–13. * (in Hungarian).

Kotlán, S., Pellérdy, L.P., 1937. Kísérleti vizsgálatok a házinyúl májcoccidiosisáról. II. Közlemények az Összehasonlítóéletes Kórtan 28, 105–120. * (in Hungarian).

Kotlán, S., Pellérdy, L.P., 1949. A survey of the species of Eimeria occurring in the domestic rabbit. Acta Veterinaria Academiae Scientiarum Hungaricae 1, 93–97. *.

Kotlán, S., Pospesch, L., 1934. A házinyúl coccidiosisának ismeretéhez. Egy új Eimeria-faj (Eimeria piriformis sp. n.) házinyúlból. Allatorvosi Lapok (Budapest) 57, 215–217. *.

Koutz, F.R., 1950. The survival of oocysts of avian coccidia in the soil. The Speculum 3, 1–4. *.

Krieg, J., 1971. Zur in vitro-Excystation und -Resistenz von Sporozoiten der kokzidienart Eimeria stiedai (Lindemann, 1965) Kisskalt & Hartmann 1907 (For the invitro excystation and resistance in sporozoites of the coccidium Eimeria stiedai [Lindemann, 1865] Kisskalt & Hartmann 1907). Inaugural Dissertation, Faculty of Veterinary Medicine, Justus Liebig-Universität Geißen, Geißen, Gemany. * (in German, English summary).

Kriegs, J.O., Zemann, A., Churakov, G., Matzke, A., Ohme, M., Zischler, H., Brosius, J., Kryger, U., Schmitz, J., 2010. Retroposon insertions provide insights into deep lagomorph evolution. Molecular Biology and Evolution 27, 2678–2681. *.

Krijgsman, B.J., 1926. Wie warden im Intestinaltractus des Wirtstieres die Sporozoiten der Coccidien aus ihren Hüllen befreit. Archiv für Protistenkunde 56, 116. *.

Kulišić, Z., Tambur, Z., Maličević, Ž., Radosavljević, M., 1998. Changes in the activity of enzymes primarily not synthesized in the liver following the infection of rabbits with intestinal coccidia. Journal of Protozoology Research 8, 1–9. *.

Kulišić, Z., Tambur, Z., Maličević, Ž., Aleksić-Bakrač, N., Mišić, Z., 2006. White blood cell differential count in rabbits artificially infected with intestinal coccidia. Journal of Protozoology Research 16, 42–50. *.

Kumar, P.N., Rajavel, A.R., Natarajan, R., Jambulingam, P., 2007. DNA barcodes can distinguish species of Indian mosquitoes (Diptera: Culicidae). Journal of Medical Entomology 44, 1–7. *.

Kummel, B., 1970. History of the Earth: An Introduction to Historical Geology. W.H. Freeman and Company, San Francisco. *.

Kutzer, E., Frey, H., 1976. Die Parasiten der Feldhasen (Lepus europaeus) in Österreich. Berliner München Tierärztliche Wochenschrift 89, 480–483. *.

Kvičerova, J., Ptáčková, P., Modrý, D., 2007. Endogenous development, pathogenicity and host specificity of Eimeria cahirinensis Couch, Blaustein, Duszynski, Shenbrot and Nevo, 1997 (Apicomplexa: Eimeriidae) from Acomys dimidiatus (Cretzschmar 1826) (Rodentia: Muridae) from the Near East. Parasitology Research 100, 219–226. *.

Kvičerova, J., Pakandl, M., Hypša, V., 2008. Phylogenetic relationships among Eimeria spp. (Apicomplexa, Eimeriidae) infecting rabbits: evolutionary significance of biological and morphological features. Parasitology 135, 443–452. *.

Lampio, T., 1946. Riistantaudit Suomessa vv. 1924–1933 (Game diseases in Finland 1924–1933). Suomen Riista 1, 93–143. *.

Learmouth, J.J., Ionas, G., Ebbertt, K.A., Kwan, E.S., 2004. Genetic characterization and transmission cycles of Cryptosporidium species isolated from humans in New Zealand. Applied Environmental Microbiology 70, 3973–3978.

Lebas, F., Coudert, P., Rouvier, R.D., Rochambeau, H., 1986. The rabbit husbandry in health and production. Animal Production and Health No. 21, FAO, Rome, Italy. *.

Lechleitner, R.R., 1959. Some parasites and infectious diseases in a black-tailed jackrabbit population in the Sacramento Valley, California. California Fish and Game 45, 83–91. *.

Le Dune, E.K., 1936. New York State Conservation Department Annual Report 26, 310–311. *.

Lee, C.D., 1934. The pathology of coccidiosis in the dog. Journal of the American Veterinary Medical Association 85, 760–781. *.

Léger, L., 1898. Essai sur la classification des Coccidies et description de quelques espèces nouvelles ou peu connues (Essay on the classification of Coccidia and the description of some new or little known species). Annales de Musee Naturaliste (Marseilles) 1, 71–123. (in French).

Leighton, F.A., Gajadhar, A.A., 2001. Besnoitia spp. and besnoitiosis. In: Samuel, W.M., Pybus, M.J., Kocan, A.A. (Eds.), Parasitic Diseases of Wild Mammals, second ed. Iowa State University Press, Ames, Iowa, pp. 468–478. *.

Leland, M.M., Hubbard, G.B., Dubey, J.P., 1992. Clinical toxoplasmosis in domestic rabbits. Laboratory Animal Science 42, 318–319.

Lepp, D.L., Todd, K.S., Samuel, W.M., 1972. Four new species of Eimeria (Protozoa: Eimeriidae) from the pika Ochotona princeps from Alberta and O. pallasi from Kazakhstan. Journal of Protozoology 19, 192–195. *.

Lepp, D.L., Todd, K.S., Samuel, W.M., 1973. Eimeria banffensis n. sp. (Protozoa: Eimeriidae) from the pika Ochotona princeps from Alberta. Transactions of the American Microscopical Society 92, 305–307. *.

Leuckart, R., 1879. Die Parasiten des Menschen und die von ihnen Herrührenden Krankheiten. Tomo I.C.F. Winter. Leipzig und Wien. *.

Levine, N.D., 1971. Coccidia of rabbits and hares. Journal of Protozoology 18 (Suppl.), 13. *.

Levine, N.D., 1973a. Introduction, history, and taxonomy. In: Hammond, D.L., Long, P.L. (Eds.), The Coccidia. *Eimeria, Isospora, Toxoplasma* and Related Genera. University Park Press, Baltimore, Maryland, pp. 1–43. *.

Levine, N.D., 1973b. Protozoan Parasites of Domestic Animals and of Man, second ed. Burgess Publishing Company, Minneapolis, Minnesota. *.

Levine, N.D., 1974. Historical aspects of research on coccidiosis. In: Proceedings of the Symposium on Coccidia and Related Organisms. Guelph, Ontario, 1973, Univeristy of Guelph, Guelph, Ontario, Canada, pp. 1–10. *.

Levine, N.D., 1988. The protozoan phylum Apicomplexa, vol. 2. CRC Press, Boca Raton, Florida. *.

Levine, N.D., Ivens, V., 1972. Coccidia of the Leporidae. Journal of Protozoology 19, 572–581. *.

Levine, N.D., Tadros, W., 1980. Named species and hosts of *Sarcocystis* (Protozoa: Apicomplexa: Sarcocystidae). Systematic Parasitology 2, 41–59. *.

Levine, P.P., 1939. The effect of sulfanilamide on the course of experimental avian coccidiosis. Cornell Veterinarian 29, 309–320. *.

Levit, A.V., Orlov, G.I., Dymkova, N.D., 1984. Sarcosporidians in the Altai pika (*Ochotona alpina*). In: Institute of Zoology, Kazakh Academy of Sciences (Ed.), Sarcosporidians of animals in Kazakhstan. Publishing House "Nauka," Alma-Ata. * (in Russian).

Leysen, E., Neirynck, A., Deleersnijder, W., Coudert, P., Peeters, J., Revets, H., Hamers, R., 1989. Identification of sporozoite surface antigens of *Eimeria stiedae* and *Eimeria magna*. In: Coccidia and Intestinal Coccidiomorphs. Tours, France, Vth International Coccidiosis Conference, pp. 509–514. 17-20 October, 1989. *.

Li, M.-H., Ooi, H.-K., 2009. Fecal occult blood manifestation of intestinal *Eimeria* spp. infection in rabbit. Veterinary Parasitology 161, 327–329. *.

Li, W.-H., Gouy, M., Sharp, P.M., O'hUigin, C., Yang, Y.-W., 1990. Molecular phylogeny of Rodentia, Lagomorpha, Primates, Artiodactyla, and Carnivora and molecular clocks. Proceedings of the National Academy of Sciences, USA 87, 6703–6707. *.

Li, M.-H., Huang, H.I., Ooi, H.-K., 2010. Prevalence, infectivity and oocyst sporulation time of rabbit-coccidia in Taiwan. Tropical Biomedicine 27, 424–429. *.

Licois, D., 2009. Letter to the Editor: Comments on the article of Ming-Hsien Li and Hong-Kean Ooi "Fecal occult blood manifestation of intestinal *Eimeria* spp. infection in rabbit.". Veterinary Parasitology 164, 363–364. *.

Licois, D., Coudert, P., 1980a. Action de la Robenidine sur l'excretion des oocystes de differentes especes de coccidies du lapin (Action of Robenidine on excretion of oocysts of different species of coccidia of the rabbit). In: Proceedings II World's Rabbit Congress, April, 1980, Barcelona, Spain, pp. 285–289. (in French).

Licois, D., Coudert, P., 1980b. Attempt to suppress immunity in rabbits immunized against *Eimeria intestinalis*. Annales de Recherches Vétérinaires 11, 273–278. *.

Licois, D., Coudert, P., 1980c. Effect of the anitcoccidial Robenidine on oocyst production in rabbits infected with different strains of coccidia. Recueil de Medecine Veterinaire 156, 391–394.

Licois, D., Mongin, P., 1980. Hypothèse sur la pathogénie de la diarrhea chez le lapin à partir de l'ètude des contenus intestinaux (An hypothesis of the pathogenesis of diarrhea in the rabbit based on a study of intestinal contents). Reproduction Nutrition Dévelopmemt 20, 1209–1216. * (in French, English summary).

Licois, D., Coudert, P., Mongin, P., 1978a. Changes in hydromineral metabolism in diarrhoeic rabbits. 1. A study of the changes in water metabolism. Annales de Recherches Vétérinaires 9, 1–10. *.

Licois, D., Coudert, P., Mongin, P., 1978b. Changes in hydromineral metabolism in diarrhoeic rabbits. 2. Study of the modifications of electrolyte metabolism. Annales de Recherches Vétérinaires 9, 453–464. *.

Licois, D., Coudert, P., Guillot, J.F., Renault, L., 1982. Diarrhee experimentale du lapin: Etude de la pathologie due a des coccidies intestinales (*E. intestinalis*) et a des *Escherichia coli* (Experimental rabbit diarrhea: Study of the disease due to intestinal coccidia [*E. intestinalis*] and *Escherichia coli*). Journees de la Recherchie Cunicole, 8–9. December, 1982 (Paris). Communication No. 27. (in French, no English summary).

Licois, D., Coudert, P., Bahangia, S., 1989. Some biological characteristics of a precocious line of *E. intestinalis*. In: Yvoré, P. (Ed.), Proceedings of the Vth International Coccidiosis Conference: Coccidia and intestinal coccidiomorphs, October 17–20 October. Les colloques de I'Inra, Paris 49, Tours, France, pp. 503–508. *.

Licois, D., Coudert, P., Boivin, M., Drouet-Viard, F., Provôt, F., 1990. Selection and characterization of a precocious line of *Eimeria intestinalis*, an intestinal rabbit coccidium. Parasitology Research 76, 192–198. *.

Licois, D., Coudert, P., Drouet-Viard, F., Boivin, M., 1992a. *Eimeria media*: selection and characterization of a precocious line. Journal of Applied Rabbit Research 15, 1423–1432. *.

Licois, D., Coudert, P., Drouet-Viard, F., Boivin, M., 1992b. *Eimeria perforans* and *E. coecicola* multiplication rate and effect of the acquired protection on the oocyst output. Journal of Applied Rabbit Research 15, 1433–1439. *.

Licois, D., Coudert, P., Bahagia, S., Rossi, G.L., 1992c. Endogenous development of *Eimeria intestinalis* in rabbits (*Oryctolagus cuniculus*). Journal of Parasitology 78, 1041–1048.

Lindemann, K., 1865. Weiteres über Gregarinen. Bulletin de la Société Impériale des Naturalistes de Moscow 38, 381–387. *.

Lindsay, D.S., Blagburn, B., 1990. Cryptosporidiosis in birds. In: Dubey, J.P., Speer, C.A., Fayer, R. (Eds.), Cryptosporidiosis in Man and Animals. CRC Press, Boca Raton, Florida, pp. 133–148.

Lindsay, D.S., Todd Jr., K.S., 1993. Coccidia of mammals. In: Parasitic Protozoa, vol. 4. Academic Press, New York, USA. 89–131. *.

Lindsay, D.S., Blagburn, B.L., Sundermann, C.A., 1989. Morphometric comparison of the oocysts of *Cryptosporidium meleagridis* and *Cryptosporidium baileyi* from birds. Proceedings of the Helminthological Society of Washington 56, 91–92. *.

Linnaeus, C., 1753. Species plantarum exhibentes plantas rite cognitas, ad genera relatas cum differentiis specificis, nominibus trivialibus, synonymis selectis, locis natalibus, secundum systema sexuale digestas..., vol. 2. Laurentii Salini, Holmiae. *.

Linnaeus, C., 1758. Systema naturae per regna tria naturae, secunelum, classes, ordines, genera, species, cum characteribus, differentiis, synonymis locis, Editio decimal, reformata, Tomus I. Laurentii Salvii, Holmiae. *.

Lipscomb, D., Platnick, N.I., Wheeler, Q.D., 2003. The intellectual content of taxonomy: a comment on DNA taxonomy. Trends in Ecology and Evolution 18, 65–66. *.

Litvenkova, E.A., 1969. Coccidia of wild animals in Byelorussia. In: Progress in Protozoology. 3rd International Congress on Protozoology, Leningrad, pp. 340–341. *.

Lizcano Herrera, J., Romero Rodríguez, J., 1969. Epizootiologie de coccidiopatias de interes veterinario en la provincial de Granada. Revista de Ibérica de Parasitologia 29, 427–432. *.

Lubimov, M.P., 1934. In: Manteufel, P. (Ed.), Biology of the Hares and Squirrels and their Diseases, Moscow and Leningrad (cited in Pellérdy, L.P. 1954. Acta Veterinaria Academiae Scientiarium Hungaricae 4, 481–487). *.

Lucus, A., Laroche, M., Durand, J., 1959. Les agents de la coccidiose du lièvre en France. Recueil de Médecine Vétérinaire de l'Ecole d'Alfort 135, 305–310. *.

Lukešová, D., Hejličk, K., Kejíková, M., Kejík, P., Punčochár, P., 1984. Sarcocystosis and toxoplasmosis in breeding rabbits. Veterinárství 34, 120–122. * (in Czech).

Lund, E.E., 1949. Considerations in the practical control of intestinal coccidiosis of domestic rabbits. Annals of the New York Academy of Science 52, 611–620. *.

Lund, E.E., 1950. A survey of intestinal parasites in domestic rabbits in six counties in southern California. Journal of Parasitology 36, 13–19. *.

Lund, E.E., 1951. Mortality among hutch-raised domestic rabbits. U.S. Department of Agriculture Circular #883, 14. *.

Luoma, J.R., 1991. Taxonomy, lacking in prestige, may be nearing a renaissance. The New York Times, Tuesday, December 10, Sec. B, 6–7. *.

Lyman, R.L., 1991. Late quaternary biography of the pygmy rabbit (*Brachylagus idahoensis*) in eastern Washington. Journal of Mammalogy 72, 110–117. *.

Lyman, R.L., 2004. Biogeographic and conservation implications of late quaternary pygmy rabbits (*Brachylagus idahoensis*) in eastern Washington. Western North American Naturalist 64, 1–6. *.

Lynch, A.J., Duszynski, D.W., Cook, J.A., 2007. Species of Coccidia (Apicomplexa: Eimeriidae) infecting pikas from Alaska, U.S.A. and Northeastern Siberia, Russia. Journal of Parasitology 93, 1230–1234. *.

Ma, P., Soave, R., 1983. Three-step stool examination for cryptosporidiosis in 10 homosexual men with protracted watery diarrhea. Journal of Infectious Diseases 147, 824–828. *.

Mace, G.M., 2004. The role of taxonomy in species conservation. Philosophical Transactions of the Royal Society of London B 359, 711–719. *.

MacKenzie, W.R., Hoxie, N.J., Proctor, M.E., Gradus, S., Blair, K.A., Peterson, D.E., Kazmierczak, J.J., Addiss, D.G., Fox, K.R., Rose, J.B., 1994. A massive outbreak in Milwaukee of *Cryptosporidium* infection transmitted through the public water supply. New England Journal of Medicine 331, 161–167. *.

MacPherson, J.M., Gajadhar, A.A., 1993. Differentiation of seven *Eimeria* species by random amplified polymorphic DNA. Veterinary Parasitology 45, 257–266. *.

Madsen, H., 1938. The coccidia of the east Greenland hares with a revision of the coccidia of hares and rabbits. Meddleser om Grønland udgivne af Kommisionen for videnskabelige undersøgelser i Grønland (Commission for Scientific Study in Greenland) 116, 1–38. *.

Mandal, A.K., 1976. Coccidia of Indian vertebrates. Records of the Zoological Survey of India 70, 39–120. *.

Mandal, A.K., 1987. Fauna of India and the Adjacent Countries: Protozoa, Sporozoa: Eucoccidiida, Eimeriidae. Zoological Survey of India. SRI Aurobindo Press, Calcutta, India. *.

Manz, W., 1867. Beitrag zur Kenntnis der Miescherschen Schläuche. Archiv für Mikroskopische Anatomie 3, 345–356. Plate 20, Figure 5. *.

Markus, M.B., 1974. Earthworms and coccidian oocysts. Annals of Tropical Medicine and Parasitology 68, 247–248.

Markus, M.B., 1980. Flies as natural transport hosts of *Sarcocystis* and other coccidia. Journal of Parasitology 66, 361–362. *.

Marotel, G., 1912. Discussion of paper by Besnoit and Robin. Bull.. et Mém. de la Société des Sciences Veterinaires de Lyon et de la Société de Médecine Vétérinaire de Lyon et du Sud-est 15, 196–217. *.

Marotel, G., Guilhon, J., 1941. Recherches sur la coccidiose du lapin. Recueil de Médecine Vétérinaire 117, 321–328. *.

Marotel, G., Guilhon, J., 1942. Note au sujet des coccidies du lapin. Recueil de Médecine Vétérinaire 118, 270. *.

Marquardt, W.C., 1973. Host and site specificity. In: Hammond, D.M., Long, P.M. (Eds.). The Coccidia. University Park Press, Baltimore, Maryland, pp. 23–43. *.

Martínez Fernández, A., Andrés Rodríguez, J., Cordero del Campillo, M., Aller Gancedo, B., 1969. Validez y extension de la especie *Eimeria perforans* (Leuckart, 1879) Sluiter y Swellengrebel, 1912 parasito intestinal del conejo. Anales de Facultad Veterinaria (León) 15, 73–85. *.

Martínez Fernández, A., de Andres Rodriguez, J., Cordero del Campillo, M., Aller Gancedo, B., 1970. Validez y extension de la especie *Eimeria perforans* (Leuckart, 1879) Sluiter y Swellengrebel, 1912, parasite intestinal del conejo. Revista de Ibérica de Parasitologia 30, 299–310. *.

Mason, R.W., 1980. The discovery of *Besnoitia wallacei* in Australia and the identification of a free-living intermediate host. Zeitschrift für Parasitenknde 61, 173–178. *.

Matschoulsky [Machul'skiĭ], S.N., 1941. Koktsidiozy promyslovykh zhivotnykh v Buryat-Mongol'skoi ASSR (Coccidiosis of animals in Buryat-Mongol, SSR). Works of Buryat Mongol Zooveterinary Institute (Ulan-Ude) 2, 134–142. * (in Russian).

Matschoulsky [Machul'skiĭ], S.N., 1947a. About coccidia in fur animals in Buryat-Mongolskoi, ASSR. Trudy Buryat-Mongol Zooveterinary Instituta (Ulan-Udé) 3, 78. * (in Russian).

Matschoulsky [Machul'skiĭ], S.N, 1947b. Sarcosporidiosis of wild animals in Buryat-Mongolia. Trudy Buryat-Mongol Zooveterinary Instituta (Ulan-Udé) 3, 87–92. * (in Russian).

Matschoulsky [Machul'skiĭ], S.N, 1949. K voprosu o kiktsidoze gryzunov yuzhnykh raĭonov Bury-Mongoliskoĭ ASSR (About coccidia in rodents of southern areas of Buryat-Mongol ASSR). Trudy Buryat-Mongol Zooveterinary Instituta (Ulan-Udé) 5, 40–56. * (in Russian).

Matsubayashi, H., 1934. Studies of the life history and classification of *Eimeria* of the rabbit. Keio-Igaku 14, 513–560. * (in Japanese, English summary).

Matsui, T., Morii, T., Iijima, T., Kobayashi, F., Fujino, T., 1989. Transformation of oocysts from several coccidian species by heat treatment. Parasitology Research 75, 264–267. *.

Matthee, C.A., van Vuuren, B.J., Bell, D., Robinson, T.J., 2004. A molecular supermatrix of the rabbits and hares (Leporidae) allows for the identification of five intercontinental exchanges during the Miocene. Systematic Biology 53, 433–447. *.

Mayberry, L.F., Marquardt, W.C., 1973. Transmission of *Eimeria separata* from the normal host, *Rattus*, to the mouse, *Mus musculus*. Journal of Parasitology 59, 198–199. *.

Mayberry, L.F., Marquardt, W.C., Nash, D.J., Plan, B., 1982. Genetic dependent transmission of *Eimeria separata* from *Rattus* to three strains of *Mus musculus*, an abnormal host. Journal of Parasitology 68, 1124–1126. *.

Mayden, R.I., 1997. A hierarchy of species concepts: The denounement in the saga of the species problem. In: Claridge, M.A., Dawah, H.A., Wilson, M.R. (Eds.). Species the Units of Diversity. Chapman and Hall, London, pp. 381–424. *.

Mayr, E., 1942. Systematics and the origin of species. Columbia University Press, New York, New York.

Mbuthia, P.G., Gathumbi, P.K., Bwangamoi, O., Wasike, P.N., 1993. Natural besnoitiosis in a rabbit. Veterinary Parasitology 45, 191–198. *.

McDonald, V., Shirley, M.W., Bellatti, M.A., 1986. *Eimeria maxima*: Characteristics of attenuated lines obtained by selection for precocious development in the chicken. Experimental Parasitology 61, 192–200. *.

McKenna, M.C., 1982. Lagomorph interrelationships. Ecobios 15, 213–223. *.

McKenna, P.B., Charleston, W.A.G., 1980. Coccidia (Protozoa: Sporozoasida) of cats and dogs. III. The occurrence of a species of *Besnoitia* in cats. New Zealand Veterinary Journal 28, 120–123. *.

McMichael, A.J., 1993. Planetary Overload: Global Environmental Change and the Health of the Human Species. Cambridge University Press, Cambridge, United Kingdom. *.

Meisel, J.L., Perera, D.R., Meligro, C., Rubin, C.E., 1976. Overwhelming watery diarrhea associated with a *Cryptosporidium* in an immunosuppressed patient. Gastroenterology 70, 1156–1160. *.

Mehlhorn, H., Scholtyseck, E., 1974. Elektronenmikroskopische Untersuchungen an Cystenstadien von *Sarcocystis tenella* aus der Oesophagus-Muskulatur des Schafes. Parasitology Research 41, 291–310. *.

Meng, J., Wyss, A.R., 2005. Glires (Lagomorpha, Rodentia). In: Rose, K.D., Archibald, J.D. (Eds.), The Rise of Placental Mammals. Origins and Relationships of the Major Extant Clades. Johns Hopkins University Press, Baltimore, Maryland, pp. 145–158. *.

Merdivenci, A., 1963. Türkiye'de evcil ve yabani tavsanlarda *Eimeria* enfeksiyonlari. Türkiye Biyolojik Derneği 13, 26–35. *.

Metcalf, M.M., 1929. Parasites and the aid they give in problems of taxonomy, geographical distribution and paleogeography. Smithsonian Miscellaneous Collections 81, 1–36. *.

Metelkin, A., 1935. The role of flies in the spread of coccidiosis in animals and man. Medical Parasitology and Parasitic Diseases, Moscow 4, 75–82. * (in Russian, English summary).

Miescher, F., 1843. Über eigenthümliche Schläuche in den Muskeln einer Hausmaus. Bericht der Verhandlungen der naturforschenden Gesellschaften, Basel 5, 198–202. *.

Millius, S., 2011. The bunny that ruled Minorca. Science News 179, 18. *.

Ming-Hsien, L., Hoong-Kean, O., 2009. Faecal occult blood manifestation of intestinal Eimeria spp. infection in rabbit. Veterinary Parasitology 161, 327–329.

Mirza, M.Y., 1970. Incidence and distribution of coccidia (Sporozoa: Eimeriidae) in mammals from Baghdad area. MS thesis. University of Baghdad, Iraq. *.

Missiaen, P., Smith, T., Guo, D.-Y., Bloch, J.I., Gingerich, P.D., 2006. Asian gliriform origin for arctostylopid mammals. Naturwissenschaften 93, 407–411. *.

Møller, T., 1958. Toxoplasmosis cuniculi, Verificering af diagnosen. Patologisk-anatomiske og serologiske undersøgelser. Nordisk Veterinaermedicin 10, 1–56. *.

Monné, L., Hönig, G., 1954. On the properties of the shells of the coccidian oocysts. Arkiv för Zoologi 7, 251–256.

Monnerot, M., Vigne, J.D., Biju-Duval, C., Casane, D., Callou, C., Hardy, C., Mougel, F., Soriguer, R., Dennebouy, N., Mounolou, J.C., 1994. Rabbit and man: genetic and historic approach. Genetics Selection Evolution 26 (Suppl. 1), 167a–182a. *.

Moreno-Montañez, T., Becerra Martell, C., Navarrete Lopez-Cozar, I., 1979. Contribucion al conocimiento de los parasitos de la liebre Lepus capensis. Revista de Ibérica de Parasitologia 39, 383–393. *.

Morgan, B.B., Waller, E.F., 1940. A survey of the parasites of the Iowa cottontail (Sylvilagus floridanus mearnsii). Journal of Wildlife Management 4, 21–26. *.

Morgan, U.M., Xiao, L., Limor, J., Gelis, S., Raidal, S.R., Fayer, R., Lal, A., Elliot, A., Thompson, R.C., 2000. Cryptosporidium meleagridis in an Indian ring-necked parrot (Psittacula krameri). Australian Veterinary Journal 78, 182–183. *.

Morgan, U.M., Monis, P.T., Xiao, L., Limor, J., Sulaiman, I., Raidal, S., O'Donoghue, P., Gasser, R., Murray, A., Fayer, R., Blagburn, B.L., Lal, A.A., Thompson, R.C., 2001. Molecular and phylogenetic characterisation of Cryptosporidium from birds. International Journal for Parasitology 31, 289–296. *.

Morgan-Ryan, U.M., Fall, A., Ward, L.A., Hijawi, N., Sulaiman, I., Fayer, R., Thompson, R.C., Olson, M., Lai, A., Xiao, L., 2002. Cryptosporidium hominis n. sp. (Apicomplexa: Cryptosporidiidae) from Homo sapiens. Journal of Eukaryotic Microbiology 49, 433–440. *.

Mosevich, T.N., Cheissin, E.M., 1961. Certain findings on the electron microscopic study of merozoites of Eimeria intestinalis from the rabbit intestine. Tsitologiya (Cytology) 3, 34–39. *.

Mosier, D.A., Cimon, K.Y., Kuhls, T.L., Oberst, R.D., Simons, K.R., 1997. Experimental cryptosporidiosis in adult and neonatal rabbits. Veterinary Parasitology 69, 163–169. *.

Munday, B.L., Mason, R.W., Hartley, W.J., Presidente, P.J.A., Obendorf, D., 1978. Sarcocystis and related organisms in Australian wildlife. I. Survey findings in mammals. Journal of Wildlife Diseases 14, 417–433. *.

Munday, B.L., Smith, D.D., Frenkel, J.K., 1980. Sarcocystis and related organisms in Australian wildlife. IV. Studies on Sarcocystis cuniculi in European rabbits (Oryctolagus cuniculus). Journal of Wildlife Diseases 16, 201–204. *.

Mundin, M.J.S., Barbon, E., 1990. Freqüência e identificação de coccidios intestinais em coelhos domésticos (Oryctolagus cuniculus) em Umberlândia, Minas Gerais (Frequency and identification of intestinal coccidia in domestic rabbits (Oryctolagus cuniculus) in Umberlândia, Minas Gerais). Arquivos Brasileiros de Medicina Veterinaria Zootecnia 42, 529–538. * (in Portuguese, English summary).

Musaev, M.A., Veisov, A.M., 1965. The coccidia of rodents in the U.S.S.R. Izvestiya Akademii Nauk Azerbaidzhanskoi SSR, Baku, 1–154. * (in Russian).

Musongong, G.A., Fakae, B.B., 1999. Prevalence of Eimeria stiedai infection in outbred domestic rabbits (Orytolagus cuniculus) in eastern Nigeria. Revue D Élevage et de Médicine vét. Pays Tropical 52, 117–118. *.

Mykytowycz, R., 1956. A survey of endoparasites of the wild rabbit, Oryctolagus cuniculus (L.), in Australia. CSIRO Wildlife Research 1, 19–25. *.

Mykytowycz, R., 1962. Epidemiology of coccidiosis (Eimeria spp.) in an experimental population of the Australian wild rabbit, Oryctolagus cuniculus (L). Parasitology 52, 375–395. *.

Niak, A., 1967. Eimeria in laboratory rabbits in Teheran. Veterinary Record 81, 549. *.

Naumov, S.P., 1939. Fluctuation in numbers among hares. Voprosy Ekologiya i Biotsenologiya Leningrad 5–6, 40–82. * (in Russian).

Neuman, M., 1962a. An outbreak of besnoitiosis in cattle. Refuah Veterinarit 19, 106–115. *.

Neuman, M., 1962b. The experimental infection of the gerbil (Meriones tristrami shawii) with Besnoitia besnoiti. Refuah Veterinarit 19, 184–188. *.

Neuman, M., Nobel, T.A., 1981. Observations on the pathology of besnoitiosis in experimental animals. Zentralblatt für Veterinärmedizin Medicine, Reihe (Series) B 28, 345–354. *.

Neuman, M., Nobel, T.A., Perelman, B.Z., 1979. The neuropathogenicity of Besnoitia besnoiti (Marotel 1912) in experimental animals. Journal of Protozoology 26, 51A (abstract 146). *.

Ng'ang'a, C.J., Munyua, W.K., Kanyari, P.W., 1994. Recovery and identification of Besnoitia and other coccidia from cat faeces around Kabete in Kenya. Bulletin of Animal Health Production Africa 42, 187–191. *.

Nickel, S., Gottwald, A., 1979. Beiträge zur Parasitenfauna der DDR. 3. Mitteilung. Endoparasiten der Feldhasen (Lepus europaeus). Angewandt Parasitologie 20, 57–62. *.

Niedźwiadek, S., Ramisz, A., Balicka, A., Bielański, P., 1990. Wpływ salinomycyny na wystepowanie kokcydiozy u królików (The influence of salinomycine on the occurrence of coccidiosis of rabbits). Riczniki Naukowe Zootechniki Monografie I Rozprawy 28, 261–269. * (in Polish, English summary).

Nieschulz, O., 1923. Über Hasenkokzidien (Eimeria leporis n. sp). Deutsche Tierärztliche Wochenschrift 31, 245–247. *.

Nime, F.A., Burek, J.D., Page, D.L., Yardley, J.H., 1976. Acute enterocolitis in a human being infected with the protozoan Cryptosporidium. Gastroenterology 70, 592–598. *.

Nolan, M.J., Jex, A.R., Haydon, S.R., Stevens, M.A., Gasser, R.B., 2010. Molecular detection of Cryptosporidium cuniculus in rabbits in Australia. Infection. Genetics and Evolution 10, 1179–1187. *.

Norton, C.C., Catchpole, J., Rose, M.E., 1977. Eimeria stiedai in rabbits: the presence of an oocyst residuum. Parasitology 75, 1–7. *.

Norton, C.C., Catchpole, J., Joyner, L.P., 1979. Redescriptions of Eimeria irresidua Kessel & Jankiewicz, 1931 and E. flavescens Marotel & Guilhon, 1941 from the domestic rabbit. Parasitology 79, 231–248. *.

Nowak, R.M., 1991. Walker's Mammals of the World, fifth ed., vol. I. The Johns Hopkins University Press, Baltimore, Maryland, pp. 539–560. *.

Nowak, R.M., 1999. Walker's Mammals of the World, sixth ed., vol. I.. The John Hopkins University Press, Baltimore, Maryland, pp. 1–793. *.

Nowell, F., Higgs, S., 1989. Eimeria species infecting wood mice (genus Apodemus) and the transfer of two species to Mus musculus. Parasitology 98, 329–336. *.

Odening, K., Stolte, M., Walter, G., Bockhardt, I., Jakob, W., 1994a. Sarcocysts (Sarcocystis sp.: Sporozoa) in the European badger Meles meles. Parasitology 108, 421–424. *.

Odening, K., Stolte, M., Walter, G., Bockhardt, I., 1994b. The European badger (Carnivora: Mustelidae) as intermediate host of further three Sarcocystis species (Sporozoa). Parasite (Paris) 1, 23–30. *.

Odening, K., Wesemeier, H.-H., Pinkowski, M., Walter, G., Sedlaczek, J., Bockhardt, I., 1994c. European hare and European rabbit (Lagomorpha) as intermediate hosts of Sarcocystis species (Sporozoa) in central Europe. Acta Protozoologica 33, 177–189. *.

Odening, K., Wesemeier, H.-H., Walter, G., Bockhardt, I., 1995a. Ultrastructure of sarcocysts from equids. Acta Parasitologica 40, 12–20. *.

Odening, K., Wesemeier, H.-H., Walter, G., Bockhardt, I., 1995b. On the morphological diagnostics and host specificity of the Sarcocystis species of some domesticated and wild Bovini (cattle, banteng and bison). Applied Parasitology 36, 161–178. *.

Odening, K., Wesemeier, H.-H., Bockhardt, I., 1996. On the sarcocysts of two further Sarcocystis species being new for the European hare. Acta Protozoologica 35, 69–72. *.

Ogedengbe, J.D., 1991. Prevalence of rabbit coccidial infection and pathogenicity of isolated Eimeria stiedai for local rabbits in Zaria. Master's Thesis, Ahmadu Bello University, Zaria, Kaduna State, northern Nigeria, Africa [Abstract only available]. *.

Oliveira, U.C., Fraga, J.S., Licois, D., Pakandl, M., Gruber, A., 2011. Development of molecular assays for the identification of the 11 Eimeria species of the domestic rabbit (Oryctolagus cuniculus). Veterinary Parasitology 176, 275–280. *.

Oncel, T., Gulegen, E., Senlik, B., Bakirci, S., 2011. Intestinal coccidiosis in Angora rabbits (Oryctolagus cuniculus) caused by Eimeria intestinalis, Eimeria perforans and Eimeria coecicola. YYU Veteriner Fakultesi Dergisi 22, 27–29. *.

Ong, C.S., Eisler, D.L., Goh, S.H., Tomblin, J., Awad-El-Kariem, F.M., Beard, C.B., Xiao, L., Sulaiman, I., Lai, A., Fyfe, M., King, A., Bowie, W.R., Isacc-Renton, J.L., 1999. Molecular epidemiology of cryptosporidiosis outbreaks and transmission in British Columbia, Canada. American Journal of Tropical Medicine and Hygiene 61, 63–69. *.

Osipovskiy, A.I., 1955. Inheritance of resistance to coccidiosis by rabbits. Zhurnal Obshchei Biologii 16, 64–68. * (in Russian).

Owen, D., 1970. Life cycle of Eimeria stiedae. Nature 227, 304. *.

Pakandl, M., 1986a. Efficacy of salinomycin, monensin and lasalocid against spontaneous Eimeria infection in rabbits. Folia Parasotology 33, 195–198.

Pakandl, M., 1986b. Two morphological types of oocyst of rabbit coccidia Eimeria media Kessel, 1929. Folia Parasitologica 33, 297–300. *.

Pakandl, M., 1988. Description of Eimeria vejdovskyi sp. n. and redescription of Eimeria media Kessel, 1929 from the rabbit. Folia Parasitologica (Praha) 35, 1–9. *.

Pakandl, M., 1989. Life cycle of Eimeria coecicola Cheissin, 1947. Folia Parasitologica (Praha) 36, 97–105. *.

Pakandl, M., 1990. Some remarks on the prevalence and species composition of hare coccidia. Folia Parasitologica (Praha) 37, 35–42. *.

Pakandl, M., Coudert, P., 1999. Life cycle of *Eimeria vejdovskyi* Pakandl, 1988: electron microscopy study. Parasitology Research 85, 850–854. *.

Pakandl, M., Jelínková, A., 2006. The rabbit coccidium *Eimeria piriformis*: selection of a precocious line and life-cycle study. Veterinary Parasitology 137, 353–354. *.

Pakandl, M., Coudert, P., Licois, D., 1993. Migration of sporozoites and merogony of *Eimeria coecicola* in gut-associated lymphoid tissue. Parasitology Research 79, 593–598. *.

Pakandl, M., Gaca, K., Drouet-Viard, F., Coudert, P., 1996a. *Eimeria coecicola*: endogenous development in gut-associated lymphoid tissue. Parasitology Research 82, 347–351. *.

Pakandl, M., Gaca, K., Licois, D., Coudert, P., 1996b. *Eimeria media* Kessel 1929: comparative study of endogenous development between precocious and parental strains. Veterinary Research 27, 465–472. *.

Pakandl, M., Eid-Ahmed, N., Licois, D., Coudert, P., 1996c. *Eimeria magna* Pérard 1925 study of the endogenous development of parental and precocious strains. Veterinary Parasitology 65, 213–222. *.

Pakandl, M., Černík, F., Coudert, P., 2003. The rabbit coccidium *Eimeria flavescens* Marotel and Guilhon 1941 an electron microscopic study of its life cycle. Parasitology Research 91, 304–311. *.

Pantenburg, B., Cabada, M.M., White Jr., A.C., 2009. Treatment of cryptospiridiosis. Experimental Reviews in Anti Infective Therapy 7, 385–391. *.

Parker, B.P., Duszynski, D.W., 1986. Polymorphism of eimerian oocysts: a dilemma posed by working with some naturally infected hosts. Journal of Parasitology 72, 602–604. *.

Pastuszko, J., 1961a. The occurrence of Eimeriinae Wenyon in the hare in Poland (Wystepowanie Eimeriinae Wenyon u zajecy w Polsce). Acta Parastiologica Polonica (Warszawa) 9, 23–32. * (in English).

Pastuszko, J., 1961b. About the specific independence of *Eimeria* sp. parasitizing in rabbits and hares (W sprawie odrebnosci gatunkow rodzaju *Eimeria* pasoztujacych u krolikow I zajecy). Wiadomosci Parazytologiczne. Polskie Towarzystow Parazytologiczne (Warszawa) 7, 305–307. * (in English and Polish).

Pastuszko, J., 1963. Kokcydiozy królików w Polsce (Coccidiosis in rabbits in Poland). Polskie Archiwum Weterynaryjne Earszawa 8, 129–140. * (in Polish with English summary).

Pavlásek, L., Lávicka, M., Tůmová, E., Skrivan, M., 1996. Spontánní kryptosporidiová nákaza u odstavených králíčat (Natural *Cryptosporidium* infection in rabbits after weaning). Veterinary Medicine—Czech (Prague) 41, 361–366. * (in Czech, English summary).

Pedraza-Diaz, S., Amar, C., McLauchlin, J., 2000. The identification and characterisation of an unusual genotype of *Cryptosporidium* from human faeces as *Cryptosporidium meleagridis*. FEMS Microbiology Letters 189, 189–194. *.

Peeters, J.E., 1988. Recent advances in intestinal pathology of rabbits and further perspectives. In: Holdas, S. (Ed.), Proceedings of the 4th World Rabbit Congress. World Rabbit Science Associatiion, Budapest, Hungary, 10–14 October, pp. 293–315.

Peeters, J.E., Geeroms, R., 1986. Efficacy of toltrazuril against intestinal and hepatic coccidiosis in rabbits. Veterinary Parasitology 22, 21–35. *.

Peeters, J.E., Halen, P., Meulemans, G., 1979. Efficacy of Robenidine in the prevention of rabbit coccidiosis. British Veterinary Journal 135, 349–354.

Peeters, J.E., Geeroms, R., Froyman, R., Halen, P., 1981. Coccidiosis in rabbits: a field study. Research in Veterinary Science 30, 328–334. *.

Peeters, J.E., Geeroms, R., Molderez, J., Halen, P., 1982. Activity of Clopidol/Methylbenzoquate, Robenidine and Salinomycin against hepatic coccidiosis in rabbits. Zentralblatt fur Veterinarmedizin Reihe B 29, 207–218. *.

Peeters, J.E., Charlier, G.J., Antoine, O., Mammerick, M., 1984. Clinical and pathological changes after *Eimeria intestinalis* infection in rabbits. Zentralblatt für Veterinärmedizin Reihe B 31, 9–24. *.

Peeters, J.E., Charlier, G.J., Dussart, P., 1986. Pouvoir pathogene de *Cryptosporidium* sp. Chez les lapereaux avant et appres sevrage. Journees de la Recherche Cunicole 37, 1–9. *.

Peeters, J.E., Geeroms, R., Norton, C.C., 1987. *Eimeria magna*: resistance against robenidine in the rabbit. Veterinary Research 121, 545–546. *.

Pellérdy, L., 1953. Beiträge zur Kenntnis der Darmkokzidiose des Kaninchens. Die endogene Entwicklung von *Eimeria piriformis*. Acta Veterinaria Academiae Scientiarium Hungaricae 3, 365–377. *.

Pellérdy, L., 1954a. Beiträge zur Specifität der Coccidien des Hasen und Kaninchens. Acta Veterinaria Academiae Scientiarium Hungaricae 4, 481–487. *.

Pellérdy, L., 1954b. *Eimeria agnosta* n. sp. from the rabbit. Acta Veterinaria Academiae Scientiarium Hungaricae 4, 259–261. *.

Pellérdy, L.P., 1956. On the status of the *Eimeria* species of *Lepus europaeus* and related species. Acta Veterinaria Academiae Scientiarum Hungaricae 6, 451–467. *.

Pellérdy, L.P., 1965. Coccidia and coccidiosis. Akadémiai Kiadó, Budapest. *.

Pellérdy, L.P., 1969a. Attempts to alter the host specificity of Eimeriae by parenteral infection experiments. Acta Veterinaria (Brno) 38, 43–46. *.

Pellérdy, L.P., 1969b. Parenteral infection experiments with *E. stiedai* (Lindemann, 1865). Acta Veterinaria Academiae Scientiarium Hungaricae 19, 171–182. *.

Pellérdy, L.P., 1974. Coccidia and coccidiosis, second ed. Joint Publication of Verlag Paul Parey, Berlin and Hamburg and Akadémiai Kiadó, Budapest. *.

Pellérdy, L.P., Babos, A., 1953. Studies on the endogenous development and pathological significance of *Eimeria media*. (Untersuchungen über die endogene Entwicklung sowie pathologische Bedeutung von *Eimeria media*). Acta Veterinaria Academiae Scientiarum Hungaricae 3, 173–188. * (in German, no English summary).

Pellérdy, L.P., Dürr, U., 1970. Zum endogenen Entwicklungszyclus von *Eimeria stiedai* (Lindemann, 1865; Kisskalt & Hartmann, 1907). Acta Veterinaria Academiae Scientiarum Hungaricae 20, 227–244. * (in German, no English summary).

Pellérdy, L., Hönich, M., Sugár, L., 1974. Studies on the development of *Eimeria leporis* (Protozoa: Sporozoa) and on its pathogenicity for the hare (*Lepus europaeus* Pall.). Acta Veterinaria Academiae Scientiarum Hungaricae 24, 163–175. *.

Penzhorn, B.L., Knapp, S.E., Speer, C.A., 1994. Enteric coccidia in free-ranging American bison (*Bison bison*) in Montana. Journal of Wildlife Diseases 30, 267–269. *.

Pérard, C., 1924a. Recherches sur les coccidies et les coccidioses du lapin. Comptes Rendus de l'Academie des Sciences (Paris) 178, 2131–2134. *.

Pérard, C., 1924b. Recherches sur les coccidies et les coccidioses du lapin. Annales de l'Institute Pasteur, Paris 38, 953–976. *.

Pérard, C., 1925a. Recherches sur les coccidies et les coccidioses du lapin. II. Contribution a l'etude de la biologie des oöcysts des coccidies (Research on coccidia and coccidiosis of the rabbit. II. Contribution to the study of the biology of coccidia oocysts). Annales de l'Institute Pasteur, Paris 39. 505–542. (in French, English summary).

Pérard, C., 1925b. Recherches sur les coccidies et les coccidioses du lapin. III. Étude de la multiplication endogène. (Identification d'une 3 espèce de coccidie du lapin: *Eimeria magna* n. sp.) (Research on coccidia and coccidiosis of the rabbit. III. Study of endogenous multiplication. [Identification of a 3rd species of rabbit coccidia: *Eimeria magna* n. sp]). Annales de l'Institute Pasteur, Paris 39, 952–961. * (in French, English summary).

Pérez-Suárez, G., Palacios, F., Boursot, P., 1994. Speciation and paraphyly in western Mediterranean hares (*L. castroviejoi, L. europaeus, L. granatensis,* and *L. capensis*) revealed by mitochondrial DNA phylogeny. Biochemical Genetics 32, 423–436. *.

Peteshev, V.M., Galuzo, I.G., Polomoshnov, A.P., 1974. Koshki—definitivnye Khoziaeva besnoitii (*Besnoitia besnoiti*) [Cats—Definitive host of *Besnoitia* (*Besnoitia besnoiti*)]. Isvestiya Akademii Nauka Kazakhskoi SSR. Seria Biologicheskiya 1, 33–38. * (in Russian).

Pfeiffer, L., 1890. Vergleichende Untersuchungen über Schwärmsporen und Dauersporen bei den Coccidieninfektionen und bei Intermittens. Fortschritte der Medizin (Munchen) 8, 939–951. *.

Pfeiffer, L., 1891. Die Protozoen als Krankheitserreger, second ed. Fischer, Jena. *.

Pfeiffer, R., 1892. Beiträge zur Protozoen-Forschung. 1 Hft. Die Coccidien-Krankheit der Kaninchen Berlin, Hirschwald. *.

Pick, P., 1974. Ultraviolett-Bestrahlung und Photoreaktivierung bei oocysten von *Eimeria stiedai* (Ultraviolet radiation and photoreactivation in oocysts of *Eimeria stiedai*). Inaugural Dissertation, Faculty of Veterinary Medicine, Justus Liebig-Universität Geißen, Geißen, Germany. (in German, no English summary).

Pimental, D., Tort, M., D'Anna, L., Krawic, A., Berger, J., Rossman, J., Mugo, F., Doon, N., Shriberg, M., Howard, E., Lee, S., Talbot, J., 1998. Ecology of increasing disease. Bioscience 48, 817–826. *.

Pohjola, S., Jokipii, L., Jokipii, A.M.M., 1984. Dimethylsulphoxide-Ziehl-Neelsen staining technique for the detection of cryptosporidial oocysts. Veterinary Record 115, 442–443. *.

Pokorný, J., Hübner, J., Zástěra, M., 1981. Izolace kmenů *Toxoplasmi gondii* z některých domácichi volně žijococj zvirat (Isolation of strains of *Toxoplasma gondii* of some domestic as well as free living animals). Ceskoslovenska Epidemiologie, Microbiologie, Immunologie 10, 323–329. *.

Polozowski, A., 1993. Kokcydioza królików I jej zapobieganie (Coccidiosis of rabbits and its control). Wiadomości Parazytologiczne 39, 13–28. * (in Polish, English summary).

Pols, J.W., 1954a. The artificial transmission of *Globidium besnoiti* Marotel, 1912, to cattle and rabbits. South African Veterinary Medical Association 25, 37–44. *.

Pols, J.W., 1954b. Preliminary notes on the behaviour of *Globidium besnoiti* Marotel, 1912, in the rabbit. South African Veterinary Medical Association 25, 45–48. *.

Pols, J.W., 1960. Studies on bovine besnoitiosis with special reference to the aetiology. Onderstepoort Journal of Veterinary Research 28, 265–356. *.

Prasad, H., 1960. Studies on the coccidia of some mammals of the families Bovidae, Cervidae and Camelidae. Zeitschrift für Parasitnekunde 20, 390–400. *.

Procunier, J.D., Fernando, M.A., Barta, J.M., 1993. Species and strain differentiation of *Eimeria* spp. of the domestic fowl using DNA polymorphisms amplified by arbitrary primers. Parasitology Research 79, 98–102. *.

Qiang, J., Luo, Z.-X., Yuan, C.-X., Wible, J.R., Zhang, J.-P., Georgi, J.A., 2002. The earliest known eutherian mammal. Nature 416, 816–822. *.

Quesenberry, K.E., Carpenter, J.W., 2010. Ferrets, Rabbits, and Rodents: Clinical Medicine and Surgery, third ed. Saunders (Elsevier). *.

Raczkowski, J.M., Wenzel, J.W., 2007. Biodiversity studies and their foundation in taxonomic scholarship. Bioscience 57, 974–979. *.

Railliet, A., Lucet, A., 1891a. Développement experimental des coccidies de l'épithélium intestinal du lapin et de la poule (Experimental development of coccidia in the intestinal epithelium of the rabbit and chicken). Comptes Rendus des Seances Socété de Biologie (Paris) 43, 820–823. * (in French).

Railliet, A., Lucet, A., 1891b. Note sur quelques espèces de coccidies encore peu étudiées (Note on some species of coccidia yet little studied). Bulletin de la Societe Zoologique de France 16, 246–250. * (in French).

Rausch, R.L., 1961. Notes on the collard pika, *Ochotona collaris* (Nelson) in Alaska. Murrelet 42, 22–24. *.

Ray, H.N., 1945. Presidential address to the Section of Zoology and Entomology of the 32nd Indian Science Congress, Nagpur. Proceedings of the Indian Science Congress Association, Part 1, 136–149. *.

Ray, H.N., Banik, D.C., 1965a. On a new coccidium, *Eimeria oryctolagi* n. sp., from the domestic rabbit *Oryctolagus* (*Lepus*) *cuniculus*. Section of Zoology and Entomology of the 32nd Indian Science Congress, Nagpur. Proceedings of the Indian Science Congress Association, Part III (Abstract, No. 72), 465. *.

Ray, H.N., Banik, D.C., 1965b. On a new coccidium, *Eimeria oryctolagi* n. sp., from the domestic rabbit *Oryctolagus* (*Lepus*) *cuniculus*. Indian Journal of Animal Health 4, 21–23. *.

Razavi, S.M., Oryan, A., Rakhshandehroo, E., Moshiri, A., Mootabi, A.A., 2010. *Eimeria* species in wild rabbits (*Oryctolagus cuniculus*) in Fars province, Iran. Tropical Biomedicine 27, 470–475. *.

Reaka-Kudla, M.L., Wilson, D.E., Wilson, E.O. (Eds.), 1997. Biodiversity II: Understanding and protecting our biological resources. Joseph Henry Press (an imprint of National Academy Press), Washington, D.C., USA and Oxford, UK. *.

Real, L.A., 1996. Sustainability and the ecology of infectious diseases. Bioscience 46, 88–97. *.

Reduker, D.W., Duszynski, D.W., Yates, T.L., 1987. Evolutionary relationships among *Eimeria* spp. (Apicomplexa) infecting Cricetid rodents. Canadian Journal of Zoology 65, 722–735. *.

Rehg, J.E., Lawton, G.W., Pakes, S.P., 1979. *Cryptosporidium cuniculus* in the rabbit (*Oryctolagus cuniculus*). Laboratory Animal Science 29, 656–660. *.

Reid, W.M., 1975. Progress in the control of coccidiosis with anticoccidials and planned immunization. American Journal of Veterinary Research 36, 593–596. *.

Remak, R., 1845. Diagnostische und pathogenetische Untersuchungen in der Klinik des Herrn Geh. Raths Dr. Schönlein, Berlin, p. 239. *.

Renaud, F., Clayton, D., De Meeüs, T., 1996. Biodiversity and evolution in host-parasite associations. Biodiversity and Conservation 5, 963–974. *.

Ride, W.D.L., Cogger, H.G., Dupis, C., Kraus, O., Minellie, A., Thompson, F.C., Tuggs, P.K. (Eds.), 2000. International Code of Zoological Nomenclature, fourth ed. Published by The International Trust for Zoological Nomenclature 1999, The Natural History Museum, Cromwell Road, London, United Kingdom. *.

Rieck, W., 1956. Studies on the proliferation of the brown hare. (Untersuchungen über die Vermehrung des Feldhasen). Zeitschrift für Jagdwissenschaft 2, 49–90. * (in German).

Ritchie, J., 1926. Note on coccidiosis of Brown hare (*Lepus europaeus*). Transactions of the Proceedings of the Perth Society of Natural Science 8, 156–157. *.

Rivolta, S., 1878. Della gregarinosi dei polli e dell' ordinamento delle gregarine e dei psorospermi degli animali domestici. Giornale de Anatomia, Fisiologia, Patologia di Animali, Pisz 10, 200–235. *.

Roberts, L.S., Janovy Jr., J., 2009. Foundations of parasitology, eighth ed. McGraw-Hill, New York, p. 369. *.

Robertson, A., 1933. Coccidiosis of the hare. Journal of Tropical Medicine and Hygiene 36, 143–148. *.

Robinson, G., Chalmers, R.M., 2010. The European rabbit (*Oryctolagus cuniculus*) a source of zoonotic cryptosporidiosis. Zoonoses Public Health 57, 1–13. *.

Robinson, T.J., Osterhoff, D.R., 1983. Protein variation and its systematic implications for the South African Leporidae (Mammalia: Lagomorpha). Animal Blood Groups Biochemical Genetics 14, 139–149. *.

Robinson, T.J., Yang, F., Harrison, W.R., 2002. Chromosome painting refines the history of genome evolution in hares and rabbits (order Lagomorpha). Cytogenetic Genome Research 96, 223–227. *.

Robinson, G., Wright, S., Elwin, K., Hadfield, S.T., Katzer, F., Bartley, P.M., Hunter, P.R., Nath, M., Innes, E.A., Chalmers, R.M., 2010. Re-description of *Cryptosporidium cuniculus* Inman and Takeuchi, 1979 (Apicomplexa: Cryptosporidiidae): morphology, biology and phylogeny. International Journal of Parasitology 40, 1539–1548. *.

Romero-Rodriguez, J., 1976. Contribucion al conocimiento de las coccidiopatias de los Lagomorpha: Estudio de los Protozoa-Eimeriidae, *E. leporis* y *E. europaea*, parasitas de *Lepus granatensis* (= *L. capansis*). Revista de Ibérica de Parasitologia 36, 131–134. *.

Rommel, M., Schwerdifeger, A., Blewaska, S., 1981. The *Sarcocystis muris*-infection as a model for research on the chemotherapy of acute sarcocystosis of domestic animals. Zentralblatt für Bakteriologie, Parasitenkunde, Infektionskrankheiten und Hygiene. Erste Abteilung Originale A 250, 268—276. *.

Rose, K.D., DeLeon, V.B., Missiaen, P., Rana, R.S., Sahni, A., Singh, L., Smith, T., 2008. Early Eocene lagomorph (Mammalia) from western India and the early diversification of Lagomorpha. Proceedings of the Royal Society B 275, 1203—1208. *.

Rose, M.E., 1958. The life cycle and development of *Eimeria stiedai* (Lindemann, 1865). Ph.D. Thesis, University of Cambridge. *.

Rose, M.E., Millard Jr., B.J., 1985. Host specificity in eimerian coccidia: Development of *Eimeria vermiformis* of the mouse, *Mus musculus* in *Rattus norvegicus*. Parasitology 90, 557—563. *.

Rossignol, J.F., Ayoub, A., Ayers, M.S., 2001. Treatment of diarrhea caused by *Cryptosporidium parvum*: a prospective randomized, double-blind, placebo-controlled study of Nitazoxanide. Journal of Infectious Disease 184, 103—106. *.

Rougeot, J., 1981. Origine et histoire du Lapin. Ethnozootechnie 27, 1—9. *.

Rutherford, R.L., 1943. The life-cycle of four intestinal coccidia of the domestic rabbit. Journal of Parasitology 29, 10—32. *.

Ryan, M.J., Sundberg, J.P., Sauerschell, R.J., Todd, K.S., 1986. *Cryptosporidium* in a wild cottontail rabbit (*Sylvilagus floridanus*). Journal of Wildlife Diseases 22, 267. *.

Ryan, U., Xiao, L., Read, C., Zhou, L., Lal, A.A., Pavlasek, I., 2003. Identification of novel *Cryptosporidium* genotypes from the Czech Republic. Applied Environmental Microbiology 69, 4302—4307. *.

Ryff, K.L., Bergstrom, R.C., 1975. Bovine coccidia in American bison. Journal of Wildlife Diseases 11, 412—414. *.

Ryley, J.F., Robinson, T.E., 1976. Life-cycle studies with *Eimeria magna* Pérard, 1925. Zeitschrift für Parasitenkunde 50, 257—275. *.

Ryšavý, B., 1954. Přispěvek k poznáni kokeidií našich i dovezených obratlovců (Contribution to the knowledge of our and imported vertebrates). Czechoslovak Parasitology 1, 131—174. * (in Czech).

Saiki, R.K., Gelfand, D.H., Stoffel, S., Scharf, S.J., Higuchi, R., Horn, G.T., Mullis, K.B., Erlich, H.A., 1988. Primedirected enzymatic amplification of DNA with a thermostable DNA polymerase. Science 239, 487—491. *.

Samoil, H.P., Samuel, W.M., 1977a. Description of nine species of *Eimeria* (Protozoa, Eimeriidae) in the snowshoe hare, *Lepus americanus*, of central Alberta. Canadian Journal of Zoology 55, 1671—1683. *.

Samoil, H.P., Samuel, W.M., 1977b. Experimental study of *Eimeria robertsoni* (Protozoa, Eimeriidae) in the snowshoe hare, *Lepus americanus*. Journal of Parasitology 63, 203—205. *.

Samoil, H.P., Samuel, W.M., 1981. Use of coccidia as indicators of phylogenetic relationships of members of the order Lagomorpha. In: Chapman, J.A., Pursley, D. (Eds.), Proceedings of the First Worldwide Furbearer Conference pp. 7—29. Baltimore, Maryland. *.

Santos, M.J., Lima., J.D., 1987. Freqüência de oocitos de *Eimeria* spp. em coelhos domésticos (*Oryctolagus cuniculus*) em quarto municípios de Minas Gerais. Arquivo Brasileiro de Medicina Veterinaria e Zootecnia 39, 93—102. *.

Santos-Mundin, M.J., Barbon, E., 1990. Freqüência e identificação de coccidios intestinais em coelhos domésticos (*Oryctolagus cuniculus*) em Umberlândia, Minas Gerais. Arquivo Brasileiro de Medicina Veterinaria e Zootecnia 42, 529—536. *.

Sanyal, P.K., Srivastava, C.P., 1988. Clinico pathological studies in experimental *Eimeria media* infection in domestic rabbit (*Oryctolagus cuniculus*). Indian Journal of Animal Research 22, 107—110. *.

Sanyal, P.K., Sharma, S.C., 1990. Clinicopathology of hepatic coccidiosis in rabbits. Indian Journal of Animal Science 60, 924—928. *.

Schipani, V., 2011. Notebook: Character flaws? The Scientist 25, 20—22. *.

Schneider, A., 1875. Contributions l'histoire des gregarines d' invertébrés de Paris et de Roscoff. Archives de Zoologie Expérimentale et Gènerale 4, 493—604. *.

Scholtyseck, E., 1962. Electron microscope studies of *Eimeria perforans* (Sporozoa). Journal of Protozoology 9, 407—414. *.

Scholtyseck, E., 1963. Elektronmikroskopische Untersuchungen über die Wechselwirkung zwischen dem Zellparasiten *Eimeria perforans* und seiner Wirtzelle. Zeitschrift für Zellforschung und Mikroskopische Anatomie 61, 220—230. *.

Scholtyseck, E., 1964. Elektronenmikroskopisch-cytochemischer Nachweis von Glykogen bein *Eimeria perforans*. Zeitschrift für Zellforschung und Mikroskopische Anatomie 64, 688—707. *.

Scholtyseck, E., 1965a. Elektronenmikroskopische Untersuchungen über die Schizogonie bei Coccidien (*Eimeria perforans* und *E. stiedae*). Zeitschrift für Parasitenkunde 26, 50—62. *.

Scholtyseck, E., 1965b. Die Mikrogametenentwicklung von *Eimeria perforans*. Zeitschrift für Zellforschung 66, 625—642. *.

Scholtyseck, E., 1973. Die Deutung von Endodyogenie bei Coccidien und Schizogonie und anderen sporozoen. Zeitschrift für Parasitenkunde 42, 87—104. *.

Scholtyseck, E., Ratanavichien, A., 1976. Endodyogenie in *Eimeria stiedai* merozoiten. Transactions of the American Microscopical Society 95, 553–556. *.

Scholtyseck, E., Spiecker, D., 1964. Vergleichende elektronenmikroskopische Untersuchungen an den Entwicklungsstadien von *Eimeria perforans* (Sporozoa) (Comparative electron microscopic studies on the developmental stages of *Eimeria perforans* [Sporozoa]). Zeitschrift für Parasitenkunde 24, 546–560. * (in German, English summary).

Scholtyseck, E., Piekarski, G., 1965. Elektronenmikroskopische Untersuchungen an Merozoiten von Eimerien (*Eimeria perforans* und *E. stiedae*) und *Toxoplasma gondii*. Zeitschrift für Parasitenkunde 26, 91–115. *.

Scholtyseck, E., Hammond, D.M., Ernst, J.V., 1966. Fine structure of the macrogametes of *Eimeria perforans, E. stiedai, E. bovis* and *E. auburnensis*. Journal of Parasitology 52, 975–987. *.

Scholtyseck, E., Rommel, A., Heller, G., 1969. Licht- und elektronenmikroskopisch Untersuchungen zur Bildung der Oocystenhülle bei Eimerien (*Eimeria perforans, E. stiedae* und *E. tenella*) (Light and electron microscope investigations on the formation of the oocyst wall in *Eimeria* species [*Eimeria perforans, E. stiedae*, and *E. tenella*]). Zeitschruft für Parasitenkunde 31, 289–298. * (in German, English summary).

Scholtyseck, E., Heller, G., Pellérdy, L.P., 1970a. Nukleolusähnliche feinstrukturen im cytoplasm der makrogameten von *Eimeria perforans* (Sporozoa, Coccidia) (Nucleolus-like ultrastructures in the cytoplasm of the macrogametes of *Eimeria perforans* [Sporozoa, Coccidia]). Acta Veterinaria Akademia Scientiarum Hungaricae 20, 429–434. * (in German, no English summary).

Scholtyseck, E.H., Mehlhorn, H., Friedhoff, K., 1970b. The fine structure of the conoid of sporozoa and related organisms. Zeitschrift für Parasitenkunde 34, 68–94. *.

Scholtyseck, E., Melhorn, H., Hammond, D.M., 1971. Fine structure of the macrogametes and oocysts of coccidia and related organisms. Zeitschrift für Parasitenkunde 37, 1–43. *.

Scholtyseck, E., Melhorn, H., Hammond, D.M., 1972. Electron microscope studies of microgametogenesis in coccidia and related groups. Zeitschrift für Parasitenkunde 38, 95–131. *.

Scholtyseck, E., Entzeroth, R., Pellérdy, L.P., 1979. Übertragung von *Eimeria stiedai* aus kaninchen (*Oryctolagus cuniculus*) auf feldhasen (*Lepus europaeus*) (Transmission of *Eimeria stiedai* from rabbit [*Oryctolagus cuniculus*] to hares [*Lepus europaeus*]). Acta Veterinaria Academiae Scientiarum Hungaricae 27, 365–373. * (in German, no English summary).

Schrecke, W., 1969. Experimentelle Coccidieninfektionen bei neugeborenen Tieren. Dissertation, Giessen. *.

Schrecke, W., Dürr, U., 1970. Excystations- und Infektionsversuche mit Kokzidienoocysten bei neugeborenen Tieren (Excystations and infection experiments with coccidia oocysts in newborn animals). Zentralblatt für Bakteriologie. I. Abteilung Originale 215, 252–258. *.

Scott, M.E., 1988. The impact of infection and disease on animal populations: inplications for conservation biology. Conservation Biology 2, 40–56. * (in German, English summary).

Seddon, H.R., 1966. Coccidiosis. In: Diseases of domestic animals in Australia, Part 4, second ed. Protozoan and virus diseases. Commonwealth of Australia, Department of Health, pp. 37–50. *.

Sedlák, K., Literák, I., Faldyna, M., Toman, M., Benák, J., 2000. Fatal toxoplasmosis in brown hares (*Lepus europaeus*): possible reasons of their high susceptibility to the infection. Veterinary Parasitology 93, 13–28. *.

Sénaud, J., Černá, Z., 1969. Etude ultrastructurale des métrozoites et de la schizogonie des Coccidies (Eimeriina): *Eimeria magna* (Perard 1925) de l'intestin des lapins et *E. tenella* (Railliet et Lucet, 1891) des coecums des poulets. Journal of Protozoology 16, 155–165. *.

Sénaud, J., Chobotar, B., Scholtyseck, E., 1976. Role of the micropore in nutrition of the Sporozoa. Ultrastructural study of *Plasmodium cathemerium, Eimeria ferrisi, E. stiedai, Besnoitia jellisoni*, and *Frenkelia* sp. Tropenmedizin und Parasitologie 27, 145–159. *.

Seville, R.S., Stanton, N.L., 1993. Eimerian guilds (Apicomplexa: Eimeriidae) in Richardson's (*Spermophilus richardsonii*) and Wyoming (*Spermophilus elegans*) ground squirrels. Journal of Parasitology 79, 973–975. *.

Shazly, M., Muborak, M., Al-Rasheid, K.A.S., Al-Ghamdy, A.O., Bashtar, A.R., 2005. Light and electron microscopic studies of *Eimeria magna* infecting the domestic rabbit, *Oryctolagus cuniculus* from Saudi Arabia. I. Asexual developmental cycles. Saudi Journal of Biological Sciences 12, 1–9. *.

Sherkov, Sh., N., Khalacheva, M., Kostova, T., Malchevski, M., Arnaudov, D., 1986. Etiopathogenesis and epizootiology of coccidiosis in rabbits. Veterinarno-Meditsinski Nauki 23, 11–17. *.

Shi, K., Jian, F., Lv, C., Ning, C., Zhang, L., Ren, X., Dearen, T.K., Li, N., Qi, M., Xiao, L., 2010. Prevalence, genetic characteristics, and zoonotic potential of *Cryptosporidium* species causing infections in farm rabbits in China. Journal of Clinical Microbiology 48, 3263–3266. *.

Shiibashi, T., Imai, T., Sato, Y., Abe, N., Yukawa, M., Nqami, S., 2006. *Cryptosporidium* infections in juvenile pet rabbits. Journal of Veterinary Medical Science 68, 281–282. *.

Shimizu, K., 1958. Studies on toxoplasmosis. I. An outbreak of toxoplasmosis among hares (*Lepus timidus ainu*) in

Sapporo. Japanese Journal of Veterinary Science 6, 157—171. *.

Sibalić, S, 1949. Effect of temperature and moisture on the exogenous development of *Eimeria perforans* and *Eimeria stiediae*. (Translated from Serbo-Croatian, Department of Agriculture and the National Science Foundation, NOLIT Publishing House, Terazije, Belgrade, Yugoslavia, 1963. Available from the Office of Technical Services, U.S. Department of Commerce, Washington 25, District of Columbia, USA). Arhiv Bioloških nauka (Beograd) 1, 253—257. *.

Siddall, M.E., Budinoff, R.B., 2005. DNA-barcoding evidence for widespread introductions of a leech from the South American *Helobdella triserialis* complex. Conservation Genetics 6, 467—472. *.

Simond, P.L., 1897. L'evolution des sporozoaires du genre *Coccidium*. Annales de l'Institute Pasteur, Paris 11. 545. *.

Simpson, G.G., 1945. The principles of classification and a classification of mammals. Bulletin of the American Museum of Natural History 85, 1—350. *.

Šlapeta, J.R., Modrý, D., Votýpka, J., Jirků, M., Lukeš, J., Koudela, B., 2003. Evolutionary relationships among cyst-forming coccidia *Sarcocystis* spp. (Alveolata: Apicomplexa: Coccidea) in endemic African tree vipers and perspective for evolution of heteroxenous life cycle. Molecular Phylogenetics and Evolution 27, 464—475. *.

Slater, R.L., Quisenberry, M.A., Fitzgerald, P.R., 1969. Pathway and timing of invasion of sporozoites of *Eimeria stiediae* (Lindemann, 1865). Journal of Parasitology 55 (Suppl.), 53 (abstract 108). *.

Slavin, D., 1955. *Cryptosporidium meleagridis* (sp. nov.). Journal of Comparative Pathology 65, 262—270. *.

Slimen, H.G., Suchentrunk, F., Elgaaied, A.B.A., 2008. On shortcomings of using mtDNA sequence divergence for the systematics of hares (genus *Lepus*): an example from cape hares. Mammalian Biology 73, 25—32. *.

Sluiter, C.P., Swellengrebel, N.H., 1912. In: De dierlijke parasieten van den mensch en van onze Huisdieren. Schletema & Holkema's Boekhandel, Amsterdam, The Netherlands, pp. 81—133. *.

Smetana, H., 1933a. Coccidiosis of the liver of rabbits. I. Experimental study on the excystation of oocysts of *Eimeria stiedae*. Archives of Pathology 15, 175—192. *.

Smetana, H., 1933b. Coccidiosis of the liver of rabbits. II. Experimental study on the mode of infection of the liver by sporozoites of *Eimeria stiedae*. Archives of Pathology 15, 330—339. *.

Smetana, H., 1933c. Coccidiosis of the liver of rabbits. III. Experimental study of the histogenesis of coccidiosis of the liver. Archives of Pathology 15, 516—636. *.

Smith, D.D., Frenkel, J.K., 1978. Cockroaches as vectors of *Sarcocystis muris* and other coccidia in the laboratory. Journal of Parasitology 64, 315—319. *.

Smith, D.D., Frenkel, J.K., 1995. Prevalence of antibodies to *Toxoplasma gondii* in wild mammals of Missouri and east central Kansas: biologic and ecologic considerations of transmission. Journal of Wildlife Diseases 31, 15—21. *.

Smith, H.V., Cacciò, S.M., Tait, A., McLauchlin, J., Thompson, R.C., 2006. Tools for investigating the environmental transmission of *Cryptosporidium* and *Giardia* infections in humans. Trends in Parasitology 22, 160—167. *.

Smith, R.P., Chalmers, R.M., Mueller-Doblies, D., Clifton-Hadley, F.A., Elwin, K., Watkins, J., Paiba, G.A., Hadfield, S.J., Giles, M., 2010. Investigation of farms linked to human patients with cryptosporidiosis in England and Wales. Preventive Veterinary Medicine 94, 9—17. *.

Snigirevskaya, E.S., 1968. The occurrence of micropore in schizonts, microgametocytes and macrogametes of *Eimeria intestinalis*. Acta Protozoologica 5, 381—386. *.

Snigirevskaya, E.S., 1969a. Changes in the ultrastructure of the nucleus and of the kinetic apparatus during microgametogenesis in the coccidians *Eimeria intestinalis* and *E. magna*. Progress in Protozoology, 3rd International Congress on Protozoology, Leningrad, p 73 (abstract). *.

Snigirevskaya, E.S., 1969b. Changes in some ultrastructures during microgametogenesis in rabbit coccidia *Eimeria intestinalis* and *E. magna*. Cytologija S.S.R. 11, 382—385. * (in Russian).

Snigirevskaya, E.S., 1969c. Electron microscopic study of the macrogametes of *Eimeria intestinalis* (coccidia). Cytologija S.S.R. 11, 700—706. * (in Russian).

Snigirevskaya, E.S., 1969d. Electron microscopic study of the schizogony process in *Eimeria intestinalis*. Acta Protozoologica 7, 57—79.

Snigirevskaya, E.S., 1972. The ultrastructure of the pellicle of intracellular parasitic protozoa—Coccidia. Tsitologiya (Cytology) 10, 1285—1290. * (in Russian, English summary).

Snigirevskaya, E.S., Cheissin, E.M., 1968. Role of the micropore in nutrition of the endogenous developmental stages of *Eimeria intestinalis*. Tsitologiya (Cytology) 10, 940—944.

Soveri, T., Valtonen., M., 1983. Endoparasites of hares (*Lepus timidus* L. and *L. europaeus* Pallas) in Finland. Journal of Wildlife Diseases 19, 337—341. *.

Speer, C.A., 1979. Further studies on the development of gamonts and oocysts of *Eimeria magna* in cultured cells. Journal of Parasitology 65, 591—598. *.

Speer, C.A., Danforth, H.D., 1976. Fine structural aspects of microgametogenesis of *Eimeria magna* in rabbits and in kidney cell cultures. Journal of Protozoology 23, 109—115. *.

Speer, C.A., Hammond, D.M., 1971. Development of first- and second-generation schizonts of *Eimeria magna* from rabbits in cell cultures. Zeitschrift für Parasitenkunde 37, 336–353. *.

Speer, C.A., Hammond, D.M., 1972a. Development of gametocytes and oocysts of *Eimeria magna* from rabbits in cell culture. Proceedings of the Helminthological Society of Washington 39, 114–118. *.

Speer, C.A., Hammond, D.M., 1972b. Motility of macrogamonts of *Eimeria magna* coccidia in cell culture. Science 178, 763–765. *.

Speer, C.A., Hammond, D.M., Elsner, Y.Y., 1973a. Further asexual development of *Eimeria magna* merozoites in cell cultures. Journal of Parasitology 59, 613–623. *.

Speer, C.A., Hammond, D.M., Youssef, N.N., Danforth, H.D., 1973b. Fine structural aspects of macrogametogenesis in *Eimeria magna*. Journal of Protozoology 20, 274–281. *.

Springer, M.S., Murphy, W.J., Eizirik, E., O'Brien, S.J., 2003. Placental mammal diversification and the Cretaceous-Tertiary boundary. Proceedings of the National Academy of Sciences, USA 100, 1056–1061. *.

Sreter, T., Varga, I., 2000. Cryptosporidiosis in birds—a review. Veterinary Parasitology 87, 261–279. *.

Sreter, T., Egyed, Z., Szell, Z., Kovacs, G., Nikolausz, M., Marialigeti, K., Varga, I., 2000. Morphologic, host specificity, and genetic characterization of a European *Cryptosporidium andersoni* isolate. Journal of Parasitology 86, 1244–1249. *.

Sroka, J., Zwoliński, J., Dutkiewicz, J., Tós-Lut, J., Latuszynska, J., 2003. Toxoplasmosis in rabbits confirmed by strain isolation: a potential risk of infection among agricultural workers. Annals of Agricultural and Environmental Medicine 10, 125–128. *.

Steindel, M., Neto, E.M., de Menezes, C.L.P., Romanha, A.J., Simpson, A.J.G., 1993. Random amplified polymorphic DNA analysis of *Trypanosoma cruzi* strains. Molecular Biochemistry and Parasitology 60, 71–80. *.

Stieda, L., 1865. Ueber die Psorospermien der Kaninchenleber und ihre Entwicklung. Archiv für Pathologische Anatomie 32, 132. *.

Stingl, H., 1974. Das extrahepatische Gallengangsystem des Kaninchens und seine Veränderungen unter dem Einfluß einer experimentellen *Eimeria stiedai*-infektion (The extrahepatic bile duct system of the rabbit and its alterations under the influence of an experimental *Eimeria stiedai* infection). Inaugural Dissertation, Faculty of Veterinary Medicine, Justus Liebig-Universität Geißen, Geißen, Germany * (in German, English summary).

Stockdale, H.D., Spencer, J.A., Blagburn, B.L., 2008. Prophylaxis and Chemotherapy. In: Fayer, R., Xiao, L. (Eds.), *Cryptosporidium* and Cryptosporidiosis. CRC Press, Boca Raton, Florida, pp. 255–287. *.

Stodart, E.A., 1968a. Coccidiosis in wild rabbits, *Oryctolagus cuniculus* (L.), at four sites in different climatic regions in eastern Australia. I. Relationship with the age of the rabbit. Australian Journal of Zoology 16, 69–85. *.

Stodart, E.A., 1968b. Coccidiosis in wild rabbits, *Oryctolagus cuniculus* (L.), at four sites in different climatic regions in eastern Australia. II. The relationship of oocyst output to climate and some aspects of the rabbits' physiology. Australian Journal of Zoology 16, 619–628. *.

Stodart, E.A., 1971. Coccidiosis in wild rabbits, *Oryctolagus cuniculus* (L.), at a site on the coastal plain in eastern Australia. Australian Journal of Zoology 19, 287–292. *.

Stolte, M., Odening, K., Bockhardt, I., 1996. The raccoon as intermediate host of three *Sarcocystis* species in Europe. Journal of the Helminthological Society of Washington 63, 145–149. *.

Stork, N.E., 1997. Measuring global biodiversity and its decline. In: Reaka-Kudia, M.L., Wilson, D.E., Wilson, E.O. (Eds.), Biodiversity II: Understanding and protecting our biological resources. Joseph Henry Press, Washington, D.C., USA, pp. 41–68. *.

Streun, A., Coudert, P., Rossi, G.L., 1979. Characterization of *Eimeria* species. II. Sequential morphologic study of the endogenous cycle of *Eimeria perforans* (Leuckart, 1879; Sluiter and Swellengrebel, 1912) in experimentally infected rabbits. Zeitschrift für Parasitenkunde 60, 37–53. *.

Sturdee, A.P., Chalmers, R.M., Bull, S.A., 1999. Detection of *Cryptosporidium* oocysts in wild mammals of mainland Britain. Veterinary Parasitology 80, 273–280. *.

Su, C., Nei, M., 1999. Fifty-million year old polymorphism at an immunoglobulin variable region gene locus in the rabbit evolutionary lineage. Proceedings of the National Academy of Sciences, USA 96, 9710–9715. *.

Sugár, L., 1978. *Eimeria babatica* sp. n. (Protozoa: Coccidia) from Europaean hare (*Lepus europaeus* Pallas) in Hungary. Parasitologia Hungarica 11, 13–15. *.

Sugár, L., 1979. *Eimeria macrosculpta* sp. n. (Protozoa: Coccidia) from European hare (*Lepus europaeus* Pallas) in Hungary. Parasitologia Hungarica 12, 9–10. *.

Sugár, L., Murai, É., Mészáros, F., 1978. Über die Endoparasiten der wildebenden Leporidae Ungarns. Parasitologia Hungarica 11, 63–85. *.

Sulaiman, I.M., Morgan, U.M., Thompson, R.C., Lal, A.A., Xiao, L., 2000. Phylogenetic relationships of *Cryptosporidium* parasites based on the 70-kilodalton heat shock protein (HSP70) gene. Applied Environmental Microbiology 66, 2385–2391. *.

Sulaiman, I.M., Lal, A.A., Xiao, L., 2002. Molecular phylogeny and evolutionary relationships of *Cryptosporidium* parasites at the actin locus. Journal of Parasitolology 88, 388–394. *.

Svanbaev, S.K., 1958. K poznaniyu fauny kiktsidiĭ gryzunov tsentral'nogo Kazakhstana (Coccidia of rodents of Central Kazakhstan). Trudy Instituta Zoologii Akadamii Nauk Kazakh SSR 9, 183—186. *.

Svanbaev, S.K., 1979. Koktsidii Dikikh Zhivotnykh Kazakhstana (Coccidia of Wild Animals in Kazakhstan). Izdatelstvo "Nauka" Kazakhskoi SSR, Alma-Ata, Kazakhstan, USSR.

Tacconi, G., Piergili-Fioretti, D., Moretti, A., Nobilini, N., Pasquali, P., 1995. Coccidia in hare (Lepus europaeus) reared in Umbria, Italy: Bioepidemiological study. Journal of Protozoology Research 5, 77—85. *.

Tadros, W., Laarman, J.J., 1976. Sarcocystis and related coccidian parasites: a brief general review, together with a discussion on some biological aspects and their life cycles and a new proposal for their classification. Acta Leidensia 44, 1—107. *.

Tadros, W., Laarman, J.J., 1977a. The cat Felis catus as the final host of Sarcocystis cuniculi Brumpt, 1913, of the rabbit Oryctolagus cuniculus. Proceedings, koninklijke Nederlandse Akademie der Wetenschappen, Serie C 80, 351—352. *.

Tadros, W., Laarman, J.J., 1977b. Establishment of the domestic cat as the definitive host of Sarcocystis cuniculi of the common European rabbit Oryctolagus cuniculus. Proceedings of the Fifth International Congress on Protozoology, New York City, 26 June-2 July, 1977, pp. 188—189. *.

Tadros, W., Laarman, J.J., 1978. A comparative study of the light and electron microscopic structure of the walls of the muscle cysts of several species of Sarcocystis eimeriid coccidia. Proceedings, koninklijke Nederlandse Akademie der Wetenschappen, Serie C 81, 469—491. *.

Tadros, W., Laarman, J.J., 1982. In: Lumsden, W.H.R., Muller, R., Baker, J.R. (Eds.), Advances in Parasitology. Current concepts on the biology, evolution and taxonomy of tissue cyst-forming eimeriid coccidia, vol. 20. Academic Press, New York, pp. 293—468. *.

Tambur, Z., Kulišić, Z., Maličević, Ž., Mihailović, M., 1998a. The influence of intestinal coccidia upon the activity of liver enzymes. Acta Veterinaria (Brno) 48, 139—146. *.

Tambur, Z., Kulišić, Z., Maličević, Ž., Mihailović, M., 1998b. Influence of intestinal coccidia infection of rabbits upon plasma and fecal protein levels, and plasma and urinary urea and creatinine levels. Acta Veterinaria (Brno) 48, 147—156. *.

Tambur, Z., Kulišić, Z., Maličević, Ž., Mihailović, M., 1998c. The influence of intestinal coccidia of rabbits upon plasma and urine electrolyte concentrations. Acta Veterinaria (Brno) 48, 225—234. *.

Tambur, Z., Kulišić, Z., Maličević, Ž., Mihailović, M., 1999. Blood glucose, plasma osmolarity and urea and creatinine clearance in rabbits artificially infected with intestinal coccidia. Acta Veterinaria (Brno) 49, 171—176. *.

Tasan, E., Özer, E., 1989. Elazığ ve Tunceli yörelerindeki yabani tavsanlarda Eimeria (Protozoa, Eimeriidae) 'larin bulunusu üzerinde bir çalisma (A study on the occurrence of Eimeria [Protozoa, Eimeriidae] in hares in the vicinities of Elazığ and Tunceli). Doğa Turk Veterinerlik Ve Hayvancilik Dergisi 13, 60—65. * (in Turkish).

Terracciano, G., Mancianti, F., Marconcini, A., 1988. Indagine parassitologica nella lepre (Parasitological survey in the hare). SUMMA 3, 217—220. * (in Italian).

Tibayrenc, M., Neubauer, K., Barnabé, C., Guerrini, F., Skarecky, D., Ayala, F.J., 1993. Genetic characterization of six parasitic protozoa: parity between random-primer DNA typing and multilocus enzyme electrophoresis. Proceedings of the National Academy of Sciences USA 90, 1335—1339. *.

Todd Jr., K.S., Hammond, D.M., 1968a. Life cycle and host specificity of Eimeria callospermophili Henry, 1932 from the Uinta ground squirrel Spermophilus armatus. Journal of Protozoology 15, 1—8. *.

Todd Jr., K.S., Hammond, D.M., 1968b. Life cycle and host specificity of Eimeria larimerensis Vetterling, 1964, from the Uinta ground squirrel Spermophilus armatus. Journal of Protozoology 15, 268—275. *.

Todd Jr., K.S., Lepp, D.L., 1972. Completion of the life cycle of Eimeria vermiformis Ernst, Chobotar, and Hammond, 1971 from the mouse Mus musculus, in dexamethasone-treated rats, Rattus norvegicus. Journal of Parasitology 58, 400—401. *.

Todd Jr., K.S., Lepp, D.L., Trayser, C.V., 1971. Development of the asexual cycle of Eimeria vermiformis Ernst, Chobotar and Hammond, 1971, from the mouse, Mus musculus, in dexamethasone-treated rats, Rattus norvegicus. Journal of Parasitology 57, 1137—1138. *.

Toula, F.H., Ramadan, H.H., 1998. Studies on coccidia species of genus Eimeria from domestic rabbit (Oryctolagus cuniculus domesticus L.) in Jeddah, Saudia Arabia. Journal of the Egyptian Society of Parasitology 28, 691—698. *.

Tsunoda, K., 1952. Eimeria matsubayashii sp. nov., a new species of rabbit coccidium. Experimental Reports of the Government Experiment Station of Animal Hygiene, Tokyo 25, 109—119. * (in Japanese with English summary).

Tyzzer, E.E., 1902. Coccidium infection of the rabbit's liver. Journal of Medical Research 7, 235—254. *.

Tyzzer, E.E., 1907. A sporozoan found in the gastric glands of the common mouse. Proceedings of the Society for Experimental Biology and Medicine 5, 12—13. *.

Tyzzer, E.E., 1910. An extracellular coccidium, Cryptosporidium muris (gen. et sp. nov.), of the gastric glands of the common mouse. Journal of Medical Research 23, 487—509. *.

Tyzzer, E.E., 1912. Cryptosporidium parvum (sp. nov.), a coccidium found in the small intestine of the common mouse. Archive für Protistenkunde 26, 394—412. *.

Tyzzer, E.E., 1929. Coccidiosis in gallinaceous birds. American Journal of Hygiene 10, 269–383. *.

Unger, C., 1977. Vorkommen und Verbreitung der Muskelsarkosporidien bei jagdbaren Wildtieren in einem jagdgebiet. Doctoral Thesis Humboldt University, Veterinary Medicine and Parasitology, Berlin. *.

Upton, S.J., McAllister, C.T., Brillhart, D.B., Duszynski, D.W., Wash, C.D., 1992. Cross-transmission studies with Eimeria arizonensis-like oocysts (Apicomplexa) in New World rodents in the genera Baiomys, Neotoma, Onychomys, Peromyscus and Reithrodontomys (Muridae). Journal of Parasitology 78, 406–413. *.

USFWS, 2001. Emergency rule to list the Columbia Basin District population segment of the pygmy rabbit (Brachylagus idahoensis) as endangered. Federal Register 66, 59734 (November 30, 2001). *.

USFWS, 2003. Final rule to list the Columbia Basin District population segment of the pygmy rabbit (Brachylagus idahoensis) as endangered. Federal Register 68, 10388 (March 5, 2003). *.

Valigurová, A., Hofmannová, L., Koudela, B., Vávra, J., 2007. An ultrastructural comparison of the attachment sites between Gregarina steini and Cryptosporidium muris. Journal of Eukaryotic Microbiology 54, 495–510. *.

Valigurová, A., Jirků, M., Koudela, B., Gelnar, M., Modrý, D., Šlapeta, J., 2008. Cryptosporidia: epicellular parasites embraced by the host cell membrane. International Journal for Parasitology 38, 913–922. *.

van Praag, E., 2011. Protozoal enteritis: Coccidiosis. http://www.medirabbit.com/EN/GI_diseases/ Protozoal_diseases/Cocc_en.htm 1–5. *.

Vande Vusse, F.J., 1965. Sarcocystis infections in relation to age of Iowa cottontails (Protozoa: Sarcocystidae). Iowa Academy of Sciences 72, 524–528. *.

Varga, I., 1976. A mezeinyúl kísérleti fertőzése Eimeria stiedaivel (Experimental transmission of Eimeria stiedai to the hare). Magyar Állatorvosok Lapja (November), 726–730 (in Hungarian, English summary) and in Acta Veterinaria Academiae Scientiarum Hungaricae 26, 105–112. * (in English).

Veisov, A.M., 1982. Materialy k izucheniyu koktsidii krolikov v Azerbaidzhanskoi SSR. v: Zayanchkauskas i dr., red. Kishechnye Prosteishie. Institut Zoologii i Parazitologii, Akademiya Nauk Litovskoi CCP, Vil'nyus (Materials for the study of the coccidia of rabbits in the Azerbaijan SSR. In Intestinal Protozoa. Zayanchkauskas et al., eds. Institute of Zoology and Parasitology, Academy of Sciences of the Lithuanian SSR, Vilnius), pp. 30–35. *.

Venturini, L., Petruccelli, M., Piscopo, M., Únzaga, J.M., Venturini, M.C., Bacigalupe, D., Basso, W., Dubey, J.P., 2002. Natural Besnoitia sp. infection in domestic rabbits from Argentina. Veterinary Parasitology 107, 273–278. *.

Vetterling, J.M., Takeuchi, A., Madden, P.A., 1971. Ultrastructure of Cryptosporidium wrari from the guinea pig. Journal of Protozoology 18, 248–260. *.

Vila-Viçosa, M.J., Caeiro, V., 1997. Contribuição para o estudo dos Eimeriidae poche (sic), 1913 parasitas do Oryctolagus cuniculus (L.), Lepus capensis (L.), Ovis aries (L.) (Contribution to the study of Eimeriidae poche (sic), 1913 parasites from Oryctolagus cuniculus (L.), Lepus capensis (L.), Ovis aries (L.); presented at the III Reunião Annual da Sociedade Portuguesa de Parasitologia, 3–4 October, 1996). Veterinária Técnica (October, 1997), 30–46. * (in Portuguese, English summary).

Vitovec, J., Pakandl, M., 1989. The pathogenicity of rabbit coccidium Eimeria coecicola Cheissin, 1947. Folia Parasitologica 36, 289–293. *.

von Braunschweig, A., 1965. Liebercoccidiose beim Hasen. Zeitschrift für Jagdwissenschaft 11, 54. *.

von Wasielewski, T., 1904. II. Kapitel. Über bau und entwicklung des Erregers der Kaninchen-Coccidiose, Eimeria cuniculi. In: Studien und Mikrophotogramme zur Kenntnis der Pathogenen Protozoen (Studies und photographic notes of the protozoan pathogens). Verlag von Johann Ambrosius Barth, Leipzig, pp. 13–68. * (in German).

Voorhies, M.R., Timperley, C.L., 1997. A new Pronotolagus (Lagomorpha: Leporidae) and other leporids from the Valentine Railway Quarries (Barstovian, Nebraska), and the archaeolagine-leporine transition. Journal of Vertebrate Paleontology 17, 725–737. *.

Votýpka, J., Hypša, V., Jirků, M., Flegr, J., Vávra, J., Lukeš, J., 1998. Molecular phylogenetic relatedness of Frankelia spp. (Protozoa: Apicomplexa) to Sarcocystis falcatula Stiles, 1893: is the genus Sarcocystis paraphyletic? Journal of Eukaryotic Microbiology 45, 137–142. *.

Wagenbach, G.E., Burns, W.C., 1969. Structure and respiration of sporulating E. stiedae and E. tenella oocysts. Journal of Protozoology 16, 257–263. *.

Waitumbi, J.N., Murphy, N.B., 1993. Inter- and intra-species differentiation of trypanosomes by genomic fingerprinting with arbitrary primers. Molecular Biochemistry and Parasitology 58, 181–186. *.

Wallace, G.D., Frenkel, J.K., 1975. Besnoitia species (Protozoa, Sporozoa, Toxoplasmatidae): Recognition of cyclic transmission by cats. Science 188, 369–371. *.

Wallace, G.D., Marshall, I., Marshall, M., 1972. Cats, rats, and toxoplasmosis on a small Pacific island. American Journal of Epidemiology 95, 475–482. *.

Waller, E.F., Morgan, B.B., 1941. The pathology of Eimeria leporis (Coccidia) in the cottontail rabbit. American Journal of Hygiene 34, 83–85. *.

Ward, R.D., Zemlak, T.S., Innes, B.H., Last, P.R., Hebert, P.D.N., 2005. DNA barcoding Australia's fish species. Philosophical Transactions of the Royal Society B 360, 1847–1857. *.

Waworuntu, F.K., 1924. Bijdrage tot de kennis van het konijnen coccidium. Schotanus & Jens, Utrecht. 1—96. *.

Weisbroth, S.H., Scher, S., 1975. Fatal intussusception associated with intestinal coccidiosis (*Eimeria perforans*) in a rabbit. Laboratory Animal Science 25, 79—81. *.

Welsh, J., McClelland, M., 1990. Fingerprinting genomes using arbitrary primers. Nucleic Acids Research 18, 7213—7218. *.

Wenyon, C.M., 1926. Protozoology, Wm. Wood and Co, New York. Vol 2, 563—880. *.

Wheeler, Q.D., 2004. Taxonomic triage and the poverty of phylogeny. Philosophical Transactions of the Royal Society B 359, 571—583. *.

Wheeler, Q.D., Cracraft, J., 1997. Chapter 28. Taxonomic preparedness: Are we ready to meet the biodiversity challenge? In: Reaka-Kudla, M.I., Wilson, D.E., Wilson, E.O. (Eds.), Biodiversity II. Understanding and Protecting Our Biological Resources, Joseph Henry Press, Washington, D.C, pp. 435—446. *.

Wheeler, Q.D., Raven, P.H., Wilson, E.O., 2004. Taxonomy: Impediment or expedient? Science 303, 285. *.

White, J.A., 1991. North American Leporidae (Mammalia, Lagomorpha) from Late Miocene (Clarendonian) to late Pliocene (Blancan). Journal of Vertebrate Paleontology 1, 67—89. *.

White Jr., A.C., Chappell, C.L., Hayat, C.S., Kimball, K.T., Flanigan, T.P., Goodgame, R.W., 1994. Paromomycin for cryptosporidiosis in AIDS: a prospective, double-blind trial. Journal of Infectious Disease 170, 419 424. *.

Wiggins, J.P., Rothenbacher, H., 1979. *Eimeria azul* sp.n. (Protozoa: Eimeriidae) from the eastern cottontail, *Sylvilagus floridanus*, in Pennsylvania. Journal of Parasitology 65, 393—394. *.

Wiggins, J.P., Cosgrove, M., Rothenbacher, H., 1980. Gastrointestinal parasites of the eastern cottontail (*Sylvilagus floridanus*) in central Pennsylvania. Journal of Wildlife Diseases 16, 541—544. *.

Wilber, P.G., Duszynski, D.W., Upton, S.J., Seville, R.S., Corliss, J.O., 1998. A revision of the taxonomy and nomenclature of the *Eimeria* (Apicomplexa: Eimeriidae) from rodents in the Tribe Marmotini (Sciuridae) Systematic Parasitology 39, 113—135. *.

Williams, J.G.K., Kubelik, A.R., Livak, K.J., Rafalski, J.A., Tingey, S.V., 1990. DNA polymorphisms amplified by arbitrary primers are useful as genetic markers. Nucleic Acids Research 18, 6531—6535. *.

Williams, R.B., Thebo, P., Marshall, R.N., Marshall, J.A., 2010. Coccidian oöcysts as type-specimens: long-term storage in aqueous potassium dichromate solution preserves DNA. Systematic Parasitology 76, 69—76. *.

Williamson, P., Day, J.G., 2007. The problem with protists: is barcoding the answer? Biologist 54, 86—90. *.

Wilson, E.O., 1992. The Diversity of Life. Harvard University Press, Cambridge, Massachusetts. *.

Wilson, E.O., 2003. The Encyclopedia of Life. Trends in Ecology and Evolution 18, 77—80. *.

Wilson, E.O., 2004. Taxonomy as a fundamental discipline. Philosophical Transactions of the Royal Society B 359, 739. *.

Wilson, E.O., 2011. Exploring a little-known planet. The Scientist 10, 64. *.

Wilson, D.E., Reeder, D.M., 2005. Mammal species of the world: A taxonomic and geographic reference, third ed., vol. 1. Johns Hopkins University Press, Baltimore, Maryland. *.

Witzmann, H., 1982. Pathomorphologische Untersuchungen zum Sarkosporidienbefall beim Feldhasen (*Lepus europaeus*). Diploma Thesis Humboldt University, Veterinary Medicine and Pathology, Berlin. *.

Witzmann, H., Ippen, R., Henne, D., 1983. Untersuchungen zum Sarkosporidienbefall beim Feldhasen (*Lepus europaeus*). Verh. ber. Erkrank. Zootiere (Berlin) 25, 315—319. *.

Xiao, L.H., Morgan, U.M., Limor, J., Escalante, A., Arrowood, M., Shulaw, W., Thompson, R.C.A., Fayer, R., Lal, A.A., 1999. Genetic diversity within *Cryptosporidium parvum* and related *Cryptosporidium* species. Applied Environmental Microbiology 65, 3386—3391. *.

Xiao, L., Limor, J., Morgan, U.M., Sulaiman, I.M., Thompson, R.C., Lal, A.A., 2000. Sequence differences in the diagnostic target region of the oocyst wall protein gene of *Cryptosporidium* parasites. Applied Environmental Microbiology 66, 5499—5502. *.

Xiao, L., Bern, C., Limor, J., Sulaiman, I., Roberts, J., Checkley, W., Cabrera, L., Gilman, R.H., Lal, A.A., 2001. Identification of 5 types of *Cryptosporidium* parasites in children in Lima. Peruvian Journal of Infectious Disease 183, 492—497. *.

Xiao, L., Sulaiman, I.M., Ryan, U.M., Zhou, L., Atwill, E.R., Tischler, M.L., Zhang, X., Fayer, R., Lal, A.A., 2002. Host adaptation and host-parasite co-evolution in *Cryptosporidium*: implications for taxonomy and public health. International Journal for Parasitology 32, 1773—1785. *.

Yakhchali, M., Tehrani, A., 2007. Eimeriidosis and pathological findings in New Zealand white rabbits. Journal of Biological Sciences 7, 1488—1491. *.

Yakimoff, W.L., 1933a. Bemerkungen über die Coccidiose der Kaninchen. Landwirtschaft Pelztierzucht 4, 113—114. *.

Yakimoff, W.L., 1933b. Über die Grösse der Oocysten der Coccidien des Kaninchens und ihre Bedeutung bei der Differentiation der verschiedenen Arten. Archiv für Protiskendue 80, 370—377. *.

Yakimoff, W.L., 1934. *Eimeria exigua* n. sp., eine neue Kaninchenkokzidie. Zentralblatt für Bakteriologie. I. Abteilung Originale 131, 18—24. *.

Yakimoff, W.L., 1936. Neue Kokzidie der Hasen. Wiener Tierartliche Monatsschrift 23, 490—491. *.

Yakimoff, W.L., Gousseff, W.F., 1938. The coccidia of mice (*Mus musculus*). Parasitology 30, 1—3. *.

Yakimoff, W.L., Iwanoff-Gobzem, P.S., 1931. Zur Frage der Infektion der Tier emit heterogenen Kokzidien. Zentralblatt für Bakteriologie. I. Abteilung Originale 122, 319—328. *.

Yakimoff, W.L., Polueketoff, A.M., Rastecaieff, E.F., 1931. Zür Frage der Hasenkokzidiose in Russland (On the question of rabbit coccidiosis in Russia). Zeitschrift Für Infektionskrankheiten. Parasitäre Krankheiten und Hygiene Der Haustiere 39, 311—319. *.

Yakimoff, W.L., Matschoulsky, S.N., Spartansky, O.A., 1936. New coccidia of hares (Neue Koksidie der Hasen). Wiener Tierärztliche Monatsschrift 23, 490—491. *.

Yamada, M., Yukawa, M., Sckikawa, H., Kenmatsu, M., Mochizuki, K., 1993. Studies on the morphology of *Sarcocystis* in thoroughbred horses in Japan. Journal of Protozoological Research 3, 14—19. *.

Yi-Fan, C., Run-Roung, Y., Jian-Hua, W., Jiang-Hui, B., Duszynski, D.W., 2009. *Eimeria* spp. (Apicomplexa: Eimeriidae) from the Plateau pika, *Ochotona curzoniae*, from Haibei Area, Qinghai Province, China, with the description of two new species. Journal of Parasitology 95, 1192—1196. *.

Yu, N., Zheng, C.L., Zhang, Y.P., Li, W.H., 2000. Molecular systematics of pikas (Genus *Ochotona*) inferred from mitochondrial DNA sequences. Molecular Phylogenetics and Evolution 16, 85—95. *.

Zachos, J., Pagani, M., Sloan, L., Thomas, E., Billups, K., 2001. Trends, rhythms, and aberrations in global climate 65 Mya to present. Science 292, 686—693. *.

Zarzara, C., Stanescu, V., Marcoci, M., Popovici, A., 1989. Incidence of hepatic and intestinal coccidiosis in farmed rabbits. Lucrarile Institutului de Cercetari Veterinare si Biopreparate Pasteur 18, 141—145.

Zhang, X.C., Hu, L.S., Li, D.C., 1992. Study on some biologic properties of *Cryptosporidium*. Bulletin of the P.L.A. Veterinary College 12, 114—118. * (in Chinese).

Zhao, X., Duszynski, D.W., 2001a. Phylogenetic relationships among rodent *Eimeria* species determined by plastid ORF470 and nuclear 18S rDNA sequences. International Journal of Parasitology 31, 715—719. *.

Zhao, X., Duszynski, D.W., 2001b. Molecular phylogenies suggest the oocyst residuum can be used to distinguish two independent lineages of *Eimeria* spp. in rodents. Parasitology Research 87, 638—643. *.

Zhu, G., Keithly, J.S., Philippe, H., 2000. What is the phylogenetic position of *Cryptosporidium*? International Journal of Systematic and Evolutionary Microbiology 50, 1673—1681. *.

Zimmer, C., 2012. King cosmos. Playboy 59, 152—156. 180—181. *.

Index

Note: Page numbers with "f" denote figures; "t" tables.

Printed and bound by CPI Group (UK) Ltd, Croydon, CR0 4YY

03/10/2024

01040323-0001